Isaac Todhunter

A history of the mathematical theory of probability from the time

of Pascal to that of Laplace

Isaac Todhunter

A history of the mathematical theory of probability from the time of Pascal to that of Laplace

ISBN/EAN: 9783742835680

Manufactured in Europe, USA, Canada, Australia, Japa

Cover: Foto ©Thomas Meinert / pixelio.de

Manufactured and distributed by brebook publishing software (www.brebook.com)

Isaac Todhunter

A history of the mathematical theory of probability from the time

of Pascal to that of Laplace

Cambridge and London:

MACMILLAN AND CO.

1865.

PREFACE.

THE favourable reception which has been granted to my *History of the Calculus of Variations during the Nineteenth Century* has encouraged me to undertake another work of the same kind. The subject to which I now invite attention has high claims to consideration on account of the subtle problems which it involves, the valuable contributions to analysis which it has produced, its important practical applications, and the eminence of those who have cultivated it.

The nature of the problems which the Theory of Probability contemplates, and the influence which this Theory has exercised on the progress of mathematical science and also on the concerns of practical life, cannot be discussed within the limits of a Preface; we may however claim for our subject all the interest which illustrious names can confer, by the simple statement that nearly every great mathematician within the range of a century and a half will come before us in the course of the history. To mention only the most distinguished in this distinguished roll—we shall find here—Pascal and Fermat, worthy to be associated by kindred genius and character—De Moivre with his rare powers of analysis, which seem to belong only to a later epoch, and which justify the honour in which he was held by Newton—Leibnitz and the eminent school of which he may be considered the founder, a school including the Bernoullis and Euler—D'Alembert, one of the most conspicuous of those who brought on the French revolution, and Condorcet, one of the most illustrious of its victims—Lagrange and Laplace who survived until the present century, and may be regarded as rivals at that time for the supremacy of the mathematical world.

I will now give an outline of the contents of the book.

The first Chapter contains an account of some anticipations of the subject which are contained in the writings of Cardan, Kepler and Galileo.

The second Chapter introduces the Chevalier de Méré who having puzzled himself in vain over a problem in chances, fortunately turned for help to Pascal: the Problem of Points is discussed in the correspondence between Pascal and Fermat, and thus the Theory of Probability begins its career.

The third Chapter analyses the treatise in which Huygens in 1659 exhibited what was then known of the subject. Works such as this, which present to students the opportunity of becoming acquainted with the speculations of the foremost men of the time, cannot be too highly commended; in this respect our subject has been fortunate, for the example which was afforded by Huygens has been imitated by James Bernoulli, De Moivre and Laplace—and the same course might with great advantage be pursued in connexion with other subjects by mathematicians in the present day.

The fourth Chapter contains a sketch of the early history of the theory of Permutations and Combinations; and the fifth Chapter a sketch of the early history of the researches on Mortality and Life Insurance. Neither of these Chapters claims to be exhaustive; but they contain so much as may suffice to trace the connexion of the branches to which they relate with the main subject of our history.

The sixth Chapter gives an account of some miscellaneous investigations between the years 1670 and 1700. Our attention is directed in succession to Caramuel, Sauveur, James Bernoulli, Leibnitz, a translator of Huygens's treatise whom I take to be Arbuthnot, Roberts, and Craig—the last of whom is notorious for an absurd abuse of mathematics in connexion with the probability of testimony.

The seventh Chapter analyses the *Ars Conjectandi* of James Bernoulli. This is an elaborate treatise by one of the greatest mathematicians of the age, and although it was unfortunately left incomplete, it affords abundant evidence of its author's ability and of his interest in the subject. Especially we may notice the famous theorem which justly bears the name of James Bernoulli, and which places the Theory of Probability in a more commanding position than it had hitherto occupied.

The eighth Chapter is devoted to Montmort. He is not to be compared for mathematical power with James Bernoulli or De Moivre; nor does he seem to have formed a very exalted idea of the true dignity and importance of the subject. But he was enthusiastically devoted to it; he spared no labour himself, and his influence direct or indirect stimulated the exertions of Nicolas Bernoulli and of De Moivre.

The ninth Chapter relates to De Moivre, containing a full analysis of his *Doctrine of Chances*. De Moivre brought to bear on the subject mathematical powers of the highest order; these powers are especially manifested in the results which he enunciated respecting the great problem of the Duration of Play. Unfortunately he did not publish demonstrations, and Lagrange

himself more than fifty years later found a good exercise for his
analytical skill in supplying the investigations; this circumstance
compels us to admire De Moivre's powers, and to regret the loss
which his concealment of his methods has occasioned to mathematics, or at least to mathematical history.

De Moivre's *Doctrine of Chances* formed a treatise on the
subject, full, clear and accurate; and it maintained its place as a
standard work, at least in England, almost down to our own day.

The tenth Chapter gives an account of some miscellaneous
investigations between the years 1700 and 1750. These investigations are due to Nicolas Bernoulli, Arbuthnot, Browne, Mairan,
Nicole, Buffon, Ham, Thomas Simpson and John Bernoulli.

The eleventh Chapter relates to Daniel Bernoulli, containing
an account of a series of memoirs published chiefly in the volumes
of the Academy of Petersburg; the memoirs are remarkable for
boldness and originality, the first of them contains the celebrated
theory of Moral Expectation.

The twelfth Chapter relates to Euler; it gives an account of
his memoirs, which relate principally to certain games of chance.

The thirteenth Chapter relates to D'Alembert; it gives a full
account of the objections which he urged against some of the
fundamental principles of the subject, and of his controversy with
Daniel Bernoulli on the mathematical investigation of the gain to
human life which would arise from the extirpation of one of the
most fatal diseases to which the human race is liable.

The fourteenth Chapter relates to Bayes; it explains the method by which he demonstrated his famous theorem, which may
be said to have been the origin of that part of the subject which
relates to the probabilities of causes as inferred from observed
effects.

The fifteenth Chapter is devoted to Lagrange; he contributed
to the subject a valuable memoir on the theory of the errors of
observations, and demonstrations of the results enunciated by De
Moivre respecting the Duration of Play.

The sixteenth Chapter contains notices of miscellaneous investigations between the years 1750 and 1780. This Chapter brings
before us Kaestner, Clark, Mallet, John Bernoulli, Beguelin,
Michell, Lambert, Buffon, Fuss, and some others. The memoir
of Michell is remarkable; it contains the famous argument for the
existence of design drawn from the fact of the closeness of certain
stars, like the Pleiades.

The seventeenth Chapter relates to Condorcet, who published a
large book and a long memoir upon the Theory of Probability.
He chiefly discussed the probability of the correctness of judgments determined by a majority of votes; he has the merit of first

submitting this question to mathematical investigation, but his own results are not of great practical importance.

The eighteenth Chapter relates to Trembley. He wrote several memoirs with the main design of establishing by elementary methods results which had been originally obtained by the aid of the higher branches of mathematics; but he does not seem to have been very successful in carrying out his design.

The nineteenth Chapter contains an account of miscellaneous investigations between the years 1780 and 1800. It includes the following names; Borda, Malfatti, Bicquilley, the writers in the mathematical portion of the *Encyclopédie Méthodique*, D'Anieres, Waring, Prevost and Lhuilier, and Young.

The twentieth Chapter is devoted to Laplace; this contains a full account of all his writings on the subject of Probability. First his memoirs in chronological order, are analysed, and then the great work in which he embodied all his own investigations and much derived from other writers. I hope it will be found that all the parts of Laplace's memoirs and work have been carefully and clearly expounded; I would venture to refer for examples to Laplace's method of approximation to integrals, to the Problem of Points, to James Bernoulli's theorem, to the problem taken from Buffon, and above all to the famous method of Least Squares. With respect to the last subject I have availed myself of the guidance of Poisson's luminous analysis, and have given a general investigation, applying to the case of more than one unknown element. I hope I have thus accomplished something towards rendering the theory of this important method accessible to students.

In an Appendix I have noticed some writings which came under my attention during the printing of the work too late to be referred to their proper places.

I have endeavoured to be quite accurate in my statements, and to reproduce the essential elements of the original works which I have analysed. I have however not thought it indispensable to preserve the exact notation in which any investigation was first presented. It did not appear to me of any importance to retain the specific letters for denoting the known and unknown quantities of an algebraical problem which any writer may have chosen to use. Very often the same problem has been discussed by various writers, and in order to compare their methods with any facility it is necessary to use one set of symbols throughout, although each writer may have preferred his peculiar set. In fact by exercising care in the choice of notation I believe that my exposition of contrasted methods has gained much in brevity and clearness without any sacrifice of real fidelity.

I have used no symbols which are not common to all mathematical literature, except $\lfloor n$ which is an abbreviation for the product $\cdot 1 . 2, \ldots n$, frequently but not universally employed : some such symbol is much required, and I do not know of any which is preferable to this, and I have accordingly introduced it in all my publications.

There are three important authors whom I have frequently cited whose works on Probability have passed through more than one edition, Montmort, De Moivre, and Laplace : it may save trouble to a person who may happen to consult the present volume if I here refer to pages 79, 136, and 495 where I have stated which editions I have cited.

Perhaps it may appear that I have allotted too much space to some of the authors whose works I examine, especially the more ancient ; but it is difficult to be accurate or interesting if the narrative is confined to a mere catalogue of titles : and as experience shews that mathematical histories are but rarely undertaken, it seems desirable that they should not be executed on a meagre and inadequate scale.

I will here advert to some of my predecessors in this department of mathematical history ; and thus it will appear that I have not obtained much assistance from them.

In the third volume of Montucla's *Histoire des Mathematiques* pages 380—426 are devoted to the Theory of Probability and the kindred subjects. I have always cited this volume simply by the name *Montucla*, but it is of course well known that the third and fourth volumes were edited from the author's manuscripts after his death by La Lande. I should be sorry to appear ungrateful to Montucla ; his work is indispensable to the student of mathematical history, for whatever may be its defects it remains without any rival. But I have been much disappointed in what he says respecting the Theory of Probability ; he is not copious, nor accurate, nor critical. Hallam has characterised him with some severity, by saying in reference to a point of mathematical history, "Montucla is as superficial as usual :" see a note in the second Chapter of the first volume of the *History of the Literature of Europe.*

There are brief outlines of the history involved or formally incorporated in some of the elementary treatises on the Theory of Probability : I need notice only the best, which occurs in the Treatise on Probability published in the Library of Useful Knowledge. This little work is anonymous, but is known to have been written by Lubbock and Drinkwater.; the former is now Sir John Lubbock, and the latter changed his name to Drinkwater-Bethune : see Professor De Morgan's *Arithmetical Books...* page 106, a letter by him in the *Assurance Magazine,* Vol. IX. page 238, and another letter by him in the *Times,* Dec. 16, 1862. The treatise is interesting and valuable, but I have not been able to agree uniformly with the historical statements which it makes or implies.

A more ambitious work bears the title *Histoire du Calcul des Probabilités depuis ses origines jusqu'à nos jours par Charles Gouraud...* Paris, 1848. This consists of 148 widely printed octavo pages; it is a popular narrative entirely free from mathematical symbols, containing however some important specific references. Exact truth occasionally suffers for the sake of a rhetorical style unsuitable alike to history and to science; nevertheless the general reader will be gratified by a lively and vigorous exhibition of the whole course of the subject. M. Gouraud recognises the value of the purely mathematical part of the Theory of Probability, but will not allow the soundness of the applications which have been made of these mathematical formulæ to questions involving moral or political considerations. His history seems to be a portion of a very extensive essay in three folio volumes containing 1929 pages written when he was very young in competition for a prize proposed by the French Academy on a subject entitled *Théorie de la Certitude;* see the *Rapport* by M. Franck in the *Séances et Travaux de l'Académie des Sciences morales et politiques,* Vol. x. pages 372, 382, and Vol. xi. page 139. It is scarcely necessary to remark that M. Gouraud has gained distinction in other branches of literature since the publication of his work which we have here noticed.

There is one history of our subject which is indeed only a sketch but traced in lines of light by the hand of the great master himself: Laplace devoted a few pages of the introduction to his celebrated work to recording the names of his predecessors and their contributions to the Theory of Probability. It is much to be regretted that he did not supply specific references throughout his treatise, in order to distinguish carefully between that which he merely transmitted from preceding mathematicians and that which he originated himself.

It is necessary to observe that in cases where I point out a similarity between the investigations of two or more writers I do not mean to imply that these investigations could not have been made independently. Such coincidences may occur easily and naturally without any reason for imputing unworthy conduct to those who succeed the author who had the priority in publication. I draw attention to this circumstance because I find with regret that from a passage in my former historical work an inference has been drawn of the kind which I here disclaim. In the case of a writer like Laplace who agrees with his predecessors, not in one or two points but in very many, it is of course obvious that he must have borrowed largely, and we conclude that he supposed the erudition of his contemporaries would be sufficient to prevent them from ascribing to himself more than was justly due.

It will be seen that I have ventured to survey a very extensive field of mathematical research. It has been my aim to estimate carefully and impartially the character and the merit of the numerous memoirs and works which I have examined; my criticism has been intentionally close and searching, but I trust never irreverent nor unjust. I have sometimes explained fully the errors which I detected; sometimes, when the detailed exposition of the error would have required more space than the matter deserved, I have given only a brief indication which may be serviceable to a student of the original production itself. I have not hesitated to introduce remarks and developments of my own whenever the subject seemed to require them. In an elaborate German review of my former publication on mathematical history it was suggested that my own contributions were too prominent, and that the purely historical character of the work was thereby impaired; but I have not been induced to change my plan, for I continue to think that such additions as I have been able to make tend to render the subject more intelligible and more complete, without disturbing in any serious degree the continuity of the history. I cannot venture to expect that in such a difficult subject I shall be quite free from error either in my exposition of the labours of others, or in my own contributions; but I hope that such failures will not be numerous nor important. I shall receive most gratefully intimations of any errors or omissions which may be detected in the work.

I have been careful to corroborate my statements by exact quotations from the originals, and these I have given in the languages in which they were published, instead of translating them ; the course which I have here adopted is I understand more agreeable to foreign students into whose hands the book may fall. I have been careful to preserve the historical notices and references which occurred in the works I studied; and by the aid of the *Table of Contents*, the *Chronological List*, and the *Index*, which accompany the present volume, it will be easy to ascertain with regard to any proposed mathematician down to the close of the eighteenth century, whether he has written anything upon the Theory of Probability.

I have carried the history down to the close of the eighteenth century; in the case of Laplace, however, I have passed beyond this limit: but by far the larger part of his labours on the Theory of Probability were accomplished during the eighteenth century, though collected and republished by him in his celebrated work in the early part of the present century, and it was therefore convenient to include a full account of all his researches in the present volume. There is ample scope for a continuation of the work which should conduct the history through the period which has elapsed since the close of the eighteenth century; and I have already made some progress in the analysis of the rich materials. But when I consider the time and labour expended on the present volume, although reluctant to abandon a long cherished design, I feel far less sanguine than once I did that I shall have the leisure to arrive at the termination I originally ventured to propose to myself.

Although I wish the present work to be regarded principally as a history, yet there are two other aspects under which it may solicit the attention of students. It may claim the title of a comprehensive treatise on the Theory of Probability, for it assumes in the reader only so much knowledge as can be gained from an elementary book on Algebra, and introduces him to almost every process and every species of problem which the literature of the subject can furnish; or the work may be considered more specially as a commentary on the celebrated treatise of Laplace,— and perhaps no mathematical treatise ever more required or more deserved such an accompaniment.

My sincere thanks are due to Professor De Morgan, himself conspicuous among cultivators of the Theory of Probability, for the kind interest which he has taken in my work, for the loan of scarce books, and for the suggestion of valuable references. A similar interest was manifested by one prematurely lost to science, whose mathematical and metaphysical genius, attested by his marvellous work on the *Laws of Thought*, led him naturally and rightfully in that direction which Pascal and Leibnitz had marked with the unfading lustre of their approbation; and who by his rare ability, his wide attainments, and his attractive character, gained the affection and the reverence of all who knew him.

I. TODHUNTER.

CAMBRIDGE,
May, 186̃5.

CONTENTS.

CHAPTER I.

CARDAN. KEPLER. GALILEO.

1. THE practice of games of chance must at all times have directed attention to some of the elementary considerations of the Theory of Probability. Libri finds in a commentary on the *Divina Commedia* of Dante the earliest indication of the different probability of the various throws which can be made with three dice. The passage from the commentary is quoted by Libri; it relates to the first line of the sixth canto of the *Purgatorio*. The commentary was published at Venice in 1477. See Libri, *Histoire des Sciences Mathématiques en Italie*, Vol. II. p. 188.

2. Some other intimations of traces of our subject in older writers are given by Gouraud in the following passage, unfortunately without any precise reference.

Les anciens paraissent avoir entièrement ignoré cette sorte de calcul. L'érudition moderne en a, il est vrai, trouvé quelques traces dans un pöeme en latin barbare intitulé : *De Vetula*, œuvre d'un moine du Bas-Empire, dans un commentaire de Dante de la fin du XV⁰ siècle, et dans les écrits de plusieurs mathématiciens italiens du moyen âge et de la renaissance, Pacioli, Tartaglia, Peverone ;......Gouraud, *Histoire du Calcul des Probabilités*, page 3.

3. A treatise by Cardan entitled *De Ludo Aleœ* next claims our attention. This treatise was published in 1663, in the first volume of the edition of Cardan's collected works, long after Cardan's death, which took place in 1576.

1

Montmort says, " Jerôme Cardan a donné un Traité De Ludo
Aleæ; mais on n'y trouve que de l'érudition et des réflexions
morales." *Essai d'Analyse...* p. XL.   Libri says, " Cardan a écrit
un traité spécial de *Ludo Aleæ*, où se trouvent résolues plusieurs
questions d'analyse combinatoire." *Histoire*, Vol. III. p. 176. The
former notice ascribes too little and the latter too much to
Cardan.

4. Cardan's treatise occupies fifteen folio pages, each containing
two columns; it is so badly printed as to be scarcely intelligible.
Cardan himself was an inveterate gambler ; and his treatise may
be best described as a gambler's manual.   It contains much miscellaneous matter connected with gambling, such as descriptions of
games and an account of the precautions necessary to be employed
in order to guard against adversaries disposed to cheat : the
discussions relating to chances form but a small portion of the
treatise.

5.   As a specimen of Cardan's treatise we will indicate the
contents of his thirteenth Chapter.   He shews the number of
cases which are favourable for each throw that can be made with
two dice.   Thus two and twelve can each be thrown in only one
way.   Eleven can be thrown in two ways, namely, by six appearing on either of the two dice and five on the other.   Ten can be
thrown in three ways, namely, by five appearing on each of the
dice, or by six appearing on either and four on the other.   And
so on.

Cardan proceeds, "Sed in Ludo fritilli undecim puncta adjicere
decet, quia una Alea potest ostendi."...The meaning apparently is,
that the person who throws the two dice is to be considered to
have thrown a given number when one of the dice *alone* exhibits
that number, as well as when the number is made up by the sum
of the numbers on the two dice.   Hence, for six or any smaller
number eleven more favourable cases arise besides those already
considered.

Cardan next exhibits correctly the number of cases which are
favourable for each throw that can be made with three dice.   Thus
three and eighteen can each be thrown in only one way ; four and

seventeen can each be thrown in three ways; and so on. Cardan also gives the following list of the number of cases *in Fritillo* :

1	2	3	4	5	6	7	8	9	10	11	12
108	111	115	120	126	133	33	36	37	36	33	26

Here we have corrected two misprints by the aid of Cardan's verbal statements. It is not obvious what the table means. It might be supposed, in analogy with what has already been said, that if a person throws three dice he is to be considered to have thrown a given number when one of the dice *alone* exhibits that number, or when two dice together exhibit it as their sum, as well as when all the three dice exhibit it as their sum : and this would agree with Cardan's remark, that for numbers higher than twelve the favourable cases are the same as those already given by him for three dice. But this meaning does not agree with Cardan's table; for with this meaning we should proceed thus to find the cases favourable for an ace: there are 5^3 cases in which no ace appears, and there are 6^3 cases in all, hence there are $6^3 - 5^3$ cases in which we have an ace or aces, that is 91 cases, and not 108 as Cardan gives.

The connexion between the numbers in the ordinary mode of using dice and the numbers which Cardan gives appears to be the following. Let n be the number of cases which are favourable to a given throw in the ordinary mode of using three dice, and N the number of cases favourable to the same throw in Cardan's mode; let m be the number of cases favourable to the given throw in the ordinary mode of using *two* dice. Then for any throw not less than thirteen, $N = n$; for any throw between seven and twelve, both inclusive, $N = 3m + n$; for any throw not greater than six, $N = 108 + 3m + n$. There is only one deviation from this law ; Cardan gives 26 favourable cases for the throw twelve, and our proposed law would give $3 + 25$, that is 28.

We do not, however, see what simple mode of playing with three dice can be suggested which shall give favourable cases agreeing in number with those determined by the above law.

6. Some further account of Cardan's treatise will be found

in the *Life of Cardan*, by Henry Morley, Vol. I. pages 92—95.
Mr Morley seems to misunderstand the words of Cardan which he
quotes on his page 92, in consequence of which he says that
Cardan "lays it down coolly and philosophically, as one of his first
axioms, that dice and cards ought to be played for money." In
the passage quoted by Mr Morley, Cardan seems rather to admit
the propriety of moderation in the stake, than to assert that there
must be a stake; this moderation Cardan recommends elsewhere,
as for example in his second Chapter. Cardan's treatise is briefly
noticed in the article *Probability* of the *English Cyclopædia*.

7. Some remarks on the subject of chance were made by
Kepler in his work *De Stella Nova in pede Serpentarii*, which was
published in 1606. Kepler examines the different opinions on the
cause of the appearance of a new star which shone with great
splendour in 1604, and among these opinions the Epicurean notion
that the star had been produced by the fortuitous concurrence
of atoms. The whole passage is curious, but we need not reproduce it, for it is easily accessible in the reprint of Kepler's works
now in the course of publication; see *Joannis Kepleri Astronomi
Opera Omnia edidit Dr Ch. Frisch*, Vol. II. pp. 714—716. See
also the *Life of Kepler* in the *Library of Useful Knowledge*, p. 13.
The passage attracted the attention of Dugald Stewart; see his
Works edited by Hamilton, Vol. I. p. 617.

A few words of Kepler may be quoted as evidence of the
soundness of his opinions; he shows that even such events as
throws of dice do not happen without a cause. He says,

Quare hoc jactu Venus cecidit, illo canis? Nimirum lusor hac vice
tessellam alio latere arripuit, aliter manu condidit, aliter intus agitavit,
alio impetu animi manusve projecit, aliter interflavit aura, alio loco
alvei impegit. Nihil hic est, quod sua causa sic caruerit, si quis ista
subtilia posset consectari.

8. The next investigation which we have to notice is that by
Galileo, entitled *Considerazione sopra il Giuco dei Dadi*. The date
of this piece is unknown; Galileo died in 1642. It appears that
a friend had consulted Galileo on the following difficulty: with
three dice the number 9 and the number 10 can each be produced
by six different combinations, and yet experience shows that the

number 10 is oftener thrown than the number 9. Galileo makes a careful and accurate analysis of all the cases which can occur, and he shows that out of 216 possible cases 27 are favourable to the appearance of the number 10, and 25 are favourable to the appearance of the number 9.

The piece will be found in Vol. XIV. pages 293—296, of *Le Opere* .... *di Galileo Galilei*, Firenze, 1855. From the *Bibliografia Galileiana* given in Vol. XV. of this edition of Galileo's works we learn that the piece first appeared in the edition of the works published at Florence in 1718 : here it occurs in Vol. III. pages 119—121.

9. Libri in his *Histoire des Sciences Mathématiques en Italie*, Vol. IV. page 288, has the following remark relating to Galileo : ..."l'on voit, par ses lettres, qu'il s'était longtemps occupé d'une question délicate et non encore résolue, relative à la manière de compter les erreurs en raison géométrique ou en proportion arithmétique, question qui touche également au calcul des probabilités et à l'arithmétique politique." Libri refers to Vol. II. page 55, of the edition of Galileo's works published at Florence in 1718 ; there can, however, be no doubt, that he means Vol. III. The letters will be found in Vol. XIV. pages 231—284 of *Le Opere...di Galileo Galilei*, Firenze, 1855 ; they are entitled *Lettere intorno la stima di un cavallo*. We are informed that in those days the Florentine gentlemen, instead of wasting their time in attention to ladies, or in the stables, or in excessive gaming, were accustomed to improve themselves by learned conversation in cultivated society. In one of their meetings the following question was proposed ; a horse is really worth a hundred crowns, one person estimated it at ten crowns and another at a thousand ; which of the two made the more extravagant estimate ? Among the persons who were consulted was Galileo ; he pronounced the two estimates to be equally extravagant, because the ratio of a thousand to a hundred is the same as the ratio of a hundred to ten. On the other hand, a priest named Nozzolini, who was also consulted, pronounced the higher estimate to be more extravagant than the other, because the excess of a thousand above a hundred is greater than that of a hundred above ten. Various letters of

Galileo and Nozzolini are printed, and also a letter of Benedetto
Castelli, who took the same side as Galileo; it appears that Galileo
had the same notion as Nozzolini when the question was first
proposed to him, but afterwards changed his mind. The matter
is discussed by the disputants in a very lively manner, and some
amusing illustrations are introduced. It does not appear, however,
that the discussion is of any scientific interest or value, and the
terms in which Libri refers to it attribute much more importance
to Galileo's letters than they deserve. The Florentine gentlemen
when they renounced the frivolities already mentioned might have
investigated questions of greater moment than that which is here
brought under our notice.

CHAPTER II.

PASCAL AND FERMAT.

10. THE indications which we have given in the preceding Chapter of the subsequent Theory of Probability are extremely slight; and we find that writers on the subject have shewn a justifiable pride in connecting the true origin of their science with the great name of Pascal. Thus,

Elle doit la naissance à deux Géomètres français du dix-septième siècle, si fécond en grands hommes et en grandes découvertes, et peut-être de tous les siècles celui qui fait le plus d'honneur à l'esprit humain. Pascal et Fermat se proposèrent et résolurent quelques problèmes sur les probabilités...Laplace, *Théorie...des Prob.* 1st edition, page 3.

Un problème relatif aux jeux de hasard, proposé à un austère janséniste par un homme du monde a été l'origine du calcul des probabilités. Poisson, *Recherches sur la Prob.* page 1.

The problem which the Chevalier de Méré (a reputed gamester) proposed to the recluse of Port Royal (not yet withdrawn from the interests of science by the more distracting contemplation of the "greatness and the misery of man"), was the first of a long series of problems, destined to call into existence new methods in mathematical analysis, and to render valuable service in the practical concerns of life." Boole, *Laws of Thought,* page 243.

11. It appears then that the Chevalier de Méré proposed certain questions to Pascal; and Pascal corresponded with Fermat on the subject of these questions. Unfortunately only a portion of the correspondence is now accessible. Three letters

of Pascal to Fermat on this subject, which were all written in 1654, were published in the *Varia Opera Mathematica D. Petri de Fermat...*Tolosæ, 1679, pages 179—188. These letters are reprinted in Pascal's works; in the edition of Paris, 1819, they occur in Vol. IV. pages 360—388. This volume of Pascal's works also contains some letters written by Fermat to Pascal, which are not given in Fermat's works; two of these relate to Probabilities, one of them is in reply to the second of Pascal's three letters, and the other apparently is in reply to a letter from Pascal which has not been preserved; see pages 385—388 of the volume.

We will quote from the edition of Pascal's works just named. Pascal's first letter indicates that some previous correspondence had occurred which we do not possess; the letter is dated July 29, 1654. He begins,

Monsieur, L'impatience me prend aussi-bien qú'à vous; et quoique je sois encore au lit, je ne puis m'empêcher de vous dire que je reçus hier au soir, de la part de M. de Carcavi, votre lettre sur les partis, que j'admire si fort, que je ne puis vous le dire. Je n'ai pas le loisir de m'étendre; mais en un mot vous avez trouvé les deux partis des dés et des parties dans la parfaite justesse : j'en suis tout satisfait ; car je ne doute plus maintenant que je ne sois dans la vérité, après la rencontre admirable où je me trouve avec vous. J'admire bien davantage la méthode des parties que celle des dés ; j'avois vu plusieurs personnes trouver celle des dés, comme M. le chevalier de Meré, qui est celui qui m'a proposé ces questions, et aussi M. de Roberval; mais M. de Meré n'avoit jamais pu trouver la juste valeur des parties, ni de biais pour y arriver : de sorte que je me trouvois seul qui eusse connu cette proportion.

Pascal's letter then proceeds to discuss the problem to which it appears from the above extract he attached the greatest importance. It is called in English the Problem of Points, and is thus enunciated : two players want each a given number of points in order to win; if they separate without playing out the game, how should the stakes be divided between them?

The question amounts to asking what is the probability which each player has, at any given stage of the game, of winning the game. In the discussion between Pascal and Fermat it is supposed that the players have equal chances of winning a single point.

12. We will now give an account of Pascal's investigations on the Problem of Points; in substance we translate his words.

The following is my method for determining the share of each player, when, for example, two players play a game of three points and each player has staked 32 pistoles.

Suppose that the first player has gained two points and the second player one point; they have now to play for a point on this condition, that if the first player gains he takes all the money which is at stake, namely 64 pistoles, and if the second player gains each player has two points, so that they are on terms of equality, and if they leave off playing each ought to take 32 pistoles. Thus, if the first player gains, 64 pistoles belong to him, and if he loses, 32 pistoles belong to him. If, then, the players do not wish to play this game, but to separate without playing it, the first player would say to the second "I am certain of 32 pistoles even if I lose this game, and as for the other 32 pistoles perhaps I shall have them and perhaps you will have them; the chances are equal. Let us then divide these 32 pistoles equally and give me also the 32 pistoles of which I am certain." Thus the first player will have 48 pistoles and the second 16 pistoles.

Next, suppose that the first player has gained two points and the second player none, and that they are about to play for a point; the condition then is that if the first player gains this point he secures the game and takes the 64 pistoles, and if the second player gains this point the players will then be in the situation already examined, in which the first player is entitled to 48 pistoles, and the second to 16 pistoles. Thus if they do not wish to play, the first player would say to the second "If I gain the point I gain 64 pistoles; if I lose it I am entitled to 48 pistoles. Give me then the 48 pistoles of which I am certain, and divide the other 16 equally, since our chances of gaining the point are equal." Thus the first player will have 56 pistoles and the second player 8 pistoles.

Finally, suppose that the first player has gained one point and

the second player none. If they proceed to play for a point the
condition is that if the first player gains it the players will be in
the situation first examined, in which the first player is entitled to
56 pistoles ; if the first player loses the point each player has then
a point, and each is entitled to 32 pistoles. Thus if they do not
wish to play, the first player would say to the second "Give me
the 32 pistoles of which I am certain and divide the remainder of
the 56 pistoles equally, that is, divide 24 pistoles equally." Thus
the first player will have the sum of 32 and 12 pistoles, that is
44 pistoles, and consequently the second will have 20 pistoles.

13. Pascal then proceeds to enunciate two general results
without demonstrations. We will give them in modern notation.

(1) Suppose each player to have staked a sum of money
denoted by A ; let the number of points in the game be $n + 1$, and
suppose the first player to have gained n points and the second
player none. If the players agree to separate without playing
any more the first player is entitled to $2A - \dfrac{A}{2^n}$.

(2) Suppose the stakes and the number of points in the game
as before, and suppose that the first player has gained one point
and the second player none. If the players agree to separate
without playing any more, the first player is entitled to

$$A + A \frac{1 \cdot 3 \cdot 5 \ldots (2n-1)}{2 \cdot 4 \cdot 6 \ldots \quad 2n} .$$

Pascal intimates that the second theorem is difficult to prove.
He says it depends on two propositions, the first of which is purely
arithmetical and the second of which relates to chances. The
first amounts in fact to the proposition in modern works on
Algebra which gives the sum of the co-efficients of the terms in
the Binomial Theorem. The second consists of a statement of
the value of the first player's chance by means of combinations,
from which by the aid of the arithmetical proposition the value
above given is deduced. The demonstrations of these two results
may be obtained from a general theorem which will be given later
in the present Chapter; see Art. 23. Pascal adds a table which

exhibits a complete statement of all the cases which can occur in a game of six points.

14. Pascal then proceeds to another topic. He says

Je n'a pas le temps de vous envoyer la démonstration d'une difficulté qui étonnoit fort M. de Meré : car il a très-bon esprit, mais il n'est pas géomètre ; c'est, comme vous savez, un grand défaut; et même il ne comprend pas qu'une ligne mathématique soit divisible à l'infini, et croit fort bien entendre qu'elle est composée de points en nombre fini, et jamais je n'ai pu l'en tirer ; si vous pouviez le faire, on le rendroit parfait. Il me disoit donc qu'il avoit trouvé fausseté dans les nombres par cette raison.

The difficulty is the following. If we undertake to throw a six with one die the odds are in favour of doing it in four throws, being as 671 to 625 ; if we undertake to throw two sixes with two dice the odds are not in favour of doing it in twenty-four throws. Nevertheless 24 is to 36, which is the number of cases with two dice, as 4 is to 6, which is the number of cases with one die. Pascal proceeds

Voilà quel étoit son grand scandale, qui lui faisoit dire hautement que les propositions n'étoient pas constantes, et que l'arithmétique se démentoit. Mais vous en verrez bien aisément la raison, par les principes où vous êtes.

15. In Pascal's letter, as it is printed in Fermat's works, the name de Méré is not given in the passage we have quoted in the preceding article ; a blank occurs after the M. It seems, however, to be generally allowed that the blank has been filled up correctly by the publishers of Pascal's works : Montmort has no doubt on the matter ; see his p. XXXII. See also Gouraud, p. 1 ; Lubbock and Drinkwater, p. 41. But there is certainly some difficulty. For in the extract which we have given in Art. 11, Pascal states that M. de Méré could solve one problem, *celle des dés*, and seems to imply that he failed only in the Problem of Points. Montucla says that the Problem of Points was proposed to Pascal by the Chevalier de Méré, "qui lui en proposa aussi quelques autres sur le jeu de dés, comme de déterminer en combien de coups on peut parier d'amener une rafle, &c. Ce chevalier, plus bel esprit que

géomètre ou analyste, résolut à la vérité ces dernières, qui ne sont pas bien difficiles; mais il échoua pour le précédent, ainsi que Roberval, à qui Pascal le proposa." p. 384. These words would seem to imply that, in Montucla's opinion, M. de Méré was not the person alluded to by Pascal in the passage we have quoted in Article 14. We may remark that Montucla was not justified in suggesting that M. de Méré must have been an indifferent mathematician, because he could not solve the Problem of Points; for the case of Roberval shews that an eminent mathematician at that time might find the problem too difficult.

Leibnitz says of M. de Méré, "Il est vrai cependant que le Chevalier avoit quelque génie extraordinaire, même pour les Mathématiques;" and these words seem intended seriously, although in the context of this passage Leibnitz is depreciating M. de Méré. Leibnitii, *Opera Omnia, ed. Dutens,* Vol. II. part 1. p. 92.

In the *Nouveaux Essais,* Liv. IV. Chap. 16, Leibnitz says, "Le Chevalier de Méré dont les *Agréments* et les autres ouvrages ont été imprimés, homme d'un esprit pénétrant et qui étoit joueur et philosophe."

It must be confessed that Leibnitz speaks far less favourably of M. de Méré in another place, *Opera,* Vol. v. p. 203. From this passage, and from a note in the article on *Zeno* in Bayle's Dictionary, to which Leibnitz refers, it appears that M. de Méré maintained that a magnitude was not infinitely divisible: this assists in identifying him with Pascal's friend who would have been perfect had it not been for this single error.

On the whole, in spite of the difficulty which we have pointed out, we conclude that M. de Méré really was the person who so strenuously asserted that the propositions of Arithmetic were inconsistent with themselves; and although it may be unfortunate for him that he is now known principally for his error, it is some compensation that his name is indissolubly associated with those of Pascal and Fermat in the history of the Theory of Probability.

16. The remainder of Pascal's letter relates to other mathematical topics. Fermat's reply is not extant ; but the nature of it may be inferred from Pascal's next letter. It appears that Fermat

sent to Pascal a solution of the Problem of Points depending on combinations.

Pascal's second letter is dated August 24th, 1654. He says that Fermat's method is satisfactory when there are only two players, but unsatisfactory when there are more than two. Here Pascal was wrong as we shall see. Pascal then gives an example of Fermat's method, as follows. Suppose there are two players, and that the first wants two points to win and the second three points. The game will then certainly be decided in the course of four trials. Take the letters a and b and write down all the combinations that can be formed of four letters. These combinations are the following, 16 in number:

a	a	a	a	a	b	a	a	b	a	a	a	b	b	a	a
a	a	a	b	a	b	a	b	b	a	a	b	b	b	a	b
a	a	b	a	a	b	b	a	b	a	b	a	b	b	b	a
a	a	b	b	a	b	b	b	b	a	b	b	b	b	b	b

Now let A denote the player who wants two points, and B the player who wants three points. Then in these 16 combinations every combination in which a occurs twice or oftener represents a case favourable to A, and every combination in which b occurs three times or oftener represents a case favourable to B. Thus on counting them it will be found that there are 11 cases favourable to A, and 5 cases favourable to B; and as these cases are all equally likely, A's chance of winning the game is to B's chance as 11 is to 5.

17. Pascal says that he communicated Fermat's method to Roberval, who objected to it on the following ground. In the example just considered it is supposed that four trials will be made; but this is not necessarily the case; for it is quite possible that the first player may win in the next two trials, and so the game be finished in two trials. Pascal answers this objection by stating, that although it is quite possible that the game *may* be finished in two trials or in three trials, yet we are at liberty to conceive that the players agree to have four trials, because, even if the game be decided in fewer than four trials, no difference will be

made in the decision by the superfluous trial or trials. Pascal puts this point very clearly.

In the context of the first passage quoted from Leibnitz in Art. 15, he refers to " les belles pensées *de Alea*, de Messieurs *Fermat, Pascal* et *Huygens*, où Mr. Roberval ne pouvoit ou ne vouloit rien comprendre."

The difficulty raised by Roberval was in effect reproduced by D'Alembert, as we shall see hereafter.

18. Pascal then proceeds to apply Fermat's method to an example in which there are *three* players. Suppose that the first player wants one point, and each of the other players two points. The game will then be certainly decided in the course of three trials. Take the letters *a*, *b*, *c* and write down all the combinations which can be formed of three letters. These combinations are the following, 27 in number:

a	a	a	b	a	a	c	a	a
a	a	b	b	a	b	c	a	b
a	a	c	b	a	c	c	a	c
a	b	a	b	b	a	c	b	a
a	b	b	b	b	b	c	b	b
a	b	c	b	b	c	c	b	c
a	c	a	b	c	a	c	c	a
a	c	b	b	c	b	c	c	b
a	c	c	b	c	c	c	c	c

Let *A* denote the player who wants one point, and *B* and *C* the other two players. By examining the 27 cases, Pascal finds 13 which are exclusively favourable to *A*, namely, those in which *a* occurs twice or oftener, and those in which *a*, *b*, and *c* each occur once. He finds 3 cases which he considers equally favourable to *A* and *B*, namely, those in which *a* occurs once and *b* twice; and similarly he finds 3 cases equally favourable to *A* and *C*. On the whole then the number of cases favourable to *A* may be considered to be $13 + \frac{3}{2} + \frac{3}{2}$, that is 16. Then Pascal finds 4 cases which are exclusively favourable to *B*, namely those represented by *bbb*, *çbb*, *bcb*, and *bbc;* and thus on the whole the number of cases favourable to B may be considered to be $4 + \frac{3}{2}$, that is $5\frac{1}{2}$. Similarly the number of cases favourable to C may be considered to be $5\frac{1}{2}$. Thus it would appear that the chances of A, B, and C are respectively as 16, $5\frac{1}{2}$, and $5\frac{1}{2}$.

Pascal, however, says that by his own method he had found that the chances are as 17, 5, and 5. He infers that the difference arises from the circumstance that in Fermat's method it is assumed that three trials will necessarily be made, which is not assumed in his own method. Pascal was wrong in supposing that the true result could be affected by assuming that three trials would necessarily be made; and indeed, as we have seen, in the case of two players, Pascal himself had correctly maintained against Roberval'that a similar assumption was legitimate.

19. A letter from Pascal to Fermat is dated August 29th, 1654. Fermat refers to the Problem of Points for the case of three players; he says that the proportions 17, 5, and 5 are correct for the example which we have just considered. This letter, however, does not seem to be the reply to Pascal's of August 24th, but to an earlier letter which has not been preserved.

On the 25th of September Fermat writes a letter to Pascal, in which Pascal's error is pointed out. Pascal had supposed that such a combination as acc represented a case equally favourable to A and C; but, as Fermat says, this case is exclusively favourable to A, because here A gains one point before C gains one; and as A only wanted one point the game is thus decided in his favour. When the necessary correction is made, the result is, that the chances of A, B, and C are as 17, 5, and 5, as Pascal had found by his own method.

Fermat then gives another solution, for the sake of Roberval, in which he does not assume that three trials will necessarily be made; and he arrives at the same result as before.

In the remainder of his letter Fermat enunciates some of his memorable propositions relating to the Theory of Numbers.

Pascal replied on October 27th, 1654, to Fermat's letter, and said that he was entirely satisfied.

20. There is another letter from Fermat to Pascal which is not dated. It relates to a simple question which Pascal had proposed to Fermat. A person undertakes to throw a six with a die in eight throws ; supposing him to have made three throws without success, what portion of the stake should he be allowed to take on condition of giving up his fourth throw? The chance of success is $\frac{1}{6}$, so that he should be allowed to take $\frac{1}{6}$ of the stake on condition of giving up his throw. But suppose that we wish to estimate the value of the fourth throw *before any throw is made*. The first throw is worth $\frac{1}{6}$ of the stake; the second is worth $\frac{1}{6}$ of what remains, that is $\frac{5}{36}$ of the stake ; the third throw is worth $\frac{1}{6}$ of what now remains, that is $\frac{25}{216}$ of the stake ; the fourth throw is worth $\frac{1}{6}$ of what now remains, that is $\frac{125}{1296}$ of the stake. .

It seems possible from Fermat's letter that Pascal had not distinguished between the two cases ; but Pascal's letter, to which Fermat's is a reply, has not been preserved, so that we cannot be certain on the point.

21. We see then that the Problem of Points was the principal question discussed by Pascal and Fermat, and it was certainly not exhausted by them. For they confined themselves to the case in which the players are supposed to possess equal skill; and their methods would have been extremely laborious if applied to any examples except those of the most simple kind. Pascal's method seems the more refined ; the student will perceive that it depends on the same principles as the modern solution of the problem by the aid of the Calculus of Finite Differences ; see Laplace, *Théorie...des Prob.* page 210.

Gouraud awards to Fermat's treatment of the problem an amount of praise which seems excessive, whether we consider that treatment absolutely or relatively in comparison with Pascal's ; see his page 9.

22. We have next to consider Pascal's *Traité du triangle arithmétique*. This treatise was printed about 1654, but not published until 1665 ; see Montucla, p. 387. The treatise will be found in the fifth volume of the edition of Pascal's works to which we have already referred.

The *Arithmetical Triangle* in its simplest form consists of the following table:

```
1   1   1   1   1   1   1   1   1   1 ...
1   2   3   4   5   6   7   8   9 ...
1   3   6   10  15  21  28  36 ...
1   4   10  20  35  56  84 ...
1   5   15  35  70  126 ...
1   6   21  56  126 ...
1   7   28  84 ...
1   8   36 ...
1   9 ...
1 ...
```

In the successive horizontal rows we have what are now called the *figurate numbers*. Pascal distinguishes them into orders. He calls the simple units 1, 1, 1, 1,... which form the first row, numbers of the first order; he calls the numbers 1, 2, 3, 4,... which form the second row, numbers of the second order; and so on. The numbers of the third order 1, 3, 6, 10,... had already received the name of *triangular* numbers; and the numbers of the fourth order 1, 4, 10, 20,... the name of *pyramidal* numbers. Pascal says that the numbers of the fifth order 1, 5, 15, 35,... had not yet received an express name, and he proposes to call them *triangulo-triangulaires*.

In modern notation the nth term of the rth order is

$$\frac{n(n+1) \dots (n+r-2)}{\lfloor r-1}.$$

Pascal constructs the *Arithmetical Triangle* by the following definition; each number is the sum of that immediately above it and that immediately to the left of it. Thus

$$10 = 4 + 6, \qquad 35 = 20 + 15, \qquad 126 = 70 + 56, \dots$$

The properties of the numbers are developed by Pascal with great skill and distinctness. For example, suppose we require the sum of the first n terms of the rth order : the sum is equal to the number of the combinations of $n + r - 1$ things taken r at a time, and Pascal establishes this by an inductive proof.

2

23. Pascal applies his *Arithmetical Triangle* to various subjects; among these we have the Problem of Points, the Theory of Combinations, and the Powers of Binomial Quantities. We are here only concerned with the application to the first subject.

In the *Arithmetical Triangle* a line drawn so as to cut off an equal number of units from the top horizontal row and the extreme left-hand vertical column is called a *base*.

The bases are numbered, beginning from the top left-hand corner. Thus the *tenth* base is a line drawn through the numbers 1, 9, 36, 84, 126, 126, 84, 36, 9, 1. It will be perceived that the rth base contains r numbers.

Suppose then that A wants m points and that B wants n points. Take the $(m+n)^{th}$ base; the chance of A is to the chance of B as the sum of the first n numbers of the base, beginning at the highest row, is to the sum of the last m numbers. Pascal establishes this by induction.

Pascal's result may be easily shewn to coincide with that obtained by other methods. For the terms in the $(m+n)^{th}$ base are the coefficients in the expansion of $(1+x)^{m+n-1}$ by the Binomial Theorem. Let $m+n-1=r$; then Pascal's result amounts to saying that the chance of A is proportional to

$$1+r+\frac{r(r-1)}{1.2}+\ldots+\frac{r(r-1)\ldots(r-n+2)}{\underline{|n-1}},$$

and the chance of B proportional to

$$1+r+\frac{r(r-1)}{1.2}+\ldots+\frac{r(r-1)\ldots(r-m+2)}{\underline{|m-1}}.$$

This agrees with the result now usually given in elementary treatises; see *Algebra*, Chapter LIII.

24. Pascal then notices some particular examples. (1) Suppose that A wants one point and B wants n points. (2) Suppose that A wants $n-1$ points and B wants n points. (3) Suppose that A wants $n-2$ points and B wants n points. An interesting relation holds between the second and third examples, which we will exhibit.

Let M denote the number of cases which are favourable to A, and N the number of cases which are favourable to B. Let $r = 2n - 2$.

In the second example we have

$$M + N = 2^r,$$

$$M - N = \frac{\lfloor r}{\lfloor n - 1 \ \lfloor n - 1} = \lambda \text{ say.}$$

Then if $2S$ denote the whole sum at stake, A is entitled to $\frac{2S}{2^r} \cdot \frac{2^r + \lambda}{2}$, that is to $\frac{S}{2^r} (2^r + \lambda)$; so that he may be considered to have recovered his own stake and to have won the fraction $\frac{\lambda}{2^r}$ of his adversary's stake.

In the third example we have

$$M + N = 2^{r-1},$$

$$M - N = \frac{2 \lfloor r - 1}{\lfloor n - 1 \ \lfloor n - 2} = \frac{2(n-1) \lfloor r - 1}{\lfloor n - 1 \ \lfloor n - 1} = \frac{2 \lambda (n-1)}{r} = \lambda.$$

Thus we shall find that A may be considered to have recovered his own stake, and to have won the fraction $\frac{\lambda}{2^{r-1}}$ of his adversary's stake.

Hence, comparing the second and third examples, we see that if the player who wins the first point also wins the second point, his advantage when he has gained the second point is double what it was when he had gained the first point, whatever may be the number of points in the game.

25. We have now analysed all that has been preserved of Pascal's researches on our subject. It seems however that he had intended to collect these researches into a complete treatise. A letter is extant addressed by him *Celeberrimæ Matheseos Academiæ Parisiensi;* this Academy was one of those voluntary associations which preceded the formation of formal scientific societies : see Pascal's Works, Vol. IV. p. 356. In the letter Pascal enumerates various treatises which he had prepared and which he hoped to

publish, among which was to be one on chances. His language shews that he had a high opinion of the novelty and importance of the matter he proposed to discuss ; he says,

Novissima autem ac penitùs intentatæ materiæ tractatio, scilicet de *compositione aleæ in ludis ipsi subjectis,* quod gallico nostro idiomate dicitur (*faire les partis des jeux*): ubi anceps fortuna æquitate rationis ita reprimitur ut utrique lusorum quod jure competit exactè semper assignetur. Quod quidem eo fortiùs ratiocinando quærendum, quo minùs tentando investigari possit : ambigui enim sortis eventus fortuitæ contingentiæ potiùs quam naturali necessitati meritò tribuuntur. Ideò res hactenus erravit incerta ; nunc autem quæ experimento rebellis fuerat, rationis dominium effugere non potuit : eam quippe tantâ securitate in artem per geometriam reduximus, ut certitudinis ejus particeps facta, jam audacter prodeat ; et sic matheseos demonstrationes cum aleæ incertitudine jungendo, et quæ contraria videntur conciliando, ab utrâque nominationem suam accipiens stupendum hunc titulum jure sibi arrogat : *aleæ geometria.*

But the design was probably never accomplished. The letter is dated 1654; Pascal died in 1662, at the early age of 39.

26. Neglecting the trifling hints which may be found in preceding writers we may say that the Theory of Probability really commenced with Pascal and Fermat; and it would be difficult to find two names which could confer higher honour on the subject.

The fame of Pascal rests on an extensive basis, of which mathematical and physical science form only a part; and the regret which we may feel at his renunciation of the studies in which he gained his earliest renown may be diminished by reflecting on his memorable *Letters,* or may be lost in deeper sorrow when we contemplate the fragments which alone remain of the great work on the evidences of religion that was to have engaged the efforts of his maturest powers.

The fame of Fermat is confined to a narrower range; but it is of a special kind which is without a parallel in the history of science. Fermat enunciated various remarkable propositions in the theory of numbers. Two of these are more important than the rest; one of them after baffling the powers of Euler and Lagrange finally yielded to Cauchy, and the other remains still unconquered. The interest which attaches to the propositions is increased by the uncertainty which subsists as to whether Fermat himself had succeeded in demonstrating them.

The French government in the time of Louis Philippe assigned a grant of money for publishing a new edition of Fermat's works; but unfortunately the design has never been accomplished. The edition which we have quoted in Art. 11 has been reprinted in facsimile by Friedlander at Berlin in 1861.

27. At the time when the Theory of Probability started from the hands of Pascal and Fermat, they were the most distinguished mathematicians of Europe. Descartes died in 1650, and Newton and Leibnitz were as yet unknown; Newton was born in 1642, and Leibnitz in 1646. Huygens was born in 1629, and had already given specimens of his powers and tokens of his future eminence; but at this epoch he could not have been placed on the level of Pascal and Fermat. In England Wallis, born in 1616, and appointed Savilian professor of geometry at Oxford in 1649, was steadily rising in reputation, while Barrow, born in 1630, was not appointed Lucasian professor of mathematics at Cambridge until 1663.

It might have been anticipated that a subject interesting in itself and discussed by the two most distinguished mathematicians of the time would have attracted rapid and general attention; but such does not appear to have been the case. The two great men themselves seem to have been indifferent to any extensive publication of their investigations; it was sufficient for each to gain the approbation of the other. Pascal finally withdrew from science and the world; Fermat devoted to mathematics only the leisure of a laborious life, and died in 1665.

The invention of the Differential Calculus by Newton and Leibnitz soon offered to mathematicians a subject of absorbing interest; and we shall find that the Theory of Probability advanced but little during the half century which followed the date of the correspondence between Pascal and Fermat.

CHAPTER III.

HUYGENS.

28. WE have now to speak of a treatise by Huygens entitled *De Ratiociniis in Ludo Aleæ.* This treatise was first printed by Schooten at the end of his work entitled *Francisci à Schooten Exercitationum Mathematicarum Libri quinque;* it occupies pages 519...534 of the volume. The date 1658 is assigned to Schooten's work by Montucla, but the only copy which I have seen is dated 1657.

Schooten had been the instructor of Huygens in mathematics; and the treatise which we have to examine was communicated by Huygens to Schooten written in their vernacular tongue, and Schooten translated it into Latin.

It appears from a letter written by Schooten to Wallis, that Wallis had seen and commended Huygens's treatise; see Wallis's *Algebra,* 1693, p. 833.

Leibnitz commends it. Leibnitii *Opera Omnia, ed. Dutens,* Vol. VI. part 1, p. 318.

29. In his letter to Schooten which is printed at the beginning of the treatise Huygens refers to his predecessors in these words: Sciendum verò, quod jam pridem inter præstantissimos totâ Galliâ Geometras calculus hic agitatus fuerit, ne quis indebitam mihi primæ inventionis gloriam hac in re tribuat. Huygens expresses a very high opinion of the importance and interest of the subject he was bringing under the notice of mathematicians.

30. The treatise is reprinted with a commentary in James Bernoulli's *Ars Conjectandi,* and forms the first of the four parts

of which that work is composed. Two English translations of the treatise have been published; one which has been attributed to Motte, but which was probably by Arbuthnot, and the other by W. Browne.

31. The treatise contains fourteen propositions. The first proposition asserts that if a player has equal chances of gaining a sum represented by a or a sum represented by b, his expectation is $\frac{1}{2}(a+b)$. The second proposition asserts that if a player has equal chances of gaining a or b or c, his expectation is $\frac{1}{3}(a+b+c)$. The third proposition asserts that if a player has p chances of gaining a and q chances of gaining b, his expectation is $\dfrac{pa+qb}{p+q}$.

It has been stated with reference to the last proposition: "Elementary as this truth may now appear, it was not received altogether without opposition." *Lubbock and Drinkwater*, p. 42. It is not obvious to what these words refer; for there does not appear to have been any opposition to the elementary principle, except at a much later period by D'Alembert.

32. The fourth, fifth, sixth, and seventh propositions discuss simple cases of the Problem of Points, when there are *two* players; the method is similar to Pascal's, see Art. 12. The eighth and ninth propositions discuss simple cases of the Problem of Points when there are *three* players; the method is similar to that for two players.

33. Huygens now proceeds to some questions relating to dice. In his tenth proposition he investigates in how many throws a player may undertake to throw a six with a single die. In his eleventh proposition he investigates in how many throws a player may undertake to throw twelve with a pair of dice. In his twelfth proposition he investigates how many dice a player must have in order to undertake that in one throw two sixes at least may appear. The thirteenth proposition consists of the following problem. *A* and *B* play with two dice; if a seven is thrown, *A* wins; if a ten is thrown, *B* wins; if any other number is thrown, the stakes are divided: compare the chances of *A* and *B*. They are shewn to be as 13 is to 11.

34.   The fourteenth proposition consists of the following problem. *A* and *B* play with two dice on the condition that *A* is to have the stake if he throws six before *B* throws seven, and that *B* is to have the stake if he throws seven before *A* throws six; *A* is to begin, and they are to throw alternately; compare the chances of *A* and *B*.

We will give the solution of Huygens. Let *B*'s chance be worth x, and the stake a, so that $a - x$ is the worth of *A*'s chance; then whenever it is *A*'s turn to throw x will express the value of *B*'s chance, but when it is *B*'s own turn to throw his chance will have a different value, say y. Suppose then *A* is about to throw; there are 36 equally likely cases; in 5 cases *A* wins and *B* takes nothing, in the other 31 cases *A* loses and *B*'s turn comes on, which is worth y by supposition. So that by the third proposition of the treatise the expectation of *B* is $\dfrac{5 \times 0 + 31y}{36}$, that is, $\dfrac{31y}{36}$. Thus

$$x = \frac{31y}{36}.$$

Now suppose *B* about to throw, and let us estimate *B*'s chance. There are 36 equally likely cases; in 6 cases *B* wins and *A* takes nothing; in the other 30 cases *B* loses and *A*'s turn comes on again, in which case *B*'s chance is worth x by supposition. So that the expectation of *B* is $\dfrac{6a + 30x}{36}$. Thus

$$y = \frac{6a + 30x}{36}.$$

From these equations it will be found that $x = \dfrac{31a}{61}$, and thus $a - x = \dfrac{30a}{61}$, so that *A*'s chance is to *B*'s chance as 30 is to 31.

35.   At the end of his treatise Huygens gives five problems without analysis or demonstration, which he leaves to the reader. Solutions are given by Bernoulli in the *Ars Conjectandi*. The following are the problems.

(1) *A* and *B* play with two dice on this condition, that *A* gains if he throws six, and *B* gains if he throws seven. *A* first has one

throw, then B has two throws, then A two throws, and so on until one or the other gains. Shew that A's chance is to B's as 10355 to 12276.

(2) Three players A, B, C take twelve balls, eight of which are black and four white. They play on the following condition; they are to draw blindfold, and the first who draws a white ball wins. A is to have the first turn, B the next, C the next, then A again, and so on. Determine the chances of the players.

Bernoulli solves this on three suppositions as to the meaning; first he supposes that each ball is replaced after it is drawn; secondly he supposes that there is only one set of twelve balls, and that the balls are not replaced after being drawn; thirdly he supposes that each player has his own set of twelve balls, and that the balls are not replaced after being drawn.

(3) There are forty cards forming four sets each of ten cards; A plays with B and undertakes in drawing four cards to obtain one of each set. Shew that A's chance is to B's as 1000 is to 8139.

(4) Twelve balls are taken, eight of which are black and four are white. A plays with B and undertakes in drawing seven balls blindfold to obtain three white balls. Compare the chances of A and B.

(5) A and B take each twelve counters and play with three dice on this condition, that if eleven is thrown A gives a counter to B, and if fourteen is thrown B gives a counter to A; and he wins the game who first obtains all the counters. Shew that A's chance is to B's as 244140625 is to 282429536481.

36. The treatise by Huygens continued to form the best account of the subject until it was superseded by the more elaborate works of James Bernoulli, Montmort, and De Moivre. Before we speak of these we shall give some account of the history of the theory of combinations, and of the inquiries into the laws of mortality and the principles of life insurance, and notices of various miscellaneous investigations.

CHAPTER IV.

ON COMBINATIONS.

37. THE theory of combinations is closely connected with the theory of probability; so that we shall find it convenient to imitate Montucla in giving some account of the writings on the former subject up to the close of the seventeenth century.

38. The earliest notice we have found respecting combinations is contained in Wallis's Algebra as quoted by him from a work by William Buckley; see Wallis's Algebra 1693, page 489. Buckley was a member of King's College, Cambridge, and lived in the time of Edward the Sixth. He wrote a small tract in Latin verse containing the rules of Arithmetic. In Sir John Leslie's *Philosophy of Arithmetic* full citations are given from Buckley's work; in Dr. Peacock's *History of Arithmetic* a citation is given; see also De Morgan's *Arithmetical Books from the invention of Printing*...

Wallis quotes twelve lines which form a *Regula Combinationis*, and then explains them. We may say briefly that the rule amounts to assigning the whole number of combinations which can be formed of a given number of things, when taken one at a time, or two at a time, or three at a time,... and so on until they are taken all together. The rule shews that the mode of proceeding was the same as that which we shall indicate hereafter in speaking of Schooten; thus for four things Buckley's rule gives, like Schooten's, $1 + 2 + 4 + 8$, that is 15 combinations in all.

By some mistake or misprint Wallis apparently overestimates the age of Buckley's work, when he says "... in Arithmetica sua, versibus scripta ante annos plus minus 190;" in the ninth Chapter of the Algebra the date of about 1550 is assigned to Buckley's death.

39. We must now notice an example of combinations which is of historical notoriety although it is very slightly connected with the theory.

A book was published at Antwerp in 1617 by Erycius Puteanus under the title, *Erycii Puteani Pietatis Thaumata in Bernardi Bauhusii è Societate Jesu Proteum Parthenium.* The book consists of 116 quarto pages, exclusive of seven pages, not numbered, which contain an Index, Censura, Summa Privilegii, and a typographical ornament.

It appears that Bernardus Bauhusius composed the following line in honour of the Virgin Mary:

Tot tibi sunt dotes, Virgo, quot sidera cælo.

This verse is arranged in 1022 different ways, occupying 48 pages of the work. First we have 54 arrangements commencing *Tot tibi;* then 25 arrangements commencing *Tot sunt;* and so on. Although these arrangements are sometimes ascribed to Puteanus, they appear from the dedication of the book to be the work of Bauhusius himself; Puteanus supplies verses of his own and a series of chapters in prose which he calls *Thaumata,* and which are distinguished by the Greek letters from A to Ω inclusive. The number 1022 is the same as the number of the stars according to Ptolemy's Catalogue, which coincidence Puteanus seems to consider the great merit of the labours of Bauhusius; see his page 82.

It is to be observed that Bauhusius did not profess to include all the possible arrangements of his line; he expressly rejected those which would have conveyed a sense inconsistent with the glory of the Virgin Mary. As Puteanus says, page 103,

Dicere horruit Vates:

Sidera tot cælo, Virgo, quot sunt tibi Dotes,

imò in hunc sensum producere Proteum recusavit, ne laudem imminueret. Sic igitur contraxit versuum numerum; ut Dotium augeret.

40. The line due to Bauhusius on account of its numerous arrangements seems to have attracted great attention during the following century; the discussion on the subject was finally settled

by James Bernoulli in his *Ars Conjectandi*, where he thus details
the history of the problem.

...Quemadmodùm cernere est in hexametro à Bernh. Bauhusio Jesuitâ
Lovaniensi in laudem Virginis Deiparæ constructo:

Tot tibi sunt Dotes, Virgo, quot sidera cœlo;

quem dignum peculiari operâ duxerunt plures Viri celebres. Erycius
Puteanus in libello, quem Thaumata Pietatis inscripsit, variationes ejus
utiles integris 48 paginis enumerat, easque numero stellarum, quarum
vulgò 1022 recensentur, accommodat, omissis scrupulosiùs illis, quæ dicere videntur, tot sidera cœlo esse, quot Mariæ dotes; nam Mariæ
dotes esse multo plures. Eundem numerum 1022 ex Puteano repetit
Gerh. Vossius, cap. 7, de Scient. Mathemat. Prestetus Gallus in primâ
editione Element. Mathemat. pag. 358. Proteo huic 2196 variationes
attribuit, sed factâ revisione in alterâ edit. tom. pr. pag. 133. numerum
earum dimidio fere auctum ad 3276 extendit. Industrii Actorum Lips.
Collectores m. Jun. 1686, in recensione Tractatûs Wallisiani de Algebrâ,
numerum in quæstione (quem Auctor ipse definire non fuit ausus) ad
2580 determinant. Et ipse postmodùm Wallisius in edit. latinâ operis
sui Oxon. anno 1693. impressâ, pagin. 494. eundem ad 3096 profert.
Sed omnes adhuc à vero deficientes, ut delusam tot Virorum post
adhibitas quoque secundas curas in re levi perspicaciam meritò mireris.
Ars Conjectandi, page 78.

James Bernoulli seems to imply that the two editions of
Wallis's Algebra differ in their enumeration of the arrangements
of the line due to Bauhusius; but this is not the case: the two
editions agree in investigation and in result.

James Bernoulli proceeds to say that he had found that there
could be 3312 arrangements without breaking the law of metre;
this excludes spondaic lines but includes those which have no
cæsura. The analysis which produces this number is given.

41. The earliest treatise on combinations which we have observed is due to Pascal. It is contained in the work on the
Arithmetical Triangle which we have noticed in Art. 22; it will
also be found in the fifth volume of Pascal's works, Paris 1819,
pages 86—107.

The investigations of Pascal on combinations depend on his
Arithmetical Triangle. The following is his principal result; we
express it in modern notation.

Take an *Arithmetical Triangle* with r numbers in its base; then the sum of the numbers in the pth horizontal row is equal to the multitude of the combinations of r things taken p at a time. For example, in Art 22 we have a triangle with 10 numbers in its base; now take the numbers in the 8th horizontal column; their sum is $1 + 8 + 36$, that is 45; and there are 45 combinations of 10 things taken 8 at a time. Pascal's proof is inductive. It may be observed that *multitudo* is Pascal's word in the Latin of his treatise, and *multitude* in the French version of a part of the treatise which is given in pages 22—30 of the volume.

From this he deduces various inferences such as the following. Let there be n things; the sum of the multitude of the combinations which can be formed, one at a time, two at a time,... , up to n at a time, is $2^n - 1$.

At the end Pascal considers this problem. Datis duobus numeris inæqualibus, invenire quot modis minor in majore combinetur. And from his *Arithmetical Triangle* he deduces in effect the following result; the number of combinations of r things taken p at a time is

$$\frac{(p+1)\,(p+2)\,(p+3)\,...\,r}{\lfloor r - p}.$$

After this problem Pascal adds,

Hoc problemate tractatum hunc absolvere constitueram, non tamen omninò sine molestia, cùm multa alia parata habeam; sed ubi tanta ubertas, vi moderanda est fames : his ergo pauca hæc subjiciam.

Eruditissimus ac mihi charisimus, D.D. de Ganières, circa combinationes, assiduo ac perutili labore, more suo, incumbens, ac indigens facili constructione ad inveniendum quoties numerus datus in alio dato combinetur, hanc ipse sibi praxim instituit.

Pascal then gives the rule; it amounts to this; the number of combinations of r things taken p at a time is

$$\frac{r\,(r-1)...\,(r-p+1)}{\lfloor p}.$$

This is the form with which we are now most familiar. It may be immediately shewn to agree with the form given before by Pascal, by cancelling or introducing factors into both numerator and denominator. Pascal however says, Excellentem hanc solutionem ipse mihi ostendit, ac etiam demonstrandam proposuit, ipsam ego sanè miratus sum, sed difficultate territus vix opus suscepi, et ipsi authori relinquendum existimavi; attamen trianguli arithmetici auxilio, sic proclivis facta est via. Pascal then establishes the correctness of the rule by the aid of his *Arithmetical Triangle;* after which he concludes thus, Hac demonstratione assecutâ, jam reliqua quæ invitus supprimebam libenter omitto, adeo dulce est amicorum memorari.

42. In the work of Schooten to which we have already referred in Art. 28 we find some very slight remarks on combinations and their applications; see pages 373—403. Schooten's first section is entitled, Ratio inveniendi electiones omnes, quæ fieri possunt, datâ multitudine rerum. He takes four letters *a, b, c, d,* and arranges them thus,

> *a.*
> *b. ab.*
> *c. ac. bc. abc.*
> *d. ad. bd. abd. cd. acd. bcd. abcd.*

Thus he finds that 15 elections can be made out of these four letters. So he adds, Hinc si per *a* designatur unum malum, per *b* unum pirum, per *c* unum prunum, et per *d* unum cerasum, et ipsa aliter atque aliter, ut supra, eligantur, electio eorum fieri poterit 15 diversis modis, ut sequitur....

Schooten next takes five letters; and thus he infers the result which we should now express by saying that, if there are *n* letters the whole number of elections is $2^n - 1$.

Hence if *a, b, c, d* are prime factors of a number, and all different, Schooten infers that the number has 15 divisors excluding unity but including the number itself, or 16 including also unity.

Next suppose some of the letters are repeated; as for example suppose we have *a, a, b,* and *c*; it is required to determine how many elections can be made. Schooten arranges the letters thus,

> *a.*
> *a. aa.*
> *b. ab. aab.*
> *c. ac. aac. bc. abc. aabc.*

We have thus $2 + 3 + 6$ elections.

Similarly if the proposed letters are a, a, a, b, b, it is found that 11 elections can be made.

In his following sections Schooten proceeds to apply these results to questions relating to the number of divisors in a number. Thus, for example, supposing a, b, c, d, to be different prime factors, numbers of the following forms all have 16 divisors, $abcd$, a^3bc, a^3b^3, a^7b, a^{15}. Hence the question may be asked, what is the least number which has 16 divisors? This question must be answered by trial; we must take the smallest prime numbers 2, 3,... and substitute them in the above forms and pick out the least number. It will be found on trial that the least number is $2^3 . 3 . 5$, that is 120. Similarly, suppose we require the least number which has 24 divisors. The suitable forms of numbers for 24 divisors are a^2bcd, a^3b^2c, a^5bc, a^5b^3, a^7b^2, $a^{11}b$ and a^{23}. It will be found on trial that the least number is $2^3 . 3^2 . 5$, that is 360.

Schooten has given two tables connected with this kind of question. (1) A table of the algebraical forms of numbers which have any given number of divisors not exceeding a hundred; and in this table, when more than one form is given in any case, the first form is that which he has found by trial will give the least number with the corresponding number of divisors. (2) A table of the least numbers which have any assigned number of divisors not exceeding a hundred. Schooten devotes ten pages to a list of all the prime numbers under 10,000.

43. A dissertation was published by Leibnitz in 1666, entitled *Dissertatio de Arte Combinatoria*; part of it had been previously published in the same year under the title of *Disputatio arithmetica de complexionibus*. The dissertation is interesting as the earliest work of Leibnitz connected with mathematics; the connexion however is very slight. The dissertation is contained in the second volume of the edition of the works of Leibnitz by Dutens; and in the first volume of the second section of the mathematical works of Leibnitz edited by Gerhardt, Halle, 1858. The dissertation is also included in the collection of the philosophical writings of Leibnitz edited by Erdmann, Berlin, 1840.

44. Leibnitz constructs a table at the beginning of his dis-

sertation similar to Pascal's *Arithmetical Triangle*, and applies it
to find the number of the combinations of an assigned set of things
taken two, three, four,... together. In the latter part of his dissertation Leibnitz shews how to obtain the number of permutations
of a set of things taken all together; and he forms the product of
the first 24 natural numbers. He brings forward several Latin
lines, including that which we have already quoted in Art. 39,
and notices the great number of arrangements which can be
formed of them.

The greater part of the dissertation however is of such a
character as to confirm the correctness of Erdmann's judgment in
including it among the philosophical works of Leibnitz. Thus,
for example, there is a long discussion as to the number of moods
in a syllogism. There is also a demonstration of the existence of
the Deity, which is founded on three definitions, one postulate,
four axioms, and one result of observation, namely, *aliquod corpus
movetur.*

45.   We will notice some points of interest in the dissertation.

(1)   Leibnitz proposes a curious mode of expression. When
a set of things is to be taken *two* at a time he uses the symbol
com2natio (combinatio); when *three* at a time he uses con3natio
(conternatio); when four at a time, con4natio, and so on.

(2)   The mathematical treatment of the subject of combinations is far inferior to that given by Pascal; probably Leibnitz
had not seen the work of Pascal. Leibnitz seems to intimate
that his predecessors had confined themselves to the combinations of things *two* at a time, and that he had himself extended
the subject so far as to shew how to obtain from his table the
combinations of things taken together more than two at a time;
generaliorem modum nos deteximus, specialis est vulgatus. He
gives the rule for the combination of things *two* at a time, namely,
that which we now express by the formula $\dfrac{n(n-1)}{2}$; but he does
not give the similar rule for combinations three, four,... at a time,
which is contained in Pascal's work.

(3)   After giving his table, which is analogous to the *Arith-*

metical Triangle, he adds, "Adjiciemus hic Theoremata quorum τὸ ὅτι ex ipsa tabula manifestum est, τὸ διότι ex tabulæ fundamento." The only theorem here that is of any importance is that which we should now express thus : if *n* be prime the number of combinations of *n* things taken *r* at a time is divisible by *n*.

(4) A passage in which Leibnitz names his predecessors may be quoted. After saying that he had partly furnished the matter himself and partly obtained it from others, he adds,

Quis illa primus detexerit ignoramus. Schwenterus *Delic.* l. 1, Sect. 1, prop. 32, apud Hieronymum Cardanum, Johannem Buteonem et Nicolaum Tartaleam, extare dicit. In Cardani tamen Practica Arithmetica quæ prodiit Mediolani anno 1539, nihil reperimus. Inprimis dilucide, quicquid dudum habetur, proposuit Christoph. Clavius in Com. supra Joh. de Sacro Bosco Sphær. edit. Romæ forma 4ta anno 1785. p. 33. seqq.

With respect to Schwenter it has been observed,

Schwenter probably alluded to Cardan's book, " De Proportionibus," in which the figurate numbers are mentioned, and their use shown in the extraction of roots, as employed by Stifel, a German algebraist, who wrote in the early part of the sixteenth century. *Lubbock and Drinkwater*, page 45.

(5) Leibnitz uses the symbols + − = in their present sense ; he uses ⌢ for multiplication and ⌣ for division. He uses the word *productum* in the sense of a sum : thus he calls 4 the productum of 3 + 1.

46. The dissertation shews that at the age of twenty years the distinguishing characteristics of Leibnitz were strongly developed. The extent of his reading is indicated by the numerous references to authors on various subjects. We see evidence too that he had already indulged in those dreams of impossible achievements in which his vast powers were uselessly squandered. He vainly hoped to produce substantial realities by combining the precarious definitions of metaphysics with the elementary truisms of logic, and to these fruitless attempts he gave the aspiring titles of *universal science*, *general science*, and *philosophical calculus*. See *Erdmann*, pages 82—91, especially page 84.

3

47. *A discourse of combinations, alternations, and aliquot parts* is attached to the English edition of Wallis's Algebra published in 1685. In the Latin edition of the Algebra, published in 1693, this part of the work occupies pages 485—529.

In referring to Wallis's Algebra we shall give the pages of the Latin edition ; but in quoting from him we shall adopt his own English version. The English version was reprinted by Maseres in a volume of reprints which was published at London in 1795 under the title of *The Doctrine of Permutations and Combinations, being an essential and fundamental part of the Doctrine of Chances.*

48. Wallis's first Chapter is *Of the variety of Elections, or Choise, in taking or leaving One or more, out of a certain Number of things proposed.* He draws up a Table which agrees with Pascal's *Arithmetical Triangle,* and shews how it may be used in finding the number of combinations of an assigned set of things taken two, three, four, five,... at a time. Wallis does not add any thing to what Pascal had given, to whom however he does not refer ; and Wallis's clumsy parenthetical style contrasts very unfavourably with the clear bright stream of thought and language which flowed from the genius of Pascal. The chapter closes with an extract from the Arithmetic of Buckley and an explanation of it ; to this we have already referred in Art. 38.

49. Wallis's second Chapter is *Of Alternations, or the different change of Order, in any Number of things proposed..* Here he gives some examples of what are now usually called permutations ; thus if there are four letters *a, b, c, d,* the number of permutations when they are taken all together is $4 \times 3 \times 2 \times 1$. Wallis accordingly exhibits the 24 permutations of these four letters. He forms the product of the first twenty-four natural numbers, which is the number of the permutations of twenty-four things taken all together.

Wallis exhibits the 24 permutations of the letters in the word *Roma* taken all together ; and then he subjoins, "Of which (in Latin) these seven are only useful ; *Roma, ramo, oram, mora, maro, armo, amor.* The other forms are useless ; as affording no (Latin) word of known signification."

Wallis then considers the case in which there is some repetition among the quantities of which we require the permutations. He takes the letters which compose the word *Messes*. Here if there were no repetition of letters the number of permutations of the letters taken all together would be $1 \times 2 \times 3 \times 4 \times 5 \times 6$, that is 720; but as Wallis explains, owing to the occurrence of the letter *e* twice, and of the letter *s* thrice, the number 720 must be divided by $2 \times 2 \times 3$, that is by 12. Thus the number of permutations is reduced to 60. Wallis exhibits these permutations and then subjoins, " Of all which varieties, there is none beside *messes* itself, that affords an useful Anagram." The chapter closes with Wallis's attempt at determining the number of arrangements of the verse

Tot tibi sunt dotes, virgo, quot sidera cælo.

The attempt is followed by these words, " I will not be positive, that there may not be some other Changes : (and then, those may be added to these :) Or, that most of these be twice repeated, (and if so, those are to be abated out of the Number :) But I do not, at present, discern either the one and other."

Wallis's attempt is a very bad specimen of analysis ; it involves both the errors he himself anticipates, for some cases are omitted and some counted more than once. It seems strange that he should have failed in such a problem considering the extraordinary powers of abstraction and memory which he possessed ; so that as he states, he extracted the square root of a number taken at random with 53 figures, in tenebris decumbens, sola fretus memoria. See his Algebra, page 450.

50. Wallis's third Chapter is *Of the Divisors and Aliquot parts, of a Number proposed.* This Chapter treats of the resolution of a number into its prime factors, and of the number of divisors which a given number has, and of the least numbers which have an assigned number of divisors.

51. Wallis's fourth Chapter is *Monsieur Fermat's Problems concerning Divisors and Aliquot Parts.* It contains solutions of two problems which Fermat had proposed as a challenge to Wallis and the English mathematicians. The problems relate to what is now called the Theory of Numbers.

52. Thus the theory of combinations is not applied by Wallis
in any manner that materially bears upon our subject. In fact
the influence of Fermat seems to have been more powerful than
that of Pascal; and the Theory of Numbers more cultivated than
the Theory of Probability.

The judgment of Montmort seems correct that nothing of any
importance in the Theory of Combinations previous to his own
work had been added to the results of Pascal. Montmort, on his
page XXXV, names as writers on the subject Prestet, Tacquet, and
Wallis. I have not seen the works of Prestet and Tacquet;
Gouraud refers to Prestet's *Nouveaux éléments de mathématiques*,
2e éd., in the following terms, Le père Prestet, enfin, fort habile
géomètre, avait expliqué avec infiniment de clarté, en 1689, les
principaux artifices de cet art ingénieux de composer et de varier
les grandeurs. *Gouraud*, page 23.

CHAPTER V.

MORTALITY AND LIFE INSURANCE.

53. THE history of the investigations on the laws of mortality and of the calculations of life insurances is sufficiently important and extensive to demand a separate work ; these subjects were originally connected with the Theory of Probability but may now be considered to form an independent kingdom in mathematical science : we shall therefore confine ourselves to tracing their origin.

54. According to Gouraud the use of tables of mortality was not quite unknown to the ancients: after speaking of such a table as unknown until the time of John de Witt he subjoins in a note,

Inconnue du moins des modernes. Car il paraîtrait par un passage du Digeste, *ad legem Falcidiam*, xxxv. 2, 68, ˙que les Romains n'en ignoraient pas absolument l'usage. Voyez à ce sujet M. V. Leclerc, *Des Journaux chez les Romains*, p. 198, et une curieuse dissertation : *De probabilitate vitæ ejusque usu forensi*, etc., d'un certain Schmelzer (Goettingue, 1787, in-8). *Gouraud*, page 14.

55. The first name which is usually mentioned in connexion with our present subject is that of John Graunt : I borrow a notice of him from Lubbock and Drinkwater, page 44. After referring to the registers of the annual numbers of deaths in London which began to be kept in 1592, and which with some

intermissions between 1594 and 1603 have since been regularly continued, they proceed thus,

They were first intended to make known the progress of the plague; and it was not till 1662 that Captain Graunt, a most acute and intelligent man, conceived the idea of rendering them subservient to the ulterior objects of determining the population and growth of the metropolis; as before his time, to use his own words, "most of them who constantly took in the weekly bills of mortality, made little or no use of them than so as they might take the same as a text to talk upon in the next company; and withal, in the plague time, how the sickness increased or decreased, that so the rich might guess of the necessity of their removal, and tradesmen might conjecture what doings they were like to have in their respective dealings." Graunt was careful to publish with his deductions the actual returns from which they were obtained, comparing himself, when so doing, to "a silly schoolboy, coming to say his lesson to the world (that peevish and tetchic master,) who brings a bundle of rods, wherewith to be whipped for every mistake he has committed." Many subsequent writers have betrayed more fear of the punishment they might be liable to on making similar disclosures, and have kept entirely out of sight the sources of their conclusions. The immunity they have thus purchased from contradiction could not be obtained but at the expense of confidence in their results.

These researches procured for Graunt the honour of being chosen a fellow of the Royal Society, ...

Gouraud says in a note on his page 16,

...John Graunt, homme sans géométrie, mais qui ne manquait ni de sagacité ni de bon sens, avait, dans une sorte de traité d'Arithmétique politique intitulé : *Natural and political observations...made upon the bills of mortality*, etc., rassemblé ces différentes listes, et donné même (*ibid.* chap. XI.) un calcul, à la vérité fort grossier, mais du moins fort original, de la mortalité probable à chaque âge d'un certain nombre d'individus supposés nés viables tous au même instant.

See also the *Athenæum* for October 31st, 1863, page 537.

56.   The names of two Dutchmen next present themselves, Van Hudden and John de Witt. Montucla says, page 407,

Le problème des rentes viagères fut traité par Van Hudden, qui quoique géomètre, ne laissa pas que d'être bourguemestre d'Amsterdam, et par le célèbre pensionnaire d'Hollande, Jean de Witt, un des pre-- miers promoteurs de la géométrie de Descartes. J'ignore le titre de l'écrit de Hudden, mais celui de Jean de Witt étoit intitulé : *De vardye van de lif-renten na proportie van de los-renten,* ou *la Valeur des rentes viagères en raison des ventes libres ou remboursables* (La Haye, 1671).

Ils étoient l'un et l'autre plus à portée que personne d'en sentir l'importance et de se procurer les dépouillemens nécessaires de registres de mortalité; aussi Leibnitz, passant en Hollande quelques années après, fit tout son possible pour se procurer l'écrit de Jean de Witt, mais il ne peut y parvenir; il n'étoit cependant pas absolument perdu, car M. Nicolas Struyck *(Inleiding tot het algemeine geography,* &c. Amst. 1740, *in* 4o. p. 345) nous apprend qu'il en a eu un exemplaire entre les mains; il nous en donne un précis, par lequel on voit combien Jean de Witt raisonnoit juste sur cette matière.

Le chevalier Petty, Anglois, qui s'occupa beaucoup de calculs politiques, entrevit le problême, mais il n'étoit pas assez géomètre pour le traiter fructueusement, en sorte que, jusqu'à Halley, l'Angleterre et la France qui empruntèrent tant et ont tant emprunté depuis, le firent comme des aveugles ou comme de jeunes débauchés.

57. With respect to Sir William Petty, to whom Montucla refers, we may remark that his writings do not seem to have been very important in connexion with our present subject. Some account of them is given in the article *Arithmétique Politique* of the original French *Encyclopédie;* the article is reproduced in the *Encyclopédie Méthodique.* Gouraud speaks of Petty thus in a note on his page 16,

Après Graunt, le chevalier W. Petty, dans différents essais d'économie politique, où il y avait, il est vrai, plus d'imagination que de jugement, s'était, de 1682 à 1687, occupé de semblables recherches.

58. With respect to Van Hudden to whom Montucla also refers we can only add that his name is mentioned with approbation by Leibnitz, in conjunction with that of John de Witt, for his researches on annuities. See *Leibnitii Opera Omnia, ed. Dutens,* Vol. II. part 1, page 93 ; Vol. VI. part 1, page 217.

59. With respect to the work of John de Witt we have some notices in the correspondence between Leibnitz and James Bernoulli; but these notices do not literally confirm Montucla's

statement respecting Leibnitz: see *Leibnizens Mathematische Schriften herausgegeben von C. I. Gerhardt,* Erste Abtheilung. Band III. Halle 1855. James Bernoulli says, page 78,

Nuper in Menstruis Excerptis Hanoverae impressis citatum inveni Tractatum quendam mihi ignotum Pensionarii de Wit von Subtiler Ausrechnung des valoris der Leib-Renten. Fortasse is quaedam huc facientia habet; quod si sit, copiam ejus mihi alicunde fieri percuperem.

In his reply Leibnitz says, page 84,

Pensionarii de Wit libellus exiguus est, ubi aestimatione illa nota utitur a possibilitate casuum aequalium aequali et hinc ostendit reditus ad vitam sufficientes pro sorte a Batavis solvi. Ideo Belgice scripserat, ut aequitas in vulgus appareret.

In his next letter, page 89, James Bernoulli says that De Witt's book will be useful to him; and as he had in vain tried to obtain it from Amsterdam he asks for the loan of the copy which Leibnitz possessed. Leibnitz replies, page 93,

Pensionarii Wittii dissertatio, vel potius Scheda impressa de reditibus ad vitam, sane brevis, extat quidem inter chartas meas, sed cum ad Te mittere vellem, reperire nondum potui. Dabo tamen operam ut nanciscare, ubi primum domi crucre licebit alicubi latitantem.

James Bernoulli again asked for the book, page 95. Leibnitz replies, page 99,

Pensionarii Wittii scriptum nondum satis quaerere licuit inter chartas; non dubito tamen, quin sim tandem reperturus, ubi vacaverit. Sed vix aliquid in eo novum Tibi occurret, cum fundamentis iisdem ubique insistat, quibus cum alii viri docti jam erant usi, tum Paschalius in Triangulo Arithmetico, et Hugenius in diss. de Alea, nempe ut medium Arithmeticum inter aeque incerta sumatur; quo fundamento etiam rustici utuntur, cum praediorum pretia aestimant, et rerum fiscalium curatores, cum reditus praefecturarum Principis medios constituunt, quando se offert conductor.

In the last of his letters to James Bernoulli which is given, Leibnitz implies that he has not yet found the book; see page 103.

We find from pages 767, 769 of the volume that Leibnitz attempted to procure a copy of De Witt's dissertation by the aid of John Bernoulli, but without success.

These letters were written in the years 1703, 1704, 1705.

60. The political fame of John de Witt has overpowered that which he might have gained from science, and thus his mathematical attainments are rarely noticed. We may therefore add that he is said to have published a work entitled *Elementa linearum curvarum*, Leyden 1650, which is commended by Condorcet; see Condorcet's *Essai...d'Analyse...* page CLXXXIV.

61. We have now to notice a memoir by Halley, entitled *An estimate of the Degrees of the Mortality of Mankind, drawn from curious Tables of the Births and Funerals at the City of Breslaw; with an Attempt to ascertain the Price of Annuities upon Lives.*

This memoir is published in Vol. XVII. of the *Philosophical Transactions*, 1693; it occupies pages 596—610.

This memoir is justly celebrated as having laid the foundations of a correct theory of the value of life annuities.

62. Halley refers to the bills of mortality which had been published for London and Dublin; but these bills were not suitable for drawing accurate deductions.

First, In that the *Number* of the People was wanting. Secondly, That the *Ages* of the People dying was not to be had. And Lastly, That both *London* and *Dublin* by reason of the great and casual Accession of *Strangers* who die therein, (as appeared in both, by the great Excess of the *Funerals* above the *Births*) rendered them incapable of being Standards for this purpose; which requires, if it were possible, that the People we treat of should not at all be changed, but die where they were born, without any Adventitious Increase from Abroad, or Decay by Migration elsewhere.

63. Halley then intimates that he had found satisfactory data in the Bills of Mortality for the city of Breslau for the years 1687, 88, 89, 90, 91; which "had then been recently communicated by Neumann (probably at Halley's request) through Justell, to the Royal Society, in whose archives it is supposed that copies of the original registers are still preserved." *Lubbock and Drinkwater*, page 45.

64. The Breslau registers do not appear to have been published themselves, and Halley gives only a very brief introduction

to the table which he deduced from them. Halley's table is in the
following form:

1	1000
2	855
3	798
4	760

......

The left-hand number indicates ages and the right-hand number the corresponding number of persons alive. We do not feel
confident of the meaning of the table. Montucla, page 408, understood that out of 1000 persons born, 855 attain to the age of one
year, then 798 out of these attain to the age of two years, and
so on.

Daniel Bernoulli understood that the number of infants born
is not named, but that 1000 are supposed to reach one year, then
855 out of these reach two years, and so on. *Hist. de l'Acad. ...
Paris*, 1760.

65. Halley proceeds to shew the use of his table in the calculation of annuities. To find the value of an annuity on the life of
a given person we must take from the table the chance that he
will be alive after the lapse of n years, and multiply this chance
by the present value of the annual payment due at the end of
n years; we must then sum the results thus obtained for all values
of n from 1 to the extreme possible age for the life of the given
person. Halley says that "This will without doubt appear to
be a most laborious Calculation." He gives a table of the value
of an annuity for every fifth year of age up to the seventieth.

66. He considers also the case of annuities on joint lives, or
on one of two or more lives. Suppose that we have two persons,
an elder and a younger, and we wish to know the probability
of one or both being alive at the end of a given number of years.
Let N be the number in the table opposite to the present age of
the younger person, and R the number opposite to that age increased by the given number of years; and let $N = R + Y$, so that
Y represents the number who have died out of N in the given
number of years. Let n, r, y denote similar quantities for the
elder age. Then the chance that both will be dead at the end

of the given number of years is $\dfrac{Yy}{Nn}$; the chance that the younger

will be alive and the elder dead is $\dfrac{Ry}{Nn}$; and so on.

Halley gives according to the fashion of the time a geometrical illustration.

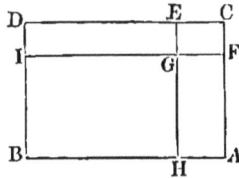

Let AB or CD represent N, and DE or BH represent R, so that EC or HA represents Y. Similarly AC, AF, CF may represent n, r, y. Then of course the rectangle $ECFG$ represents Yy, and so on.

In like manner, Halley first gives the proposition relating to three lives in an algebraical form, and then a geometrical illustration by means of a parallelepiped. We find it difficult in the present day to understand how such simple algebraical propositions could be rendered more intelligible by the aid of areas and solids.

67. On pages 654—656 of the same volume of the *Philosophical Transactions* we have *Some further Considerations on the Breslaw Bills of Mortality. By the same Hand, &c.*

68. De Moivre refers to Halley's memoir, and republishes the table; see De Moivre's *Doctrine of Chances*, pages 261, 345.

CHAPTER VI.

MISCELLANEOUS INVESTIGATIONS

BETWEEN THE YEARS 1670 AND 1700.

69. THE present chapter will contain notices of various contributions to our subject, which were made between the publication of the treatise by Huygens and of the more elaborate works by James Bernoulli, Montmort, and De Moivre.

70. A Jesuit named John Caramuel published in 1670, under the title of *Mathesis Biceps*, two folio volumes of a course of Mathematics ; it appears from the list of the author's works at the beginning of the first volume that the entire course was to have comprised four volumes.

There is a section called *Combinatoria* which occupies pages 921—1036, and part of this is devoted to our subject.

Caramuel gives first an account of combinations in the modern sense of the word; there is nothing requiring special attention here : the work contains the ordinary results, not proved by general symbols but exhibited by means of examples. Caramuel refers often to Clavius and Izquierdus as his guides.

After this account of combinations in the modern sense Caramuel proceeds to explain the *Ars Lulliana*, that is the method of affording assistance in reasoning, or rather in disputation, proposed by Raymond Lully.

71. Afterwards we have a treatise on chances under the title of *Kybeia, quæ Combinatoriæ genus est, de Alea, et Ludis Fortunæ*

serio disputans. This treatise includes a reprint of the treatise of Huygens, which however is attributed to another person. Caramuel says, page 984,

Dum hoc Syntagma Perillustri Domino N. Viro cruditissimo communicarem, ostendit etiam mihi ingeniosam quamdam de eodem argumento Diatribam, quam à Christiano Severino Longomontano fuisse scriptam putabat, et, quia est curiosa, et brevis, debuit huic Quæstioni subjungi...

In the table of contents to his work, page XXVIII, Caramuel speaks of the tract of Huygens as

Diatribe ingeniose à Longomontano, ut putatur, de hoc eodem argumento scripta : nescio an evulgata.

Longomontanus was a Danish astronomer who lived from 1562 to 1647.

72. Nicolas Bernoulli speaks very severely of Caramuel. He says Un Jesuite nommé Caramuel, que j'ai cité dans ma These... mais comme tout ce qu'il donne n'est qu'un amas de paralogismes, je ne le compte pour rien. *Montmort,* p. 387.

By his *These* Nicolas Bernoulli probably means his *Specimina Artis conjectandi...,* which will be noticed in a subsequent Chapter, but Caramuel's name is not mentioned in that essay as reprinted in the *Acta Erud....Suppl.*

John Bernoulli in a letter to Leibnitz speaks more favourably of Caramuel ; see page 715 of the volume cited in Art. 59.

73. Nicolas Bernoulli has exaggerated the Jesuit's blunders. Caramuel touches on the following points, and correctly : the chances of the throws with two dice ; simple cases of the Problem of Points for two players ; the chance of throwing an ace once at least in two throws, or in three throws ; the game of *Passe-dix.*

He goes wrong in trying the Problem of Points for three players, which he does for two simple cases ; and also in two other problems, one of which is the fourteenth of Huygens's treatise, and the other is of exactly the same kind.

Caramuel's method with the fourteenth problem of Huygens's treatise is as follows. Suppose the stake to be 36 ; then A's chance at his first throw is $\frac{5}{36}$, and $\frac{5}{36} \times 36 = 5$; thus taking 5 from 36 we may consider 31 as left for B. Now B's chance of success in a single throw is $\frac{6}{36}$; thus $\frac{6}{36} \times 31$, that is $5\frac{1}{6}$, may be considered the value of his first throw.

Thus Caramuel assigns 5 to A and $5\frac{1}{6}$ to B, as the value of their first throws respectively; then the remaining $25\frac{5}{6}$ he proposes to divide equally between A and B. This is wrong: he ought to have continued his process, and have assigned to A for his second throw $\frac{5}{36}$ of the $25\frac{5}{6}$, and then to B for his second throw $\frac{6}{36}$ of the remainder; and so on. Thus he would have had for the shares of each player an infinite geometrical progression, and the result would have been correct.

It is strange that Caramuel went wrong when he had the treatise of Huygens to guide him; it seems clear that he followed this guidance in the discussion of the Problem of Points for *two* players, and then deserted it.

74. In the *Journal des Sçavans* for Feb. 1679, Sauveur gave some formulæ without demonstration relating to the advantage of the Banker at the game of *Bassette*. Demonstrations of the formulæ will be found in the *Ars Conjectandi* of James Bernoulli, pages 191—199. I have examined Sauveur's formulæ as given in the Amsterdam edition of the *Journal*. There are six series of formulæ; in the first five, which alone involve any difficulty, Sauveur and Bernoulli agree: the last series is obtained by simply subtracting the second from the fifth, and in this case by mistake or misprint Sauveur is wrong. Bernoulli seems to exaggerate the discrepancy when he says, Quòd si quis D.ni Salvatoris Tabellas cum hisce nostris contulerit, deprehendet illas in quibusdam locis, præsertim ultimis, nonnihil emendationis indigere. Montucla, page 390, and Gouraud, page 17, seem also to think Sauveur more inaccurate than he really is.

An *éloge* of Sauveur by Fontenelle is given in the volume for 1716 of the *Hist. de l'Acad....Paris*. Fontenelle says that Bassette was more beneficial to Sauveur than to most of those who played at it with so much fury; it was at the request of the Marquis of Dangeau that Sauveur undertook the investigation of the chances of the game. Sauveur was in consequence introduced at court, and had the honour of explaining his calculations to the King and Queen. See also *Montmort*, page XXXIX.

75. James Bernoulli proposed for solution two problems in chances in the *Journal des Sçavans* for 1685. They are as follows:

1. *A* and *B* play with a die, on condition that he who first throws an ace wins. First *A* throws once, then *B* throws once, then *A* throws twice, then *B* throws twice, then *A* throws three times, then *B* throws three times, and so on until ace is thrown.

2. Or first *A* throws once, then *B* twice, then *A* three times, then *B* four times, and so on.

The problems remained unsolved until James Bernoulli himself gave the results in the *Acta Eruditorum* for 1690. Afterwards in the same volume Leibnitz gave the results. The chances involve infinite series which are not summed.

James Bernoulli's solutions are reprinted in the collected edition of his works, Geneva, 1744; see pages 207 and 430. The problems are also solved in the *Ars Conjectandi*, pages 52—56.

76. Leibnitz took great interest in the Theory of Probability and shewed that he was fully alive to its importance, although he cannot be said himself to have contributed to its advance. There was one subject which especially attracted his attention, namely that of games of all kinds; he himself here found an exercise for his inventive powers. He believed that men had nowhere shewn more ingenuity than in their amusements, and that even those of children might usefully engage the attention of the greatest mathematicians. He wished to have a systematic treatise on games, comprising first those which depended on numbers alone, secondly those which depended on position, like chess, and lastly those which depended on motion, like billiards. This he considered would be useful in bringing to perfection the art of invention, or

as he expresses it in another place, in bringing to perfection the art of arts, which is the art of thinking.

See *Leibnitii Opera Omnia, ed. Dutens*, Vol. v. pages 17, 22, 28, 29, 203, 206. Vol. vi. part 1, 271, 304. *Erdmann*, page 175.

See also *Opera Omnia, ed. Dutens*, Vol. vi. part 1, page 36, for the design which Leibnitz entertained of writing a work on estimating the probability of conclusions obtained by arguments.

77. Leibnitz however furnishes an example of the liability to error which seems peculiarly characteristic of our subject. He says, *Opera Omnia, ed. Dutens*, Vol. vi. part 1, page 217,

...par exemple, avec deux dés, il est aussi faisable de jetter douze points, que d'en jetter onze; car l'un et l'autre ne se peut faire que d'une seule manière; mais il est trois fois plus faisable d'en jetter sept; car cela se peut faire en jettant six et un, cinq et deux, quatre et trois; et une combinaison ici est aussi faisable que l'autre.

It is true that eleven can only be made up of six and five; but the six may be on *either* of the dice and the five on the other, so that the chance of throwing eleven with two dice is twice as great as the chance of throwing twelve: and similarly the chance of throwing seven is six times as great as the chance of throwing twelve.

78. A work entitled *Of the Laws of Chance* is said by Montucla to have appeared at London in 1692; he adds mais n'ayant jamais rencontré ce livre, je ne puis en dire davantage. Je le soupçonne, néanmoins de Benjamin Motte, depuis secrétaire de la société royale. *Montucla*, page 391.

Lubbock and Drinkwater say respecting it, page 43,

This essay, which was edited, and is generally supposed to have been written by Motte, the secretary of the Royal Society, contains a translation of Huyghens's treatise, and an application of his principles to the determination of the advantage of the banker at pharaon, hazard, and other games, and to some questions relating to lotteries.

A similar statement is made by Galloway in his *Treatise on Probability*, page 5.

79. It does not appear however that there was any fellow of the Royal Society named Motte; for the name does not occur

in the list of fellows given in Thomson's *History of the Royal Society.*

I have no doubt that the work is due to Arbuthnot. For there is an English translation of Huygens's treatise by W. Browne, published in 1714; in his Advertisement to the Reader Browne says, speaking of Huygens's treatise,

Besides the Latin Editions it has pass'd thro', the learned Dr Arbuthnott publish'd an English one, together with an Application of the General Doctrine to some particular Games then most in use; which is so intirely dispers'd Abroad, that an Account of it is all we can now meet with.

This seems to imply that there had been no other translation except Arbuthnot's; and the words "an Application of the General Doctrine to some particular Games then most in use" agree very well with some which occur in the work itself: "It is easy to apply this method to the Games that are in use amongst us." See page 28 of the fourth edition.

Watt's *Bibliotheca Britannica,* under the head Arbuthnot, places the work with the date 1692.

80. I have seen only one copy of this book, which was lent to me by Professor De Morgan. The title page is as follows:

Of the laws of chance, or, a method of calculation of the hazards of game, plainly demonstrated, and applied to games at present most in use; which may be easily extended to the most intricate cases of chance imaginable. The fourth edition, revis'd by John Ham. By whom is added, a demonstration of the gain of the banker in any circumstance of the game call'd Pharaon; and how to determine the odds at the Ace of Hearts or Fair Chance; with the arithmetical solution of some questions relating to lotteries; and a few remarks upon Hazard and Backgammon. London. Printed for B. Motte and C. Bathurst, at the *Middle-Temple* Gate in *Fleet-street,* M.DCC.XXXVIII.

81. I proceed to describe the work as it appears in the fourth edition.

The book is of small octavo size; it may be said to consist of two parts. The first part extends to page 49; it contains a translation of Huygens's treatise with some additional matter. Page 50 is blank; page 51 is in fact a title page containing a reprint

4

of part of the title we have already given, namely from "a demonstration" down to "Backgammon."

The words which have been quoted from Lubbock and Drinkwater in Art. 78, seem not to distinguish between these two
parts. There is nothing about the " advantage of the banker
at Pharaon" in the first part; and the investigations which are
given in the second part could not, I believe, have appeared so
early as 1692: they seem evidently taken from De Moivre. De
Moivre says in the second paragraph of his preface,

I had not at that time read anything concerning this Subject, but
Mr. Huygens's Book, de Ratiociniis in Ludo Aleæ, and a little English Piece (which was properly a Translation of it) done by a very ingenious Gentleman, who, tho' capable of carrying the matter a great
deal farther, was contented to follow his Original; adding only to it
the computation of the Advantage of the Setter in the Play called
Hazard, and some few things more.

82. The work is preceded by a Preface written with vigour
but not free from coarseness. We will give some extracts, which
show that the writer was sound in his views and sagacious in
his expectations.

It is thought as necessary to write a Preface before a Book, as
it is judg'd civil, when you invite a Friend to Dinner to proffer him
a Glass of Hock beforehand for a Whet: And this being maim'd
enough for want of a Dedication, I am resolv'd it shall not want an
Epistle to the Reader too. I shall not take upon me to determine,
whether it is lawful to play at Dice or not, leaving that to be disputed
betwixt the Fanatick Parsons and the Sharpers; I am sure it is lawful
to deal with Dice as with other Epidemic Distempers; ......

A great part of this Discourse is a Translation from Mons. Huygens's Treatise, De ratiociniis in ludo Aleæ; one, who in his Improvements of Philosophy, has but one Superior, and I think few or no
equals. The whole I undertook for my own Divertisement, next to
the Satisfaction of some Friends, who would now and then be wran-
gling about the Proportions of Hazards in some Cases that are here
decided. All it requir'd was a few spare Hours, and but little Work
for the Brain; my Design in publishing it, was to make it of more
general Use, and perhaps persuade a raw Squire, by it, to keep his
Money in his Pocket; and if, upon this account, I should incur the

Clamours of the Sharpers, I do not much regard it, since they are a sort of People the World is not bound to provide for.......

...It is impossible for a Die, with such determin'd force and direction, not to fall on such a determin'd side, and therefore I call that Chance which is nothing but want of Art;......

The Reader may here observe the Force of Numbers, which can be successfully applied, even to those things, which one would imagine are subject to no Rules. There are very few things which we know, which are not capable of being reduc'd to a Mathematical Reasoning; and when they cannot, it's a sign our Knowledge of them is very small and confus'd; and where a mathematical reasoning can be had, it's as great folly to make use of any other, as to grope for a thing in the dark, when you have a Candle standing by you. I believe the Calculation of the Quantity of Probability might be improved to a very useful and pleasant Speculation, and applied to a great many Events which are accidental, besides those of Games;......

...There is likewise a Calculation of the Quantity of Probability founded on Experience, to be made use of in Wagers about any thing; it is odds, if a Woman is with Child, but it shall be a Boy; and if you would know the just odds, you must consider the Proportion in the Bills that the Males bear to the Females: The Yearly Bills of Mortality are observ'd to bear such Proportion to the live People as 1 to 30, or 26; therefore it is an even Wager, that one out of thirteen dies within a Year (which may be a good reason, tho' not the true, of that foolish piece of Superstition), because, at this rate, if 1 out of 26 dies, you are no loser. It is but 1 to 18 if you meet a Parson in the Street, that he proves to be a Non-Juror, because there is but 1 of 36 that are such.

83. Pages 1 to 25 contain a translation of Huygens's treatise including the five problems which he left unsolved. Respecting these our author says

The Calculus of the preceding Problems is left out by Mons. Huygens, on purpose that the ingenious Reader may have the satisfaction of applying the former method himself; it is in most of them more laborious than difficult: for Example, I have pitch'd upon the second and third, because the rest can be solv'd after the same Method.

Our author solves the second problem in the first of the three senses which it may bear according to the *Ars Conjectandi*, and he arrives at the same result as James Bernoulli on page 58 of the *Ars Conjectandi*. Our author adds,

I have suppos'd here the Sense of the Problem to be, that when any one chus'd a Counter, he did not diminish their number; but if he miss'd of a white one, put it in again, and left an equal hazard to him who had the following choice; for if it be otherwise suppos'd, A's share will be $\frac{55}{123}$, which is less than $\frac{9}{19}$.

This result $\frac{55}{123}$ however is wrong in either of the other two senses which James Bernoulli ascribes to the problem, for which he obtains $\frac{77}{165}$ and $\frac{101}{125}$ respectively as the results; see Art. 35.

84. Then follow some other calculations about games. We have some remarks about the Royal-Oak Lottery which are analogous to those made on the Play of the Royal Oak by De Moivre in the Preface to his *Doctrine of Chances*.

A table is given of the number of various throws which can be made with three dice. Pages 34—39 are taken from Pascal; they seem introduced abruptly, and they give very little that had not already occurred in the translation of Huygens's treatise.

85. Our author touches on Whist; and he solves two problems about the situation of honours. These solutions are only approximate, as he does not distinguish between the dealers and their adversaries. And he also solves the problem of comparing the chances of two sides, one of which is at eight and the other at nine; the same remark applies to this solution. He makes the chances as 9 to 7; De Moivre by a stricter investigation makes them nearly as 25 to 18. See *Doctrine of Chances*, page 176.

86. Our author says on page 43,

All the former Cases can be calculated by the Theorems laid down by Monsieur Huygens; but Cases more compos'd require other Principles; for the easy and ready Computation of which, I shall add one Theorem more, demonstrated after Monsieur Huygens's method.

The theorem is : "if I have p Chances for a, q Chances for b, and r Chances for c, then my hazard is worth $\dfrac{ap + bq + cr}{p + q + r}$." Our author demonstrates this, and intimates that it may be extended to the case when there are also s Chances for d, &c.

Our author then considers the game of Hazard. He gives an investigation similar to that in De Moivre, and leading to the same results; see *Doctrine of Chances*, page 160.

87. The first part of the book concludes thus :

All those Problems suppose Chances, which are in an equal probability to happen; if it should be suppos'd otherwise, there will arise variety of Cases of a quite different nature, which, perhaps, 'twere not unpleasant to consider : I shall add one Problem of that kind, leaving the Solution to those who think it merits their pains.

In Parallelipipedo cujus latera sunt ad invicem in ratione a, b, c: Invenire quotâ vice quivis suscipere potest, ut datum quodvis planum, v. g. ab jaciat.

The problem was afterwards discussed by Thomas Simpson ; it is Problem XXVII. of his *Nature and Laws of Chance*.

88. It will be convenient to postpone an account of the second part of the book until after we have examined the works of De Moivre.

89. We next notice *An Arithmetical Paradox, concerning the Chances of Lotteries*, by the Honourable Francis Roberts, Esq. ; Fellow of the R. S.

This is published in Vol. XVII. of the *Philosophical Transactions*, 1693 ; it occupies pages 677—681.

Suppose in one lottery that there are three blanks, and three prizes each of 16 pence ; suppose in another lottery that there are four blanks, and two prizes each of 2 shillings. Now for one drawing, in the first lottery the expectation is $\frac{1}{2}$ of 16 pence, and in the second it is $\frac{1}{3}$ of 2 shillings ; so that it is 8 pence in each case. The paradox which Roberts finds is this ; suppose that a gamester pays a shilling for the chance in one of these lotteries ; then although, as we have just seen, the expectations are *equal*, yet the odds against him are 3 to 1 in the first lottery, and only 2 to 1 in the second.

The paradox is made by Roberts himself, by his own arbitrary definition of *odds*.

Supposing a lottery has a blanks and b prizes, and let each prize be r shillings; and suppose a gamester gives a shilling for one drawing in the lottery; then Roberts says the odds against him are formed by the product of $\frac{a}{b}$ and $\frac{1}{r-1}$, that is, the odds are as a to $b\,(r-1)$. This is entirely arbitrary.

The mere algebra of the paper is quite correct, and is a curious specimen of the mode of work of the day.

The author is doubtless the same whose name is spelt Robartes in De Moivre's Preface.

90. I borrow from Lubbock and Drinkwater an account of a work which I have not seen; it is given on their page 45.

It is not necessary to do more than mention an essay, by Craig, on the probability of testimony, which appeared in 1699, under the title of "Theologiæ Christianæ Principia Mathematica." This attempt to introduce mathematical language and reasoning into moral subjects can scarcely be read with seriousness; it has the appearance of an insane parody of Newton's Principia, which then engrossed the attention of the mathematical world. The author begins by stating that he considers the mind as a movable, and arguments as so many moving forces, by which a certain velocity of suspicion is produced, &c. He proves gravely, that suspicions of any history, transmitted through the given time (*cæteris paribus*), vary in the duplicate ratio of the times taken from the beginning of the history, with much more of the same kind with respect to the estimation of equable pleasure, uniformly accelerated pleasure, pleasure varying as any power of the time, &c. &c.

It is stated in biographical dictionaries that Craig's work was reprinted at Leipsic in 1755, with a refutation by J. Daniel Titius; and that some *Animadversiones* on it were published by Peterson in 1701.

Prevost and Lhuilier notice Craig's work in a memoir published in the *Mémoires de l'Acad....Berlin*, 1797. It seems that Craig concluded that faith in the Gospel so far as it depended on oral tradition expired about the year 800, and that so far as it depended on written tradition it would expire in the year 3150. Peterson by adopting a different law of diminution concluded that faith would expire in 1789.

See *Montmort*, page XXXVIII.; also the *Athenæum* for Nov. 7th, 1863, page 611.

91. *A Calculation of the Credibility of Human Testimony* is contained in Vol. XXI. of the *Philosophical Transactions;* it is the volume for 1699 : the essay occupies pages 359—365. The essay is anonymous ; Lubbock and Drinkwater suggest that it may be by Craig.

The views do not agree with those now received.

First suppose we have *successive* witnesses. Let a report be transmitted through a series of n witnesses, whose credibilities are $p_1, p_2, \ldots p_n$: the essay takes the product $p_1 p_2 \ldots p_n$ as representing the resulting probability.

Next, suppose we have *concurrent* witnesses. Let there be two witnesses ; the first witness is supposed to leave an amount of uncertainty represented by $1 - p_1$; of this the second witness removes the fraction p_2, and therefore leaves the fraction $(1 - p_1)(1 - p_2)$: thus the resulting probability is $1 - (1 - p_1)(1 - p_2)$. Similarly if there are three concurrent testimonies the resulting probability is $1 - (1 - p_1)(1 - p_2)(1 - p_3)$; and so on for a greater number.

The theory of this essay is adopted in the article *Probabilité* of the original French *Encyclopédie*, which is reproduced in the *Encyclopédie Méthodique:* the article is unsigned, so that we must apparently ascribe it to Diderot. The same theory is adopted by Bicquilley in his work *Du Calcul des Probabilités*.

CHAPTER VII.

JAMES BERNOULLI.

92. WE now propose to give an account of the *Ars Conjectandi* of James Bernoulli.

James Bernoulli is the first member of the celebrated family of this name who is associated with the history of Mathematics. He was born 27th December, 1654, and died 16th August, 1705. For a most interesting and valuable account of the whole family we may refer to the essay entitled *Die Mathematiker Bernoulli... von Prof. Dr. Peter Merian*, Basel, 1860.

93. Leibnitz states that at his request James Bernoulli studied the subject. Feu Mr. *Bernoulli* a cultivé cette matière sur mes exhortations; Leibnitii *Opera Omnia, ed. Dutens*, Vol. VI. part 1, page 217. But this statement is not confirmed by the correspondence between Leibnitz and James Bernoulli, to which we have already referred in Art. 59. It appears from this correspondence that James Bernoulli had nearly completed his work before he was aware that Leibnitz had heard any thing about it. Leibnitz says, page 71,

Audio a Te doctrinam de aestimandis probabilitatibus (quam ego magni facio) non parum esse excultam. Vellem aliquis varia ludendi genera (in quibus pulchra hujus doctrinae specimina) mathematice tractaret. Id simul amoenum et utile foret nec Te aut quocunque gravissimo Mathematico indignum.

James Bernoulli in reply says, page 77,

Scire libenter velim, Amplissime Vir, a quo habeas, quod Doctrina de probabilitatibus aestimandis a me excolatur. Verum est me a pluribus retro annis hujusmodi speculationibus magnopere delectari, ut vix putem, quemquam plura super his meditatum esse. Animus etiam erat, Tractatum quendam conscribendi de hac materia; sed saepe per integros annos seposui, quia naturalis meus torpor, quem accessoria valetudinis meae infirmitas immane quantum auxit, facit ut aegerrime ad scribendum accedam; et saepe mihi optarem amanuensem, qui cogitata mea leviter sibi indicata plene divinare, scriptisque consignare posset. Absolvi tamen jam maximam Libri partem, sed deest adhuc praecipua, qua artis conjectandi principia etiam ad civilia, moralia et oeconomia applicare doceo...

James Bernoulli then proceeds to speak of the celebrated theorem which is now called by his name.

Leibnitz in his next letter brings some objections against the theorem; see page 83: and Bernoulli replies; see page 87. Leibnitz returns to the subject; see page 94: and Bernoulli briefly replies, page 97,

Quod Verisimilitudines spectat, et earum augmentum pro aucto scil. observationum numero, res omnino se habet ut scripsi, et certus sum Tibi placituram demonstrationem, cum publicavero.

94. The last letter from James Bernoulli to Leibnitz is dated 3rd June, 1705. It closes in a most painful manner. We here see him, who was perhaps the most famous of all who have borne his famous name, suffering under the combined sorrow arising from illness, from the ingratitude of his brother John who had been his pupil, and from the unjust suspicions of Leibnitz who may be considered to have been his master:

Si rumor vere narrat, redibit certe frater meus Basileam, non tamen Graecam (cum ipse sit ἀναλφάβητος) sed meam potius stationem (quam brevi cum vita me derelicturum, forte non vane, existimat) occupaturus. De iniquis suspicionibus, quibus me immerentem onerasti in Tuis penultimis, alias, ubi plus otii nactus fuero. Nunc vale et fave etc.

95. The *Ars Conjectandi* was not published until eight years after the death of its author. The volume of the *Hist. de l'Acad....Paris* for 1705, published in 1706, contains Fontenelle's *Éloge* of James Bernoulli. Fontenelle here gave a brief notice, derived from Hermann, of the contents of the *Ars Conjectandi* then unpublished. A brief notice is also give in another *Éloge* of

James Bernoulli which appeared in the *Journal des Sçavans*
for 1706: this notice is attributed to Saurin by Montmort; see his
page IV.

References to the work of James Bernoulli frequently occur in
the correspondence between Leibnitz and John Bernoulli; see the
work cited in Art. 59, pages 367, 377, 836, 845, 847, 922, 923,
925, 931.

96. The *Ars Conjectandi* was published in 1713. A preface
of two pages was supplied by Nicolas Bernoulli, the son of a
brother of James and John. It appears from the preface that
the fourth part of the work was left unfinished by its author; the
publishers had desired that the work should be finished by John
Bernoulli, but the numerous engagements of this mathematician
had been an obstacle. It was then proposed to devolve the task
on Nicolas Bernoulli, who had already turned his attention to
the Theory of Probability. But Nicolas Bernoulli did not consider himself adequate to the task; and by his advice the work
was finally published in the state in which its author had left it;
the words of Nicolas Bernoulli are, Suasor itaque fui, ut Tractatus
iste qui maxima ex parte jam impressus erat, in eodem quo eum
Auctor reliquit statu cum publico communicaretur.

The *Ars Conjectandi* is not contained in the collected edition
of James Bernoulli's works.

97. The *Ars Conjectandi*, including a treatise on infinite series,
consists of 306 small quarto pages besides the title leaf and the
preface. At the end there is a dissertation in French, entitled
Lettre à un Amy, sur les Parties du Jeu de Paume which occupies 35 additional pages. Montucla speaks of this letter as the
work of an anonymous author; see his page 391: but there can
be no doubt that it is due to James Bernoulli, for to him Nicolas
Bernoulli assigns it in the preface to the *Ars Conjectandi*, and
in his correspondence with Montmort. See *Montmort*, page 333.

98. The *Ars Conjectandi* is divided into four parts. The
first part consists of a reprint of the treatise of Huygens *De Ratiociniis in Ludo Aleæ*, accompanied with a commentary by James
Bernoulli. The second part is devoted to the theory of permutations and combinations. The third part consists of the solution

of various problems relating to games of chance. The fourth part proposed to apply the Theory of Probability to questions of interest in morals and economical science. We may observe that instead of the ordinary symbol of equality, =, James Bernoulli uses ∞, which Wallis ascribes to Des Cartes; see Wallis's *Algebra*, 1693, page 138.

99. A French translation of the first part of the *Ars Conjectandi* was published in 1801, under the title of *L'Art de Conjecturer, Traduit du Latin de Jacques Bernoulli; Avec des Observations, Éclaircissemens et Additions. Par L. G. F. Vastel,...* Caen. 1801. The second part of the *Ars Conjectandi* is included in the volume of reprints which we have cited in Art. 47; Maseres in the same volume gave an English translation of this part.

100. The first part of the *Ars Conjectandi* occupies pages 1—71; with respect to this part we may observe that the commentary by James Bernoulli is of more value than the original treatise by Huygens. The commentary supplies other proofs of the fundamental propositions and other investigations of the problems; also in some cases it extends them. We will notice the most important additions made by James Bernoulli.

101. In the Problem of Points with two players, James Bernoulli gives a table which furnishes the chances of the two players when one of them wants any number of points not exceeding nine, and the other wants any number of points not exceeding seven; and, as he remarks, this table may be prolonged to any extent; see his page 16.

102. James Bernoulli gives a long note on the subject of the various throws which can be made with two or more dice, and the number of cases favourable to each throw. And we may especially remark that he constructs a large table which is equivalent to the theorem we now express thus: the number of ways in which m can be obtained by throwing n dice is equal to the co-efficient of x^m in the development of $(x + x^2 + x^3 + x^4 + x^5 + x^6)^n$ in a series of powers of x. See his page 24.

103. The tenth problem is to find in how many trials one may undertake to throw a six with a common die. James Bernoulli gives a note in reply to an objection which he suggests might be urged against the result; the reply is perhaps only intended as a popular illustration: it has been criticized by Prevost in the *Nouveaux Mémoires de l'Acad....Berlin* for 1781.

104. James Bernoulli gives the general expression for the chance of succeeding m times at least in n trials, when the chance of success in a single trial is known. Let the chances of success and failure in a single trial be $\dfrac{b}{a}$ and $\dfrac{c}{a}$ respectively: then the required chance consists of the terms of the expansion of $\left(\dfrac{b}{a}+\dfrac{c}{a}\right)^n$ from $\left(\dfrac{b}{a}\right)^n$ to the term which involves $\left(\dfrac{b}{a}\right)^m \left(\dfrac{c}{a}\right)^{n-m}$, both inclusive.

This formula involves a solution of the Problem of Points for two players of unequal skill; but James Bernoulli does not explicitly make the application.

105. James Bernoulli solves four of the five problems which Huygens had placed at the end of his treatise; the solution of the fourth problem he postpones to the third part of his book as it depends on combinations.

106. Perhaps however the most valuable contribution to the subject which this part of the work contains is a method of solving problems in chances which James Bernoulli speaks of as his own, and which he frequently uses. We will give his solution of the problem which forms the fourteenth proposition of the treatise of Huygens: we have already given the solution of Huygens himself; see Art. 34.

Instead of two players conceive an infinite number of players each of whom is to have one throw in turn. The game is to end as soon as a player whose turn is denoted by an odd number throws a six, or a player whose turn is denoted by an even number throws a seven, and such player is to receive the whole sum at stake. Let b denote the number of ways in which six can be thrown, c the number of ways in which six can fail; so that $b = 5$, and $c = 31$; let e denote the number of ways in which seven can be thrown, and f the number of ways in which seven can fail, so that $e = 6$, and $f = 30$; and let $a = b + c = e + f$.

Now consider the expectations of the different players; they are as follows:

I.   II.   III.   IV.   V.   VI.   VII.   VIII. ...

$$\frac{b}{a}, \quad \frac{ce}{a^2}, \quad \frac{bcf}{a^3}, \quad \frac{c^2ef}{a^4}, \quad \frac{bc^2f^2}{a^5}, \quad \frac{c^3ef^2}{a^6}, \quad \frac{bc^3f^3}{a^7}, \quad \frac{c^4ef^3}{a^8} \cdots$$

For it is obvious that $\dfrac{b}{a}$ expresses the expectation of the first player. In order that the second player may win, the first throw must fail and the second throw must succeed; that is there are ce favourable cases out of a^2 cases, so the expectation is $\dfrac{ce}{a^2}$. In order that the third player may win, the first throw must fail, the second throw must fail, and the third throw must succeed; that is there are cfb favourable cases out of a^3 cases, so the expectation is $\dfrac{bcf}{a^3}$. And so on for the other players. Now let a single player, A, be substituted in our mind in the place of the first, third, fifth,...; and a single player, B, in the place of the second, fourth, sixth.... We thus arrive at the problem proposed by Huygens, and the expectations of A and B are given by two infinite geometrical progressions. By summing these progressions we find that A's expectation is $\dfrac{ab}{a^2 - cf}$, and B's expectation is $\dfrac{ce}{a^2 - cf}$; the proportion is that of 30 to 31, which agrees with the result in Art. 34.

107. The last of the five problems which Huygens left to be solved is the most remarkable of all; see Art. 35. It is the first example on the *Duration of Play*, a subject which afterwards exercised the highest powers of De Moivre, Lagrange, and Laplace. James Bernoulli solved the problem, and added, without a demonstration, the result for a more general problem of which that of Huygens was a particular case; see *Ars Conjectandi* page 71.

Suppose A to have m counters, and B to have n counters; let their chances of winning in a single game be as a to b; the loser in each game is to give a counter to his adversary: required the chance of each player for winning all the counters of his adversary. In the case taken by Huygens m and n were equal.

It will be convenient to give the modern form of solution of the problem.

Let u_x denote A's chance of winning all his adversary's counters when he has himself x counters. In the next game A must either win or lose a counter; his chances for these two contingencies are $\dfrac{a}{a+b}$ and $\dfrac{b}{a+b}$ respectively: and then his chances of winning all his adversary's counters are u_{x+1} and u_{x-1} respectively. Hence

$$u_x = \frac{a}{a+b} u_{x+1} + \frac{b}{a+b} u_{x-1}.$$

This equation is thus obtained in the manner exemplified by Huygens in his fourteenth proposition; see Art. 34.

The equation in Finite Differences may be solved in the ordinary way; thus we shall obtain

$$u_x = C_1 + C_2 \left(\frac{b}{a}\right)^x,$$

where C_1 and C_2 are arbitrary constants. To determine these constants we observe that A's chance is zero when he has no counters, and that it is unity when he has all the counters. Thus u_x is equal to 0 when x is 0, and is equal to 1 when x is $m+n$. Hence we have

$$0 = C_1 + C_2, \quad 1 = C_1 + C_2 \left(\frac{b}{a}\right)^{m+n};$$

therefore

$$C_1 = -C_2 = \frac{a^{m+n}}{a^{m+n} - b^{m+n}}.$$

Hence

$$u_x = \frac{a^{m+n} - a^{m+n-x} b^x}{a^{m+n} - b^{m+n}}.$$

To détermine A's chance at the beginning of the game we must put $x = m$; thus we obtain

$$u_m = \frac{a^n (a^m - b^m)}{a^{m+n} - b^{m+n}}.$$

In precisely the same manner we may find B's chance at any stage of the game; and his chance at the beginning of the game will be

$$\frac{b^m (a^n - b^n)}{a^{m+n} - b^{m+n}}.$$

It will be observed that the sum of the chances of A and B at the beginning of the game is *unity*. The interpretation of this result is that one or other of the players must eventually win all the counters; that is, the play must terminate. This might have been expected, but was not assumed in the investigation.

The formula which James Bernoulli here gives will next come before us in the correspondence between Nicolas Bernoulli and Montmort; it was however first published by De Moivre in his *De Mensura Sortis*, Problem IX., where it is also demonstrated.

108. We may observe that Bernoulli seems to have found, as most who have studied the subject of chances have also found, that it was extremely easy to fall into mistakes, especially by attempting to reason without strict calculation. Thus, on his page 15, he points out a mistake into which it would have been easy to fall, *nisi nos calculus aliud docuisset*. He adds,

Quo ipso proin monemur, ut cauti simus in judicando, ¯nec ratiocinia nostra super quâcunque statim analogiâ in rebus deprehensâ fundare suescamus; quod ipsum tamen etiam ab iis, qui vel maximè sapere videntur, nimis frequenter fieri solet.

Again, on his page 27,

Quæ quidem eum in finem hîc adduco, ut palàm fiat, quàm parùm fidendum sit ejusmodi ratiociniis, quæ corticem tantùm attingunt, nec in ipsam rei naturam altiùs penetrant; tametsi in toto vitæ usu etiam apud sapientissimos quosque nihil sit frequentius.

Again, on his page 29, he refers to the difficulty which Pascal says had been felt by M. de * * * *, whom James Bernoulli calls Anonymus quidam cæterà subacti judicii Vir, sed Geometriæ expers. James Bernoulli adds,

Hâc enim qui imbuti sunt, ejusmodi ἐναντιοφανείαι minimè morantur, probè conscii dari innumera, quæ admoto calculo aliter se habere comperiuntur, quàm initio apparebant; ideoque sedulò cavent, juxtà id quod semel iterumque monui, ne quicquam analogiis temerè tribuant.

109. The second part of the *Ars Conjectandi* occupies pages 72—137: it contains the doctrine of Permutations and Combinations. James Bernoulli says that others have treated this subject before him, and especially Schooten, Leibnitz, Wallis and Prestet ; and so he intimates that his matter is not entirely new. He continues thus, page 73,

...tametsi quædam non contemnenda de nostro adjecimus, inprimis demonstrationem generalem et facilem proprietatis numerorum figuratorum, cui cætera pleraque innituntur, et quam nemo quod sciam ante nos dedit eruitve.

110. James Bernoulli begins by treating on permutations; he proves the ordinary rule for finding the number of permutations of a set of things taken all together, when there are no repetitions among the set of things and also when there are. He gives a full analysis of the number of arrangements of the verse Tot tibi sunt dotes, Virgo, quot sidera cœli ; see Art. 40. He then considers combinations; and first he finds the total number of ways in which a set of things can be taken, by taking them one at a time, two at a time, three at a time,...He then proceeds to find what we should call the number of combinations of n things taken r at a time; and here is the part of the subject in which he added most to the results obtained by his predecessors. He gives a figure which is substantially the same as Pascal's *Arithmetical Triangle;* and he arrives at two results, one of which is the well-known form for the nth term of the rth order of figurate numbers, and the other is the formula for the sum of a given number of terms of the series of figurate numbers of a given order ; these results are expressed definitely in the modern notation as we now have them in works on Algebra. The mode of proof is more laborious, as might be expected. Pascal as we have seen in Arts. 22 and 41, employed without any scruple, and indeed rather with approbation, the method of induction : James Bernoulli however says, page 95,... modus demonstrandi per inductionem parùm scientificus est.

James Bernoulli names his predecessors in investigations on figurate numbers in the following terms on his page 95 :

Multi, ut hoc in transitu notemus, numerorum figuratorum contemplationibus vacârunt (quos inter Faulhaberus et Remmelini Ulmenses, Wallisius, Mercator in Logarithmotechniâ, Prestetus, aliique)...

111. We may notice that James Bernoulli gives incidentally on his page 89 a demonstration of the Binomial Theorem for the case of a positive integral exponent. Maseres considers this to be the first demonstration that appeared; see page 233 of the work cited in Art. 47.

112. From the summation of a series of figurate numbers James Bernoulli proceeds to derive the summation of the powers of the natural numbers. He exhibits definitely Σn, Σn^2, Σn^3,... up to Σn^{10}; he uses the symbol $\int$ where we in modern books use Σ. He then extends his results by induction without demonstration, and introduces for the first time into Analysis the coefficients since so famous as the *numbers of Bernoulli*. His general formula is that

$$\Sigma n = \frac{n^{c+1}}{c+1} + \frac{n}{2} + \frac{c}{2} An^{c-1} + \frac{c\,(c-1)\,(c-2)}{2.3.4} Bn^{c-3}$$

$$+ \frac{c\,(c-1)\,(c-2)\,(c-3)\,(c-4)}{2.3.4.5.6} Cn^{-5}$$

$$+ \frac{c\,(c-1)\,(c-2)\,(c-3)\,(c-4)\,(c-5)\,(c-6)}{2.3.4.5.6.7.8} Dn^{c-7} + \dots$$

where $\quad A = \frac{1}{6}$, $B = -\frac{1}{30}$, $C = \frac{1}{42}$, $D = -\frac{1}{30}$, $\dots$

He gives the numerical value of the sum of the tenth powers of the first thousand natural numbers; the result is a number with thirty-two figures. He adds, on his page 98,

E quibus apparet, quàm inutilis censeuda sit opera Ismaëlis Bullialdi, quam conscribendo tam spisso volumini Arithmeticæ suæ Infinitorum impendit, ubi nihil præstitit aliud, quàm ut primarum tantum sex potestatum summas (partem ejus quod unicâ nos consecuti sumus paginâ) immenso labore demonstratas exhiberet.

For some account of Bulliald's *spissum volumen*, see Wallis's *Algebra*, Chap. LXXX.

113. James Bernoulli gives in his fourth Chapter the rule now well known for the number of the combinations of n things

taken c at a time. He also draws various simple inferences from the rule. He digresses from the subject of this part of his book to resume the discussion of the Problem of Points; see his page 107. He gives two methods of treating the problem by the aid of the theory of combinations. The first method shews how the table which he had exhibited in the first part of the *Ars Con-* *jectandi* might be continued and the law of its terms expressed; the table is a statement of the chances of A and B for winning the game when each of them wants an assigned number of points. Pascal had himself given such a table for a game of six points; an extension of the table is given on page 16 of the *Ars Con-* *jectandi*, and now James Bernoulli investigates general expressions for the component numbers of the table. From his investigation he derives the result which Pascal gave for the case in which one player wants one point more than the other player. James Bernoulli concludes this investigation thus; Ipsa solutio Pascaliana, quæ Auctori suo tantopere arrisit.

James Bernoulli's other solution of the Problem of Points is much more simple and direct, for here he *does* make the application to which we alluded in Art. 104. Suppose that A wants m points and B wants n points; then the game will certainly be decided in $m + n - 1$ trials. As in each trial A and B have equal chances of success the whole number of possible cases is 2^{m+n-1}. And A wins the game if B gains no point, or if B gains just one point, or just two points,... or any number up to $n - 1$ inclusive. Thus the number of cases favourable to A is

$$1 + \mu + \frac{\mu(\mu-1)}{2} + \frac{\mu(\mu-1)(\mu-2)}{\underline{3}} + ... + \frac{\mu(\mu-1)...(\mu-n+2)}{\underline{n-1}},$$

where $\mu = m + n - 1$.

Pascal had in effect advanced as far as this; see Art. 23: but the formula is more convenient than the *Arithmetical Triangle*.

114. In his fifth Chapter James Bernoulli considers another question of combinations, namely that which in modern treatises is enunciated thus: to find the number of homogeneous products of the rth degree which can be formed of n symbols. In his sixth Chapter he continues this subject, and makes a slight reference to

the doctrine of the number of divisors of a given number; for more information he refers to the works of Schooten and Wallis, which we have already examined; see Arts. 42, 47.

115. In his seventh Chapter James Bernoulli gives the formula for what we now call the number of permutations of n things taken c at a time. In the remainder of this part of his book he discusses some other questions relating to permutations and combinations, and illustrates his theory by examples.

116. The third part of the *Ars Conjectandi* occupies pages 138—209; it consists of twenty-four problems which are to illustrate the theory that has gone before in the book. James Bernoulli gives only a few lines of introduction, and then proceeds to the problems, which he says,

...nullo ferè habito selectu, prout in adversariis reperi, proponam, præmissis etiam vel interspersis nonnullis facilioribus, et in quibus nullus combinationum usus apparet.

117. The fourteenth problem deserves some notice. There are two cases in it, but it will be sufficient to consider one of them. A is to throw a die, and then to repeat his throw as many times as the number thrown the first time. A is to have the whole stake if the sum of the numbers given by the latter set of throws exceeds 12; he is to have half the stake if the sum is equal to 12; and he is to have nothing if the sum is less than 12. Required the value of his expectation. It is found to be $\frac{15295}{31104}$, which is rather less than $\frac{1}{2}$. After giving the correct solution James Bernoulli gives another which is plausible but false, in order, as he says, to impress on his readers the necessity of caution in these discussions. The following is the false solution.

A has a chance equal to $\frac{1}{6}$ of throwing an ace at his first trial; in this case he has only one throw for the stake, and that throw may give him with equal probability any number between 1 and 6 inclusive, so that we may take $\frac{1}{6}(1+2+3+4+5+6)$, that is $3\frac{1}{2}$, for his mean throw. We may observe that $3\frac{1}{2}$ is the Arithmetical mean between 1 and 6. Again A has a chance equal to $\dfrac{1}{6}$ of throwing a two at his first trial; in this case he has two throws for the stake, and these two throws may give him any number between 2 and 12 inclusive; and the probability of the number 2 is the same as that of 12, the probability of 3 is the same as that of 11, and so on; hence as before we may take $\dfrac{1}{2}(2+12)$, that is 7, for his mean throw. In a similar way if three, four, five, or six be thrown at the first trial, the corresponding means of the numbers in the throws for the stake will be respectively $10\frac{1}{2}$, 14, $17\frac{1}{2}$, and 21. Hence the mean of all the numbers is

$$\dfrac{1}{6}\{3\tfrac{1}{2}+7+10\tfrac{1}{2}+14+17\tfrac{1}{2}+21\},\text{ that is } 12\tfrac{1}{4};$$

and as this number is greater than 12 it might appear that the odds are in favour of A.

A false solution of a problem will generally appear more plausible to a person who has originally been deceived by it than to another person who has not seen it until after he has studied the accurate solution. To some persons James Bernoulli's false solution would appear simply false and not plausible; it leaves the problem proposed and substitutes another which is entirely different. This may be easily seen by taking a simple example. Suppose that A instead of an equal chance for any number of throws between one and six inclusive, is restricted to one *or* six throws, and that each of these two cases is equally likely. Then, as before, we may take $\dfrac{1}{2}\{3\tfrac{1}{2}+21\}$, that is $12\tfrac{1}{4}$ as the mean throw. But it is obvious that the odds are against him; for if he has only one throw he cannot obtain 12, and if he has six throws he will not necessarily obtain 12. The question is not *what is the mean number* he will obtain, but *how many throws* will give him 12 or more, and *how many* will give him less than 12.

James Bernoulli seems not to have been able to make out more than that the second solution must be false because the first is unassailable; for after saying that from the second solution we might suppose the odds to be in favour of A, he adds, Hujus autem contrarium ex priore solutione, quæ sua luce radiat, apparet; ...

The problem has been since considered by Mallet and by Fuss, who agree with James Bernoulli in admitting the plausibility of the false solution.

118. James Bernoulli examines in detail some of the games of chance which were popular in his day. Thus on pages 167 and 168 he takes the game called *Cinq et neuf.* He takes on pages 169—174 a game which had been brought to his notice by a stroller at fairs. According to James Bernoulli the chances were against the stroller, and so as he says, istumque proin hoc aleæ genere, ni præmia minuat, non multum lucrari posse. We might desire to know more of the stroller who thus supplied the occasion of an elaborate discussion to James Bernoulli, and who offered to the public the amusement of gambling on terms unfavourable to himself.

James Bernoulli then proceeds to a game called *Trijaques.* He considers that, it is of great importance for a player to maintain a serene composure even if the cards are unfavourable, and that a previous calculation of the chances of the game will assist in securing the requisite command of countenance and temper. As James Bernoulli speaks immediately afterwards of what he had himself formerly often observed in the game, we may perhaps infer that *Trijaques* had once been a favourite amusement with him.

119. The nineteenth problem is thus enunciated,

In quolibet Aleæ genere, si ludi Oeconomus seu Dispensator *(le Banquier du Jeu)* nonnihil habeat prærogativæ in eo consistentis, ut paulo major sit casuum numerus quibus vincit quàm quibus perdit; et major simul casuum numerus, quibus in officio Oeconomi pro ludo sequenti confirmatur, quàm quibus œconomia in collusorem transfertur. Quæritur, quanti privilegium hoc Oeconomi sit æstimandum?

The problem is chiefly remarkable from the fact that James Bernoulli candidly records two false solutions which occurred to him before he obtained the true solution.

120. The twenty-first problem relates to the game of *Bassette;*

James Bernoulli devotes eight pages to it, his object being to estimate the advantage of the banker at the game. See Art. 74.

The last three problems which James Bernoulli discusses arose from his observing that a certain stroller, in order to entice persons to play with him, offered them among the conditions of the game one which was apparently to their advantage, but which on investigation was shewn to be really pernicious; see his pages 208, 209.

121. The fourth part of the *Ars Conjectandi* occupies pages 210—239; it is entitled *Pars Quarta, tradens usum et applicationem præcedentis Doctrinæ in Civilibus, Moralibus et Oeconomicis.* It was unfortunately left incomplete by the author; but nevertheless it may be considered the most important part of the whole work. It is divided into five Chapters, of which we will give the titles.

I. *Præliminaria quædam de Certitudine, Probabilitate, Necessitate, et Contingentia Rerum.*

II. *De Scientia et Conjectura. De Arte Conjectandi. De Argumentis Conjecturarum. Axiomata quædam generalia huc pertinentia.*

III. *De variis argumentorum generibus, et quomodo eorum pondera æstimentur ad supputandas rerum probabilitates.*

IV. *De duplici Modo investigandi numeros casuum. Quid sentiendum de illo, qui instituitur per experimenta. Problema singulare eam in rem propositum, &c.*

V. *Solutio Problematis præcedentis.*

122. We will briefly notice the results of James Bernoulli as to the probability of arguments. He distinguishes arguments into two kinds, *pure* and *mixed.* He says, *Pura* voco, quæ in quibusdam casibus ita rem probant, ut in aliis nihil positivè probent: *Mixta,* quæ ita rem probant in casibus nonnullis, ut in cæteris probent contrarium rei.

Suppose now we have three arguments of the *pure* kind leading to the same conclusion; let their respective probabilities be

$1 - \dfrac{c}{a}$, $1 - \dfrac{f}{d}$, $1 - \dfrac{i}{g}$. Then the resulting probability of the conclusion is $1 - \dfrac{cfi}{adg}$. This is obvious from the consideration that any one of the arguments would establish the conclusion, so that the conclusion fails only when all the arguments fail.

Suppose now that we have in addition two arguments of the *mixed* kind: let their respective probabilities be $\dfrac{q}{q+r}$, $\dfrac{t}{t+u}$. Then James Bernoulli gives for the resulting probability

$$1 - \frac{cfiru}{adg\,(ru + qt)}.$$

But this formula is inaccurate. For the supposition $q = 0$ amounts to having one argument *absolutely decisive against* the conclusion, while yet the formula leaves still a certain probability *for* the conclusion. The error was pointed out by Lambert; see Prevost and Lhuilier, *Mémoires de l'Acad....Berlin* for 1797.

123. The most remarkable subject contained in the fourth part of the *Ars Conjectandi* is the enunciation and investigation of what we now call *Bernoulli's Theorem*. It is introduced in terms which shew a high opinion of its importance:

Hoc igitur est illud Problema, quod evulgandum hoc loco proposui, postquam jam per vicennium pressi, et cujus tum novitas, tum summa utilitas cum pari conjuncta difficultate omnibus reliquis hujus doctrinæ capitibus pondus et pretium superaddere potest. *Ars Conjectandi*, page 227. See also De Moivre's *Doctrine of Chances*, page 254.

We will now state the purely algebraical part of the theorem. Suppose that $(r + s)^{nt}$ is expanded by the Binomial Theorem, the letters all denoting integral numbers and t being equal to $r + s$. Let u denote the sum of the greatest term and the n preceding terms and the n following terms. Then by taking n large enough the ratio of u to the sum of all the remaining terms of the expansion may be made as great as we please.

If we wish that this ratio should not be less than c it will be sufficient to take n equal to the greater of the two following expressions,

$$\frac{\log c + \log (s-1)}{\log (r+1) - \log r} \left(1 + \frac{s}{r+1}\right) - \frac{s}{r+1},$$

and
$$\frac{\log c + \log (r-1)}{\log (s+1) - \log s} \left(1 + \frac{r}{s+1}\right) - \frac{r}{s+1}.$$

James Bernoulli's demonstration of this result is long but perfectly satisfactory; it rests mainly on the fact that the terms in the Binomial series increase continuously up to the greatest term, and then decrease continuously. We shall see as we proceed with the history of our subject that James Bernoulli's demonstration is now superseded by the use of Stirling's Theorem.

124.   Let us now take the application of the algebraical result to the Theory of Probability.   The greatest term of $(r+s)^{nt}$, where $t = r+s$ is the term involving $r^{nr} s^{ns}$.   Let r and s be proportional to the probability of the happening and failing of an event in a single trial.   Then the sum of the $2n+1$ terms of $(r+s)^{nt}$ which have the greatest term for their middle term corresponds to the probability that in nt trials the number of times the event happens will lie between $n(r-1)$ and $n(r+1)$, both inclusive; so that the ratio of the number of times the event happens to the whole number of trials lies between $\frac{r+1}{t}$ and $\frac{r-1}{t}$.   Then, by taking for n the greater of the two expressions in the preceding article, we have the odds of c to 1, that the ratio of the number of times the event happens to the whole number of trials lies between $\frac{r+1}{t}$ and $\frac{r-1}{t}$.

As an example James Bernoulli takes

$$r = 30, \quad s = 20, \quad t = 50.$$

He finds for the odds to be 1000 to 1 that the ratio of the number of times the event happens to the whole number of trials shall lie between $\frac{31}{50}$ and $\frac{29}{50}$, it will be sufficient to make 25550 trials; for the odds to be 10000 to 1, it will be sufficient to make 31258 trials; for the odds to be 100000 to 1, it will be sufficient to make 36966 trials; and so on.

125. Suppose then that we have an urn containing white balls and black balls, and that the ratio of the number of the former to the latter is *known to be* that of 3 to 2. We learn from the preceding result that if we make 25550 drawings of a single ball, replacing each ball after it is drawn, the odds are 1000 to 1 that the white balls drawn lie between $\frac{31}{50}$ and $\frac{29}{50}$ of the whole number drawn. This is the *direct* use of James Bernoulli's theorem. But he himself proposed to employ it *inversely* in a far more important way. Suppose that in the preceding illustration we do not know anything beforehand of the ratio of the white balls to the black; but that we have made a large number of drawings, and have obtained a white ball R times, and a black ball S times: then according to James Bernoulli we are to infer that the ratio of the white balls to the black balls in the urn is approximately $\frac{R}{S}$. To determine the precise numerical estimate of the probability of this inference requires further investigation: we shall find as we proceed that this has been done in two ways, by an inversion of James Bernoulli's theorem, or by the aid of another theorem called Bayes's theorem; the results approximately agree. See Laplace, *Théorie...des Prob....* pages 282 and 366.

126. We have spoken of the *inverse* use of James Bernoulli's theorem as the most important; and it is clear that he himself was fully aware of this. This use of the theorem was that which Leibnitz found it difficult to admit, and which James Bernoulli maintained against him; see the correspondence quoted in Art. 59, pages 77, 83, 87, 94, 97.

127. A memoir on infinite series follows the *Ars Conjectandi*, and occupies pages 241—306 of the volume; this is contained in the collected edition of James Bernoulli's works, Geneva, 1744: it is there broken up into parts and distributed through the two volumes of which the edition consists.

This memoir is unconnected with our subject, and we will therefore only briefly notice some points of interest which it presents.

128.   James Bernoulli enforces the importance of the subject in the following terms, page 243,

Cæterum quantæ sit necessitatis pariter et utilitatis hæc serierum contemplatio, ei sane ignotum esse non poterit, qui perspectum habuerit, ejusmodi series sacram quasi esse anchoram, ad quam in maxime arduis et desperatæ solutionis Problematibus, ubi omnes alias humani ingenii vires naufragium passæ, velut ultimi remedii loco confugiendum est.

129.   The principal artifice employed by James Bernoulli in this memoir is that of subtracting one series from another, thus obtaining a third series.   For example,

let $\qquad S = 1 + \dfrac{1}{2} + \dfrac{1}{3} + \ldots + \dfrac{1}{n+1}$,

then $\qquad S = \quad 1 + \dfrac{1}{2} + \dfrac{1}{3} + \ldots + \dfrac{1}{n} + \dfrac{1}{n+1}$;

therefore $0 = -1 + \dfrac{1}{1 \cdot 2} + \dfrac{1}{2 \cdot 3} + \dfrac{1}{3 \cdot 4} + \ldots + \dfrac{1}{n(n+1)} + \dfrac{1}{n+1}$,

therefore $\qquad \dfrac{1}{1 \cdot 2} + \dfrac{1}{2 \cdot 3} + \dfrac{1}{3 \cdot 4} + \ldots + \dfrac{1}{n(n+1)} = 1 - \dfrac{1}{n+1}$.

Thus the sum of n terms of the series, of which the rth term is $\dfrac{1}{r(r+1)}$, is $\dfrac{n}{n+1}$.

130.   James Bernoulli says that his brother first observed that the sum of the infinite series $\dfrac{1}{1} + \dfrac{1}{2} + \dfrac{1}{3} + \dfrac{1}{4} + \ldots$ is infinite ; and he gives his brother's demonstration and his own ; see his page 250.

131.   James Bernoulli shews that the sum of the infinite series $\dfrac{1}{1} + \dfrac{1}{2^2} + \dfrac{1}{3^2} + \dfrac{1}{4^2} + \ldots$ is finite, but confesses himself unable to give the sum.   He says, page 254, Si quis inveniat nobisque communicet, quod industriam nostram elusit hactenus, magnas de nobis gratias feret.   The sum is now known to be $\dfrac{\pi^2}{6}$; this result is due to Euler: it is given in his *Introductio in Analysin Infinitorum*, 1748, Vol. I. page 130.

132. James Bernoulli seems to be on more familiar terms with infinity than mathematicians of the present day. On his page 262 we find him stating, correctly, that the sum of the infinite series $\frac{1}{\sqrt{1}} + \frac{1}{\sqrt{2}} + \frac{1}{\sqrt{3}} + \frac{1}{\sqrt{4}} + \dots$ is infinite, for the series is greater than $\frac{1}{1} + \frac{1}{2} + \frac{1}{3} + \frac{1}{4} + \dots$ He adds that the sum of all the odd terms of the first series is to the sum of all the even terms as $\sqrt{2} - 1$ is to 1; so that the sum of the odd terms would appear to be *less* than the sum of the even terms, which is impossible. But the paradox does not disturb James Bernoulli, for he adds,

...cujus ἐναντιοφανείας rationem, etsi ex infiniti natura finito intellectui comprehendi non posse videatur, nos tamen satis perspectam habemus.

133. At the end of the volume containing the *Ars Conjectandi* we have the *Lettre à un Amy, sur les Parties du Jeu de Paume,* to which we have alluded in Art. 97.

The nature of the problem discussed may be thus stated. Suppose A and B two players; let them play a set of games, say five, that is to say, the player gains the set who first wins five games. Then a certain number of sets, say four, make a match. It is required to estimate the chances of A and B in various states of the contest. Suppose for example that A has won two sets, and B has won one set; and that in the set now current A has won two games and B has won one game. The problem is thus somewhat similar in character to the Problem of Points, but more complicated. James Bernoulli discusses it very fully, and presents his result in the form of tables. He considers the case in which the players are of unequal skill; and he solves various problems arising from particular circumstances connected with the game of tennis to which the letter is specially devoted.

On the second page of the letter is a very distinct statement of the use of the celebrated theorem known by the name of Bernoulli; see Art. 123.

134. One problem occurs in this *Lettre à un Amy...* which it may be interesting to notice.

Suppose that A and B engage in play, and that each in turn

by the laws of the game has an advantage over his antagonist. Thus
suppose that A's chance of winning in the 1st, 3rd, 5th... games is
always p, and his chance of losing q; and in the 2nd, 4th, 6th...
games suppose that A's chance of winning is q and his chance of
losing p. The chance of B is found by taking that of A from
unity; so that B's chance is p or q according as A's is q or p.

Now let A and B play, and suppose that the stake is to be
assigned to the player who first wins n games. There is however to
be this peculiarity in their contest: If each of them obtains $n-1$
games it will be necessary for one of them to win two games in
succession to decide the contest in his favour; if each of them
wins one of the next two games, so that each has scored n games,
the same law is to hold, namely, that one must win two games in
succession to decide the contest in his favour; and so on.

Let us now suppose that $n = 2$, and estimate the advantage of
A. Let x denote this advantage, S the whole sum to be gained.

Now A may win the first and second games; his chance for
this is pq, and then he receives S. He may win the first game,
and lose the second; his chance for this is p^2. He may lose the
first game and win the second; his chance for this is q^2. In the
last two cases his position is neither better nor worse than at first;
that is he may be said to receive x.

Thus $$x = pq\,S + (p^2 + q^2)\,x\,;$$

therefore $$x = \frac{pq\,S}{1 - p^2 - q^2} = \frac{pq\,S}{2\,pq} = \frac{S}{2}\,.$$

Hence of course B's advantage is also $\dfrac{S}{2}$. Thus the players
are on an equal footing.

James Bernoulli in his way obtains this result. He says that
whatever may be the value of n, the players are on an equal footing; he verifies the statement by calculating numerically the
chances for $n = 2, 3, 4$ or 5, taking $p = 2q$. See his pages 18, 19.

Perhaps the following remarks may be sufficient to shew that
whatever n may be, the players must be on an equal footing. By
the peculiar law of the game which we have explained, it follows
that the contest is not decided until one player has gained at least
n games, and is at least two games in advance of his adversary.

Thus the contest is either decided in an *even* number of games, or else in an odd number of games in which the victor is at least three games in advance of his adversary: in the last case no advantage or disadvantage will accrue to either player if they play one more game and count it in. Thus the contest may be conducted without any change of probabilities under the following laws: the number of games shall be *even*, and the victor gain not less than n and be at least two in advance of his adversary. But since the number of games is to be *even* we see that the two players are on an equal footing.

135. Gouraud has given the following summary of the merits of the *Ars Conjectandi*; see his page 28 :

Tel est ce livre de l'*Ars conjectandi*, livre qui, si l'on considère le temps où il fut composé, l'originalité, l'étendue et la pénétration d'esprit qu'y montra son auteur, la fécondité étonnante de la constitution scientifique qu'il donna au Calcul des probabilités, l'influence enfin qu'il devait exercer sur deux siècles d'analyse, pourra sans exagération être regardé comme un des monuments les plus importants de l'histoire des mathématiques. Il a placé à jamais le nom de Jacques Bernoulli parmi les noms de ces inventeurs, à qui la postérité reconnaissante reporte toujours et à bon droit, le plus pur mérite des découvertes, que sans leur premier effort, elle n'aurait jamais su faire.

This panegyric, however, seems to neglect the simple fact of the date of *publication* of the *Ars Conjectandi*, which was really subsequent to the first appearance of Montmort and De Moivre in this field of mathematical investigation. The researches of James Bernoulli were doubtless the earlier in existence, but they were the later in appearance before the world ; and thus the influence which they might have exercised had been already produced. The problems in the first three parts of the *Ars Conjectandi* cannot be considered equal in importance or difficulty to those which we find investigated by Montmort and De Moivre; but the memorable theorem in the fourth part, which justly bears its author's name, will ensure him a permanent place in the history of the Theory of Probability.

CHAPTER VIII.

MONTMORT.

136. THE work which next claims attention is that of Montmort; it is entitled *Essai d'Analyse sur les Jeux de Hazards*.

Fontenelle's *Éloge de M. de Montmort* is contained in the volume for 1719 of the *Hist. de l'Acad...Paris*, which was published in 1721; from this we take a few particulars.

Pierre Remond de Montmort was born in 1678. Under the influence of his guide, master, and friend, Malebranche, he devoted himself to religion, philosophy, and mathematics. He accepted with reluctance a canonry of Nôtre-Dame at Paris, which he relinquished in order to marry. He continued his simple and retired life, and we are told that, *par un bonheur assez singulier le mariage lui rendit sa maison plus agréable*. In 1708 he published his work on Chances, where with the courage of Columbus he revealed a new world to mathematicians.

After Montmort's work appeared De Moivre published his essay *De Mensura Sortis*. Fontenelle says,

Je ne dissimulerai point qui M. de Montmort fut vivement piqué de cet ouvrage, qui lui parut avoir été entiérement fait sur le sien, et d'après le sien. Il est vrai, qu'il y étoit loué, et n'étoit-ce pas assez, dira-t-on ? mais un Seigneur de fief n'en quittera pas pour des louanges celui qu'il prétend lui devoir foi et hommage des terres qu'il tient de lui. Je parle selon sa prétention, et ne décide nullement s'il étoit en effet le Seigneur.

Montmort died of small pox at Paris in 1719. He had been engaged on a work entitled *Histoire de la Géométrie*, but had not proceeded far with it; on this subject Fontenelle has some interesting remarks. See also Montucla's *Histoire des Mathematiques*, first edition, Preface, page vii.

137. There are two editions of Montmort's work; the first appeared in 1708; the second is sometimes said to have appeared in 1713, but the date 1714 is on the title page of my copy, which appears to have been a present to 'sGravesande from the author. Both editions are in quarto; the first contains 189 pages with a preface of XXIV pages, and the second contains 414 pages with a preface and advertisement of XLII pages. The increased bulk of the second edition arises, partly from the introduction of a treatise on combinations which occupies pages 1—72, and partly from the addition of a series of letters which passed between Montmort and Nicholas Bernoulli with one letter from John Bernoulli. The name of Montmort does not appear on the title page or in the work, except once on page 338, where it is used with respect to a place.

Any reference which we make to Montmort's work must be taken to apply to the *second* edition unless the contrary is stated.

Montucla says, page 394, speaking of the second edition of Montmort's work, Cette édition, indépendamment de ses augmentations et corrections faites à la première, est remarquable par de belles gravures à la tête de chaque partie. These engravings are four in number, and they occur *also in the first edition*, and of course the impressions will naturally be finer in the earlier edition. It is desirable to correct the error implied in Montucla's statement, because the work is scarce, and thus those who merely wish for the engravings may direct their attention to the first edition, leaving the second for mathematicians.

138. Leibnitz corresponded with Montmort and his brother; and he records a very favourable opinion of the work we are now about to examine. He says, however, J'aurois souhaité les loix des Jeux un peu mieux décrites, et les termes expliqués en faveur des étrangers et de la postérité. Leibnitii *Opera Omnia, ed. Dutens*, Vol. v. pages 17 and 28.

Reference is also made to Montmort and his book in the correspondence between Leibnitz and John and Nicholas Bernoulli;

see the work cited in Art. 59, pages 827, 836, 837, 842, 846, 903, 985, 987, 989.

139. We will now give a detailed account of Montmort's work; we will take the second edition as our standard, and point out as occasion may require when our remarks do not apply to ' the first edition also.

140. The preface occupies XXIV pages. Montmort refers to the fact that James Bernoulli had been engaged on a work entitled *De arte conjectandi*, which his premature death had prevented him from completing. Montmort's introduction to these studies had arisen from the request of some friends that he would determine the advantage of the banker at the game of Pharaon; and he had been led on to compose a work which might compensate for the loss of Bernoulli's.

Montmort makes some judicious observations on the foolish and superstitious notions which were prevalent among persons devoted to games of chance, and proposes to check these by shewing, not only to such persons but to men in general, that there are rules in chance, and that for want of knowing these rules mistakes are made which entail adverse results; and these results men impute to destiny instead of to their own ignorance. Perhaps however he speaks rather as a philosopher than as a gambler when he says positively on his page VIII,

On joueroit sans doute avec plus d'agrément si l'on pouvoit sçavoir à chaque coup l'esperance qu'on a de gagner, ou le risque que l'on court de perdre. On seroit plus tranquile sur les évenemens du jeu, et on sentiroit mieux le ridicule de ces plaintes continuelles ausquelles se laissent aller la plûpart des Joueurs dans les rencontres les plus communes, lorsqu'elles leur sont contraires.

141. Montmort divides his work into four parts. The first part contains the theory of combinations; the second part discusses certain games of chance depending on cards; the third part discusses certain games of chance depending on dice; the fourth part contains the solution of various problems in chances, including the five problems proposed by Huygens. To these four parts must be added the letters to which we have alluded in Art. 137.

Montmort gives his reasons for not devoting a part to the application of his subject to political, economical, and moral questions, in conformity with the known design of James Bernoulli; see his pages XIII—XX. His reasons contain a good appreciation of the difficulty that must attend all such applications, and he thus states the conditions under which we may attempt them with advantage: 1°. borner la question que l'on se propose à un petit nombre de suppositions, établies sur des faits certains; 2°. faire abstraction de toutes les circonstances ausquelles la liberté de l'homme, cet écueil perpetuel de nos connoissances, pourroit avoir quelque part. Montmort praises highly the memoir by Halley, which we have already noticed; and also commends Petty's *Political Arithmetic ;* see Arts. 57, 61.

Montmort refers briefly to his predecessors, Huygens, Pascal, and Fermat. He says that his work is intended principally for mathematicians, and that he has fully explained the various games which he discusses because, pour l'ordinaire les Sçavans ne sont pas Joueurs; see his page XXIII.

142. After the preface follows an *Avertissement* which was not in the first edition. Montmort says that two small treatises on the subject had appeared since his first edition; namely a thesis by Nicolas Bernoulli *De arte conjectandi in Jure,* and a memoir by De Moivre, *De mensura sortis.*

Montmort seems to have been much displeased with the terms in which reference was made to him by De Moivre. De Moivre had said,

Hugenius, primus quod sciam regulas tradidit ad istius generis Problematum Solutionem, quas nuperrimus autor Gallus variis exemplis pulchre illustravit ; sed non videntur viri clarissimi ea simplicitate ac generalitate usi fuisse quam natura rei postulabat : etenim dum plures quantitates incognitas usurpant, ut varias Collusorum conditiones repraesentent, calculum suum nimis perplexum reddunt ; dumque Collusorum dexteritatem semper aequalem ponunt, doctrinam hanc ludorum intra limites nimis arctos continent.

Montmort seems to have taken needless offence at these words ; he thought his own performances were undervalued, and accordingly he defends his own claims: this leads him to give a sketch

6

of the history of the Theory of Probability from its origin. He attributes to himself the merit of having explored a subject which had been only slightly noticed and then entirely forgotten for sixty years; see his page xxx.

143. The first part of Montmort's work is entitled *Traité des Combinaisons;* it occupies pages 1—72. Montmort says, on his page xxv, that he has here collected the theorems on Combinations which were scattered over the work in the first edition, and that he has added some theorems.

Montmort begins by explaining the properties of Pascal's *Arithmetical Triangle.* He gives the general expression for the term which occupies an assigned place in the *Arithmetical Triangle.* He shews how to find the sum of the squares, cubes, fourth powers,... of the first n natural numbers. He refers, on his page 20, to a book called the *New introduction to the Mathematics* written by M. Johnes, sçavant Geometre Anglois. The author here meant is one who is usually described as the *father of Sir William Jones.* Montmort then investigates the number of permutations of an assigned set of things taken in an assigned number together.

144. Much of this part of Montmort's work would however be now considered to belong rather to the chapter on Chances than to the chapter on Combinations in a treatise on Algebra. We have in fact numerous examples about drawing cards and throwing dice.

We will notice some of the more interesting points in this part. We may remark that in order to denote the number of combinations of n things taken r at a time, Montmort uses the symbol of a small rectangle with n above it and r below it.

145. Montmort proposes to establish the Binomial Theorem; see his page 32. He says that this theorem may be demonstrated in various ways. His own method will be seen from an example. Suppose we require $(a + b)^4$. Conceive that we have four counters each having two faces, one black and one white. Then Montmort has already shewn by the aid of the *Arithmetical Triangle* that if the four counters are thrown promiscuously there is *one* way in which all the faces presented will be black, *four* ways in which

three faces will be black and one white, *six* ways in which two faces will be black and two white; and so on. Then he reasons thus: we know by the rules for multiplication that in order to raise $a + b$ to the fourth power (1) we must take the fourth power of a and the fourth power of b, which is the same thing as taking the four black faces and the four white faces, (2) we must take the cube of a with b, and the cube of b with a in as many ways as possible, which is the same thing as taking the three black faces with one white face, and the three white faces with one black face, (3) we must take the square of a with the square of b in as many ways as possible, which is the same thing as taking the two black faces with the two white faces. Hence the coefficients in the Binomial Theorem must be the numbers 1, 4, 6, which we have already obtained in considering the cases which can arise with the four counters.

146. Thus in fact Montmort argues *à priori* that the coefficients in the expansion of $(a + b)^n$ must be equal to the numbers of cases corresponding to the different ways in which the white and black faces may appear if n counters are thrown promiscuously, each counter having one black face and one white face.

Montmort gives on his page 34 a similar interpretation to the coefficients of the multinomial theorem. Hence we see that he in some cases passed from theorems in Chances to theorems in pure Algebra, while we now pass more readily from theorems in pure Algebra to their application to the doctrine of Chances.

147. On his page 42 Montmort has the following problem: There are p dice each having the same number of faces; find the number of ways in which when they are thrown at random we can have a aces, b twos, c threes, ...

The result will be in modern notation

$$\frac{\lfloor p}{\lfloor a \lfloor b \lfloor c \cdots}$$

He then proceeds to a case a little more complex, namely where we are to have a of one sort of faces, b of another sort, c of a third sort, and so on, without specifying whether the a faces

are to be aces, or twos, or threes, ..., and similarly without specifying for the b faces, or the c faces, ...

He had given the result for this problem in his first edition,
page 137, where the factors B, C, D, E, F, ... must however be
omitted from his denominator; he suppressed the demonstration .
in his first edition because he said it would be long and abstruse,
and only intelligible to such persons as were capable of discovering
it for themselves.

148.   On his page 46 Montmort gives the following problem,
which is new in the second edition : There are n dice each having
f faces, marked with the numbers from 1 to f; they are thrown at
random: determine the number of ways in which the sum of the
numbers exhibited by the dice will be equal to a given number p.

We should now solve the problem by finding the coefficient
of x^p in the expansion of

$$(x + x^2 + x^3 + ... + x^f)^n,$$

that is the coefficient of x^{p-n} in the expansion of $\left(\dfrac{1-x^f}{1-x}\right)^n$, that is
in the expansion of $(1-x)^{-n} (1-x^f)^n$.   Let $p - n = s$; then the
required number is

$$\frac{n(n+1)\,...\,(n+s-1)}{\underline{|s}} - n\,\frac{n(n+1)\,...\,(n+s-f-1)}{\underline{|s-f}}$$

$$+ \frac{n(n-1)}{1.2}\,\frac{n(n+1)\,...\,(n+s-2f-1)}{\underline{|s-2f}} - \,......$$

The series is to be continued so long as all the factors which
occur are positive.   Montmort demonstrates the formula, but in a
much more laborious way than the above.

149.   The preceding formula is one of the standard results of
the subject, and we must now trace its history.   The formula was
first published by De Moivre without demonstration in the *De
Mensura Sortis*.   Montmort says, on his page 364, that it was derived
from page 141 of his first edition; but this assertion is quite un-
founded, for all that we have in Montmort's first edition, at the
place cited, is a table of the various throws which can be made
with any number of dice up to nine in number.   Montmort how-

ever shews by the evidence of a letter addressed to John Bernoulli, dated 15th November, 1710, that he was himself acquainted with the formula before it was published by De Moivre; see *Montmort*, page 307. De Moivre first published his demonstration in his *Miscellanea Analytica*, 1730, where he ably replied to the assertion that the formula had been derived from the first edition of Montmort's work; see *Miscellanea Analytica*, pages 191—197. De Moivre's demonstration is the same as that which we have given.

150. Montmort then proceeds to a more difficult question. Suppose we have three sets of cards, each set containing ten cards marked with the numbers 1, 2, ... 10. If three cards are taken out of the thirty, it is required to find in how many ways the sum of the numbers on the cards will amount to an assigned number.

In this problem the assigned number may arise (1) from three cards no two of which are of the same set, (2) from three cards two of which are of one set and the third of another set, (3) from three cards all of the same set. The first case is treated in the problem, Article 148; the other two cases are new.

Montmort here gives no general solution; he only shews how a table may be made registering all the required results.

He sums up thus, page 62: Cette methode est un peu longue, mais j'ai de la peine à croire qu'on puisse en trouver une plus courte.

The problem discussed here by Montmort may be stated thus: We require the number of solutions of the equation $x + y + z = p$, under the restriction that x, y, z shall be positive integers lying between 1 and 10 inclusive, and p a positive integer which has an assigned value lying between 3 and 30 inclusive.

151. In his pages 63—72 Montmort discusses a problem in the summation of series. We should now enunciate it as a general question of Finite Differences: to find the sum of any assigned number of terms of a series in which the Finite Differences of a certain order are zero.

In modern notation, let u_n denote the nth term and suppose that the $(m + 1)^{\text{th}}$ Finite Difference is zero.

Then it is shewn in works on Finite Differences, that

$$u_n = u_0 + n\Delta u_0 + \frac{n(n-1)}{1.2} \Delta^2 u_0 + \dots$$

$$+ \frac{n(n-1)\dots(n-m+1)}{\lfloor m} \Delta^m u_0 .$$

This formula Montmort gives, using A, B, C,... for Δu_0, $\Delta^2 u_0$, $\Delta^3 u_0$,...

By the aid of this formula the summation of an assigned number of terms of the proposed series is reduced to depend on the summation of series of which $\frac{n(n-1)\dots(n-r+1)}{\lfloor r}$ may be taken as the type of the general term; and such summations have been already effected by means of the *Arithmetical Triangle* and its properties.

152. Montmort naturally attaches great importance to this general investigation, which is new in the second edition. He says, page 65,

Ce Problême a, comme l'on voit, toute l'étendue et toute l'universalité possible, et semble ne rien laisser à désirer sur cette matiere, qui n'a encore été traitée par personne, que je sçache : j'en avois obmis la démonstration dans le Journal des Sçavans du mois de Mars 1711.

De Moivre in his *Doctrine of Chances* uses the rule which Montmort here demonstrates. In the first edition of the *Doctrine of Chances*, page 29, we are told that the "Demonstration may be had from the *Methodus Differentialis* of Sir *Isaac Newton*, printed in his *Analysis*." In the second edition of the Doctrine of Chances, page 52, and in the third edition, page 59, the origin of the rule is carried further back, namely, to the fifth Lemma of the *Principia*, Book III. See also *Miscellanea Analytica*, page 152.

De Moivre seems here hardly to do full justice to Montmort; for the latter is fairly entitled to the credit of the first explicit enunciation of the rule, even though it may be implicitly contained in Newton's *Principia* and *Methodus Differentialis*.

153. Montmort's second part occupies pages 73—172; it re-

lates to games of chance involving cards. The first game is that called *Pharaon*.

This game is described by De Moivre, and some investigations given by him relating to it. De Moivre restricts himself to the case of a common pack of cards with *four* suits; Montmort supposes the number of suits to be any number whatever. On the other hand De Moivre calculates the percentage of gain of the banker, which he justly considers the most important and difficult part of the problem; see *Doctrine of Chances*, pages IX, 77, 105.

Montmort's second edition gives the general results more compactly than the first.

154. We shall make some remarks in connection with Montmort's investigations on Pharaon, for the sake of the summation of certain series which present themselves.

155. Suppose that there are p cards in the pack, which the Banker has, and that his adversary's card occurs q times in the pack. Let u_p denote the Banker's advantage, A the sum of money which his adversary stakes. Montmort shews that

$$u_p = \frac{q(q-1)}{p(p-1)} \frac{1}{2} A + \frac{(p-q)(p-q-1)}{p(p-1)} u_{p-2},$$

supposing that $p-2$ is greater than q. That is Montmort should have this; but he puts $(pq-q^2)2A + (q^2-q)\frac{3}{2}A$, on his page 89, by mistake for $q(q-1)\frac{1}{2}A$; he gets right on his page 90. Montmort is not quite full enough in the details of the treatment of this equation. The following results will however be found on examination.

If q is even we can by successive use of the formula make u_p depend on u_q; and then it follows from the laws of the game that u_q is equal to A if q is equal to 2, and to $\frac{1}{2}A$ if q is greater than 2. Thus we shall have, if q is an even number greater than 2,

$$u_p = \frac{q\,(q-1)}{p\,(p-1)}\,\frac{1}{2}\,A\left\{1 + \frac{(p-q)\,(p-q-1)}{(p-2)\,(p-3)}\right.$$

$$+ \frac{(p-q)\,(p-q-1)\,(p-q-2)\,(p-q-3)}{(p-2)\,(p-3)\,(p-4)\,(p-5)}$$

$$\left. + \ldots\ldots + \frac{(p-q)\,(p-q-1)\ldots 1}{(p-2)\,(p-3)\ldots(q-1)}\right\}.$$

If $q = 2$ the last term within the brackets should be doubled.

Again if q is odd we can by successive use of the fundamental formula make u_p depend on u_{q+1}, and if q is greater than unity it can be shewn that $u_{q+1} = \dfrac{q-1}{q+1}\,\dfrac{A}{2}$. Thus we shall have, if q is an odd number greater than unity,

$$u_p = \frac{q\,(q-1)}{p\,(p-1)}\,\frac{1}{2}\,A\left\{1 + \frac{(p-q)\,(p-q-1)}{(p-2)\,(p-3)}\right.$$

$$+ \frac{(p-q)\,(p-q-1)\,(p-q-2)\,(p-q-3)}{(p-2)\,(p-3)\,(p-4)\,(p-5)}$$

$$\left. + \ldots\ldots + \frac{(p-q)\,(p-q-1)\ldots 2}{(p-2)\,(p-3)\ldots\ldots q}\right\}.$$

If $q = 1$ we have by a special investigation $u_p = \dfrac{A}{p}$.

If we suppose q even and $p - q$ not less than $q - 1$, or q odd and $p - q$ not less than q, some of the terms within the brackets may be simplified. Montmort makes these suppositions, and consequently he finds that the series within the brackets may be expressed as a fraction, of which the common denominator is

$$(p-2)\,(p-3)\ldots(p-q+1);$$

the numerator consists of a series, the first term of which is the same as the denominator, and the last term is

$$(q-2)\,(q-3)\ldots 2.1,\ \text{or}\ (q-1)\,(q-2)\ldots 3.2,$$

according as q is even or odd.

The matter contained in the present article was not given by Montmort in his first edition; it is due to John Bernoulli: see Montmort's, page 287.

156. We are thus naturally led to consider the summation of certain series.

Let $\phi(n, r) = \dfrac{n(n+1)(n+2)\ldots(n+r-1)}{\lfloor r}$.

so that $\phi(n, r)$ is the nth number of the $(r+1)^{\text{th}}$ order of figurate numbers.

Let $S\phi(n, r)$ stand for $\phi(n, r) + \phi(n-2, r) + \phi(n-4, r) + \ldots$, so that $S\phi(n, r)$ is the sum of the alternate terms of the series of figurate numbers of the $(r+1)^{\text{th}}$ order, beginning with the nth and going backwards. It is required to find an expression for $S\phi(n, r)$. It is known that

$$\phi(n, r) + \phi(n-1, r) + \phi(n-2, r) + \phi(n-3, r) + \ldots = \phi(n, r+1);$$

and by taking the terms in pairs it is easy to see that

$$\phi(n, r) - \phi(n-1, r) + \phi(n-2, r) - \phi(n-3, r) + \ldots = S\phi(n, r-1);$$

therefore, by addition,

$$S\phi(n, r) = \frac{1}{2}\phi(n, r+1) + \frac{1}{2}S\phi(n, r-1).$$

Hence, continuing the process, we shall have

$$S\phi(n, r) = \frac{1}{2}\phi(n, r+1) + \frac{1}{4}\phi(n, r) + \frac{1}{8}\phi(n, r-1) + \ldots$$

$$\ldots + \frac{1}{2^r}\phi(n, 2) + \frac{1}{2^r}S\phi(n, 0);$$

and we must consider $S\phi(n, 0) = \frac{1}{2}n$, if n be even, and $= \frac{1}{2}(n+1)$, if n be odd.

We may also obtain another expression for $S\phi(n, r)$. For change n into $n+1$ in the two fundamental relations, and subtract, instead of adding as before; thus

$$S\phi(n, r) = \frac{1}{2}\phi(n+1, r+1) - \frac{1}{2}S\phi(n+1, r-1).$$

Hence, continuing the process, we shall have

$$S\phi(n, r) = \frac{1}{2}\phi(n+1, r+1) - \frac{1}{4}\phi(n+2, r) + \frac{1}{8}\phi(n+3, r-1)$$

$$- \ldots - \frac{(-1)^r}{2^r}\phi(n+r, 2) + \frac{(-1)^r}{2^r}S\phi(n+r, 0).$$

157. Montmort's own solution of the problem respecting
Pharaon depends on the first mode of summation explained in Art.
156, which coincides with Montmort's own process. The fact that
in Montmort's result when q is odd, $q-1$ terms are to be taken,
and when q is even, q terms are to be taken and the last doubled,
depends on the different values we have to ascribe to $S\phi\,(n,\,0)$ according as n is even or odd; see Montmort's page 98.

Montmort gives another form to his result on his page 99;
this he obtained, after the publication of his first edition, from
Nicolas Bernoulli. It appears however that a wrong date is here
assigned to the communication of Nicolas Bernoulli; see Montmort's page 299. This form depends on the second mode of summation explained in Art. 156. It happens that in applying this
second mode of summation to the problem of Pharaon $n+r$ is
always odd; so that in Nicolas Bernoulli's form for the result
we have only one case, and not two cases according as q is even
or odd.

There is a memoir by Euler on the game of Pharaon in the
Hist. de l'Acad....Berlin for 1764, in which he expresses the advantage of the Banker in the same manner as Nicolas Bernoulli.

158. Montmort gives two tables of numerical results respecting Pharaon. One of these tables purports to be an exact exhibition of the Banker's advantage at any stage of the game, supposing
it played with an ordinary pack of 52 cards; the other table is an
approximate exhibition of the Banker's advantage. A remark may
be made with respect to the former table. The table consists of
four columns; the first and third are correct. The second column
should be calculated from the formula $\dfrac{n+2}{2n\,(n-1)}$, by putting for n
in succession 50, 48, 46, ... 4. But in the two copies of the second
edition of Montmort's book which I have seen the column is given
incorrectly; it begins with $\dfrac{3117}{350350}$ instead of $\dfrac{26}{2450}$, and of the remaining entries some are correct, but not in their simplest forms,
and others are incorrect. The fourth column should be calculated
from the formula $\dfrac{2n-5}{2\,(n-1)\,(n-3)}$, by putting for n in succession
50, 48, 46 ... 4; but there are errors and unreduced results in it;

it begins with a fraction having twelve figures in its denominator, which in its simplest form would only have four figures.

In the only copy of the first edition which I have seen these columns are given correctly; in both editions the description given in the text corresponds not to the incorrect forms but to the correct forms.

159. Montmort next discusses the game of *Lansquenet;* this discussion occupies pages 105—129. It does not appear to present any point of interest, and it would be useless labour to verify the complex arithmetical calculations which it involves. A few lines which occurred on pages 40 and 41 of Montmort's first edition are omitted in the second; while the Articles 84 and 95 of the second edition are new. Article 84 seems to have been suggested to Montmort by John Bernoulli; see Montmort's page 288 : it relates to a point which James Bernoulli had found difficult, as we have already stated in Art. 119.

160. Montmort next discusses the game of *Treize;* this discussion occupies pages 130—143. The problem involved is one of considerable interest, which has maintained a permanent place in works on the Theory of Probability.

The following is the problem considered by Montmort.

Suppose that we have thirteen cards numbered 1, 2, 3 ... up to 13; and that these cards are thrown promiscuously into a bag. The cards are then drawn out singly; required the chance that, once at least, the number on a card shall coincide with the number expressing the order in which it is drawn.

161. In his first edition Montmort did not give any demonstrations of his results; but in his second edition he gives two demonstrations which he had received from Nicolas Bernoulli; see his pages 301, 302. We will take the first of these demonstrations.

Let $a, b, c, d, e, \ldots$ denote the cards, n in number. Then the number of possible cases is $\lfloor n$. The number of cases in which a is first is $\lfloor n-1$. The number of cases in which b is second, but a not first, is $\lfloor n-1 - \lfloor n-2$. The number of cases in which c is third, but a not first nor b second, is $\lfloor n-1 - \lfloor n-2 - \{\lfloor n-2 - \lfloor n-3\}$, that is $\lfloor n-1 - 2\lfloor n-2 + \lfloor n-3$. The number of cases in which d is fourth, but neither a, b, nor c in its proper place is $\lfloor n-1 - 2\lfloor n-2 + \lfloor n-3 - \{\lfloor n-2 - 2\lfloor n-3 + \lfloor n-4\}$, that is $\lfloor n-1 - 3\lfloor n-2 + 3\lfloor n-3 - \lfloor n-4$. And generally the number of cases in which the mth card is in its proper place, while none of its predecessors is in its proper place, is

$$\lfloor n-1 - (m-1)\lfloor n-2 + \frac{(m-1)(m-2)}{1.2}\lfloor n-3$$

$$- \frac{(m-1)(m-2)(m-3)}{\lfloor 3}\lfloor n-4 + \ldots\ldots + (-1)^{m-1}\lfloor n-m.$$

We may supply a step here in the process of Nicolas Bernoulli, by shewing the truth of this result by induction. Let $\psi(m, n)$ denote the number of cases in which the mth card is the first that occurs in its right place; we have to trace the connexion between $\psi(m, n)$ and $\psi(m+1, n)$. The number of cases in which the $(m+1)^{\text{th}}$ card is in its right place while none of the cards between b and the mth card, both inclusive, is in its right place, is $\psi(m, n)$. From this number we must reject all those cases in which a is in its right place, and thus we shall obtain $\psi(m+1, n)$. The cases to be rejected are in number $\psi(m, n-1)$. Thus

$$\psi(m+1, n) = \psi(m, n) - \psi(m, n-1).$$

Hence we can shew that the form assigned by Nicolas Bernoulli to $\psi(m, n)$ is universally true.

Thus if a person undertakes that the mth card shall be the first that is in its right place, the number of cases favourable to him is $\psi(m, n)$, and therefore his chance is $\dfrac{\psi(m, n)}{\lfloor n}$.

If he undertakes that at least one card shall be in its right place, we obtain the number of favourable cases by summing $\psi(m, n)$ for all values of m from 1 to n both inclusive: the chance is found by dividing this sum by $\lfloor n$.

Hence we shall obtain for the chance that at least one card is in its right place,

We may observe that if we subtract the last expression from unity we obtain the chance that no card is in its right place. Hence if $\phi(n)$ denote the number of cases in which no card is in its right place, we obtain

$$\phi(n) = \lfloor n \left\{ \frac{1}{2} - \frac{1}{\lfloor 3} + \frac{1}{\lfloor 4} + \ldots - \frac{(-1)^{n-1}}{\lfloor n} \right\}.$$

162. The game which Montmort calls *Treize* has sometimes been called *Rencontre*. The problem which is here introduced for the first time has been generalised and discussed by the following writers: De Moivre, *Doctrine of Chances*, pages 109—117. Euler, *Hist. de l'Acad....Berlin*, for 1751. Lambert, *Nouveaux Mémoires de l'Acad. ... Berlin*, for 1771. Laplace, *Théorie ... des Prob.* pages 217—225. Michaelis, *Mémoire sur la probabilité du jeu de rencontre*, Berlin, 1846.

163. Pages 148—156 of Montmort relate to the game of *Bassette*. This is one of the most celebrated of the old games; it bears a great resemblance to Pharaon.

As we have already stated, this game was discussed by James Bernoulli, who summed up his results in the form of six tables; see Art. 119. The most important of these tables is in the fourth, which is in effect also reproduced in De Moivre's investigations. The reader who wishes to obtain a notion of the game may consult De Moivre's *Doctrine of Chances*, pages 69—77.

164. James Bernoulli and De Moivre confine themselves to the case of a common pack of cards, so that a particular card, an ace for example, cannot occur more than four times. Montmort however, considers the subject more generally, and gives formulæ for a pack of cards consisting of any number of suits. Montmort gives a general formula on his page 153 which is new in his second edition. The quantity which De Moivre denotes by y and puts equal to $\frac{1}{2}$ is taken to be $\frac{2}{3}$ by Montmort.

Montmort gives a numerical table of the advantage of the Banker at Bassette. In the second edition some fractions are left unreduced which were reduced to their lowest terms in the first edition, the object of the change being probably to allow

the law of formation to be more readily perceived. The last
fraction, given in the table was wrong in the first edition; see
Montmort's page 303. It would be advisable to multiply both
numerator and denominator of this fraction by 12 to maintain
uniformity in the table.

165. Montmort devotes his pages 157—172 to some problems respecting games which are not entirely games of chance.
He gives some preliminary remarks to shew that the complete
discussion of such games is too laborious and complex for our
powers of analysis; he therefore restricts himself to some special
problems relating to the games.

The games are not described, so that it would be difficult to
undertake an examination of Montmort's investigations. Two of
the problems, namely, those relating to the game of *Piquet*, are
given by De Moivre with more detail than by Montmort; see
Doctrine of Chances, page 179. These problems are simple exercises in combinations; and it would appear that all Montmort's
other problems in this part of his book are of a similar kind, pre-
senting no difficulty except that arising from a want of familiarity
with the undescribed games to which they belong.

166. Montmort's third part occupies pages 173—215; it
relates to games of chance involving dice. This part is almost
identically repeated from the first edition.

The first game is called *Quinquenove;* it is described, and a
calculation given of the disadvantage of a player. The second
game is called *Hazard;* this is also described, and a calculation
given of the disadvantage of the player who holds the dice. This
game is discussed by De Moivre; see his pages 160—166. The
third game is called *Esperance;* it is described and a particular
case of it with three players is calculated. The calculation is
extremely laborious, and the chances of the three players are
represented by three fractions, the common denominator being a
number of twenty figures. Then follow games called *Trois Dez,
Passe-dix, Rafle;* these are described somewhat obscurely, and
problems respecting them are solved; *Raffling* is discussed by De
Moivre; see pages 166—172 of the *Doctrine of Chances.*

167. The last game is called *Le Jeu des Noyaux*, which Montmort says the Baron de la Hontan had found to be in use among the savages of Canada ; see Montmort's pages XII and 213. The game is thus described,

On y joue avec huit noyaux noirs d'un côté et blancs de l'autre : on jette les noyaux en l'air : alors si les noirs se trouvent impairs, celui qui a jetté les noyaux gagne ce que l'autre Joueur a mis au jeu : S'ils se trouvent ou tous noirs ou tous blancs, il en gagne le double ; et hors de ces deux cas il perd sa mise.

Suppose eight dice each having only two faces, one face black and one white ; let them be thrown up at random. There are then 2^8, that is 256, equally possible cases. It will be found that there are 8 cases for one black and seven white, 56 cases for three black and five white, 28 cases for two black and six white, and 70 cases for four black and four white ; and there is only one case for all black. Thus if the whole stake be denoted by A, the chance of the player who throws the dice is

$$\frac{1}{256}\left\{ (8+8+56+56)\,A + 2\left(A + \frac{1}{2}A\right) \right\},$$

and the chance of the other player is

$$\frac{1}{256}\left\{ (28+28+70)\,A + 2\left(0 - \frac{1}{2}A\right) \right\}.$$

The former is equal to $\frac{131}{256}A$, and the latter to $\frac{125}{256}A$.

Montmort says that the problem was proposed to him by a lady who gave him almost instantly a correct solution of it; but he proceeds very rudely to depreciate the lady's solution by insinuating that it was only correct by accident, for her method was restricted to the case in which there were only *two* faces on each of the dice : Montmort then proposes a similar problem in which each of the dice has *four* faces.

Montmort should have recorded the name of the only lady who has contributed to the Theory of Probability.

168. The fourth part of Montmort's book occupies pages 216—282; it contains the solution of various problems respecting chances, and in particular of the five proposed by Huygens in 1657; see Art. 35. This part of the work extends to about double the length of the corresponding part in the first edition.

169. Montmort's solution of Huygens's first problem is similar to that given by James Bernoulli. The first few lines of Montmort's *Remarque* on his page 217 are not in his first edition; they strongly resemble some lines in the *Ars Conjectandi*, page 51. But Montmort does not refer to the latter work, either in his preface or elsewhere, although it appeared before his own second edition; the interval however between the two publications may have been very small, and so perhaps Montmort had not seen the *Ars Conjectandi* until after his own work had been completely printed.

The solution of Huygens's fifth problem is very laborious, and inferior to that given by James Bernoulli; and Montmort himself admits that he had not adopted the best method; see his page 223.

The solutions of Huygens's problems which Montmort gave in his first edition received the benefit of some observations by John Bernoulli; these are printed in Montmort's fifth part, pages 292—294, and by the aid of them the solutions in the second edition were improved: but Montmort's discussions of the problems remain still far less elaborate than those of James Bernoulli.

170. Montmort next takes two problems which amount to finding the value of an annuity, allowing compound interest. Then he proceeds to the problem of which a particular example is to find in how many throws with a single die it will be an even chance to throw a six.

171. Montmort now devotes his pages 232—248 to the Problem of Points. He reprints Pascal's letter of August 14th, 1654, to which we have alluded in Art. 16, and then he adds, page 241,

Le respect que nous avons pour la réputation et pour la mémoire de M. Pascal, ne nous permet pas de faire remarquer ici en détail toutes les fautes de raisonnement qui sont dans cette Lettre ; il nous suffira d'avertir que la cause de son erreur est de n'avoir point d'égard aux divers arrangemens des lettres.

Montmort's words seem to imply that Pascal's letter contains a large amount of error; we have, however, only the single fundamental inaccuracy which Fermat corrected, as we have shewn in Art. 19, and the inference that it was not allowable to suppose that a certain number of trials will necessarily be made; see Art. 18.

172. Montmort gives for the first time two formulæ either of which is a complete solution of the Problem of Points when there are two players, taking into account difference of skill. We will exhibit these formulæ in modern notation. Suppose that A wants m points and B wants n points; so that the game will be necessarily decided in $m + n - 1$ trials; let $m + n - 1 = r$. Let p denote A's skill, that is his chance of winning in a single trial, and let q denote B's skill; so that $p + q = 1$.

Then A's chance of winning the game is

$$p^r + rp^{r-1}q + \frac{r(r-1)}{1.2}p^{r-2}q^2 + \ldots\ldots + \frac{\lfloor r}{\lfloor m \lfloor n-1}p^m q^{n-1};$$

and B's chance of winning the game is

$$q^r + rq^{r-1}p + \frac{r(r-1)}{1.2}q^{r-2}p^2 + \ldots\ldots + \frac{\lfloor r}{\lfloor n \lfloor m-1}q^n p^{m-1}.$$

This is the first formula. According to the second formula A's chance of winning the game is

$$p^m \left\{ 1 + mq + \frac{m(m+1)}{1.2}q^2 + \ldots\ldots + \frac{\lfloor r-1}{\lfloor m-1 \lfloor n-1}q^{n-1} \right\};$$

and B's chance of winning the game is

$$q^n \left\{ 1 + np + \frac{n(n+1)}{1.2}p^2 + \ldots\ldots + \frac{\lfloor r-1}{\lfloor m-1 \lfloor n-1}p^{m-1} \right\}.$$

Montmort demonstrates the truth of these formulæ, but we need not give the demonstrations here as they will be found in elementary works; see *Algebra*, Chapter LIII.

173. In Montmort's first edition he had confined himself to the case of *equal* skill and had given only the *first* formula,

so that he had not really advanced beyond Pascal, although the
formula would be more convenient than the use of the *Arithmetical Triangle*; see Art. 23. The first formula for the case
of unequal skill was communicated to Montmort by John Bernoulli in a letter dated March 17th, 1710; see Montmort's page 295.
As we have already stated the formula was known to James
Bernoulli; see Art. 113. The second formula for the Problem of
Points must be assigned to Montmort himself, for it now appears
before us for the first time.

174. It will be interesting to make some comparison between
the two formulæ given in Art. 172.

It may be shewn that we have *identically*

$$p^r + rp^{r-1}q + \frac{r(r-1)}{1.2}p^{r-2}q^2 + \ldots + \frac{\lfloor r}{\lfloor m \rfloor \lfloor n-1 \rfloor}p^m q^{n-1}$$

$$= p^m \left\{ (p+q)^{r-m} + m(p+q)^{r-m-1}q + \frac{m(m+1)}{1.2}(p+q)^{r-m-2}q^2 + \right.$$

$$\left. \ldots + \frac{\lfloor r-1}{\lfloor m-1 \rfloor \lfloor n-1 \rfloor}q^{n-1} \right\}.$$

This may be shewn by picking out the coefficients of the
various powers of q in the expression on the right-hand side,
making use of the relations presented by the identity

$$(1-q)^{-(t-m)}(1-q)^{-m} = (1-q)^{-t}.$$

Thus we see that if $p+q$ be equal to unity the two expressions given in Art. 172 for A's chance are numerically equal.

175. If however $p+q$ be not equal to unity the two expressions given in Art. 172 for A's chance are *not* numerically equal.
If we suppose $p+q$ less than unity, we can give the following interpretation to the formulæ. Suppose that A's chance of winning
in a single trial is p, and B's chance is q, and that there is the
chance $1-p-q$ that it is a drawn contest.

Then the formula

$$p^m \left\{ 1 + mq + \frac{m(m+1)}{1.2}q^2 + \ldots + \frac{\lfloor r-1}{\lfloor m-1 \rfloor \lfloor n-1 \rfloor}q^{n-1} \right\}$$

expresses the chance that A shall win m points before either a single drawn contest occurs, or B wins n points.

This is easily seen by examining the reasoning by which the formula is established in the case when $p + q$ *is* equal to unity.

But the formula

$$p^r + rp^{r-1}q + \frac{r(r-1)}{1.2}p^{r-2}q^2 + \dots + \frac{\lfloor r}{\lfloor m \lfloor n-1}p^m q^{n-1}$$

expresses the chance that A shall win m points out of r, on the condition that r trials are to be made, and that A is not to be considered to have won if a drawn contest should occur even *after* he has won his m points.

This follows from the fact that if we expand $(p + q + 1 - p - q)^r$ in powers of p, q, $1 - p - q$, a term such as $Cp^\rho q^\sigma (1 - p - q)^\tau$ expresses the chance that A wins ρ points, B wins σ points, and τ contests are drawn.

Or we may treat this second case by using the transformation in Art. 174. Then we see that $(p + q)^{-m}$ expresses the chance that there shall be no drawn contest after the m points which A is supposed to have won; $(p + q)^{-m-1}$ expresses the chance that there shall be no drawn contest after the m points which A is supposed to have won, and the single point which B is supposed to have won; and so on.

176. Montmort thinks it might be easily imagined that the chances of A and B, if they respectively want km and kn points, would be the same as if they respectively wanted m and n points; but this he says is not the case; see his page 247. He seems to assert that as k increases the chance of the player of greater skill necessarily increases with it. He does not however demonstrate this.

We know by Bernoulli's theorem that if the number of trials be made large enough, there is a very high probability that the number of points won by each player respectively will be nearly in the ratio of his skill; so that *if the ratio of* m *to* n *be less than that of the skill of* A *to the skill of* B, we can, by increasing k, obtain as great a probability as we please that A will win km points before B wins kn points.

Montmort probably implies, though he does not state, the condition which we have put in Italics.

177. Montmort devotes his pages 248—257 to the discussion of a game of Bowls, which leads to a problem resembling the Problem of Points. The problem was started by De Moivre in his *De Mensura Sortis;* see *Montmort,* page 366, and the *Doctrine of Chances,* page 121. De Moivre had supposed the players to be of equal skill, and each to have the same number of balls ; Montmort generalised the problem by supposing players of unequal skill and having unequal numbers of balls. Thus the problem was not in Montmort's first edition.

Montmort gives on his page 256 a simple example of a solution of a problem which appears very plausible, but which is incorrect. Suppose A plays with one bowl and B with two bowls ; required their respective chances in one trial, assuming equal skill. Considering that any one of the three bowls is as likely as the others to be first, the chance of B is $\frac{2}{3}$ and that of A is $\frac{1}{3}$. But by the incorrect solution Montmort arrives at a different result. For suppose A to have delivered his bowl. Then B has the chance $\frac{1}{2}$ with his first bowl of beating A; and the chance $\frac{1}{2} \times \frac{1}{2}$ of failing with his first bowl and being successful with his second. Thus B's chance appears to be $\frac{3}{4}$. Montmort considers the error of this solution to lie in the assumption that when B has failed to beat A with his first bowl it is still an even chance that he will beat A with his second bowl : for the fact that B failed with his first bowl suggests that A's bowl has a position better than the average, so that B's chance of success with his second bowl becomes less than an even chance.

178. Montmort then takes four problems in succession of trifling importance. The first relates to a lottery which was started in Paris in 1710, in which the projector had offered to the public terms which were very disadvantageous to himself. The second is an easy exercise in combinations. The third relates to a game called *Le Jeu des Oublieux.* The fourth is an extension of Huygens's eleventh problem, and is also given in the *Ars Conjectandi,* page 34. These four problems are new in the second edition.

179. Montmort now passes to a problem of a more important
character which occupies his pages 268—277, and which is also
new in the second edition; it relates to the *Duration of Play*;
see Art. 107.

Suppose A to have m counters and B to have n counters; let
their chances of winning a single game be as a to b; the loser in
each game is to give a counter to his adversary: required the chance
that A will have won all B's counters on or before the xth game.
This is the most difficult problem which had as yet been solved
in the subject. Montmort's formula is given on his pages 268, 269.

180. The history of this problem up to the current date will
be found by comparing the following pages of Montmort's book,
275, 309, 315, 324, 344, 368, 375, 380.

It appears that Montmort worked at the problem and also
asked Nicolas Bernoulli to try it. Nicolas Bernoulli sent a
solution to Montmort, which Montmort said he admired but
could not understand, and he thought his own method of investigation and that of Nicolas Bernoulli must be very different: but
after explanations received from Nicolas Bernoulli, Montmort
came to the conclusion that the methods were the same. Before
however the publication of Montmort's second edition, De Moivre
had solved the problem in a different manner in the *De Mensura
Sortis.*

181. The general problem of the Duration of Play was studied
by De Moivre with great acuteness and success; indeed his investigation forms one of his chief contributions to the subject.

He refers in the following words to Nicolas Bernoulli and
Montmort:

Monsieur *de Monmort,* in the Second Edition of his Book of Chances,
having given a very handsom Solution of the Problem relating to the
duration of Play, (which Solution is coincident with that of Monsieur
Nicolas Bernoully, to be seen in that Book) and the demonstration of it
being very naturally deduced from our first Solution of the foregoing
Problem, I thought the Reader would be well pleased to see it transferred to this place.

Doctrine of Chances; first edition, page 122.

... the Solution of Mr *Nicolas Bernoulli* being very much crouded with Symbols, and the verbal Explication of them too scanty, I own I did not understand it thoroughly, which obliged me to consider Mr. *de Monmort's* Solution with very great attention: I found indeed that he was very plain, but to my great surpriz) I found him very erroneous; still in my Doctrine of Chances I printed that Solution, but rectified and ascribed it to Mr. *de Monmort*, without the least intimation of any alterations made by me; but as I had no thanks for so doing, I resume my right, and now print it as my own....

Doctrine of Chances; second edition page 181, third edition, page 211.

The language of De Moivre in his second and third editions would seem to imply that the solutions of Nicolas Bernoulli and Montmort are different; but they are really coincident, as De Moivre had himself stated in his first edition. The statement that Montmort's solution is very erroneous, is unjustly severe; Montmort has given his formula without proper precaution, but his example which immediately follows shews that he was right himself and would serve to guide his readers. The second edition of the *Doctrine of Chances* appeared nearly twenty years after the death of Montmort; and the change in De Moivre's language respecting him seems therefore especially ungenerous.

182. We shall not here give Montmort's general solution of the Problem of the Duration of Play; we shall have a better opportunity of noticing it in connexion with De Moivre's investigations. We will make three remarks which may be of service to any student who examines Montmort's own work.

Montmort's general statement on his pages 268, 269, might easily mislead; the example at the end of page 269 is a safer guide. If the statement were literally followed, the second line in the example would consist of as many terms as the first line, the fourth of as many terms as the third, and the sixth of as many terms as the fifth; but this would be wrong, shewing that the general statement is not literally accurate.

Montmort's explanation at the end of his page 270, and the beginning of his page 271, is not satisfactory. It is not true as he intimates, that the four letters *a* and the eleven letters *b* must be so arranged that only a single b is to come among the four letters a: we might have such an arrangement as *aaabbbbbbbbbbba*. We shall return to this point in our account of De Moivre's investigations.

On his page 272 Montmort gives a rule deduced from his formula; he ought to state that the rule assumes that the players are of equal skill: his rule also assumes that $p - m$ is an *even* number.

183. On his pages 275, 276 Montmort gives without demonstration results for two special cases.

(1) Suppose that there are two players of equal skill, and that each starts with two counters; then $1 - \dfrac{1}{2^x}$ is the chance that the match will be ended in $2x$ games at most. The result may be deduced from Montmort's general expression. A property of the Binomial Coefficients is involved which we may briefly indicate.

Let u_1, u_2, u_3, ... denote the successive terms in the expansion of $(1 + 1)^{2x}$. Let S denote the sum of the following series

$$u_x + 2u_{x-1} + u_{x-2} + 0 + u_{x-4} + 2u_{x-5} + u_{x-6} + 0 + u_{x-8} + \cdots$$

Then shall $S = 2^{2x-1} - 2^{x-1}$.

For let v_r denote the rth term in the expansion of $(1 + 1)^{2x-1}$, and w_r the rth term in the expansion of $(1 + 1)^{2x-2}$. Then

$$u_r = v_r + v_{r-1},$$
$$u_{r-1} = v_{r-1} + v_{r-2} = w_{r-1} + 2w_{r-2} + w_{r-3}.$$

Employ the former transformation in the odd terms of our proposed series, and the latter in the even terms; thus we find that the proposed series becomes

$$v_x + v_{x-1} + v_{x-2} + v_{x-3} + v_{x-4} + \cdots$$
$$+ 2 \{ w_{x-1} + 2w_{x-2} + w_{x-3} + 0 + w_{x-5} + \cdots \}.$$

The first of these two series is equal to $\dfrac{1}{2} (1 + 1)^{2x-1}$; and the second is a series of the same kind as that which we wish to sum with x changed into $x - 1$. Thus we can finish the demonstration *by induction*; for obviously

$$2 (2^{2x-3} - 2^{x-2}) + 2^{2x-2} = 2^{2x-1} - 2^{x-1}.$$

(2) Next suppose that each player starts with three counters; then $1 - \dfrac{3^x}{4^x}$ is the chance that the match will be ended in $2x + 1$ games at most. This result had in fact been given by Montmort in his first edition, page 184. It may be deduced from Montmort's general expression, and involves a property of the Binomial Coefficients which we will briefly indicate.

Let $u_1,\ u_2,\ u_3,\ \ldots$ denote the successive terms in the expansion of $(1 + 1)^{2x+1}$. Let S denote the sum of the following series

$$u_x + 2u_{x-1} + 2u_{x-2} + u_{x-3} + 0 + 0 + u_{x-6} + 2u_{x-7} + 2u_{x-8} + u_{x-9} + 0 + 0 + \ldots$$

Then shall $S = 2^{2x} - 3^x$.

If w_r denote the rth term in the expansion of $(1 + 1)^{2x-1}$ we can shew that

$$u_x + 2u_{x-1} + 2u_{x-2} + u_{x-3}$$
$$= w_x + w_{x-1} + w_{x-2} + w_{x-3} + w_{x-4} + w_{x-5}$$
$$+ 3\,(w_{x-1} + 2w_{x-2} + 2w_{x-3} + w_{x-4}).$$

By performing a similar transformation on every successive four significant terms of the original series we transform it into $\dfrac{1}{2}\,(1 + 1)^{2x-1} + 3\Sigma$, where Σ is a series like S with x changed into $x - 1$. Thus

$$S = 2^{2x-2} + 3\Sigma.$$

Hence by induction we find that $S = 2^{2x} - 3^x$.

184. Suppose the players of equal skill, and that each starts with the same *odd* number of counters, say m; let $f = \dfrac{m+1}{2}$. Then Montmort says, on his page 276, that we may wager with advantage that the match will be concluded in $3f^2 - 3f + 1$ trials. Montmort does not shew how he arrived at this approximation. The expression may be put in the form $\dfrac{3}{4}\,m^2 + \dfrac{1}{4}$. De Moivre spoke favourably of this approximation on page 148 of his first edition; he says, "Now Mr *de Montmort* having with great Sagacity discovered that Analogy, in the case of an equal and Odd number of Stakes, on supposition of an equality of Skill between the

Gamesters..." In his second and third editions De Moivre with-drew this commendation, and says respecting the rule "Which tho' near the Truth in small numbers, yet is very defective in large ones, for it may be proved that the number of Games found by his Expression, far from being above what is requisite is really below it." *Doctrine of Chances*, third edition, page 218.

De Moivre takes for an example $m = 45$; and calculates by his own mode of approximation that about 1531 games are requisite in order that it may be an even chance that the match will be concluded ; Montmort's rule would assign 1519 games. We should differ here with De Moivre, and consider that the results are rather remarkable for their near agreement than for their dis-crepancy.

The problem of the Duration of Play is fully discussed by Laplace, *Théorie...des Prob.* pages 225—238.

185. Montmort gives some numerical results for a simple problem on his page 277. Suppose in the problem of Art. 107 that the two players are of equal skill, each having originally n counters. Proceeding as in that Article, we have

$$u_x = \frac{1}{2}\left(u_{x+1} + u_{x-1}\right).$$

Hence we find $u_x = Cx + C_1$, where C and C_1 are arbitrary con-stants. To determine them we have

$$u_0 = 0, \ u_{2n} = 1 ;$$

hence finally, $$u_x = \frac{x}{2n}.$$

Montmort's example is for $n = 6$; he gave it in his first edition, page 178. He did not however appear to have observed the gene-ral law, at which John Bernoulli expressed his surprise ; see Mont-mort's page 295.

186. Montmort now proposes on pages 278—282 four pro-blems for solution ; they were originally given at the end of the first edition.

The first problem is *sur le Jeu du Treize*. It is not obvious why this problem is repeated, for Montmort stated the results on his pages 130—143, and demonstrations by Nicolas Bernoulli are given on pages 301, 302.

The second problem is *sur le Jeu appellé le Her;* a discussion respecting this problem runs through the correspondence between Montmort and Nicolas Bernoulli. See Montmort's pages 321, 334, 338, 348, 361, 376, 400, 402, 403, 409, 413. We will return to this problem in Art. 187.

The third problem is *sur le Jeu de la Ferme;* it is not referred to again in the book.

The fourth Problem is *sur le Jeu des Tas.* We will return to this problem in Art. 191.

Montmort's language in his *Avertissement,* page XXV, leads to the expectation that solutions of all the four problems will be found in the book, whereas only the first is solved, and indeed Montmort himself seems not to have solved the others; see his page 321.

187. It may be advisable to give some account of the discussion respecting the game called *Her.* The game is described by Montmort as played by several persons; but the discussion was confined to the case of two players, and we will adopt this limitation.

Peter holds a common pack of cards; he gives a card at random to Paul and takes one himself; the main object is for each to obtain a higher card than his adversary. The order of value is *ace, two, three, ... ten, Knave, Queen, King.*

Now if Paul is not content with his card he may compel Peter to change with him; but if Peter has a *King* he is allowed to retain it. If Peter is not content with the card which he at first obtained, or which he has been compelled to receive from Paul, he is allowed to change it for another taken out of the pack at random; but if the card he then draws is a *King* he is not allowed to have it, but must retain the card with which he was dissatisfied. If Paul and Peter finally have cards of the same value Paul is considered to lose.

188. The problem involved amounts to a determination of the relative chances of Peter and Paul; and this depends on their using or declining their rights of changing their cards. Montmort communicated the problem to two of his friends, namely Walde-grave, of whom we hear again, and a person who is called some-

times M. l'Abbé de Monsoury and sometimes M. l'Abbé d'Orbais. These two persons differed with Nicolas Bernoulli respecting a point in the problem; Nicolas Bernoulli asserted that in a certain contingency of the game each player ought to take a certain course out of two which were open to him; the other two persons contended that it was not certain that one of the courses ought to be preferred to the other.

Montmort himself scarcely interfered until the end of the correspondence, when he intimated that his opinion was contrary to that of Nicolas Bernoulli; it would seem that the latter intended to produce a fuller explanation of his views, but the correspondence closes without it.

189. We will give some details in order to shew the nature of the dispute.

It will naturally occur to the reader that one general principle must hold, namely, that if a player has obtained a high card it will be prudent for him to rest content with it and not to run the risk involved in changing that card for another. For example, it appears to be tacitly allowed by the disputants that if Paul has obtained an *eight*, or a higher card, he will remain content with it, and not compel Peter to change with him; and, on the other hand, if Paul has obtained a *six*, or a lower card, he will compel Peter to change. The dispute turns on what Paul should do if he has obtained a *seven*. The numerical data for discussing this case will be found on Montmort's page 339; we will reproduce them with some explanation of the process by which they are obtained.

I.   Paul has a *seven;* required his chance if he compels Peter to change.

Supposing Paul to change, Peter will know what Paul has and will know that he himself now has a *seven;* so he remains content if Paul has a *seven*, or a lower card, and takes another card if Paul has an *eight* or a higher card. Thus Paul's chance arises from the hypotheses that Peter originally had *Queen, Knave, ten, nine,* or *eight*. Take one of these cases, for example, that of the *ten*. The chance that Peter had a *ten* is $\frac{4}{51}$; then Paul takes it, and Peter gets the *seven*. There are 50 cards left and Peter takes one of these instead of his seven; 39 cards out of the 50 are favourable to Paul, namely 3 *sevens*, 4 *Kings*, 4 *nines*, 4 *eights*, 4 *sixes*, ... 4 *aces*.

Proceeding in this way we find for Paul's chance

$$\frac{4}{51} \cdot \frac{47 + 43 + 39 + 35 + 31}{50}, \text{ that is } \frac{780}{51.50}.$$

In this case Paul's chance can be estimated without speculating upon the conduct of Peter, because there can be no doubt as to what that conduct will be.

II. Paul has a *seven;* required his chance if he retains the *seven*.

The chance in this case depends upon the conduct of Peter. Now it appears to be tacitly allowed by the disputants that if Peter has a *nine* or a higher card he will retain it, and if he has a *seven* or a lower card he will take another instead. The dispute turns on what he will do if he has an *eight*.

(1) Suppose that Peter's rule is to retain an *eight*.

Paul's chance arises from the hypotheses that Peter has a *seven, six, five, four, three, two,* or *ace*, for which he proceeds to take another card.

We shall find now, by the same method as before, that Paul's chance is

$$\frac{3}{51} \cdot \frac{24}{50} + \frac{4}{51} \cdot \frac{27}{50} + \frac{4}{51} \cdot \frac{27}{50} + \frac{4}{51} \cdot \frac{27}{50} + \frac{4}{51} \cdot \frac{27}{50} + \frac{4}{51} \cdot \frac{27}{50} + \frac{4}{51} \cdot \frac{27}{50},$$

that is
$$\frac{720}{51.50}.$$

(2) Suppose that Peter's rule is to change an *eight*.

We have then to add $\frac{4}{51} \cdot \frac{24}{50}$ to the preceding result; and thus we obtain for Paul's chance $\frac{816}{51.50}.$

Thus we find that in Case I. Paul's chance is $\frac{780}{51.50}$, and that in Case II. it is either $\frac{720}{51.50}$ or $\frac{816}{51.50}$. If it be an even chance which rule Peter adopts we should take $\frac{1}{2}\left(\frac{720}{51.50}+\frac{816}{51.50}\right)$, that

is, $\frac{768}{51.50}$ as Paul's chance in Case **II**. Thus in Case **II**. Paul's chance is less than in Case **I**.; and therefore he should adopt the rule of changing when he has a *seven*. This is one of the arguments on which Nicolas Bernoulli relies.

On the other hand his opponents, in effect, deny the correctness of estimating it as an even chance that Peter will adopt either of the two rules which have been stated.

We have now to estimate the following chance. Peter has an *eight* and Paul has not compelled him to change; what is Peter's chance? Peter must argue thus:

I. Suppose Paul's rule is to change a *seven;* then he now has an *eight* or a higher card. That is, he must have one out of a certain 23 cards.

(1) If I retain my *eight* my chance of beating him arises only from the hypothesis that his card is one of the 3 *eights;* that is, my chance is $\frac{3}{23}$.

(2) If I change my *eight* my chance arises from the five hypotheses that Paul has *Queen, Knave, ten, nine,* or *eight;* so that my chance is

$$\frac{4}{23}\cdot\frac{3}{50}+\frac{4}{23}\cdot\frac{7}{50}+\frac{4}{23}\cdot\frac{11}{50}+\frac{4}{23}\cdot\frac{15}{50}+\frac{3}{23}\cdot\frac{22}{50},$$

that is $\frac{210}{23.50}$.

II. Suppose Paul's rule is to retain a *seven*. Then, as before,

(1) If I retain my *eight* my chance is $\frac{7}{27}$.

(2) If I change my *eight* my chance is

$$\frac{4}{27}\cdot\frac{3}{50}+\frac{4}{27}\cdot\frac{7}{50}+\frac{4}{27}\cdot\frac{11}{50}+\frac{4}{27}\cdot\frac{15}{50}+\frac{3}{27}\cdot\frac{22}{50}+\frac{4}{27}\cdot\frac{26}{50},$$

that is $\frac{314}{27.50}$.

190. These numerical results were accepted by the disputants. We may sum them up thus. The question is whether Paul should retain a certain card, and whether Peter should retain a certain card. If Paul knows his adversary's rule, he should adopt the contrary, namely retaining when his adversary changes, and changing when his adversary retains. If Peter knows his adversary's rule he should adopt the same, namely, retaining when his adversary retains and changing when his adversary changes.

Now Nicolas Bernoulli asserted that Paul should change, and therefore of course that Peter should. The objection to this is briefly put thus by Montmort, page 405,

En un mot, Monsieur, si je sçai que vous êtes le conseil de Pierre, il est évident que je dois moi Paul me tenir au sept; et de même si je suis Pierre, et qui je sçache que vous êtes le conseil de Paul, je dois changer au huit, auquel cas vous aurés donné un mauvais conseil à Paul.

The reader will be reminded of the old puzzle respecting the veracity of the Cretans, since Epimenides the Cretan said they were liars.

The opponents of Nicolas Bernoulli at first contended that it was indifferent for Paul to retain a *seven* or to change it, and also for Peter to retain an *eight* or to change it; and in this Montmort considered they were wrong. But in conversation they explained themselves to assert that no absolute rule could be laid down for the players, and in this Montmort considered that they were right; see his page 403.

The problem is considered by Trembley in the *Mémoires de l'Acad....Berlin*, for 1802.

191. The fourth problem which Montmort proposed for solution is *sur le Jeu des Tas*. The game is thus described, page 281,

Pour comprendre de quoi il s'agit, il faut sçavoir qu'après les reprises d'hombre un des Joueurs s'amuse souvent à partager le jeu en dix tas composés chacun de quatre cartes couvertes, et qu'ensuite retournant la premiere de chaque tas, il ôte et met à part deux à deux toutes celles qui se trouvent semblables, par exemple, deux Rois, deux valets, deux six, &c. alors il retourne les cartes qui suivent immédiatement celles qui viennent de lui donner des doublets, et il continue d'ôter et de mettre à part celles qui viennent par doublet jusqu'à ce qu'il en soit

venu à la dernicre de chaque tas, après les avoir enlevé toutes deux à
deux, auquel cas seulement il a gagné.

The game is not entirely a game of pure chance, because the
player may often have a choice of various methods of pairing and
removing cards. In the description of the game forty cards are
supposed to be used, but Montmort proposes the problem for solution generally without limiting the cards to forty. He requires
the chance the player has of winning and also the most advantageous method of proceeding. He says the game was rarely
played for money, but intimates that it was in use among ladies.

192. On his page 321 Montmort gives, without demonstration,
the result in a particular case of this problem, namely when the
cards consist of n pairs, the two cards in each pair being numbered
alike ; the cards are supposed placed at random in n lots, each of
two cards. He says that the chance the player has of winning is
$\frac{n-1}{2n-1}$. On page 334 Nicolas Bernoulli says that this formula is
correct, but he wishes to know how it was found, because he himself can only find it by induction, by putting for n in succession
2, 3, 4, 5, ...We may suppose this means that Nicolas Bernoulli verified by trial that the formula was correct in certain cases, but could
not give a general demonstration. Montmort seems to have
overlooked Nicolas Bernoulli's inquiry, for the problem is never
mentioned again in the course of the correspondence. As the result
is remarkable for its simplicity, and as Nicolas Bernoulli found the
problem difficult, it may be interesting to give a solution. It will
be observed that in this case the game is one of pure chance, as the
player never has any choice of courses open to him.

193. The solution of the problem depends on our observing
the state of the cards at the epoch at which the player loses, that
is at the epoch at which he can make no more pairs among the
cards exposed to view; the player may be thus arrested at the
very beginning of the game, or after he has already taken some
steps : at this epoch the player *is left with some number of lots,
which are all unbroken, and the cards exposed to view present no
pairs.* This will be obvious on reflection.

We must now determine (1) the whole number of possible cases, and (2) the whole number of cases in which the player is arrested at the very beginning.

(1) We may suppose that $2n$ cards are to be put in $2n$ places, and thus $\lfloor 2n$ will be the whole number of possible cases.

(2) Here we may find the number of cases by supposing that the n upper places are first filled and then the n lower places. We may put in the first place any card out of the $2n$, then in the second place any card of the $2n - 2$ which remain by rejecting the companion card to that we put in the first place, then in the third place any card of the $2n - 4$ which remain by rejecting the two companion cards, and so on. Thus the n upper places can be filled in $2^n \lfloor n$ ways. Then the n lower places can be filled in $\lfloor n$ ways. Hence we get $2^n \lfloor n \lfloor n$ cases in which the player is arrested at the very beginning.

We may divide each of these expressions by $\lfloor n$ if we please to disregard the different order in which the n lots may be supposed to be arranged. Thus the results become $\dfrac{\lfloor 2n}{\lfloor n}$ and $2^n \lfloor n$ respectively; we shall use these forms.

Let u_n denote the whole number of unfavourable cases, and let f_r denote the whole number of favourable cases when the cards consist of r pairs. Then

$$u_n = 2^n \lfloor n + \Sigma \frac{\lfloor n}{\lfloor r \lfloor n - r} f_r \lfloor n - r \, 2^{n-r},$$

the summation extending from $r = 2$ to $r = n - 1$, both inclusive.

For, as we have stated, the player loses by being left with some number of lots, all unbroken, in which the exposed cards contain no pairs. Suppose he is left with $n - r$ lots, so that he has got rid of r lots of the original n lots. The factor $\dfrac{\lfloor n}{\lfloor r \lfloor n - r}$ gives the number of ways in which r pairs can be selected from n pairs; the factor f_r gives the number of ways in which these pairs can be so arranged as to enable the player to get rid of them; the factor $\lfloor n - r \, 2^{n-r}$ gives the number of ways in which the remaining $n - r$ pairs can be distributed into $n - r$ lots without a single pair occurring among the exposed cards.

It is to be observed that the case in which $r = 1$ does not occur, from the nature of the game; for the player, if not arrested at the very beginning, will certainly be able to remove *two* pairs. We may however if we please consider the summation to extend from $r = 1$ to $r = n - 1$, since $f_r = 0$ when $r = 1$.

We have then

$$u_n = 2^n \lfloor n \left\{ 1 + \Sigma \frac{f_r}{2^r \lfloor r} \right\}.$$

The summation for u_{n-1} extends to one term less; thus we shall find that

$$u_n = 2n\,u_{n-1} + 2nf_{n-1}.$$

But

$$u_{n-1} + f_{n-1} = \frac{\lfloor 2n - 2}{\lfloor n - 1};$$

therefore

$$u_n = \frac{2n \lfloor 2n - 2}{\lfloor n - 1}.$$

Hence $\quad f_n = \dfrac{\lfloor 2n}{\lfloor n} - u_n = \dfrac{2 \lfloor 2n - 2}{\lfloor n - 2}; \quad \text{and } f_n \div \dfrac{\lfloor 2n}{\lfloor n} = \dfrac{n - 1}{2n - 1}.$

This is Montmort's result.

194. We now arrive at what Montmort calls the fifth part of his work, which occupies pages 283—414. It consists of the correspondence between Montmort and Nicolas Bernoulli, together with one letter from John Bernoulli to Montmort and a reply from Montmort. The whole of this part is new in the second edition.

John Bernoulli, the friend of Leibnitz and the master of Euler, was the third brother in the family of brothers of whom James Bernoulli was the eldest. John was born in 1667, and died in 1748. The second brother of the family was named Nicolas; his son of the same name, the friend and correspondent of Montmort, was born in 1687, and died in 1759.

195. Some of the letters relate to Montmort's first edition, and it is necessary to have access to this edition to study the letters with advantage; because although Montmort gives references to the corresponding passages in the second edition, yet

as these passages have been modified or corrected in accordance
with the criticisms contained in the letters, it is not always obvious what the original reading was.

196. The first letter is from John Bernoulli; it occupies
pages 283—298 ; the letter is also reprinted in the collected
edition of John Bernoulli's works, in four volumes, Lausanne and
Geneva, 1742; see Vol. I. page 453.

John Bernoulli gives a series of remarks on Montmort's first
edition, correcting some errors and suggesting some improvements.
He shews that Montmort did not present his discussion relating
to Pharaon in the simplest form; Montmort however did not
modify this part of his work. John Bernoulli gave a general
formula for the advantage of the Banker, and this Montmort did
adopt, as we have seen in Art. 155.

197. John Bernoulli points out a curious mistake made
by Montmort twice in his first edition; see his pages 288, 296.
Montmort had considered it practically impossible to find the
numerical value of a certain number of terms of a geometrical
progression ; it would seem that he had forgotten or never known
the common Algebraical formula which gives the sum. The
passages cited by John Bernoulli are from pages 35 and 181 of
the first edition ; but in the only copy which I have seen of the
first edition the text does not correspond with John Bernoulli's
quotations : it appears however that in each place the original page
has been cancelled and replaced by another in order to correct
the mistake.

After noticing the mistake, John Bernoulli proceeds thus in
his letter :

...mais pour le reste, vous faites bien d'employer les logarithmes,
je m'en suis servi utilement dans une pareille occasion il y a bien
douze ans, où il s'agissoit de déterminer combien il restoit de vin et
d'eau mêlé ensemble dans un tonneau, lequel étant au commencement
tout plein de vin, on en tireroit tous les jours pendant une année
une certaine mesure, en le remplissant incontinent après chaque extraction avec de l'eau pure. Vous trouverés la solution de cette question qui est assés curieuse dans ma dissertation *De Nutritione*, que Mr
Varignon vous pourra communiquer. Je fis cette question pour faire

comprendre comment on peut déterminer la quantité de vieille matière qui reste dans nos corps mêlée avec de la nouvelle qui nous vient tous les jours par la nourriture, pour réparer la perte que nos corps font insensiblement par la transpiration continuelle.

The dissertation *De Nutritione* will be found in the collected edition of John Bernoulli's works; see Vol. I. page 275.

198. John Bernoulli passes on to a remark on Montmort's discussion of the game of Treize. The remark enunciates the following theorem.

Let $$\phi(n) = 1 - \frac{1}{\lfloor 2} + \frac{1}{\lfloor 3} - \frac{1}{\lfloor 4} + \ldots + \frac{(-1)^{n+1}}{\lfloor n},$$

and let

$$\psi(n) = \phi(n) + \frac{1}{1}\phi(n-1) + \frac{1}{2}\phi(n-2) + \ldots + \frac{1}{\lfloor n-1}\phi(1);$$

then shall $$\psi(n) = \frac{1}{1} + \frac{1}{\lfloor 2} + \frac{1}{\lfloor 3} + \frac{1}{\lfloor 4} + \ldots + \frac{1}{\lfloor n}.$$

We may prove this by induction. For we may write $\psi(n)$ in the following form,

$$1\left\{1 + \frac{1}{1} + \frac{1}{\lfloor 2} + \frac{1}{\lfloor 3} + \ldots\ldots\ldots\ldots + \frac{1}{\lfloor n-1}\right\}$$

$$-\frac{1}{2}\left\{1 + \frac{1}{1} + \frac{1}{\lfloor 2} + \frac{1}{\lfloor 3} + \ldots\ldots\ldots + \frac{1}{\lfloor n-2}\right\}$$

$$+\frac{1}{\lfloor 3}\left\{1 + \frac{1}{1} + \frac{1}{\lfloor 2} + \frac{1}{\lfloor 3} + \ldots\ldots + \frac{1}{\lfloor n-3}\right\}$$

$$-\ldots\ldots$$

Hence we can shew that

$$\psi(n+1) = \psi(n) + \frac{1}{\lfloor n+1}.$$

199. John Bernoulli next adverts to the solutions which Montmort had given of the five problems proposed by Huygens; see Art. 35.

According to John Bernoulli's opinion, Montmort had not understood the second and third problems in the sense which Huygens had intended; in the fifth problem Montmort had changed the enunciation into another quite different, and yet had really solved the problem according to Huygens's enunciation. By the corrections which he made in his second edition, Montmort shewed that he admitted the justice of the objections urged against his solutions of the second and fifth problems; in the case of the third problem he retained his original opinion; see his pages 292, 305.

John Bernoulli next notices the solution of the Problem of Points, and gives a general formula, to which we have referred in Art. 173. Then he adverts to a problem which Montmort had not fully considered; see Art. 185.

200. John Bernoulli gives high praise to Montmort's work, but urges him to extend and enrich it. He refers to the four problems which Montmort had proposed for investigation; the first he considers too long to be finished in human life, and the fourth he cannot understand: the other two he thinks might be solved by great labour. This opinion seems singularly incorrect. The first problem is the easiest of all, and has been solved without difficulty; see Article 161: perhaps however John Bernoulli took it in some more general sense; see Montmort's page 308. The fourth problem is quite intelligible, and a particular case of it is simple; see Art. 193. The third and fourth problems seem to be far more intractable.

201. A letter to Montmort from Nicolas Bernoulli occupies pages 299—303. This letter contains corrections of two mistakes which occurred in Montmort's first edition. It gives without demonstration a formula for the advantage of the Banker at Pharaon, and also a formula for the advantage of the Banker at Bassette; Montmort quoted the former in the text of his second edition; see Art. 157. Nicolas Bernoulli gives a good investigation of the formulæ which occur in analysing the game of Treize; see Art. 161. He also discusses briefly a game of chance which we will now explain.

202. Suppose that a set of players A, B, C, D, ... undertake to play a set of l games with cards. A is at first the dealer, there are m chances out of $m + n$ that he retains the deal at the next game, and n chances out of $m + n$ that he loses it; if he loses the

deal the player on his right hand takes it; and so on in order.
B is on the left of A, C is on the left of B, and so on. Let the
advantages of the players when A deals be a, b, c, d, ... respectively; these advantages are supposed to depend entirely on
the situation of the players, the game being a game of pure
chance.

Let the chances of A, B, C, D, ... be denoted by z, y, x, u, ...;
and let $s = m + n$.

Then Nicolas Bernoulli gives the following values :

$$z = a + \frac{ma+nb}{s} + \frac{m^2a+2mnb+n^2c}{s^2} + \frac{m^3a+3m^2nb+3mn^2c+n^3d}{s^3} + \ldots,$$

$$y = b + \frac{nb+nc}{s} + \frac{m^2b+2mnc+n^2d}{s^2} + \frac{m^3b+3m^2nc+3mn^2d+n^2e}{s^3} + \ldots,$$

$$x = c + \frac{nc+nd}{s} + \frac{m^2c+2mnd+n^2e}{s^2} + \frac{m^3c+3m^2nd+3mn^2e+n^3f}{s^3} + \ldots,$$

$$u = d + \frac{md+ne}{s} + \frac{m^2d+2mne+n^2f}{s^2} + \frac{m^3d+3m^2ne+3mn^2f+n^3g}{s^3} + \ldots,$$

and so on.

Each of these series is to continue for l terms. If there are
not so many as l players, the letters in the set a, b, c, d, e, f, g, ...
will recur. For example, if there are only four players, then
$e = a$, $f = b$, $g = c$, ...

It is easy to see the meaning of the separate terms. Take, for
example, the value of z. A deals; the advantage directly arising
from this is a. Then there are m chances out of s that A will have
the second deal, and n chances out of s that the deal will pass on
to the next player, and thus put A in the position originally held
by B. Hence we have the term $\dfrac{ma+nb}{s}$. Again, for the third
deal; there are $(m+n)^2$, that is, s^2 possible cases; out of these
there are m^2 cases in which A will have the third deal, $2mn$ cases
in which the player on the right of A will have it, and n^2 cases in
which the player next on the right will have it. Hence we
have the term $\dfrac{m^2a + 2mnb + n^2c}{s^2}$. And so on.

Nicolas Bernoulli then gives another form for these expressions; we will exhibit that for z from which the others can be deduced.

Let $\qquad q = \dfrac{s}{n}, \quad r = \left(\dfrac{m}{s}\right)^l, \quad t = \dfrac{n}{m}.$ Then

$$z = aq\,(1-r) + bq\left\{1 - r\,[1+tl]\right\} + cq\left\{1 - r\left[1+tl+\frac{t^2l\,(l-1)}{1.2}\right]\right\}$$

$$+ dq\left\{1-r\left[1+tl+\frac{t^2l\,(l-1)}{1.2}+\frac{t^3l\,(l-1)\,(l-2)}{1.2.3}\right]\right\}$$

$$+ \dots;$$

this series is to be continued for l terms.

The way in which this transformation is effected is the following: suppose for example we pick out the coefficient of c in the value of z, we shall find it to be

$$\frac{n^2}{1.2\,s^2}\left\{1.2 + 3.2\,\frac{m}{s} + 4.3\,\frac{m^2}{s^2} + 5.4\,\frac{m^3}{s^3} + \dots\right\},$$

where the series in brackets is to consist of $l-2$ terms.

We have then to shew that this expression is equal to

$$q\left\{1 - r\left[1+tl+\frac{t^2l\,(l-1)}{1.2}\right]\right\}.$$

We will take the general theorem of which this is a particular case. Let

$$S = \frac{n^\lambda}{s^\lambda\,\lfloor\lambda}\left\{P_1 + P_2\,\frac{m}{s} + P_3\,\frac{m^2}{s^2} + \dots \text{ to } l-\lambda \text{ terms}\right\},$$

where $\qquad\qquad P_\rho = \dfrac{\lfloor\rho+\lambda-1}{\lfloor\rho-1}.$

Let $\qquad\qquad u = 1 + \dfrac{m}{s} + \dfrac{m^2}{s^2} + \dots\dots + \dfrac{m^{l-1}}{s^{l-1}};$

then $\qquad S = \dfrac{n^\lambda}{\lfloor\lambda}\,\dfrac{d^\lambda u}{dm^\lambda}.$

Now $\qquad\qquad u = \dfrac{1 - \left(\dfrac{m}{s}\right)^l}{1 - \dfrac{m}{s}} = \dfrac{1-\mu^l}{1-\mu}$ say;

thus
$$\frac{d^\lambda u}{dm^\lambda} = \frac{\lfloor \lambda}{s^\lambda} \frac{1-\mu^l}{(1-\mu)^{\lambda+1}} - \frac{\lambda}{1} \frac{\lfloor \lambda-1}{s^\lambda} \frac{l\mu^{l-1}}{(1-\mu)^\lambda}$$

$$- \frac{\lambda(\lambda-1)}{1.2} \frac{\lfloor \lambda-2}{s^\lambda} \frac{l(l-1)\mu^{l-2}}{(1-\mu)^{\lambda-1}}$$

$$- \frac{\lambda(\lambda-1)(\lambda-2)}{1.2.3} \frac{\lfloor \lambda-3}{s^\lambda} \frac{l(l-1)(l-2)\mu^{l-3}}{(1-\mu)^{\lambda-2}}$$

$$- \ldots\ldots$$

$$= \frac{s\lfloor \lambda}{n^{\lambda+1}} \left\{ 1 - r \left[1 + tl + \frac{t^2 l(l-1)}{1.2} + \frac{t^3 l(l-1)(l-2)}{1.2.3} + \ldots \right] \right\},$$

where the series between square brackets is to extend to $\lambda+1$ terms.

We may observe that by the nature of the problem we have

$$a+b+c+\ldots = 0, \text{ and also } z+y+x+\ldots = 0.$$

The problem simplifies very much if we may regard l as infinite or very great. For then let z denote the advantage of A; if A obtains the next deal we may consider that his advantage is still z; if A loses the next deal his advantage is the same as that of B originally. Thus

$$z = a + \frac{mz + ny}{s}.$$

Multiply by s and transpose ; therefore

$$z = y + aq.$$

Similarly we have

$$y = x + bq, \quad x = u + cq, \quad \ldots\ldots$$

Hence we shall obtain

$$z = \frac{q}{p} \left\{ a(p-1) + b(p-2) + c(p-3) + \ldots \right\},$$

where p denotes the number of players ; and the values of y, x, ... may be obtained by symmetrical changes in the letters.

We may also express the result thus,

$$z = -\frac{q}{p} \left\{ a + 2b + 3c + \ldots \right\}.$$

203. The next letter is from Montmort to John Bernoulli; it occupies pages 303—307. Montmort makes brief observations on the points to which John Bernouilli had drawn his attention; he suggests a problem on the Duration of Play for the consideration of Nicolas Bernoulli.

204. The next letter is from Nicolas Bernoulli to Montmort; it occupies pages 308—314.

Nicolas Bernoulli first speaks of the game of *Treize*, and gives a general formula for it; but by accident he gave the formula incorrectly, and afterwards corrected it when Montmort drew his attention to it; see Montmort's pages 315, 323.

We will here investigate the formula after the manner given by Nicolas Bernoulli for the simple case already considered in Art. 161.

Suppose there are n cards divided into p sets. Denote the cards of a set by $a, b, c, \ldots$ in order.

The whole number of cases is $\lfloor n$.

The number of ways in which a can stand first is $p \lfloor n-1$.

The number of ways in which b can stand second without a standing first is $p \lfloor n-1 - p^2 \lfloor n-2$.

The number of ways in which c can stand third without a standing first or b second is $p \lfloor n-1 - 2p^2 \lfloor n-2 + p^3 \lfloor n-3$.
And so on.

Hence the chance of winning by the first card is $\dfrac{p}{n}$; the chance of winning by the second card is $\dfrac{p}{n} - \dfrac{p^2}{n(n-1)}$; the chance of winning by the third card is $\dfrac{p}{n} - \dfrac{2p^2}{n(n-1)} + \dfrac{p^3}{n(n-1)(n-2)}$; and so on.

Hence the chance of winning by one or other of the first m cards is

$$\frac{mp}{n} - \frac{m(m-1)}{1.2} \frac{p^2}{n(n-1)} + \frac{m(m-1)(m-2)}{1.2.3} \frac{p^3}{n(n-1)(n-2)} - \ldots$$

And the entire chance of winning is found by putting $m = \dfrac{n}{p}$, so that it is

$$\frac{1}{1} - \frac{n-p}{1.2\,(n-1)} + \frac{(n-p)\,(n-2p)}{1.2.3\,(n-1)\,(n-2)}$$
$$- \frac{(n-p)\,(n-2p)\,(n-3p)}{1.2.3.4\,(n-1)\,(n-2)\,(n-3)} + \dots$$

205. Nicolas Bernoulli then passes on to another game in which he objects to Montmort's conclusion. Montmort had found a certain advantage for the first player, on the assumption that the game was to conclude at a certain stage; Nicolas Bernoulli thought that at this stage the game ought not to terminate, but that the players should change their positions. He says that the advantage for the first player should be only half what Montmort stated. The point is of little interest, as it does not belong to the theory of chances but to the conventions of the players; Montmort, however, did not admit the justice of the remarks of Nicolas Bernoulli; see Montmort's pages 309, 317, 327.

206. Nicolas Bernoulli then considers the problem on the Duration of Play which had been suggested for him by Montmort. Nicolas Bernoulli here gives the formulæ to which we have already alluded in Art. 180; but the meaning of the formulæ was very obscure, as Montmort stated in his reply. Nicolas Bernoulli gives the result which expresses the chances of each player when the number of games is unlimited; he says this may be deduced from the general formulæ, and that he had also obtained it previously by another method. See Art. 107.

207. Nicolas Bernoulli then makes some remarks on the summation of series. He exemplifies the method which is now common in elementary works on Algebra. Suppose we require the sum of the squares of the first n triangular numbers, that is, the sum of n terms of the series of which the rth term is $\left\{\frac{r\,(r+1)}{1.2}\right\}^2$. Assume that the sum is equal to

$$an^5 + bn^4 + cn^3 + dn^2 + en + f;$$

and then determine a, b, c, d, e, f by changing n into $n+1$ in the assumed identity, subtracting, and equating coefficients. This method is ascribed by Nicolas Bernoulli to his uncle John.

Nicolas Bernoulli also indicates another method; he resolves $\left\{\dfrac{r\,(r+1)}{1\,.\,2}\right\}^2$ into

$$6\,\frac{r\,(r+1)\,(r+2)\,(r+3)}{1\,.\,2\,.\,3\,.\,4} - 6\,\frac{r\,(r+1)\,(r+2)}{1\,.\,2\,.\,3} + \frac{r\,(r+1)}{1\,.\,2}\,;$$

and thus finds that the required sum is

$$6\,\frac{n\,(n+1)\,(n+2)\,(n+3)\,(n+4)}{1\,.\,2\,.\,3\,.\,4\,.\,5} - 6\,\frac{n\,(n+1)\,(n+2)\,(n+3)}{1\,.\,2\,.\,3\,.\,4}$$
$$+ \frac{n\,(n+1)\,(n+2)}{1\,.\,2\,.\,3}\,.$$

208. It seems probable that a letter from Montmort to Nicolas Bernoulli, which has not been preserved, preceded this letter from Nicolas Bernoulli. For Nicolas Bernoulli refers to the problem about a lottery, as if Montmort had drawn his attention to it; see Art. 180: and he intimates that Montmort had offered to undertake the printing of James Bernoulli's unpublished *Ars Conjectandi*. Neither of these points had been mentioned in Montmort's preceding letters as we have them in the book.

209. The next letter is from Montmort to Nicolas Bernoulli; it occupies pages 315—323. The most interesting matter in this letter is the introduction for the first time of a problem which has since been much discussed. The problem was proposed to Montmort, and also solved, by an English gentleman named Waldegrave; see Montmort's pages 318 and 328. In the problem as originally proposed only three players are considered, but we will enunciate it more generally. Suppose there are $n+1$ players; two of them play a game; the loser deposits a shilling, and the winner then plays with the third player; the loser deposits a shilling, and the winner then plays with the fourth player; and so on. The player who lost the first game does not enter again until after the $(n+1)^{th}$ player has had his turn. The process continues until one player has beaten in continued succession all the other players, and then he receives all the money which has been deposited. It is required to determine the expectation of each of the players, and also the chance that the money will be won when, or before, a certain number of games has been played. The game is sup-

posed a game of pure chance, or which is the same thing, the players are all supposed of equal skill.

Montmort himself in the case of three players states all the required results, but does not give demonstrations. In the case of four players he states the numerical probability that the money will be won in any assigned number of games between 3 and 13 inclusive, but he says that the law of the numbers which he assigns is not easy to perceive. He attempted to proceed further with the problem, and to determine the advantage of each player when there are four players, and also to determine the probability of the money being won in an assigned number of games when there are five or six players. He says however, page 320, mais cela m'a paru trop difficile, ou plûtôt j'ai manqué de courage, car je serois sûr d'en venir à bout.

210. There are references to this problem several times in the correspondence of Montmort and Nicolas Bernoulli; see Montmort's pages 328, 345, 350, 366, 375, 380, 400. Nicolas Bernoulli succeeded in solving the problem generally for any number of players; his solution is given in Montmort's pages 381—387, and is perhaps the most striking investigation in the work. The following remarks may be of service to a student of this solution.

(1) On page 386 Nicolas Bernoulli ought to have stated how many terms should be taken of the two series which he gives, namely, a number expressed by the greatest integer contained in $\frac{n+p-1}{n}$. On page 330 where he does advert to this point he puts by mistake $\frac{n+p}{n}$ instead of $\frac{n+p-1}{n}$.

(2) The expressions given for a, b, c, ... on page 386 are correct, except that given for a; the value of a is $\frac{2}{2^n}$, and not $\frac{1}{2^n}$, as the language of Nicolas Bernoulli seems to imply.

(3) The chief results obtained by Nicolas Bernoulli are stated at the top of page 329; these results agree with those afterwards given by Laplace.

211. Although the earliest *notice* of the problem occurs in the letter of Montmort's which we are now examining, yet the earliest *publication* of it is due to De Moivre; it is Problem XV. of the *De Mensura Sortis.* We shall however speak of it as *Waldegrave's Problem,* from the person whose name we have found first associated with it.

Tho problem is discussed by Laplace, *Théorie...des Prob.* page 238, and we shall therefore have to recur to it.

212. Montmort refers on page 320 to a book entitled *Traité du Jeu,* which he says he had lately received from Paris. He says it is *un Livre de morale.* He praises the author, but considers him to be wrong sometimes in his calculation of chances, and gives an example. Nicolas Bernoulli in reply says that the author of the book is Mr Barbeyrac. Nicolas Bernoulli agrees with Montmort in his general opinion respecting the book, but in the example in question he thinks Barbeyrac right and Montmort wrong. The difference in result arises from a difference in the way of understanding the rules of the game. Montmort briefly replied; see pages 332, 346.

Montmort complains of a dearth of mathematical memoirs; he says, page 322,

Je suis étonné de voir les Journeaux de Leipsic si dégarnis de morceaux de Mathematiques: ils doivent en partie leur réputation aux excellens Memoires que Messieurs vos Oncles y envoyoient souvent : les Geometres n'y trouvent plus depuis cinq ou six ans les mêmes richesses qu'autrefois, faites-en des reproches à M. votre Oncle, et permettés-moi de vous en faire aussi, *Luceat lux vestra coram hominibus.*

213. The next letter is from Nicolas Bernoulli to Montmort; it occupies pages 323—337. It chiefly relates to matters which we have already sufficiently noticed, namely, the games of Treize, Her, and Tas, and Waldegrave's Problem. Nicolas Bernoulli adverts to the letter by his uncle James on the game of Tennis, which was afterwards published at the end of the *Ars Conjectandi,* and he proposes for solution four of the problems which are considered in the letter in order to see if Montmort's results will agree with those of James Bernoulli.

Nicolas Bernoulli gives at the end of his letter an example of summation of series. He proposes to sum p terms of the series 1, 3, 6, 10, 15, 21, ... He considers the series

$$1 + 3x + 6x^2 + 10x^3 + 15x^4 + 21x^5 + \ldots$$

which he decomposes into a set of series, thus:

$$1 + 2x + 3x^2 + 4x^3 + 5x^4 + \ldots$$
$$+ \ x + 2x^2 + 3x^3 + 4x^4 + \ldots$$
$$+ \ x^2 + 2x^3 + 3x^4 + \ldots$$
$$+ \ x^3 + 2x^4 + \ldots$$
$$+ \ x^4 + \ldots$$
$$+ \ldots$$

The series in each horizontal row is easily summed to p terms; the expression obtained takes the form $\dfrac{0}{0}$ when $x = 1$, and Nicolas Bernoulli evaluates the indeterminate form, as he says, ...en. me servant de la regle de mon Oncle, que feu Monsieur le Marquis de l'Hôpital a inscré dans son Analyse des infiniment petits, ...

The investigation is very inaccurately printed.

214. The next letter is from Montmort to Nicolas Bernoulli; it occupies pages 337—347. Besides remarks on the game of Her and on Waldegrave's Problem, it contains some attempts at the problems which Nicolas Bernoulli had proposed out of his uncle's letter on the game of Tennis. But Montmort found the problems difficult to understand, and asked several questions as to their meaning.

215. Montmort gives on his page 342 the following equation as the result of one of the problems,

$$4m^3 - 8m^2 + 14m + 6 = 3^{m+1},$$

and he says that this is satisfied approximately by $m = 5\frac{57}{320}$; but there is some mistake, for the equation has no root between 5 and 6. The correct equation should apparently be

$$8m^3 - 12m^2 + 16m + 6 = 3^{m+1},$$

which has a root between 5·1 and 5·2.

216. One of the problems is the following. The skill of A, that is his chance of success in a single trial, is p, the skill of B is q. A and B are to play for victory in two games out of three, each game being for two points. In the first game B is to have a point given to him, in the second the players are to be on an equality, and in the third also B is to have a point given to him. Required the skill of each player so that on the whole the chances may be equal. A's chance of success in the first game or in the third game is p^2, and B's chance is $q^2 + 2qp$. A's chance of success in the second game is $p^3 + 3p^2q$, and B's chance is $q^3 + 3q^2p$. Hence A's chance of success in two games out of three is

$$p^2 (p^3 + 3p^2q) + p^2 (q^2 + 2qp) (p^3 + 3p^2q) + p^4 (q^3 + 3q^2p);$$

and this by supposition must equal $\frac{1}{2}$.

This agrees with Montmort's result by putting $\frac{a}{a+b}$ for p and $\frac{b}{a+b}$ for q, allowing for a mistake which was afterwards corrected; see Montmort's pages 343, 350, 352.

217. The letter closes with the following interesting piece of literary history.

Je ne sçai si vous sçavés qu'on réimprime la Recherche de la verité. Le R. P. Malbranche m'a dit que cet Ouvrage paroîtroit au commencement d'Avril. Il y aura un grand nombre d'additions sur des sujets très importans. Vous y verrés entr'autres nouveautés une Dissertation sur la cause de la pesanteur, qui apparemment fixera les doutes de tant de Sçavans hommes qui ne sçavent à quoi s'en tenir sur cette matiere. Il prouve d'une maniere invincible la necessité de ses petits tourbillons pour rendre raison de la cause de la pesanteur, de la dureté et fluidité des corps et. des principaux phenomenes touchant la lumiere et les couleurs; sa theorie s'accorde le mieux du monde avec les belles experiences que M. Newton a rapporté dans son beau Traité *De Natura Lucis et Colorum.* Je peux me glorifier auprès du Public que mes prieres ardentes et réiterées depuis plusieurs années, ont contribué à déterminer cet incomparable Philosophe à écrire sur cette

matière qui renferme toute la Physique generale. Vous verrés avec
admiration que ce grand homme a porté dans ces matieres obscures
cette netteté d'idées, cette sublimité de genie et d'invention qui bril-
lent avec tant d'éclat dans ses Traités de Metaphysique.

Posterity has not adopted the high opinion which Montmort
here expresses respecting the physical speculations of his friend
and master; Malebranche is now remembered and honoured for
his metaphysical works alone, which have gained the following
testimony from one of the greatest critics :

As a thinker, he is perhaps the most profound that France has
ever produced, and as a writer on philosophical subjects, there is not
another European author who can be placed before him.

Sir William Hamilton's *Lectures on Metaphysics*, Vol. I. page 262 ;
see also his edition of *Reid's Works*, page 266.

218. The next letter is from Montmort to Nicolas Bernoulli ;
it occupies pages 352—360. We may notice that Montmort here
claims to be the first person who called attention to the theorem
which is now given in elementary treatises on Algebra under the
following enunciation : To find the number of terms in the expansion of any multinomial, the exponent being a positive integer.
See Montmort's page 355.

219. Montmort gives in this letter some examples of the rectification of curves ; see his pages 356, 357, 359, 360. In particular
he notices one which he had himself discussed in the early days
of the Integral Calculus, when, as he says, the subject was well
known only by five or six mathematicians. This example is the
rectification of the curve called after the name of its inventor De
Beaune ; see John Bernoulli's works, Vol. I. pages 62, 63. What
Montmort gives in this letter is not intelligible by itself, but it can
be understood by the aid of the original memoir, which is in the
Journal des Sçavans, Vol. XXXI.

These remarks by Montmort on the rectification of curves are
of no great interest except to a student of the history of the Integral Calculus, and they are not free from errors or misprints.

220. Montmort quotes the following sentence from a letter written by Pascal to Fermat.

Pour vous parler franchement de la Geometrie, je la trouve le plus haut exercice de l'esprit; mais en même temps je la connois pour si inutile, que je fais peu de différence entre un homme qui n'est que Geometre et un habile Artisan; aussi je l'appelle le plus beau métier du monde; mais enfin ce n'est qu'un métier: et j'ai souvent dit qu'elle est bonne pour faire l'essai, mais non pas l'emploi de notre force.

Montmort naturally objects to this decision as severe and humiliating, and probably not that which Pascal himself would have pronounced in his earlier days.

221. The next letter is also from Montmort to Nicolas Bernoulli; it occupies pages 361—370. Montmort says he has just received De Moivre's book, by which he means the memoir *De Mensura Sortis*, published by De Moivre in the *Philosophical Transactions*; and he proceeds to analyse this memoir. Montmort certainly does not do justice to De Moivre. Montmort in fact considers that the first edition of his own work contained implicitly all that had been given in the *De Mensura Sortis;* and he seems almost to fancy that the circumstance that a problem had been discussed in the correspondence between himself and the Bernoullis was sufficient ground to deprive De Moivre of the credit of originality. The opinion of Nicolas Bernoulli was far more favourable to De Moivre; see Montmort's pages 362, 375, 378, 386.

De Moivre in his *Miscellanea Analytica* replied to Montmort, as we shall see hereafter.

222. On his page 365 Montmort gives some remarks on the second of the five problems which Huygens proposed for solution; see Art. 35.

Suppose there are three players; let a be the number of white balls, and b of black balls; let $c = a + b$. The balls are supposed not to be replaced after being drawn; then the chance of the first player is

$$\frac{a}{c} + \frac{b\,(b-1)\,(b-2)\,a}{c\,(c-1)\,(c-2)\,(c-3)} + \frac{b\,(b-1)\,\dots\,(b-5)\,a}{c\,(c-1)\,\dots\,(c-6)} + \dots$$

Montmort takes credit to himself for summing this series, so as to find its value when a and b are large numbers; but, without saying so, he assumes that $a = 4$. Thus the series becomes

$$\frac{4\,\lfloor b}{\lfloor c}\left\{\frac{\lfloor c-1}{\lfloor b}+\frac{\lfloor c-4}{\lfloor b-3}+\frac{\lfloor c-7}{\lfloor b-6}+\ldots\right\}.$$

Let $p = b + 3$, then $c = p + 1$; thus the series within brackets becomes

$$p\,(p-1)\,(p-2)+(p-3)\,(p-4)\,(p-5)$$
$$+\,(p-6)\,(p-7)\,(p-8)+\ldots$$

Suppose we require the sum of n terms of the series. The rth term is

$$(p-3r+3)\,(p-3r+2)\,(p-3r+1)\,;$$

assume that it is equal to

$$A+B\,(r-1)+\frac{C\,(r-1)\,(r-2)}{1\,.\,2}+\frac{D\,(r-1)\,(r-2)\,(r-3)}{1\,.\,2\,.\,3},$$

where A, B, C, D are to be independent of r.

We shall find that

$$A = p\,(p-1)\,(p-2),$$
$$B = -\,(9p^2 - 45p + 60),$$
$$C = 54p - 216,$$
$$D = -\,162.$$

Hence the required sum of n terms is

$$np\,(p-1)\,(p-2)-\frac{n\,(n-1)}{1\,.\,2}\,(9p^2-45p+60)$$

$$+\frac{n\,(n-1)\,(n-2)}{1\,.\,2\,.\,3}\,(54p-216)-\frac{n\,(n-1)\,(n-2)\,(n-3)}{1\,.\,2\,.\,3\,.\,4}\,162.$$

This result is sufficiently near Montmort's to shew that he must have adopted nearly the same method; he has fallen into some mistake, for he gives a different expression for the terms independent of p.

In the problem on chances to which this is subservient we should have to put for n the greatest integer in $\frac{p}{3}$.

Montmort refers on his page 364 to a letter dated June 8th, 1710, which does not appear to have been preserved.

223. The next letter is from Nicolas Bernoulli to Montmort; it occupies pages 371—375. Nicolas Bernoulli demonstrates a property of De Beaune's curve ; he also gives a geometrical rectification of the logarithmic curve ; but his results are very incorrect. He then remarks on a subject which he says had been brought to his notice in Holland, and on which a memoir had been inserted in the *Philosophical Transactions*. The subject is the argument for Divine Providence taken from the constant regularity observed in the births of both sexes. The memoir to which Bernoulli refers is by Dr John Arbuthnot ; it is in Vol. XXVII. of the *Philosophical Transactions*, and was published in 1710. Nicolas Bernoulli had discussed the subject in Holland with 'sGravesande.

Nicolas Bernoulli says that he was obliged to refute the argument. What he supposes to be a refutation amounts to this ; he examined the registers of births in London for the years from 1629 to 1710 inclusive ; he found that on the average 18 males were born for 17 females. The greatest variations from this ratio were in 1661, when 4748 males and 4100 females were born, and in 1703, when 7765 males and 7683 females were born. He says then that we may bet 300 to 1 that out of 14,000 infants the ratio of the males to the females will fall within these limits ; we shall see in Art. 225 the method by which he obtained this result.

224. The next letter is also from Nicolas Bernoulli to Montmort ; it occupies pages 375—387. It contains some remarks on the game of *Her*, and some remarks in reply to those made by Montmort on De Moivre's memoir *De Mensura Sortis*. The most important part of the letter is an elaborate discussion of Waldegrave's problem ; we have already said enough on this problem, and so need only add that Nicolas Bernoulli speaks of this discussion as that which he preferred to every thing else which he had produced on the subject ; see page 381. The approbation which he thus bestows on his own work seems well deserved.

225. The next letter is also from Nicolas Bernoulli to Montmort ; it occupies pages 388—393. It is entirely occupied with

the question of the ratio of male infants to female infants. We have already stated that Nicolas Bernoulli had refused to see any argument for Divine Providence in the fact of the nearly constant ratio. *He assumes that the probability of the birth of a male is to the probability of the birth of a female as* 18 *to* 17 ; he then shews that the chances are 43 to 1 that out of 14,000 infants the males will lie between 7037 and 7363. His investigation involves a general demonstration of the theorem of his uncle James called Bernoulli's Theorem. The investigation requires the summation of terms of a binomial series ; this is effected approximately by a process which is commenced in these words : Or comme ces termes sont furieusement grands, il faut un artifice singulier pour trouver ce rapport : voici comment je m'y suis pris.

The whole investigation bears some resemblance to that of James Bernoulli and may have been suggested by it, for Nicolas Bernoulli says at the end of it, Je me souviens que feu mon Oncle a démontré une semblable chose dans son Traité *De Arte Conjectandi,* qui s'imprime à présent à Bâle, ...

226. The next letter is from Montmort to Nicolas Bernoulli ; it occupies pages 395—400. Montmort records the death of the Duchesse d'Angoulême, which caused him both grief and trouble ; he says he cannot discuss geometrical matters, but will confine himself to literary intelligence.

He mentions a work entitled *Prémotion Physique, ou Action de Dieu sur les Creatures démontrée par raisonnement.* The anonymous author pretended to follow the method of mathematicians, and on every page were to be found such great words as *Definition, Axiom, Theorem, Demonstration, Corollary,* &c.

Montmort asks for the opinion of Nicolas Bernoulli and his uncle respecting the famous *Commercium Epistolicum* which he says M^m de la Societé Royale ont fait imprimer pour assurer à M. Newton la gloire d'avoir inventé le premier et seul les nouvelles methodes.

Montmort speaks with approbation of a little treatise which had just appeared under the title of *Mechanique du Feu.*

Montmort expresses his strong admiration of two investigations which he had received from Nicolas Bernoulli ; one of these was

the solution of Waldegrave's problem, and the other apparently
the demonstration of James Bernoulli's theorem: see Arts. 224, 225.
Montmort says, page 400,

> Tout cela étoit en verité bien difficile et d'un grand travail.
> Vous êtes un terrible homme; je croyois que pour avoir pris les de-
> vants je ne serois pas si-tôt ratrappé, mais je vois bien que je me suis
> trompé: je suis à présent bien derriere vous; et forcé de mettre toute
> mon ambition à vous suivre de loin.

227. This letter from Montmort is interesting, as it records
the perplexity in which the writer found himself between the
claims of the rival systems of natural philosophy, the Cartesian
and the Newtonian. He says, page 397,

> Dérangé comme je le suis par l'autorité de M. Newton, et d'un
> si grand nombre de sçavans Geometres Anglois, je serois presque tenté
> de renoncer pour jamais à l'étude de la Physique, et de remettre à
> sçavoir tout cela dans le Ciel; mais non, l'autorité des plus grands
> esprits ne doit point nous faire de loi dans les choses où la raison
> doit décider.

228. Montmort gives in this letter his views respecting a
History of Mathematics; he says, page 399,

> Il seroit à souhaiter que quelqu'un voulût prendre la peine de
> nous apprendre comment et en quel ordre les découvertes en Mathe-
> matiques se sont succedées les unes aux autres, et à qui nous en avons
> l'obligation. On a fait l'Histoire de la Peinture, de la Musique, de
> la Medecine, &c. Une bonne Histoire des Mathematiques, et en par-
> ticulier de la Geometrie, seroit un Ouvrage beaucoup plus curieux et
> plus utile: Quel plaisir n'auroit-on pas de voir la liaison, la connexion
> des methodes, l'enchaînement des differentes theories, à commencer
> depuis les premiers temps jusqu'au nôtre ou cette science se trouve
> portée à un si haut degré de perfection. Il me semble qu'un tel
> Ouvrage bien fait pourroit être en quelque sorte regardé comme l'his-
> toire de l'esprit humain; puisque c'est dans cette science plus qu'en
> toute autre chose, que l'homme fait connoître l'excellence de ce don
> d'intelligence que Dieu lui a accordé pour l'élever au dessus de toutes
> les autres Creatures.

Montmort himself had made some progress in the work which he here recommends; see Art. 137. It seems however that his manuscripts were destroyed or totally dispersed; see Montucla, *Histoire des Mathematiques* first edition, preface, page IX.

229. The next letter is from Nicolas Bernoulli to Montmort; it occupies pages 401, 402. Nicolas Bernoulli announces that the *Ars Conjectandi* has just been published, and says, Il n'y aura gueres rien de nouveau pour vous. He proposes five problems to Montmort in return for those which Montmort had proposed to him. He says that he had already proposed the first problem in his last letter; but as the problem does not occur before in the correspondence, a letter must have been suppressed, or a portion of it omitted.

The third problem is as follows. *A* and *B* play with a common die, *A* deposits a crown, and *B* begins to play; if *B* throws an even number he takes the crown, if he throws an odd number he deposits a crown. Then *A* throws, and takes a crown if he throws an even number, but does not deposit a crown if he throws an odd number. Then *B* throws again, and so on. Thus each takes a crown if he throws an even number, but *B* alone deposits a crown if he throws an odd number. The play is to continue as long as there is any sum deposited. Determine the advantage of *A* or *B*.

The fourth problem is as follows. *A* promises to give to *B* a crown if *B* with a common die throws six at the first throw, two crowns if *B* throws six at the second throw, three crowns if *B* throws six at the third throw; and so on.

The fifth problem generalises the fourth, *A* promises to give *B* crowns in the progression 1, 2, 4, 8, 16, ... or 1, 3, 9, 27, ... or 1, 4, 9, 16, 25, ... or 1, 8, 27, 64, ... instead of in the progression 1, 2, 3, 4, 5, as in the fourth problem.

230. The next letter is the last; it is from Montmort to Nicolas Bernoulli, and it occupies pages 403—412. It enters largely on the game of *Her*. With respect to the five problems proposed to him, Montmort says that he has not tried the first and second, that the fourth and fifth present no difficulty, but that the third is much more difficult. He says that it took him

a long time to convince himself that there would be neither advantage nor disadvantage for B, but that he had come to this conclusion, and so had Waldegrave, who had worked with him at the problem. It would seem however, that this result is obvious, for B has at every trial an equal chance of winning or losing a crown.

Montmort proposes on his page 408 a problem to Nicolas Bernoulli, but the game to which it relates is not described.

231. In the fourth problem given in Art. 229, the advantage of B is expressed by the series

$$\frac{1}{6} + \frac{2}{6^2} + \frac{3}{6^3} + \frac{4}{6^4} + \dots \text{ in infinitum.}$$

This series may be summed by the ordinary methods.

We shall see that a problem of the same kind as the fourth and fifth of those communicated by Nicolas Bernoulli to Montmort, was afterwards discussed by Daniel Bernoulli and others, and that it has become famous under the title of the *Petersburg Problem*.

232. Montmort's work on the whole must be considered highly creditable to his acuteness, perseverance, and energy. The courage is to be commended which led him to labour in a field hitherto so little cultivated, and his example served to stimulate his more distinguished successor. De Moivre was certainly far superior in mathematical power to Montmort, and enjoyed the great advantage of a long life, extending to more than twice the duration of that of his predecessor ; on the other hand, the fortunate circumstances of Montmort's position gave him that abundant leisure, which De Moivre in exile and poverty must have found it impossible to secure.

CHAPTER IX.

DE MOIVRE.

233. ABRAHAM De Moivre was born at Vitri, in Champagne, in 1667. On account of the revocation of the edict of Nantes, in 1685, he took shelter in England, where he supported himself by giving instruction in mathematics and answers to questions relating to chances and annuities. He died at London in 1754.

John Bernoulli speaks thus of De Moivre in a letter to Leibnitz, dated 26 Apr. 1710; see page 847 of the volume cited in Art. 59:

...Dominus Moyvraeus, insignis certe Geometra, qui haud dubie adhuc haeret Londini, luctans, ut audio, cum fame et miseria, quas ut depellat, victum quotidianum ex informationibus adolescentum petere cogitur. O duram sortem hominis! et parum aptam ad excitanda ingenia nobilia; quis non tandem succumberet sub tam iniquae fortunae vexationibus? vel quodnam ingenium etiam fervidissimum non algeat tandem? Miror certe Moyvracum tantis angustiis pressum ea tamen adhuc praestare, quae praestat.

De Moivre was elected a Fellow of the Royal Society in 1697; his portrait, strikingly conspicuous among those of the great chiefs of science, may be seen in the collection which adorns the walls of the apartment used for the meetings of the Society. It is recorded that Newton himself, in the later years of his life, used to reply to inquirers respecting mathematics in these words: "Go to Mr De Moivre, he knows these things better than I do." In the long list of men ennobled by genius, virtue, and misfortune, who have found an asylum in England, it would be difficult to name one who has conferred more honour on his adopted country than De Moivre.

234. Number 329 of the *Philosophical Transactions* consists entirely of a memoir entitled *De Mensura Sortis, seu, de Probabilitate Eventuum in Ludis a Casu Fortuito Pendentibus.* Autore Abr. De Moivre, R.S.S.

The number is stated to be for the months of January, February, and March 1711 ; it occupies pages 213—264 of Volume XXVII. of the Philosophical Transactions.

The memoir was afterwards expanded by De Moivre into his work entitled *The Doctrine of Chances: or, a Method of Calculating the Probabilities of Events in Play.* The first edition of this work appeared in 1718 ; it is in quarto and contains xiv + 175 pages, besides the title-leaf and a dedication. The second edition appeared in 1738 ; it is in large quarto, and contains xiv + 258 pages, besides the title-leaf and a dedication and a page of corrections. The third edition appeared in 1756, after the author's death ; it is in large quarto, and contains xii + 348 pages, besides the title-leaf and a dedication.

235. I propose to give an account of the memoir *De Mensura Sortis,* and of the third edition of the *Doctrine of Chances.* In my account of the memoir I shall indicate the corresponding parts of the *Doctrine of Chances;* and in my account of the *Doctrine of Chances* I shall give such remarks as may be suggested by comparing the third edition of the work with those which preceded it ; any reference to the *Doctrine of Chances* must be taken to apply to the *third* edition, unless the contrary is stated.

236. It may be observed that the memoir *De Mensura Sortis* is not reprinted in the abridgement of the Philosophical Transactions up to the year 1800, which was edited by Hutton, Shaw, and Pearson.

The memoir is dedicated to Francis Robartes, at whose recommendation it had been drawn up. The only works of any importance at this epoch, which had appeared on the subject, were the treatise by Huygens, and the first edition of Montmort's book. De Moivre refers to these in words which we have already quoted in Art. 142.

De Moivre says that Problems 16, 17, 18 in his memoir were proposed to him by Robartes. In the Preface to the Doctrine of Chances, which is said to have been written in 1717, the origin of the memoir is explained in the following words :

'Tis now about Seven Years, since I gave a Specimen in the *Philosophical Transactions*, of what I now more largely treat of in this Book. The occasion of my then undertaking this Subject was chiefly owing to the Desire and Encouragement of the Honourable *Francis Robartes* Esq. (now Earl of Radnor); who, upon occasion of a French Tract, called *L'Analyse des Jeux de Hazard*, which had lately been published, was pleased to propose to me some Problems of much greater difficulty than any he had found in that Book; which having solved to his Satisfaction, he engaged me to methodize those Problems, and to lay down the Rules which had led me to their Solution. After I had proceeded thus far, it was enjoined me by the Royal Society, to communicate to them what I had discovered on this Subject: and thereupon it was ordered to be published in the Transactions, not so much as a matter relating to Play, but as containing some general Speculations not unworthy to be considered by the Lovers of Truth.

237. The memoir consists of twenty-six Problems, besides a few introductory remarks which explain how probability is measured.

238. The first problem is to find the chance of throwing an ace twice or oftener in eight throws with a single die ; see *Doctrine of Chances*, page 13.

239. The second problem is a case of the Problem of Points. A is supposed to want 4 points, and B to want 6 points ; and A's chance of winning a single point is to B's as 3 is to 2 ; see *Doctrine of Chances*, page 18. It is to be remembered that up to this date, in all that had been *published* on the subject, the chances of the players for winning a single point had always been assumed equal ; see Art. 173.

240. The third problem is to determine the chances of A and B for winning a single game, supposing that A can give B two games out of three ; the fourth problem is of a similar kind, supposing

that A can give B one game out of three : see Problems I. and II. of the *Doctrine of Chances.*

241. The fifth problem is to find how many trials must be made to have an even chance that an event shall happen *once* at least. Montmort had already solved the problem ; see Art. 170.

De Moivre adds a useful approximate formula which is now one of the permanent results in the subject; we shall recur to it in noticing Problem III. of the *Doctrine of Chances,* where it is reproduced.

242. De Moivre then gives a Lemma: To find how many Chances there are upon any number of Dice, each of them of the same number of Faces, to throw any given number of points; see *Doctrine of Chances,* page 39. We have already given the history of this Lemma in Art. 149.

243. The sixth problem is to find how many trials must be made to have an even chance that an event shall happen *twice* at least. The seventh problem is to find how many trials must be made to have an even chance that an event shall happen *three* times at least, or *four* times at least, and so on. See Problems III. and IV. of the *Doctrine of Chances.*

244. The eighth problem is an example of the Problem of Points with three players; it is Problem VI. of the *Doctrine of Chances.*

245. The ninth problem is the fifth of those proposed for solution by Huygens, which Montmort had enunciated wrongly in his first edition; see Art. 199. Here we have the first *publication* of the general formula for the chance which each of two players has of ruining the other in an unlimited number of games; see Art. 107. The problem is Problem VII. of the *Doctrine of Chances.*

246. The tenth problem is Problem VIII. of the *Doctrine of Chances,* where it is thus enunciated :

Two Gamesters A and B lay by 24 Counters, and play with three Dice, on this condition; that if 11 Points come up, A shall take one

Counter out of the heap; if 14, *B* shall take out one; and he shall be reputed the winner who shall soonest get 12 Counters.

This is a very simple problem. De Moivre seems quite unnecessarily to have imagined that it could be confounded with that which immediately preceded it; for at the end of the ninth problem he says,

Maxime cavendum est ne Problemata propter speciem aliquam affinitatis inter se confundantur. Problema sequens videtur affine superiori.

After enunciating his ninth problem he says,

Problema istud a superiore in hoc differt, quod 23 ad plurimum tesserarum jactibus, ludus necessario finietur; cum ludus ex lege superioris problematis, posset in aeternum continuari, propter reciprocationem lucri et jacturæ se invicem perpetuo destruentium.

247. The eleventh and twelfth problems consist of the second of those proposed for solution by Huygens, taken in two meanings; they form Problems x. and xi. of the *Doctrine of Chances*. The meanings given by De Moivre to the enunciation coincide with the first and second of the three considered by James Bernoulli; see Arts. 35 and 199.

248. The thirteenth problem is the first of those proposed for solution by Huygens; the fourteenth problem is the fourth of the same set: see Art. 35. These problems are very simple and are not repeated in the *Doctrine of Chances*. In solving the fourth of the set De Moivre took the meaning to be that *A* is to draw three white balls *at least*. Montmort had taken the meaning to be that *A* is to draw *exactly* three white balls. John Bernoulli in his letter to Montmort took the meaning to be that *A* is to draw three white balls *at least*. James Bernoulli had considered both meanings. See Art. 199.

249. The fifteenth problem is that which we have called Waldegrave's problem; see Art. 211. De Moivre here discusses the problem for the case of three players: this discussion is repeated, and extended to the case of four players, in the *Doctrine of Chances*, pages 132—159. De Moivre was the first in *publishing* a solution of the problem.

250. The sixteenth and seventeenth problems relate to the game of bowls; see Art. 177. These problems are reproduced in a more general form in the *Doctrine of Chances*, pages 117—123. Respecting these two problems Montmort says, on his page 366,

Les Problêmes 16 et 17 ne sont que deux cas très simples d'un même Problême, c'est presque le seul qui m'ait échapé de tous ceux que je trouve dans ce Livre.

251. The eighteenth and nineteenth problems are Problems XXXIX. and XL. of the *Doctrine of Chances*, where we shall find it more convenient to notice them.

252. The remaining seven problems of the memoir form a distinct section on the Duration of Play. They occur as Problems LVIII, LX, LXI, LXII, LXIII, LXV, LXVI, of the *Doctrine of Chances;* and we shall recur to them.

253. It will be obvious from what we have here given that the memoir *De Mensura Sortis* deserves especial notice in the history of our subject. Many important results were here first *published* by De Moivre, although it is true that these results already existed in manuscript in the *Ars Conjectandi* and the correspondence between Montmort and the Bernoullis.

We proceed to the *Doctrine of Chances*.

254. The second edition of the *Doctrine of Chances* contains an *Advertisement* relating to the additions and improvements effected in the work; this is not reprinted in the third edition. The second edition has at the end a Table of Contents which neither of the others has. The third edition has the following *Advertisement:*

The Author of this Work, by the failure of his Eye-sight in extreme old age, was obliged to entrust the Care of a new Edition of it to one of his Friends; to whom he gave a Copy of the former, with some marginal Corrections and Additions, in his own hand writing. To these the Editor has added a few more, where they were thought necessary: and has disposed the whole in better Order; by restoring to their proper places some things that had been accidentally *misplaced,* and by putting all the Problems concerning *Annuities* together; as they stand in the late *improved* edition of the Treatise on that Subject. An *Appendix* of several useful Articles is likewise subjoined : the whole according to a Plan concerted with the Author, above a year before his death.

255. The following list will indicate the parts which are new in the third edition. The *Remark*, pages 30—33 ; the *Remark*, pages 48, 49; the greater part of the second *Corollary*, pages 64—66; the *Examples*, page 88; the *Scholium*, page 95 ; the *Remark*, page 116 ; the third *Corollary*, page 138 ; the second *Corollary*, page 149 ; the *Remark*, pages 151—159 ; the fourth *Corollary*, page 162 ; the second *Corollary*, pages 176—179 ; the Note at the foot of page 187 ; the *Remark*, pages 251—254.

The part on life annuities is very much changed, according to the plan laid down in the *Advertisement*.

In the second and third editions the numbers of the Problems agree up to Problem XI; Problem XII. of the third edition had been Problem LXXXIX. of the second; from Problem XII. to Problem LXIX. of the third edition inclusive, the number of each Problem exceeds by unity its number in the second edition ; Problem LXIX. of the second edition is incorporated in the third edition with Problem VI ; Problems LXX. and LXXI. are the same in the two editions, allowing for a misprint of LXXI. for LXX. in the second edition. After this the numbering differs considerably because in the second edition Problems respecting life annuities are not separated from the other Problems as they are in the third edition.

The first edition of the work was dedicated to Newton : the second was dedicated to Lord Carpenter, and the dedication of the second edition is reprinted at the beginning of the third ; the dedication to Newton is reprinted on page 329 of the third edition.

256. The first edition of the *Doctrine of Chances* has a good preface explaining the design and utility of the book and giving an account of its contents; the preface is reproduced in the other editions with a few omissions. It is to be regretted that the following paragraphs were not retained, which relate respectively to the first and second editions of Montmort's work :

However, had I allowed my self a little more time to consider it, I had certainly done the Justice to its Author, to have owned that he had not only illustrated *Huygens's* Method by a great variety of well

chosen Examples, but that he had added to it several curious things of
his own Invention.

....

Since the printing of my Specimen, *Mr. de Monmort*, Author of the
Analyse des jeux de Hazard, Published a Second Edition of that Book,
in which he has particularly given many proofs of his singular Genius,
and extraordinary Capacity; which Testimony I give both to Truth,
and to the Friendship with which he is pleased to Honour me.

The concluding paragraph of the preface to the first edition
refers to the *Ars Conjectandi*, and invites Nicolas and John Bernoulli to prosecute the subject begun in its fourth part; this
paragraph is omitted in the other editions.

We repeat that we are about to analyse the *third* edition of the
Doctrine of Chances, only noticing the previous editions in cases of
changes or additions in matters of importance.

257. The Doctrine of Chances begins with an *Introduction* of
33 pages, which explains the chief rules of the subject and illustrates them by examples; this part of the work is very much fuller
than the corresponding part of the first edition, so that our remarks
on the *Introduction* do not apply to the first edition. De Moivre
considers carefully the following fundamental theorem: suppose
that the odds for the happening of an event at a single trial are as
a to b, then the chance that the event will happen r times at least
in n trials is found by taking the first $n - r + 1$ terms of the expansion of $(a + b)^n$ and dividing by $(a+b)^n$. We know that the result
can also be expressed in another manner corresponding to the
second formula in Art. 172; it is curious that De Moivre gives
this without demonstration, though it seems less obvious than
that which he has demonstrated.

To find the chance that an event may happen *just* r times, De
Moivre directs us to subtract the chance that it will happen *at least*
$r - 1$ times from the chance that it will happen *at least r* times.
He notices, but less distinctly than we might expect, the modern
method which seems more simple and more direct, by which we
begin with finding the chance that an event shall happen *just r*
times and deduce the chance that it shall happen *at least r*
times.

258. De Moivre notices the advantage arising from employing a single letter instead of two or three to denote the probability of the happening of one event. Thus if x denote the probability of the happening of an event, $1 - x$ will denote the probability of its failing. So also y and z may denote the probabilities of the happening of two other events respectively. Then, for example,

$$x\,(1 - y)\,(1 - z)$$

will represent the probability of the first to the exclusion of the other two. De Moivre says in conclusion, "and innumerable cases of the same nature, belonging to any number of Events, may be solved without any manner of trouble to the imagination, by the mere force of a proper notation."

259. In his third edition De Moivre draws attention to the convenience of approximating to a fraction with a large numerator and denominator by continued fractions, which he calls "the Method proposed by Dr *Wallis*, *Huygens*, and others." He gives the rule for the formation of the successive convergents which is now to be found in elementary treatises on Algebra; this rule he ascribes to Cotes.

260. The *Doctrine of Chances* contains 74 problems exclusive of those relating to life annuities; in the first edition there were 53 problems.

261. We have enunciated Problems I. and II. in Art. 240. Suppose p and q to represent the chances of A and B in a single game. Problem I. means that it is an even chance that A will win three games before B wins one; thus $p^3 = \frac{1}{2}$. Hence $p = \frac{1}{\sqrt[3]{2}}$, and $q = 1 - \frac{1}{\sqrt[3]{2}}$. Problem II. means that it is an even chance that A will win three games before B wins two. Thus $p^4 + 4p^3q = \frac{1}{2}$; which must be solved by trial.

These problems are simple examples of the general formula in Art. 172.

262. Problems III, IV, and V. are included in the following

general enunciation. Suppose a the number of chances for the happening of an event in a single trial, and b the number of chances for its failing: find how many trials must be made to have an even chance that the event will happen r times at least.

For example, let $r = 1$.

Suppose x the number of trials. Then the chance that the event fails x times in succession is $\dfrac{b^x}{(a+b)^x}$. And by supposition this is equal to the chance of its happening once at least in x trials. Therefore each of these chances must be equal to $\dfrac{1}{2}$. Thus

$$\frac{b^x}{(a+b)^x} = \frac{1}{2};$$

from this equation x may be found by logarithms.

De Moivre proceeds to an approximation. Put $\dfrac{b}{a} = q$. Thus

$$x \log \left(1 + \frac{1}{q}\right) = \log 2.$$

If $q = 1$, we have $x = 1$. If q be greater than 1, we have by expanding $\log \left(1 + \dfrac{1}{q}\right)$,

$$x \left\{ \frac{1}{q} - \frac{1}{2q^2} + \frac{1}{3q^3} - \frac{1}{4q^4} + \dots \right\} = \log 2,$$

where $\log 2$ will mean the logarithm to the Napierian base. Then if q be large we have approximately

$$x = q \log 2 = \frac{7}{10} q \text{ nearly.}$$

De Moivre says, page 37,

Thus we have assigned the very narrow limits within which the ratio of x to q is comprehended; for it begins with unity, and terminates at last in the ratio of 7 to 10 very near.

But x soon converges to the limit $0.7q$, so that this value of x may be assumed in all cases, let the value of q be what it will.

The fact that this result *is* true when q is moderately large is the clement of truth in the mistake made by M. de Méré; he assumed that such a result should hold for *all* values of q: see Art. 14.

263. As another example of the general enunciation of Art. 262, let $r = 3$.

The chance that the event will happen at least 3 times in x trials is equal to the first $x - 2$ terms of the expansion of

$$\left(\frac{a}{a+b} + \frac{b}{a+b}\right)^x,$$

and this chance by hypothesis is $\frac{1}{2}$. Hence the last three terms of the expansion will also be equal to $\frac{1}{2}$, that is,

$$b^x + x b^{x-1} a + \frac{x(x-1)}{1 \cdot 2} b^{x-2} a^2 = \frac{1}{2} (a+b)^x.$$

Put $\frac{b}{a} = q$; thus $\left(1 + \frac{1}{q}\right)^x = 2 \left\{1 + \frac{x}{q} + \frac{x(x-1)}{2q^2}\right\}.$

If $q = 1$ we find $x = 5$.

If q be supposed indefinitely great, and we put $\frac{x}{q} = z$, we get

$$e^z = 2 \left(1 + z + \frac{z^2}{2}\right),$$

where e is the base of the Napierian logarithms.

By trial it is found that $z = 2 \cdot 675$ nearly. Hence De Moivre concludes that x always lies between $5q$ and $2 \cdot 675q$.

264. De Moivre exhibits the following table of results obtained in the manner shewn in the two preceding Articles.

A Table *of the Limits*.

The Value of x will always be

For a single Event, between $1q$ and $0 \cdot 693q$.
For a double Event, between $3q$ and $1 \cdot 678q$.
For a triple Event, between $5q$ and $2 \cdot 675q$.
For a quadruple Event, between $7q$ and $3 \cdot 672q$.
For a quintuple Event, between $9q$ and $4 \cdot 670q$.
For a sextuple Event, between $11q$ and $5 \cdot 668q$.
&c.

And if the number of Events contended for, as well as the number q be pretty large in respect to Unity; the number of Trials requisite for those Events to happen n times will be $\dfrac{2n-1}{2}\,q$, or barely nq.

De Moivre seems to have inferred the general result enunciated in the last sentence, from observing the numerical values obtained in the six cases which he had calculated, for he gives no further investigation.

265. In Art. 263 we have seen that De Moivre concludes that $\dfrac{x}{q}$ *always* lies between 5 and 2·675. This may appear very probable, but it is certainly not demonstrated. It is quite conceivable, in the absence of any demonstration to the contrary, that $\dfrac{x}{q}$ should at first increase with q, and so be greater than 5, and then decrease and become less than 2·675, and then increase again to its limit 2·675. The remark applies to the general proposition, whatever be the value of r, as well as to the particular example in which $r = 3$.

It would not be very easy perhaps to shew from such an equation as that in Art. 263, that x *increases continually* with q; and yet from the nature of the question we may conclude that this must be the case. For if the chance of success in a single trial is diminished, it appears obvious that the number of trials must be increased, in order to secure an even chance for the event to happen once at least.

266. On pages 39—43 of the *Doctrine of Chances*, we have the Lemma of which we have already given an account; see Art. 242.

267. Problem VI. of the *Doctrine of Chances* is an example of the Problem of Points with three players. De Moivre gives the same kind of solution as Fermat: see Arts. 16 and 18. In the third edition there is also a discussion of some simple cases according to the method which Pascal used for two players; see Art. 12. De Moivre also gives here a good rule for solving the problem for any number of players; the rule is founded on

Fermat's method, and is intended to lighten as much as possible the labour which must be incurred in applying the method to complex cases. The rule was first published in the *Miscellanea Analytica*, in 1730; it is given in the second edition of the *Doctrine of Chances* on pages 191, 192.

268. Problem VII. is the fifth of those proposed by Huygens for solution; see Art. 35. We have already stated that De Moivre generalises the problem in the same way as James Bernoulli, and the result, with a demonstration, was first published in the *De Mensura Sortis;* see Arts. 107, 245. De Moivre's demonstration is very ingenious, but not quite complete. For he finds the *ratio* of the chance that *A* will ruin *B* to the chance that *B* will ruin *A*; then he assumes in effect that in the long run one or other of the players *must* be ruined: thus he deduces the absolute values of the two chances.

See the first Appendix to Professor De Morgan's *Essay on Probabilities* in the *Cabinet Cyclopædia*.

We have spoken of Problem VIII. in Art. 246.

269. Problem IX. is as follows,

Supposing *A* and *B*, whose proportion of skill is as *a* to *b*, to play together, till *A* either wins the number *q* of Stakes, or loses the number *p* of them; and that *B* sets at every Game the sum *G* to the sum *L*; it is required to find the Advantage or Disadvantage of *A*.

This was Problem XLIII. of the first edition of the *Doctrine of Chances*, in the preface to which it is thus noticed:

The 43d Problem having been proposed to me by Mr. *Thomas Woodcock*, a Gentleman whom I infinitely respect, I attempted its Solution with a very great desire of obtaining it; and having had the good Fortune to succeed in it, I returned him the Solution a few Days after he was pleased to propose it. This Problem is in my Opinion one of the most curious that can be propos'd on this Subject; its Solution containing the Method of determining, not only that Advantage which results from a Superiority of Chance, in a Play confined to a certain number of Stakes to be won or lost by either Party, but also that which may result from an unequality of Stakes; and even compares those two Advantages together, when the Odds of Chance being on one side, the Odds of Money are on the other.

In the *Miscellanea Analytica*, page 204, the problem is again said to have been proposed by Thomas Woodcock, *spectatissimo viro*, but he is not mentioned in the second or third edition of the *Doctrine of Chances;* so that De Moivre's infinite respect for him seems to have decayed and disappeared in a finite time.

The solution of the problem is as follows:

Let R and S respectively represent the Probabilities which A and B have of winning all the Stakes of their Adversary; which Probabilities have been determined in the viith *Problem*. Let us first suppose that the Sums deposited by A and B are equal, *viz.* G, and G: now since A is either to win the sum qG, or lose the sum pG, it is plain that the Gain of A ought to be estimated by $RqG - SpG$; moreover since the Sums deposited are G and G, and that the proportion of the Chances to win one Game is as a to b, it follows that the Gain of A for each individual Game is $\dfrac{aG - bG}{a + b}$; and for the same reason the Gain of each individual Game would be $\dfrac{aG - bL}{a + b}$, if the Sums deposited by A and B were respectively L and G. Let us therefore now suppose that they are L and G; then in order to find the whole Gain of A in this second circumstance, we may consider that whether A and B lay down equal Stakes or unequal Stakes, the Probabilities which either of them has of winning all the Stakes of the other, suffer not thereby any alteration, and that the Play will continue of the same length in both circumstances before it is determined in favour of either; wherefore the Gain of each individual Game in the first case, is to the Gain of each individual Game in the second, as the whole Gain of the first case, to the whole Gain of the second; and consequently the whole Gain of the second case will be $\overline{Rq - Sp} \times \dfrac{aG - bL}{a - b}$, or restoring the values of R and S,

$$\dfrac{qa^q \times \overline{a^p - b^p} - pb^p \times \overline{a^q - b^q}}{a^{p+q} - b^{p+q}} \text{ multiplied by } \dfrac{aG - bL}{a - b}.$$

270. In the first edition of the *Doctrine of Chances*, pages 136—142, De Moivre gave a very laborious solution of the preceding Problem. To this was added a much shorter solution, communicated by Nicolas Bernoulli from his uncle. This solution was founded on an artifice which De Moivre had himself used in

the ninth problem of the *De Mensura Sortis*. De Moivre however renounces for himself the claim to the merit of the solution. This renunciation he repeats in the *Miscellanea Analytica*, page 206, where he names the author of the simple solution which we have already given. He says,

Ego vere illud ante libenter fassus sum, idque ipsum etiamnum libenter fateor, quamvis solutio Problematis mei noni causam fortasse dederit hujus solutionis, me tamen nihil juris in eam habere, eamque Cl. illius Autori ascribi æquum esse.

Septem aut octo abhinc annis D. *Stevens* Int. Templ. Socius, Vir ingenuus, singulari sagacitate præditus, id sibi propositum habens ut Problema superius allatum solveret, hac ratione solutionem facile assecutus est, quam mihi his verbis exhibuit.

Then follows the solution, after which De Moivre adds,

Doctissimus adolescens D. *Cranmer*, apud Genevenses Mathematicæ Professor dignissimus, cujus recordatio æque ac Collegæ ejus peritissimi D. *Calandrin* mihi est perjucunda, cum superiore anno Londini commoraretur, narravit mihi se ex literis D. *Nic. Bernoulli* ad se datis accepisse Cl. Virum novam solutionem hujus Problematis adeptum esse, quam prioribus autor anteponebat ; cum vero nihil de via solutionis dixerit, si mihi conjicere liceat qualis ea sit, hanc opinor eandem esse atque illam quam modo attuli.

271. We have already spoken of Problems x. and xi. in Art. 247. In his solution of Problem x. De Moivre uses the theorem for the summation of series to which we have referred in Art. 152. A corollary was added in the second edition and was expanded in the third edition, on which we will make a remark.

Suppose that A, B, and C throw *in order* a die of n faces, and that a faces are favourable to A, and b to B, and c to C, where $a + b + c = n$. Required the chances which A, B, and C have respectively of being the first to throw a corresponding face. It may be easily shewn that the chances are proportional to an^2, $(b + c) bn$, and $(b + c) (a + c) c$, respectively. De Moivre, in his third edition, page 65, seems to imply that *before the order was fixed*, the chances would be proportional to a, b, c. This must of course mean that such would be the case if there were *no order at all;* that is if the die were to be thrown and the stake awarded to A, B, or C, according as the face which appeared was one of the a, b, c respectively. If there is to be an order, but the order is as likely to be one as another, the result will be different. The chance of A for example will be one sixth of the sum arising from six possible and equally likely cases. It will be found that A's chance is

$$\frac{a\left\{6a^2 + 9a\,(b+c) + 3\,(b^2+c^2) + 8bc\right\}}{6\left\{n^3 - (b+c)\,(c+a)\,(a+b)\right\}}.$$

272. Problem XII. appeared for the first time in the second edition, page 248, with this preliminary notice. "A particular *Friend* having desired of me that to the preceding Problems I would add one more, I have thought fit to comply with his desire; the Problem was this." The problem is of no great importance; it is solved by the method often used in the *Ars Conjectandi*, which we have explained in Art. 106.

273. Problem XIII. relates to the game of *Bassette*, and Problem XIV. to the game of *Pharaon;* these problems occupy pages 69—82 of the work. We have already sufficiently noticed these games; see Arts. 154, 163. De Moivre's discussion is the same in all his three editions, except that a paragraph on page 37 of the first edition, extending from the words "Those who are ..." to the end of the page, is omitted in the following editions. The paragraph is in fact an easy example of the formulæ for the game of Bassette.

274. Problems XV. to XX. form a connected series. De Moivre solves simple examples in chances and applies his results to establish a Theory of Permutations and Combinations; in modern times we usually adopt the reverse order, establish the Theory of Permutations and Combinations first, and afterwards apply the theory in the discussion of chances. We will take an example of De Moivre's method from his Problem XV. Suppose there are six things a, b, c, d, e, f, and let two of them be taken at random; required the chance that a shall stand first, and b second. The

chance of taking a first is $\frac{1}{6}$; and there are then five things left,

and the chance of now taking b is $\frac{1}{5}$. Therefore the required

chance is $\frac{1}{30}$. Then De Moivre says,

Since the taking a in the first place, and b in the second, is but one single Case of those by which six Things may change their order, being taken two and two; it follows that the number of Changes or Permutations of six Things, taken two and two, must be 30.

275. In his Preface De Moivre says,

Having explained the common Rules of Combinations, and given a Theorem which may be of use for the Solution of some Problems relating to that Subject, I lay down a new Theorem, which is properly a contraction of the former, whereby several Questions of Chance are resolved with wonderful ease, tho' the Solution might seem at first sight to be of insuperable difficulty.

The *new Theorem* amounts to nothing more than the simplification of an expression by cancelling factors, which occur in its numerator and denominator; see *Doctrine of Chances*, pages ix. 89.

276. Problems XXI. to XXV. consist of easy applications to questions concerning Lotteries of the principles established in the Problems XV. to XX.; only the first two of these questions concerning Lotteries appeared in the first edition.

A Scholium is given on page 95 of the third edition which deserves notice. De Moivre quotes the following formula: Suppose a and n to be positive integers; then

$$\frac{1}{n}+\frac{1}{n+1}+\frac{1}{n+2}+\frac{1}{n+3}+\dots+\frac{1}{a-1}$$

$$=\log\frac{a}{n}+\frac{1}{2n}-\frac{1}{2a}+\frac{A}{2}\left(\frac{1}{n^2}-\frac{1}{a^2}\right)+\frac{B}{4}\left(\frac{1}{n^4}-\frac{1}{a^4}\right)$$

$$+\frac{C}{6}\left(\frac{1}{n^6}-\frac{1}{a^6}\right)+\dots;$$

where $\qquad A=\frac{1}{6},\quad B=-\frac{1}{30},\quad C=\frac{1}{42},\quad\dots$

As De Moivre says A, B, C, ... are "the numbers of Mr. *James Bernoulli* in his excellent Theorem for the Summing of Powers." See Art. 112. De Moivre refers for the demonstration of the formula to the Supplement to the *Miscellanea Analytica*, where the formula first appeared. We shall recur to this in speaking of the *Miscellanea Analytica*.

277. Problems XXVII. to XXXII. relate to the game of *Quadrille*; although the game is not described there is no difficulty in understanding the problems which are simple examples of the Theory of Combinations: these problems are not in the first edition.

278. Problem XXXIII. is To find at *Pharaon* how much it is that the Banker gets *per Cent.* of all the Money that is adventured. De Moivre in his Preface seems to attach great importance to this solution; but it scarcely satisfies the expectations which are thus raised. The player who stakes against the bank is in fact supposed to play merely by chance without regard to what would be his best course at any stage of the game, although the previous investigations of Montmort and De Moivre shewed distinctly that some courses were far less pernicious than others.

The Banker's adversary in De Moivre's solution is therefore rather a machine than a gambler with liberty of choice.

279. Problem XXXIV. is as follows :

Supposing A and B to play together, that the Chances they have respectively to win are as a to b, and that B obliges himself to set to A so long as A wins without interruption : what is the advantage that A gets by his hand?

The result is, supposing each to stake one,

$$\frac{a-b}{a+b}\left\{ 1 + \frac{a}{a+b} + \frac{a^2}{(a+b)^2} + \frac{a^3}{(a+b)^3} + \dots \text{ in infinitum} \right\},$$

that is,

$$\frac{a-b}{b}.$$

280. Problems XXXV. and XXXVI. relate to the game discussed by Nicolas Bernoulli and Montmort, which is called *Treize* or *Rencontre*; see Art. 162.

De Moivre treats the subject with great ingenuity and with more generality than his predecessors, as we shall now shew.

281. Problem XXXV. is thus enunciated :

Any number of Letters a, b, c, d, e, f, &c., all of them different, being taken promiscuously as it happens: to find the Probability that some of them shall be found in their places according to the rank they obtain in the Alphabet; and that others of them shall at the same time be displaced.

Let n be the number of the letters ; suppose that p specified letters are to be in their places, q specified letters out of their places, and the remaining $n - p - q$ letters free from any restriction. The chance that this result will happen is

$$\frac{1}{n(n-1)\ldots(n-p+1)} \left\{ 1 - \frac{q}{1} \frac{1}{n-p} + \frac{q(q-1)}{1.2} \frac{1}{(n-p)(n-p-1)} - \ldots \right\}.$$

This supposes that p is greater than 0; if $p = 0$, the result is

$$1 - \frac{q}{1} \frac{1}{n} + \frac{q(q-1)}{1.2} \frac{1}{n(n-1)} - \ldots$$

If we suppose in this formula $q = m - 1$, we have a result already implicitly given in Art. 161.

In demonstrating these formulæ De Moivre is content to examine a few simple cases and assume that the law which presents itself will hold universally. We will indicate his method.

The chance that a is in the first place is $\frac{1}{n}$; the chance that a is in the first place, and b in the second place is $\frac{1}{n(n-1)}$: hence the chance that a is in the first place and b not in the second place is

$$\frac{1}{n} - \frac{1}{n(n-1)}.$$

Similarly the chance that a, b, c are all in their proper places is $\frac{1}{n(n-1)(n-2)}$; subtract this from the chance that a and b are in their proper places, and we have the chance that a and b are in their proper places, and c not in its proper place : thus this chance is

$$\frac{1}{n(n-1)} - \frac{1}{n(n-1)(n-2)}.$$

De Moivre uses a peculiar notation for facilitating this process. Let $+a$ denote the chance that a is in its proper place and $-a$ the chance that it is out of it; let $+b$ denote the chance that b is in its proper place and $-b$ the chance that it is out of it; and so on. And in general let such a symbol as $+a+b+c-d-e$ denote that a, b, c are in their proper places, and d, e out of theirs.

Let $\dfrac{1}{n} = r$, $\quad \dfrac{1}{n(n-1)} = s$, $\quad \dfrac{1}{n(n-1)(n-2)} = t$,

$$\frac{1}{n(n-1)(n-2)(n-3)} = v, \ldots$$

Then we have the following results:

$$
\begin{aligned}
&+b \qquad\quad = r \\
&+b+a = s \\
\hline
&+b-a = r-s \quad\quad\quad\quad (1)
\end{aligned}
$$

$$
\begin{aligned}
&+c+b \qquad\quad = s \\
&+c+b+a = t \\
\hline
&+c+b-a = s-t \quad\quad\quad\quad (2)
\end{aligned}
$$

$$
\begin{aligned}
&+c-a \qquad\quad = r-s \quad\quad \text{by (1)}\\
&+c-a+b = \quad\; s-t \quad \text{by (2)}\\
\hline
&+c-a-b = \qquad r-2s+t \quad\quad (3)
\end{aligned}
$$

$$
\begin{aligned}
&+d+c+b \qquad\quad = t \\
&+d+c+b+a = v \\
\hline
&+d+c+b-a = t-v \quad\quad\quad\quad (4)
\end{aligned}
$$

$$
\begin{aligned}
&+d+c-a \qquad\quad = s-t \quad\quad \text{by (2)}\\
&+d+c-a+b = \quad\; t-v \quad \text{by (4)}\\
\hline
&+d+c-a-b = \qquad s-2t+v \quad\quad (5)
\end{aligned}
$$

$$
\begin{aligned}
&+d-b-a \qquad\quad = r-2s+t \quad\quad \text{by (3)}\\
&+d-b-a+c = \qquad s-2t+v \quad \text{by (5)}\\
\hline
&d-b-a-c = \qquad r-3s+3t-v \quad\quad (6)
\end{aligned}
$$

It is easy to translate into words any of these symbolical processes. Take for example that which leads to the result (2):

$$+c+b = s;$$

this means that the chance that c and b are in their proper places
is s; and this we know to be true ;

$$+ c + b + a = t,$$

this means that the chance that c, b, a are all in their proper
places is t; and this we know to be true.

From these two results we deduce that the chance that c and b
are in their proper places, and a out of its place is $s - t$; and this
is expressed symbolically thus,

$$+ c + b - a = s - t.$$

Similarly, to obtain the result (3) ; we know from the result (1)
that $r - s$ is the chance that c is in its proper place, and a out of
its proper place ; and we know from the result (2) that $s - t$ is the
chance that c and b are in their proper places, and a out of its proper place ; hence we infer that the chance that c is in its proper
place, and a and b out of their proper places is $r - 2s + t$; and this
result is expressed symbolically thus,

$$+ c - a - b = r - 2s + t.$$

282. De Moivre refers in his Preface to this process in the following terms :

In the 35th and 36th Problems, I explain a new sort of *Algebra*,
whereby some Questions relating to Combinations are solved by so easy
a Process, that their Solution is made in some measure an immediate
consequence of the Method of Notation. I will not pretend to say that
this new *Algebra* is absolutely necessary to the Solving of those Questions which I make to depend on it, since it appears that Mr. *Montmort*,
Author of the *Analyse des Jeux de Hazard*, and Mr. *Nicholas Bernoulli*
have solved, by another Method, many of the cases therein proposed :
But I hope I shall not be thought guilty of too much Confidence, if
I assure the Reader, that the Method I have followed has a degree of
Simplicity, not to say of Generality, which will hardly be attained by
any other Steps than by those I have taken.

283. De Moivre himself enunciates his result verbally ; it is of
course equivalent to the formula which we have given in Art. 281,
but it will be convenient to reproduce it. The notation being that
already explained, he says,

...then let all the quantities 1, r, s, t, v, &c. be written down with Signs alternately positive and negative, beginning at 1, if p be $= 0$; at r, if p be $= 1$; at s, if p be $= 2$; &c. Prefix to these Quantities the Coefficients of a Binomial Power, whose index is $= q$; this being done, those Quantities taken all together will express the Probability required.

284. The enunciation and solution of Problem XXXVI. are as follows:

Any given number of Letters a, b, c, d, e, f, &c., being each repeated a certain number of times, and taken promiscuously as it happens: To find the Probability that of some of those sorts, some one Letter of each may be found in its place, and at the same time, that of some other sorts, no one Letter be found in its place.

Suppose n be the number of all the Letters, l the number of times that each Letter is repeated, and consequently $\frac{n}{l}$ the whole number of Sorts: suppose also that p be the number of Sorts of which some one Letter is to be found in its place, and q the number of Sorts of which no one Letter is to be found in its place. Let now the prescriptions given in the preceding Problem be followed in all respects, saving that r must here be made $= \frac{l}{n}$, $s = \frac{l^2}{n(n-1)}$, $t = \frac{l^3}{n(n-1)(n-2)}$, &c., and the Solution of any particular case of the Problem will be obtained.

Thus if it were required to find the Probability that no Letter of any sort shall be in its place, the Probability thereof would be expressed by the Series

$$1 - qr + \frac{q(q-1)}{1.2}s - \frac{q(q-1)(q-2)}{1.2.3}t + \frac{q(q-1)(q-2)(q-3)}{1.2.3.4}v \ \&c.$$

of which the number of Terms is equal to $q + 1$.

But in this particular case q would be equal to $\frac{n}{l}$, and therefore, the foregoing Series might be changed into this, viz.

$$\frac{1}{2}\frac{n-l}{n-1} - \frac{1}{6}\frac{(n-l)(n-2l)}{(n-1)(n-2)} + \frac{1}{24}\frac{(n-l)(n-2l)(n-3l)}{(n-1)(n-2)(n-3)} \&c.$$

of which the number of Terms is equal to $\frac{n-l}{l}$.

285. De Moivre then adds some Corollaries. The following is the first of them:

From hence it follows, that the Probability of one or more Letters, indeterminately taken, being in their places, will be expressed as follows:

$$1 - \frac{1}{2}\frac{n-l}{n-1} + \frac{1}{6}\frac{(n-l)(n-2l)}{(n-1)(n-2)} - \frac{1}{24}\frac{(n-l)(n-2l)(n-3l)}{(n-1)(n-2)(n-3)} \&c.$$

This agrees with what we have already given from Nicolas Bernoulli; see Art. 204.

In the next three Corollaries De Moivre exhibits the probability that two or more letters should be in their places, that three or more should be, and that four or more should be.

286. The four Corollaries, which we have just noticed, are examples of the most important part of the Problem; this is treated by Laplace, who gives a general formula for the probability that any assigned number of letters or some greater number shall be in their proper places. *Théorie...des Prob.* pages 217—222. The part of Problems xxxv. and xxxvi. which De Moivre puts most prominently forward in his enunciations and solutions is the condition that p letters are to be in their places, q out of their places, and $n-p-q$ free from any restriction; this part seems peculiar to De Moivre, for we do not find it before his time, nor does it seem to have attracted attention since.

287. A *Remark* is given on page 116 which was not in the preceding editions of the *Doctrine of Chances*. De Moivre shews that the sum of the series

$$1 - \frac{1}{2} + \frac{1}{6} - \frac{1}{24} + \dots \text{ in infinitum,}$$

is equal to unity diminished by the reciprocal of the base of the Napierian logarithms.

288. The fifth Corollary to Problem xxxvi. is as follows:

If A and B each holding a Pack of Cards, pull them out at the same time one after another, on condition that every time two like Cards are pulled out, A shall give B a Guinea; and it were required to find what consideration B ought to give A to play on those Terms: the Answer will be one Guinea, let the number of Cards be what it will.

Altho' this be a Corollary from the preceding Solutions, yet it may more easily be made out thus; one of the Packs being the Rule whereby to estimate the order of the Cards in the second, the Probability that the two first Cards are alike is $\frac{1}{52}$, the Probability that the two second are alike is also $\frac{1}{52}$, and therefore there being 52 such alike combinations, it follows that the value of the whole is $\frac{52}{52} = 1$.

It may be interesting to deduce this result from the formulæ already given. The chance that out of n cards, p *specified cards* will be in their places, and all the rest out of their places will be obtained by making $q = n - p$ in the first formula of Art. 281. The chance that *any* p cards will be in their places, and all the rest out of their places will be obtained by multiplying the preceding result by $\dfrac{\lfloor n}{\lfloor n - p \rfloor p}$. And since in this case B receives p guineas, we must multiply by p to obtain B's advantage. Thus we obtain

$$\frac{1}{\lfloor p - 1} \left\{ 1 - 1 + \frac{1}{2} - \frac{1}{\lfloor 3} + \frac{1}{\lfloor 4} - \dots + \frac{(-1)^{n-p}}{\lfloor n - p} \right\}.$$

Denote this by $\phi(p)$; then we are to shew that the sum of the values of $\phi(p)$ obtained by giving to p all values between 1 and n inclusive is unity.

Let $\psi(n)$ denote the sum; then it may be easily shewn that

$$\psi(n + 1) - \psi(n) = 0.$$

Thus $\psi(n)$ is constant for all values of n; and it $= 1$ when $n = 1$, so that $\psi(n)$ is always $= 1$.

289. The sixth Corollary to Problem XXXVI. is as follows:

If the number of Packs be given, the Probability that any given number of Circumstances may happen in any number of Packs, will easily be found by our Method : thus if the number of Packs be k, the Probability that one Card or more of the same Suit and Name in every one of the Packs may be in the same position, will be expressed as follows,

$$\frac{1}{n^{k-2}} - \frac{1}{2\,\{n\,(n-1)\}^{k-2}} + \frac{1}{\underline{|3}\,\{n\,(n-1)\,(n-2)\}^{k-2}}$$

$$- \frac{1}{\underline{|4}\,\{n\,(n-1)\,(n-2)\,(n-3)\}^{k-2}}\ \&\text{c.}$$

Laplace demonstrates this result; see *Théorie ... des Prob.* page 224.

290. Problems XXXVII. and XXXVIII. relate to the game of Bowls ; see Arts. 177, 250.

De Moivre says, page 120,

Having given formerly the Solution of this Problem, proposed to me by the Honourable *Francis Robartes*, Esq;, in the *Philosophical Transactions* Number 329; I there said, by way of Corollary, that if the proportion of Skill in the Gamesters were given, the Problem might also be solved: since which time M. *de Monmort*, in the second Edition of a Book by him published upon the Subject of Chance, has solved this Problem as it is extended to the consideration of the Skill, and to carry his Solution to a great number of Cases, giving also a Method whereby it might be carried farther : But altho' his Solution is good. as he has made a right use of the Doctrine of Combinations, yet I think mine has a greater degree of Simplicity, it being deduced from the original Principle whereby I have demonstrated the Doctrine of Permutations and Combinations :...

291. Problems XXXIX. to XLII. form a connected set. Problem XXXIX. is as follows :

To find the Expectation of *A*, when with a Die of any given number of Faces, he undertakes to fling any number of them in any given number of Casts.

Let $p + 1$ be the number of faces on the die, n the number of casts, f the number of faces which A undertakes to fling. Then A's expectation is

$$\frac{1}{(p+1)^n} \left\{ (p+1)^n - \frac{f}{1} p^n + \frac{f(f-1)}{1 \cdot 2} (p-1)^n \right.$$

$$\left. - \frac{f(f-1)(f-2)}{\underline{3}} (p-2)^n + \dots \right\}.$$

De Moivre infers this general result from the examination of the simple cases in which f is equal to 1, 2, 3, 4 respectively.

He says in his Preface respecting this problem,

When I began for the first time to attempt its Solution, I had nothing else to guide me but the common Rules of Combinations, such as they had been delivered by Dr. *Wallis* and others; which when I endeavoured to apply, I was surprized to find that my Calculation swelled by degrees to an intolerable Bulk: For this reason I was forced to turn my Views another way, and to try whether the Solution I was seeking for might not be deduced from some easier considerations; whereupon I happily fell upon the Method I have been mentioning, which as it led me to a very great Simplicity in the Solution, so I look upon it to be an Improvement made to the Method of Combinations.

The problem has attracted much attention; we shall find it discussed by the following writers: Mallet, *Acta Helvetica*, 1772; Euler, *Opuscula Analytica*, Vol. II. 1785; Laplace, *Mémoires... par divers Savans*, 1774, *Théorie... des Prob.* page 191; Trembley, *Mémoires de l'Acad... Berlin*, 1794, 1795.

We shall recur to the problem when we are giving an account of Euler's writings on our subject.

292.   Problem XL. is as follows:

To find in how many Trials it will be probable that A with a Die of any given number of Faces shall throw any proposed number of them.

Take the formula given in Art. 291, suppose it equal to $\frac{1}{2}$, and seek for the value of n. There is no method for solving this equation exactly, so De Moivre adopts an approximation. He supposes that $p + 1$, p, $p - 1$, $p - 2$, ... are in Geometrical

Progression, which supposition he says "will very little err from
the truth, especially if the proportion of p to 1, be not very small."

Put r for $\dfrac{p+1}{p}$; thus the equation becomes

$$1-\frac{f}{1}\frac{1}{r^n}+\frac{f(f-1)}{1.2}\frac{1}{r^{2n}}-\frac{f(f-1)(f-2)}{\lfloor 3}\frac{1}{r^{3n}}+\ldots=\frac{1}{2};$$

that is
$$\left(1-\frac{1}{r^n}\right)^{\!f}=\frac{1}{2}.$$

Hence
$$\frac{1}{r^n}=1-\left(\tfrac{1}{2}\right)^{\frac{1}{f}},$$

and then n may be found by logarithms.

De Moivre says in his Preface respecting this problem,

The 40th Problem is the reverse of the preceding; It contains a
very remarkable Method of Solution, the Artifice of which consists
in changing an Arithmetic Progression of Numbers into a Geometric
one; this being always to be done when the Numbers are large, and
their Intervals small. I freely acknowledge that I have been indebted
long ago for this useful Idea, to my much respected Friend, That Excellent Mathematician Dr. *Halley*, Secretary to the *Royal Society*,
whom I have seen practise the thing on another occasion: For this
and other Instructive Notions readily imparted to me, during an un-
interrupted Friendship of five and Twenty years, I return him my
very hearty Thanks.

Laplace also notices this method of approximation in solving
the problem, and he compares its result with that furnished by his
own method; see *Théorie ... des Prob.* pages 198—200.

293. Problem XLI. is as follows:

Supposing a regular Prism having a Faces marked I, b Faces
marked II, c Faces marked III, d Faces marked IV, &c. what is the
Probability that in a certain number of throws n, some of the Faces
marked I will be thrown, as also some of the Faces marked II?

This is an extension of Problem XXXIX; it was not in the first
edition of the *Doctrine of Chances*.

Let $a+b+c+d+\ldots=s$; then the Probability required
will be

$$\frac{1}{s^n}\left[s^n-\{(s-a)^n+(s-b)^n\}+(s-a-b)^n\right].$$

11

If it be required that some of the Faces marked i, some of the Faces marked ii, and some of the Faces marked iii be thrown, the Probability required will be

$$\frac{1}{s^n}\left[s^n - \left\{(s-a)^n + (s-b)^n + (s-c)^n\right\} \right.$$
$$+ (s-a-b)^n + (s-b-c)^n + (s-c-a)^n$$
$$\left. - (s-a-b-c)^n \right].$$

And so on if other Faces are required to be thrown.

De Moivre intimates that these results follow easily from the method adopted in Problem XXXIX.

294. Problem XLII. first appeared in the second edition; it is not important.

Problem XLIII. is as follows:

Any number of Chances being given, to find the Probability of their being produced in a given order, without any limitation of the number of times in which they are to be produced.

It may be remarked that, for an approximation, De Moivre proposes to replace several numbers representing chances by a common mean value; it is however not easy to believe that the result would be very trustworthy. This problem was not in the first edition.

295. Problems XLIV. and XLV. relate to what we have called Waldegrave's Problem; see Art. 211.

In De Moivre's first edition, the problem occupies pages 77—102. De Moivre says in his preface that he had received the solution by Nicolas Bernoulli before his own was published; and that both solutions were printed in the *Philosophical Transactions*, No. 341. De Moivre's solution consists of a very full and clear discussion of the problem when there are three players, and also when there are four players; and he gives a little aid to the solution of the general problem. The last page is devoted to an explanation of a method of solving the problem which Brook Taylor communicated to De Moivre.

In De Moivre's third edition the problem occupies pages 132—159. The matter given in the first edition is here reproduced, omitting,

however, some details which the reader might be expected to fill
up for himself, and also the method of Brook Taylor. On the
other hand, the last nine pages of the discussion in the third
edition were not in the first edition; these consist of explanations
and investigations with the view of enabling a reader to determine
numerical results for any number of players, supposing that at
any stage it is required to stop the play and divide the money
deposited equitably. This part of the problem is peculiar to
De Moivre.

The discussions which De Moivre gives of the particular
cases of three players and four players are very easy and satisfactory; but as a general solution his method seems inferior to
that of Nicolas Bernoulli. We may remark that the investigation
for three players given by De Moivre will enable the student to
discover how Montmort obtained the results which he gives without demonstration for three players; see Art. 209. De Moivre
determines a player's expectation by finding first the advantage
resulting from his chance of winning the whole sum deposited, and
then his disadvantage arising from the contributions which he
may have had to make himself to the whole sum deposited; the
expectation is obtained by subtracting the second result from the
first. Montmort determined the expectation by finding, first the
advantage of the player arising from his chance of winning the
deposits of the other two players, and then the disadvantage
arising from the chance which the other two players have of
winning his deposits; the expectation is obtained by subtracting
the second result from the first.

The problem will come before us again as solved by Laplace.

296. Problem XLVI. is on the game of *Hazard*; there is no
description of the game here, but there is one given by Montmort
on his page 177; and from this description, De Moivre's solution
can be understood: his results agree with Montmort's. Problem XLVII. is also on Hazard; it relates to a point in the game
which is not noticed by Montmort, and it is only from De Moivre's
investigation itself that we can discover what the problem is,
which he is considering. With respect to this problem, De Moivre
says, page 165,

11—2

After I had solved the foregoing Problem, which is about 12 years
ago, I spoke of my Solution to Mr. *Henry Stuart Stevens*, but without communicating to him the manner of it: As he is a Gentleman
who, besides other uncommon Qualifications, has a particular Sagacity
in reducing intricate Questions to simple ones, he brought me, a few
days after, his Investigation of the Conclusion set down in my third
Corollary; and as I have had occasion to cite him before, in another
Work, so I here renew with pleasure the Expression of the Esteem
which I have for his extraordinary Talents:

Then follows the investigation due to Stevens. The above
passage occurs for the first time in the second edition, page 140;
the name however is there spelt Stephens: see also Art. 270.

Problem XLVII. is not in the first edition; on the other hand,
a table of numerical values of chances at Hazard, without accompanying explanations, is given on pages 174, 175 of the first
edition, which is not reproduced in the other editions.

297. Problems XLVIII. and XLIX. relate to the game of *Raffling*.
If three dice are thrown, some throws will present triplets, some
doublets, and some neither triplets nor doublets; in the game
of *Raffles* only those throws count which present triplets or
doublets. The game was discussed by Montmort in his
pages 207—212; but he is not so elaborate as De Moivre. Both
writers give a numerical table of chances, which De Moivre says was
drawn up by Francis Robartes, twenty years before the publication of Montmort's work; see *Miscellanea Analytica*, page 224.

Problem XLIX. was not in De Moivre's first edition, and
Problem XLVIII. was not so fully treated as in the other editions.

298. Problem L. is entitled *Of Whisk;* it occupies pages 172—179.
This is the game now called *Whist.* De Moivre determines the
chances of various distributions of the *Honours* in the game. Thus,
for example, he says that the probability that there are no Honours
on either side is $\frac{650}{1666}$; this of course means that the Honours
are equally divided. The result would be obtained by considering
two cases, namely, 1st, that in which the card turned up is an

Honour, and 2nd, that in which the card turned up is not an Honour. Thus we should have for the required probability

$$\frac{4}{13} \cdot \frac{3}{1} \cdot \frac{25.26.25}{51.50.49} + \frac{9}{13} \cdot \frac{4.3}{1.2} \cdot \frac{25.24.26.25}{51.50.49.48};$$

and this will be found equal to $\frac{650}{1666}$.

De Moivre has two Corollaries, which form the chief part of his investigation respecting Whist.

The first begins thus:

From what we have said, it will not be difficult to solve this Case at Whisk; *viz.* which side has the best, of those who have VIII of the Game, or of those who at the same time have IX?

In order to which it will be necessary to premise the following Principle.

1° That there is but 1 Chance in 8192 to get VII. by Triks.

2° That there are 13 Chances in 8192 to get VI.

3° That there are 78 Chances in 8192 to get V.

4° That there are 286 Chances in 8192 to get IV.

5° That there are 715 Chances in 8192 to get III.

6° That there are 1287 Chances in 8192 to get II.

7° That there are 1716 Chances in 8192 to get I.

All this will appear evident to those who can raise the Binomial $a + b$ to its thirteenth power.

But it must carefully be observed that the foregoing Chances express the Probability of getting so many Points by Triks, and neither more nor less.

De Moivre states his conclusion thus:

From whence it follows that without considering whether the VIII are Dealers or Eldest, there is one time with another the Odds of somewhat less than 7 to 5; and very nearly that of 25 to 18.

The second Corollary contains tables of the number of chances for any assigned number of Trumps in any hand. De Moivre says,

By the help of these Tables several useful Questions may be resolved; as 1°. If it is asked, what is the Probability that the Dealer has precisely III Trumps, besides the Trump Card? The Answer, by *Tab.* I. is $\frac{4662}{15875}$; ...

In the first edition there was only a brief notice of Whist, occupying scarcely more than a page.

299. Problems LI. to LV. are on *Piquet*. The game is not described, but there is no difficulty in understanding the problems, which are easy examples of combinations. The following *Remark* occurs on page 186 ; it was not in the first edition :

It may easily be perceived from the Solution of the preceding Problem, that the number of variations which there are in twelve Cards make it next to impossible to calculate some of the Probabilities relating to Piquet, such as that which results from the priority of Hand, or the Probabilities of a Pic, Repic or Lurch; however notwithstanding that difficulty, one may from observations often repeated, nearly estimate what those Probabilities are in themselves, as will be proved in its place when we come to treat of the reasonable conjectures which may be deduced from Experiments; for which reason I shall set down some Observations of a Gentleman who has a very great degree of Skill and Experience in that Game, after which I shall make an application of them.

The discussion of Piquet was briefer in the first than in the following editions.

300. We will give the enunciation of Problem LVI. and the beginning of the solution.

Problem LVI. Of Saving Clauses.

A has 2 Chances to beat *B*, and *B* has 1 chance to beat *A* ; but there is one Chance which intitles them both to withdraw their own Stake, which we suppose equal to *s* ; to find the Gain of *A*.

Solution.

This Question tho' easy in itself, yet is brought in to caution Beginners against a Mistake which they might commit by imagining that the Case, which intitles each Man to recover his own Stake, needs not be regarded, and that it is the same thing as if it did not exist. This I mention so much more readily, that some people who have pretended great skill in these Speculations of Chance have themselves fallen into that error.

This problem was not in the first edition. The gain of A is $\frac{1}{4} s$.

301. Problem LVII, which was not in the first edition, is as follows:

A and B playing together deposit £s apiece; A has 2 Chances to win s, and B 1 Chance to win s, whereupon A tells B that he will play with him upon an equality of Chance, if he B will set him $2s$ to $1s$, to which B assents: to find whether A has any advantage or disadvantage by that Bargain.

In the first case A's expectation is $\frac{1}{3} s$, and in the second, it is $\frac{1}{2} s$; so that he gains $\frac{1}{6} s$ by the bargain.

302. We now arrive at one of the most important parts of De Moivre's work, namely, that which relates to the Duration of Play; we will first give a full account of what is contained in the third edition of the *Doctrine of Chances,* and afterwards state how much of this was added to the investigations originally published in the *De Mensura Sortis.*

De Moivre himself regarded his labours on this subject with the satisfaction which they justly merited; he says in his Preface,

When I first began to attempt the general Solution of the Problem concerning the Duration of Play, there was nothing extant that could give me any light into that Subject; for altho' Mr *de Monmort*, in the first Edition of his Book, gives the Solution of this Problem, as limited to three Stakes to be won or lost, and farther limited by the Supposition of an Equality of Skill between the Adventurers; yet he having given no Demonstration of his Solution, and the Demonstration when discovered being of very little use towards obtaining the general Solution of the Problem, I was forced to try what my own Enquiry would lead me to, which having been attended with Success, the result of what I found was afterwards published in my *Specimen* before mentioned.

The *Specimen* is the Essay *De Mensura Sortis.*

303. The general problem relating to the Duration of Play may be thus enunciated: suppose A to have m counters, and B to have n counters; let their chances of winning in a single game be as a to b; the loser in each game is to give a counter to his adversary: required the probability that *when or before a certain number of games has been played,* one of the players will have won all the counters of his adversary. It will be seen that the words in italics constitute the advance which this problem makes beyond the more simple one discussed in Art. 107.

De Moivre's Problems LVIII. and LIX. amount to solving the problem of the Duration of Play for the case in which m and n are *equal.*

After discussing some cases in which $n = 2$ or 3, De Moivre lays down a General Rule, thus:

A General Rule for determining what Probability there is that the Play shall not be determined in a given number of Games.

Let n be the number of Pieces of each Gamester. Let also $n + d$ be the number of Games given; raise $a + b$ to the Power n, then cut off the two extream Terms, and multiply the remainder by $aa + 2ab + bb$: then cut off again the two Extreams, and multiply again the remainder by $aa + 2ab + bb$, still rejecting the two Extreams; and so on, making as many Multiplications as there are Units in $\frac{1}{2}d$; make the last Product the Numerator of a Fraction whose Denominator let be $(a+b)^{n+d}$, and that Fraction will express the Probability required,......; still observing that if d be an odd number, you write $d-1$ in its room.

For an example, De Moivre supposes $n = 4$, $d = 6$.

Raise $a + b$ to the fourth power, and reject the extremes; thus we have $4a^3b + 6a^2b^2 + 4ab^3$.

Multiply by $a^2 + 2ab + b^2$, and reject the extremes; thus we have $14a^4b^2 + 20a^3b^3 + 14a^2b^4$.

Multiply by $a^2 + 2ab + b^2$, and reject the extremes; thus we have $48a^5b^3 + 68a^4b^4 + 48a^3b^5$.

Multiply by $a^2 + 2ab + b^2$, and reject the extremes; thus we have $164a^6b^4 + 232a^5b^5 + 164a^4b^6$.

Thus the probability that the Play will not be ended in 10 games is

$$\frac{164\,a^6b^4 + 232\,a^5b^5 + 164\,a^4b^6}{(a+b)^{10}}.$$

De Moivre leaves his readers to convince themselves of the accuracy of his rule; and this is not difficult.

De Moivre suggests that the work of multiplication may be abbreviated by omitting the a and b, and restoring them at the end; this is what we now call the *method of detached coefficients*.

304. The terms which are *rejected* in the process of the preceding Article will furnish an expression for the probability that the play *will* be ended in an assigned number of games. Thus if $n = 4$ and $d = 6$, this probability will be found to be

$$\frac{a^4 + b^4}{(a+b)^4} + \frac{4a^5b + 4ab^5}{(a+b)^6} + \frac{14a^6b^2 + 14a^2b^6}{(a+b)^8} + \frac{48a^7b^3 + 48a^3b^7}{(a+b)^{10}},$$

that is, $\dfrac{a^4 + b^4}{(a+b)^4}\left\{ 1 + \dfrac{4ab}{(a+b)^2} + \dfrac{14a^2b^2}{(a+b)^4} + \dfrac{48a^3b^3}{(a+b)^6} \right\}.$

Now here we arrive at one of De Moivre's important results; he gives, *without demonstration*, general formulæ for determining those numerical coefficients which in the above example have the values 4, 14, 48. De Moivre's formulæ amount to two laws, one connecting each coefficient with its predecessors, and one giving the value of each coefficient separately. We can make the laws most intelligible by demonstrating them. We start from a result given by Laplace. He shews, *Théorie ... des Prob.*, page 229, that the chance of A for winning precisely at the $(n + 2x)^{\text{th}}$ game is the coefficient of t^{n+2x} in the expansion of

$$\frac{a^n t^n}{\left\{ \dfrac{1 + \sqrt{(1 - 4abt^2)}}{2} \right\}^n + \left\{ \dfrac{1 - \sqrt{(1 - 4abt^2)}}{2} \right\}^n},$$

where it is supposed that $a + b = 1$.

Now the denominator of the above expression is known to be equal to

$$1 - nc + \frac{n(n-3)}{1 \cdot 2}c^2 - \frac{n(n-4)(n-5)}{\lfloor 3}c^3 + \dots$$

where $c = abt^2$; see *Differential Calculus*, Chapter IX.

We can thus obtain by the ordinary doctrine of Series, a linear relation between the coefficient of t^{n+2x} and the coefficients of the preceding powers of t, namely, t^{n+2x-2}, t^{n+2x-4}, ... This is De Moivre's first law; see his page 198.

Again; we may write the above fraction in the form

$$\frac{a^n t^n}{N^n (1 + c^n N^{-2n})},$$

where

$$N = \frac{1 + \sqrt{(1 - 4abt^2)}}{2};$$

and then by expanding, we obtain

$$a^n t^n \left\{ N^{-n} - (abt^2)^n N^{-3n} + (abt^2)^{2n} N^{-5n} - \dots \right\}.$$

The coefficient of t^{2x} in N^{-n} is known to be

$$a^x b^x \frac{n \, (n + x + 1) \, (n + x + 2) \dots (n + 2x - 1)}{\lfloor x};$$

see *Differential Calculus*, Chapter IX.

Similarly we get the coefficient of t^{2x-2n} in N^{-3n}, of t^{2x-4n} in N^{-5n}, and so on.

Thus we obtain the coefficient of t^{n+2x} in the expansion of the original expression.

This is De Moivre's second law; see his page 199.

305. De Moivre's Problems LX. LXI. LXII. are simple examples formed on Problems LVIII. and LIX. They are thus enunciated:

LX. Supposing A and B to play together till such time as four Stakes are won or lost on either side; what must be their proportion of Skill, otherwise what must be their proportion of Chances for winning any one Game assigned, to make it as probable that the Play will be ended in four Games as not?

LXI. Supposing that A and B play till such time as four Stakes are won or lost: What must be their proportion of Skill to make it a Wager of three to one, that the Play will be ended in four Games?

LXII. Supposing that A and B play till such time as four Stakes are won or lost; What must be their proportion of Skill to make it an equal Wager that the Play will be ended in six Games?

306. Problems LXIII. and LXIV. amount to the general enunciation we have given in Art. 303; so that the restriction that
m and n are equal which was imposed in Problems LVIII. and
LIX. is now removed. As before De Moivre states, *without de-*
monstration, two general laws, which we will now give.

Laplace shews, *Théorie...des Prob.* page 228, that the chance
of A for winning precisely at the $(n + 2x)^{\text{th}}$ game is the coefficient
of t^{n+2x} in the expansion of

$$a^n t^n \frac{\left\{ \dfrac{1 + \sqrt{(1 - 4c)}}{2} \right\}^m - \left\{ \dfrac{1 - \sqrt{(1 - 4c)}}{2} \right\}^m}{\left\{ \dfrac{1 + \sqrt{(1 - 4c)}}{2} \right\}^{m+n} - \left\{ \dfrac{1 - \sqrt{(1 - 4c)}}{2} \right\}^{m+n}}.$$

Let $\dfrac{\sqrt{(1 - 4c)}}{2}$ be denoted by h; then the fractional expression
which multiplies $a^n t^n$ becomes by expansion, and striking out $2h$
from numerator and denominator,

$$\frac{\left(\dfrac{1}{2}\right)^{m-1} + \dfrac{m(m-1)(m-2)}{\underline{3}} \left(\dfrac{1}{2}\right)^{m-3} h^2 + \dfrac{m(m-1)(m-2)(m-3)(m-4)}{\underline{5}} \left(\dfrac{1}{2}\right)^{m-5} h^4 + \ldots}{(m+n)\left(\dfrac{1}{2}\right)^{m+n-1} + \dfrac{(m+n)(m+n-1)(m+n-2)}{\underline{3}} \left(\dfrac{1}{2}\right)^{m+n-3} h^2 + \ldots},$$

We have to arrange the denominator according to powers of
t, and to shew that it is equal to

$$1 - labt^2 + \frac{(l-1)(l-2)}{1 \cdot 2}(abt^2)^2 - \frac{(l-2)(l-3)(l-4)}{\underline{3}}(abt^2)^3 + \ldots$$

where $l = m + n - 2$.

Now, as in Art. 304, we have

$$\left\{ \frac{1 + \sqrt{(1 - 4c)}}{2} \right\}^r + \left\{ \frac{1 - \sqrt{(1 - 4c)}}{2} \right\}^r,$$

$$= 1 - rc + \frac{r(r-3)}{1 \cdot 2}c^2 - \frac{r(r-4)(r-5)}{\underline{3}}c^3 + \ldots;$$

and the left-hand member is equivalent to

$$2\left\{ \left(\frac{1}{2}\right)^r + \frac{r(r-1)}{1 \cdot 2}\left(\frac{1}{2}\right)^{r-2} h^2 + \frac{r(r-1)(r-2)(r-3)}{\underline{4}}\left(\frac{1}{2}\right)^{r-4} h^4 + \ldots \right\}.$$

Differentiate both sides with respect to t observing that $\dfrac{hdh}{dt} = -abt$. Thus,

$$2\left\{ \frac{r(r-1)}{1}\left(\frac{1}{2}\right)^{r-2} + \frac{r(r-1)(r-2)(r-3)}{\underline{3}}\left(\frac{1}{2}\right)^{r-4} h^2 + \ldots \right\},$$

$$= 2\left\{ r - \frac{r(r-3)}{1}abt^2 + \frac{r(r-4)(r-5)}{1\cdot 2}(abt^2)^2 - \ldots \right\}.$$

Now put $r = l + 3$; and we obtain the required result.

Thus a linear relation can be obtained between the coefficients of successive powers of t.

This is De Moivre's first law; see his page 205.

Again; let $N = \dfrac{1 + \sqrt{(1-4c)}}{2}$; then the original expression becomes

$$\frac{a^n t^n N^m (1 - c^m N^{-2m})}{N^{m+n}(1 - c^{m+n}N^{-2m-2n})}$$

$$= a^n t^n N^{-n}(1 - c^m N^{-2m})(1 - c^{m+n}N^{-2m-2n})^{-1}.$$

We may now proceed as in the latter part of Art. 304, to determine the coefficient of t^{n+2x}.

The result will coincide with De Moivre's second law; see his page 207.

307. Problem LXV. is a particular case of the problem of Duration of Play; m is now supposed infinite: in other words A has *unlimited capital* and we require his chance of ruining B in an assigned number of games.

De Moivre solves this problem in two ways. We will here give his first solution with the first of the two examples which accompany it.

SOLUTION.

Supposing n to be the number of Stakes which A is to win of B, and $n+d$ the number of Games; let $a+b$ be raised to the Power whose Index is $n+d$; then if d be an odd number, take so many Terms of that Power as there are Units in $\dfrac{d+1}{2}$; take also so many of the Terms next following as have been taken already, but prefix to them in an inverted order, the Coefficients of the preceding Terms. But if d be an even number, take so many Terms of the said Power as there are Units in $\frac{1}{2}d+1$; then take as many of the Terms next following

as there are Units in $\frac{1}{2}d$, and prefix to them in an inverted order the Coefficients of the preceding Terms, omitting the last of them; and those Terms taken all together will compose the Numerator of a Fraction expressing the Probability required, the Denominator of which Fraction ought to be $(a+b)^{n+d}$.

EXAMPLE I.

Supposing the number of Stakes, which A is to win, to be *Three*, and the given number of Games to be *Ten;* let $a+b$ be raised to the tenth power, viz. $a^{10}+10a^9b+45a^8bb+120a^7b^3+210a^6b^4+252a^5b^5 +210a^4b^6+120a^3b^7+45aab^8+10ab^9+b^{10}$. Then, by reason that $n=3$, and $n+d=10$, it follows that d is $=7$, and $\frac{d+1}{2}=4$. Wherefore let the Four first Terms of the said Power be taken, viz. $a^{10}+10a^9b +45a^8bb+120a^7b^3$, and let the four Terms next following be taken likewise without regard to their Coefficients, then prefix to them in an inverted order, the Coefficients of the preceding Terms: thus the four Terms following with their new Coefficients will be $120a^6b^4+45a^5b^5 +10a^4b^6+1a^3b^7$. Then the Probability which A has of winning three Stakes of B in ten Games or sooner, will be expressed by the following Fraction

$$\frac{a^{10}+10a^9b+45a^8bb+120a^7b^3+120a^6b^4+45a^5b^5+10a^4b^6+a^3b^7}{(a+b)^{10}},$$

which in the Case of an Equality of Skill between A and B will be reduced to $\frac{352}{1024}$ or $\frac{11}{32}$.

308. In De Moivre's solution there is no difficulty in seeing the origin of his *first* set of terms, but that of the *second* set of terms is not so immediately obvious. We will take his example, and account for the last four terms.

The last term is a^3b^7. There is only *one* way in which B's capital may be exhausted while A wins only three games; namely, A must win the first three games.

The next term is $10a^4b^6$. There are *ten* ways in which B's capital may be exhausted while A wins only four games. For let there be ten places; put b in any one of the first three places, and fill up the remaining places with the letters *aaaabbbbb* in this order ; or put *a* in any one of the last seven places, and fill up the remaining places with the letters *aaabbbbbb* in this order ; we thus obtain the ten admissible cases.

The next term is $45a^5b^5$. There are *forty-five* ways in which *B*'s capital may be exhausted while *A* wins only five games. For let there be ten places. Take any two of the first three places and put *b* in each, and fill up the remaining places with the letters *aaaaabb* in this order. Or take any two of the last seven places and put *a* in each, and fill up the remaining places with the letters *aaabbbbb* in this order. Or put *b* in any one of the first three places and *a* in any one of the last seven ; and fill up the remaining places with the letters *aaaabbbb* in this order. On the whole we shall obtain a number equal to the number of combinations of 10 things taken 2 at a time. The following is the general result : suppose we have to arrange *r* letters *a* and *s* letters *b*, so that in each arrangement there shall be *n* more of the letters *a* than of the letters *b* before we have gone through the arrangement ; then if *r* is less than $s+n$ the number of different arrangements is the same as the number of combinations of $r+s$ things taken $r-n$ at a time. For example, let $r=6$, $s=4$, $n=3$; then the number of different arrangements is $\dfrac{10 \times 9 \times 8}{1 \times 2 \times 3}$, that is 120.

The result which we have here noticed was obtained by Montmort, but in a very unsatisfactory manner : see Art. 182.

De Moivre's first solution of his Problem LXV. is based on the same principles as Montmort's solution of the general problem of the Duration of Play.

309. De Moivre's second solution of his Problem LXV. consists of a formula which he gives without demonstration. Let us return to the expression in Art. 306, and suppose *m* infinite. Then the chance of *A* for winning precisely at the $(n+2x)^{th}$ game is the coefficient of t^{n+2x} in the expansion of

$$\frac{a^n t^n}{\left\{ \dfrac{1 + \sqrt{(1-4c)}}{2} \right\}^n},$$

that is $\quad a^n \dfrac{n\,(n+x+1)\,(n+x+2)\,\ldots\ldots\,(n+2x-1)}{\lfloor x} \, a^x b^x \,;$

see Art. 304.

The chance of A for winning *at or before* the $(n+2x)^{th}$ game is therefore

$$a^n \left\{ 1 + nab + \frac{n\,(n+3)}{1\,.\,2}\,a^2b^2 + \ldots \right.$$

$$\left. + \frac{n\,(n+x+1)\,(n+x+2)\,\ldots\,(n+2x-1)}{\lfloor x}\,a^x b^x \right\}.$$

Laplace, *Théorie...des Prob.*, page 235.

310. De Moivre says with respect to his Problem LXV,

In the first attempt that I had ever made towards solving the general Problem of the Duration of Play, which was in the year 1708, I began with the Solution of this LXVth Problem, well knowing that it might be a Foundation for what I farther wanted, since which time, by a due repetition of it, I solved the main Problem : but as I found afterwards a nearer way to it, I barely published in my first Essay on those matters, what seemed to me most simple and elegant, still preserving this Problem by me in order to be published when I should think it proper.

De Moivre goes on to speak of the investigations of Montmort and Nicolas Bernoulli, in words which we have already quoted ; see Art. 181.

311. Dr L. Oettinger on pages 187, 188 of his work entitled *Die Wahrscheinlichkeits-Rechnung, Berlin*, 1852, objects to some of the results which are obtained by De Moivre and Laplace.

Dr Oettinger seems to intimate that in the formula, which we have given at the end of Art. 309, Laplace has omitted to lay down the condition that A has an unlimited capital ; but Laplace has distinctly introduced this condition on his page 234.

Again, speaking of De Moivre's solution of his Problem LXIV. Dr Oettinger says, Er erhält das nämliche unhaltbare Resultat, welches *Laplace* nach ihm aufstellte.

But there is no foundation for this remark ; De Moivre and

Laplace are correct. The misapprehension may have arisen from
reading only a part of De Moivre's page 205, and so assuming a
law of a series to hold universally, which he distinctly says breaks
off after a certain number of terms.

The just reputation of Dr Oettinger renders it necessary for me
to notice his criticisms, and to record my dissent from them.

312. De Moivre's Problems LXVI. and LXVII. are easy deductions from his preceding results ; they are thus enunciated :

LXVI. To find what Probability there is that in a given number
of Games A may be a winner of a certain number q of Stakes, and at
some other time B may likewise be winner of the number p of Stakes,
so that both circumstances may happen.

LXVII. To find what Probability there is, that in a given number
of Games A may win the number q of Stakes; with this farther condition, that B during that whole number of Games may never have
been winner of the number p of Stakes.

313. De Moivre now proceeds to express his results relating
to the Duration of Play in another form. He says, page 215,

The Rules hitherto given for the Solution of Problems relating to
the Duration of Play are easily practicable, if the number of Games
given is but small; but if that number is large, the work will be very
tedious, and sometimes swell to that degree as to be in some manner
impracticable : to remedy which inconveniency, I shall here give an
Extract of a paper by me produced before the Royal Society, wherein
was contained a Method of solving very expeditiously the chief Problems relating to that matter, by the help of a Table of Sines, of which
I had before given a hint in the first Edition of my *Doctrine of Chances*,
pag. 149, and 150.

The paper produced before the Royal Society does not appear
to have been published in the *Philosophical Transactions;* probably we have the substance of it in the *Doctrine of Chances*.

De Moivre proceeds according to the announcement in the
above extract, to express his results relating to the Duration of
Play by the help of Trigonometrical Tables; in Problem LXVIII. he
supposes the players to have equal skill, and in Problem LXIX. he
supposes them to have unequal skill.

The demonstrations of the formulæ are to be found in the *Miscellanea Analytica*, pages 76—83, and in the *Doctrine of Chances*, pages 230—234. De Moivre supposes the players to start with the *same* number of counters; but he says on page 83 of the *Miscellanea Analytica*, that solutions similar but somewhat more complex could be given for the case in which the original numbers of counters were *different*. This has been effected by Laplace in his discussion of the whole problem.

314. De Moivre's own demonstrations depend on his doctrine of Recurring Series; by this doctrine De Moivre could effect what we should now call the integration of a linear equation in Finite Differences: the equation in this case is that furnished by the first of the two laws which we have explained in Arts. 304, 306. Certain trigonometrical formulæ are also required; see *Miscellanea Analytica*, page 78. One of these, De Moivre says, constat ex Æquationibus ad circulum vulgo notis; the following is the property: in elementary works on Trigonometry we have an expansion of cos $n\theta$ in descending powers of cos θ; now cos $n\theta$ vanishes when $n\theta$ is any odd multiple of $\dfrac{\pi}{2}$, and therefore the equivalent expansion must also vanish. The other formulæ which De Moivre uses are in fact deductions from the general theorem which is called *De Moivre's property of the Circle;* they are as follows; let $\alpha = \dfrac{\pi}{2n}$, then we have

$$1 = 2^{n-1} \sin \alpha \sin 3\alpha \sin 5\alpha \dots \sin (2n\alpha - \alpha) \, ;$$

also if n be even we have

$$\cos n\phi = 2^{n-1} \{\sin^2 \alpha - \sin^2 \phi\} \{\sin^2 3\alpha - \sin^2 \phi\} \dots$$

$$\dots \{\sin^2 (n-3) \alpha - \sin^2 \phi\} \{\sin^2 (n-1) \alpha - \sin^2 \phi\} :$$

see *Plane Trigonometry*, Chap. XXIII.

De Moivre uses the first of these formulæ; and also a formula which may be deduced from the second by differentiating with respect to ϕ, and after differentiation putting ϕ equal to α, or 3α, or 5α, ...

315. De Moivre applies his results respecting the Duration

of Play to test the value of an approximation proposed by Montmort; we have already referred to this point in Art. 184.

316. It remains to trace the history of De Moivre's investigations on this subject.

The memoir *De Mensura Sortis* contains the following Problems out of those which appear in the *Doctrine of Chances*, LVIII, LX, LXII, LXIII, the first solution of LXV, LXVI. The first edition of the *Doctrine of Chances* contains all that the third does, except the Problems LXVIII. and LXIX; these were added in the second edition. As we proceed with our history we shall find that the subject engaged the attention of Lagrange and Laplace, the latter of whom has embodied the researches of his predecessors in the *Théorie...des Prob.* pages 225—238.

317. With one slight exception noticed in Art. 322, the remainder of the *Doctrine of Chances* was not in the first edition but was added in the second edition.

318. The pages 220—229 of the *Doctrine of Chances*, form a digression on a subject, which is one of De Moivre's most valuable contributions to mathematics, namely that of *Recurring Series*. He says, page 220,

The Reader may have perceived that the Solution of several Problems relating to Chance depends upon the Summation of Series; I have, as occasion has offered, given the Method of summing them up; but as there are others that may occur, I think it necessary to give a summary View of what is most requisite to be known in this matter; desiring the Reader to excuse me, if I do not give the Demonstrations, which would swell this Tract too much; especially considering that I have already given them in my *Miscellanea Analytica*.

319. These pages of the *Doctrine of Chances* will not present any difficulty to a student who is acquainted with the subject of Recurring Series, as it is now explained in works on Algebra; De Moivre however gives some propositions which are not usually reproduced in the present day.

320. One theorem may be noticed which is enunciated by De Moivre, on his page 224, and also on page 167 of the *Miscellanea Analytica*.

The general term of the expansion of $(1-r)^{-p}$ in powers of r is $\dfrac{p\,(p+1)\,\ldots\,(p+n-1)}{\lfloor n}\,r^n$; the sum of the first n terms of the expansion is equivalent to the following expression

$$\frac{1-r^n-nr^n\,(1-r)-\dfrac{n\,(n+1)}{1\,.\,2}\,r^n\,(1-r)^2-\ldots-\dfrac{\lfloor n+p-2}{\lfloor n-1\,\lfloor p-1}\,r^n(1-r)^{p-1}}{(1-r)^p}\,.$$

This may be easily shewn to be true when $n=1$, and then, by induction, it may be shewn to be generally true. For

$$r^{n+1}=r^n\left\{1-(1-r)\right\},$$

so that

$$r^{n+1}+(n+1)\,r^{n+1}\,(1-r)+\frac{(n+1)\,(n+2)}{1\,.\,2}\,r^{n+1}\,(1-r)^2+\ldots$$

$$=r^n\left\{1-(1-r)\right\}+(n+1)\,r^n\,(1-r)\left\{1-(1-r)\right\}$$

$$+\frac{(n+1)\,(n+2)}{1\,.\,2}\,r^n\,(1-r)^2\left\{1-(1-r)\right\}+\ldots$$

$$=r^n+nr^n\,(1-r)+\frac{n\,(n+1)}{1\,.\,2}\,r^n(1-r)^2+\ldots+\frac{\lfloor n+p-2}{\lfloor n-1\,\lfloor p-1}\,r^n\,(1-r)^{p-1}$$

$$-\frac{\lfloor n+1+p-2}{\lfloor n\,\lfloor p-1}\,r^n\,(1-r)^p.$$

Thus the additional term obtained by changing n into $n+1$ is $\dfrac{\lfloor n+p-1}{\lfloor n\,\lfloor p-1}\,r^n$ as it should be; so that if De Moivre's theorem is true for any value of n, it is true when n is changed into $n+1$.

321. Another theorem may be noticed; it is enunciated by De Moivre on his page 229. Having given the scales of relation of two Recurring Series, it is required to find the scale of relation of the Series arising from the product of corresponding terms.

For example, let $u_n r^n$ be the general term in the expansion according to powers of r of a proper Algebraical fraction of which the denominator is $1-fr+gr^2$; and let $v_n a^n$ be the general term in the expansion according to powers of a of a proper Algebraical fraction of which the denominator is $1 - ma + pa^2$. We have to find the scale of relation of the Series of which the general term is $u_n v_n (ra)^n$.

We know by the ordinary theory of decomposing Recurring Series into Geometrical Progressions that

$$u_n r^n \times v_n a^n = r^n a^n (R_1 \rho_1{}^n + R_2 \rho_2{}^n)(A_1 \alpha_1{}^n + A_2 \alpha_2{}^n),$$

where ρ_1 and ρ_2 are the reciprocals of the roots of the equation

$$1 - fr + gr^2 = 0,$$

and α_1 and α_2 are the reciprocals of the roots of the equation

$$1 - ma + pa^2 = 0 ;$$

and R_1, R_2, A_1, A_2 are certain constants.

Thus $u_n v_n = R_1 A_1 (\rho_1 \alpha_1)^n + R_1 A_2 (\rho_1 \alpha_2)^n$
$$+ R_2 A_1 (\rho_2 \alpha_1)^n + R_2 A_2 (\rho_2 \alpha_2)^n ;$$

this shews that the required scale of relation will involve four terms besides unity. The four quantities $\rho_1 \alpha_1$, $\rho_1 \alpha_2$, $\rho_2 \alpha_1$, $\rho_2 \alpha_2$ will be the reciprocals of the roots of the equation in z which is found by eliminating r and a from

$$1 - fr + gr^2 = 0, \quad 1 - ma + pa^2 = 0, \quad ra = z ;$$

this equation therefore is

$$1 - fmz + (pf^2 + gm^2 - 2gp) z^2 - fgmpz^3 + g^2 p^2 z^4 = 0.$$

Thus we have determined the required scale of relation ; for the denominator of the fraction which by expansion produces $u_n v_n (ra)^n$ as its general term will be

$$1 - fmra + (pf^2 + gm^2 - 2gp) r^2 a^2 - fgmpr^3 a^3 + g^2 p^2 r^4 a^4.$$

De Moivre adds, page 229,

But it is very observable, that if one of the differential Scales be the Binomial $1 - a$ raised to any Power, it will be sufficient to raise the other differential Scale to that Power, only substituting ar for r, or leaving the Powers of r as they are, if a be restrained to Unity; and that Power of the other differential Scale will constitute the differential Scale required.

This is very easily demonstrated. For suppose that one scale of relation is $(1-a)^t$; then by forming the product of the corresponding terms of the two Recurring Series, we obtain for the general term

$$\frac{\lfloor t+n-1}{\lfloor t-1 \rfloor \lfloor n} a^n \left\{ R_1\rho_1{}^n + R_2\rho_2{}^n + R_3\rho_3{}^n + \ldots \right\}$$

This shews that the general term will be the coefficient of r^n in the expansion of

$$\frac{R_1}{(1-ra\rho_1)^t} + \frac{R_2}{(1-ra\rho_2)^t} + \frac{R_3}{(1-ra\rho_3)^t} + \ldots;$$

and by bringing these fractions to a common denominator, we obtain De Moivre's result.

322. De Moivre applies his theory of Recurring Series to demonstrate his results relating to the Duration of Play, as we have already intimated in Art. 313; and to illustrate still further the use of the theory he takes two other problems respecting play. These problems are thus enunciated :

LXX. M and N, whose proportion of Chances to win one Game are respectively as a to b, resolve to play together till one or the other has lost 4 Stakes: two Standers by, R and S, concern themselves in the Play, R takes the side of M, and S of N, and agree betwixt them, that R shall set to S, the sum L to the sum G on the first Game, $2L$ to $2G$ on the second, $3L$ to $3G$ on the third, $4L$ to $4G$ on the fourth, and in case the Play be not then concluded, $5L$ to $5G$ on the fifth, and so increasing perpetually in Arithmetic Progression the Sums which they are to set to one another, as long as M and N play; yet with this farther condition, that the Sums, set down by them R and S, shall at the end of each Game be taken up by the Winner, and not left upon the Table to be taken up at once upon the Conclusion of the Play: it is demanded how the Gain of R is to be estimated before the Play begins.

LXXI. If M and N, whose number of Chances to win one Game are respectively as a to b, play together till four Stakes are won or lost on either side; and that at the same time, R and S whose number of Chances to win one Game are respectively as c to d, play also together till five Stakes are won or lost on either side; what is the Probability that the Play between M and N will be ended in fewer Games, than the Play between R and S.

The particular case of 'Problem LXXI. in which $a = b$, and $c = d$, was given in the first edition of the *Doctrine of Chances*, page 152.

323. Problems LXXII. and LXXIII. are important; it will be sufficient to enunciate the latter.

A and *B* playing together, and having a different number of Chances to win one Game, which number of Chances I suppose to be respectively as a to b, engage themselves to a Spectator *S*, that after a certain number of Games is over, *A* shall give him as many Pieces as he wins Games, over and above $\dfrac{a}{a+b} n$, and *B* as many as he wins Games, over and above the number $\dfrac{b}{a+b} n$; to find the Expectation of *S*.

Problem LXXII. is a particular case of Problem LXXIII. obtained by supposing a and b to be equal.

These two problems first appeared in the *Miscellanea Analytica*, pages 99—101. We there find the following notice respecting Problem LXXII:

Cum aliquando labente Anno 1721, Vir Clarissimus *Alex. Cuming* Eq. Au. Regiæ Societatis Socius, quæstionem infra subjectam mihi proposuisset, solutionem problematis ei postero die tradideram.

After giving the solution De Moivre proceeds to Problem LXXIII. which he thus introduces :.

Eodem procedendi modo, solutum fuerat Problema sequens ab eodem Cl. viro etiam propositum, ejusdem generis ac superius sed multo latius patens.

We will give a solution of Problem LXXIII; De Moivre in the *Doctrine of Chances* merely states the result.

Let $n = c\,(a + b)$; consider the expectation of *S* so far as it depends on *A*. The chance that *A* will win all the games is $\dfrac{a^n}{(a+b)^n}$, and in this case he gives cb to *S*. The chance that *A* will win $n-1$ games is $\dfrac{na^{n-1}b}{(a+b)^n}$, and in this case he gives $cb - 1$ to *S*. And so on.

Thus we have to sum the series

$$a^n bc + na^{n-1}b \,(bc - 1) + \frac{n\,(n-1)}{1 \cdot 2}\, a^{n-2}b^2\,(bc - 2) + \dots \,,$$

the series extending so long as the terms in brackets are positive.

We have

$$a^n bc - na^{n-1}b = a^{n-1}b\,(ac - n) = -\, a^{n-1}b\,bc \,;$$

thus the first *two* terms amount to

$$(n - 1)\, a^{n-1}b\,bc.$$

Now combine this with $-\dfrac{n\,(n-1)}{1 \cdot 2}\, a^{n-2}b^2 2$; we get

$$(n - 1)\, a^{n-2}b^2\,(ac - n),\ \text{that is } -(n - 1)\, a^{n-2}b^2\,bc \,;$$

thus the first *three* terms amount to

$$\frac{(n - 1)\,(n - 2)}{1 \cdot 2}\, a^{n-2}b^2 bc.$$

This process may be carried on for any number of terms; and we shall thus obtain for the sum of bc terms

$$\frac{(n - 1)\,(n - 2)\,\dots\,(n - bc + 1)}{\lfloor bc - 1}\, a^{n-bc+1}b^{bc-1}bc.$$

This may be expressed thus

$$\frac{\lfloor n}{n\,\lfloor bc\,\lfloor ac}\, a^{ac}b^{bc}ac\,bc,$$

which is equivalent to De Moivre's result. The expectation of S from B will be found to be the same as it is from A.

324. When the chances of A and B for winning a single game are in the proportion of a to b we know, from Bernoulli's theorem, that there is a high probability that in a large number of trials the number of games won by A and B respectively will be nearly in the ratio of a to b. Accordingly De Moivre passes naturally from his Problem LXXIII. to investigations which in fact amount to what we have called the inverse use of Bernoulli's theorem; see Art. 125. De Moivre says,

...I'll take the liberty to say, thát this is the hardest Problem that can be proposed on the Subject of Chance, for which reason I have reserved it for the last, but I hope to be forgiven if my Solution is not fitted to the capacity of all Readers; however I shall derive from it some Conclusions that may be of use to every body: in order thereto, I shall here translate a Paper of mine which was printed *November* 12, 1733, and communicated to some Friends, but never yet made public, reserving to myself the right of enlarging my own Thoughts, as occasion shall require.

Then follows a section entitled *A Method of approximating the Sum of the Terms of the Binomial* $(a+b)^n$ *expanded into a Series, from whence are deduced some practical Rules to estimate the Degree of Assent which is to be given to Experiments.* This section occupies pages 243—254 of the *Doctrine of Chances;* we shall find it convenient to postpone our notice of it until we examine the *Miscellanea Analytica.*

325.   De Moivre's Problem LXXIV. is thus enunciated:

To find the Probability of throwing a Chance assigned a given number of times without intermission, in any given number of Trials.

It was introduced in the second edition, page 243, in the following terms:

When I was just concluding this Work, the following Problem was mentioned to me as very difficult, for which reason I have considered it with a particular attention.

De Moivre does not demonstrate his results for this problem; we will solve the problem in the modern way.

Let a denote the chance for the event in a single trial, b the chance against it; let n be the number of trials, p the number of times without intermission for which the event is required to happen. We shall speak of this as a *run of p*.

Let u_n denote the probability of having the required run of p in n trials; then

$$u_{n+1} = u_n + (1 - u_{n-p})\, ba^p:$$

for in $n+1$ trials we have *all* the favourable cases which we have in n trials, and some more, namely those in which after having failed in $n-p$ trials, we fail in the $(n-p+1)^{th}$ trial, and then have a run of p.

Let $u_n = 1 - v_n$, and substitute in the equation; thus

$$v_{n+1} = v_n - ba^p v_{n-p}.$$

The Generating Function of v_n will therefore be

$$\frac{\phi(t)}{1 - t + ba^p t^{p+1}},$$

where $\phi(t)$ is an arbitrary function of t which involves no powers of t higher than t^p.

The Generating Function of u_n is therefore

$$\frac{1}{1-t} - \frac{\phi(t)}{1 - t + ba^p t^{p+1}};$$

we may denote this by

$$\frac{\psi(t)}{(1-t)(1 - t + ba^p t^{p+1})},$$

where $\psi(t)$ is an arbitrary function of t which involves no powers of t higher than t^{p+1}. Now it is obvious that $u_n = 0$ if n be less than p, also $u_p = a^p$, and $u_{p+1} = a^p + ba^p$.

Hence we find that

$$\psi(t) = a^p t^p (1 - at),$$

so that the Generating Function of u_n is

$$\frac{a^p t^p (1 - at)}{(1-t)(1 - t + ba^p t^{p+1})}.$$

The coefficient of t^n in the expansion of this function will therefore be obtained by expanding

$$\frac{a^p (1 - at)}{1 - t + ba^p t^{p+1}},$$

and taking the coefficients of all the powers of t up to that of t^{n-p} inclusive.

It may be shewn that De Moivre's result agrees with this after allowing for a slight mistake. He says we must divide unity by $1 - x - ax^2 - a^2 x^3 - \ldots - a^{p-1} x^p$, take $n - p + 1$ terms of the series, multiply by $\frac{a^p}{(a+b)^p}$, and finally put $x = \frac{b}{a+b}$. The mistake here is that in the series $1 - x - ax^2 - a^2 x^3 - \ldots - a^{p-1} x^p$ instead of a we ought to read $\dfrac{a}{b}$. De Moivre is correct in an example which he gives on his page 255. Let $\dfrac{a}{b} = c$, then according to De Moivre's rule corrected we have to expand

$$\frac{1}{1 - x \dfrac{1 - c^p x^p}{1 - cx}} \frac{a^p}{(a+b)^p}, \quad \text{that is} \quad \frac{1 - cx}{1 - x(1+c) + c^p x^{p+1}} \frac{a^p}{(a+b)^p}.$$

This will be seen to agree with our result remembering that we took $a + b = 1$.

De Moivre himself on his page 256 practically gives this form to his result by putting

$$1 - x \frac{1 - c^p a^p}{1 - cx} \quad \text{for} \quad 1 - x - cx^2 - c^2 x^3 - \ldots - c^{p-1} x^p.$$

De Moivre gives without demonstration on his page 259 an approximate rule for determining the number of trials which must be made in order to render the chance of a run of p equal to one half.

De Moivre's Problem LXXIV. has been extended by Condorcet, *Essai...de l'Analyse...* pages 73—86, and by Laplace, *Théorie...des Prob.* pages 247—253.

326.   De Moivre's pages 261—328 are devoted to *Annuities on Lives;* an Appendix finishes the book, occupying pages 329—348 : this also relates principally to annuities, but it contains a few notes on the subject of Probability. As we have already stated in Art. 53, we do not profess to give an account of the investigations relating to mortality and life insurance.

We may remark that there is an Italian translation of De Moivre's treatise on Annuities, with notes and additions ; the title is *La Dottrina degli Azzardi...de Abramo Moivre: Trasportata dall' Idioma Inglese,...dal Padre Don Roberto Gaeta...sotto l'assistenza del Padre Don Gregorio Fontana...In Milano* 1776. This translation does not discuss the general Theory of Probability, but only annuities on lives and similar subjects.

In the *Advertisement* to the second edition of the *Doctrine of Chances*, page XIII, De Moivre says,

There is in the World a Gentleman of an older Date, who in the year 1726 did assure the Public that he could calculate the Values of Lives if he would, but that he would not,...

De Moivre proceeds to make some sarcastic remarks; a manuscript note in my copy says that the person here meant was "John Smart of Guildhall, who in that year published Tables of Interest, Discount, Annuities, &c. 4to."

327. We have now to notice De Moivre's work entitled *Miscellanea Analytica de Seriebus et Quadraturis*...London, 1730.

This is a quarto volume containing 250 pages, a page of Errata, a Supplement of 22 pages, and two additional pages of Errata; besides the title page, dedication, preface, index, and list of subscribers to the work.

We have already had occasion to refer to the *Miscellanea Analytica* as supplying matter bearing on our subject; we now however proceed to examine a section of the work which is entirely devoted to controversy between Montmort and De Moivre. This section is entitled *Responsio ad quasdam Criminationes;* it occupies pages 146—229, and is divided into seven Chapters.

328. In the first Chapter the design of the section is explained. De Moivre relates the history of the publication of Montmort's first edition, of the memoir *De Mensura Sortis*, and of Montmort's second edition. De Moivre sent a copy of the *De Mensura Sortis* to Montmort, who gave his opinion of the memoir in a letter to Nicolas Bernoulli, which was published in the second edition of Montmort's book; see Art. 221. De Moivre states briefly the animadversions of Montmort, distributing them under nine heads.

The publication of Montmort's second edition however does not seem to have produced any quarrel between him and De Moivre; the latter returned his thanks for the present of a copy of the work, and after this a frequent interchange of letters took place between the two mathematicians. In 1715 Montmort visited England, and was introduced to Newton and other distinguished men; he was also admitted as a member of the Royal Society. De Moivre sent to Montmort a copy of the *Doctrine of Chances* when it was published, and about two years afterwards Montmort died.

De Moivre quotes the words of Fontenelle which we have already given in Art. 136, and intimates that these words induced him to undertake a comparison between his own labours and those of Montmort, in order to vindicate his own claims. As the *Doctrine of Chances* was written in English it was not readily accessible to all who would take an interest in the dispute; and this led De Moivre to devote a section to the subject in his *Miscellanea Analytica*.

329. The second Chapter of the *Responsio...* is entitled *De Methodo Differentiarum, in qua exhibetur Solutio Stirlingiana de media Coefficiente Binomii.* The general object is to shew that in the summation of series De Moivre had no need for any of Montmort's investigations. De Moivre begins by referring to a certain theorem which we have noticed in Art. 152; he gives some examples of the use of this theorem. He also adverts to other methods of summation.

Montmort had arrived at a very general result in the summation of series. Suppose $u^n r^n$ to denote the nth term of a series, where u_n is such that $\Delta^m u_n$ is zero, m being any positive integer; then Montmort had succeeded in summing any assigned number of terms of the series. De Moivre shews that the result can be easily obtained by the method of Differences, that is by the method which we have explained in Art. 151.

The investigations by Montmort on the summation of series to which De Moivre refers were published in Vol. XXX. of the *Philosophical Transactions*, 1717.

This Chapter of the *Responsio...* gives some interesting details respecting Stirling's Theorem including a letter from Stirling himself.

330. The third Chapter of the *Responsio...* is entitled *De Methodo Combinationum;* the fourth *De Permutationibus;* the fifth *Combinationes et Permutationes ulterius consideratæ:* these Chap-

ters consist substantially of translations of portions of the *Doctrine of Chances*, and so do not call for any remark. The sixth Chapter is entitled *De Numero Punctorum in Tesseris;* it relates entirely to the formula of which we have given the history in Art. 149.

331. The seventh Chapter of the *Responsio...* is entitled *Solutiones variorum Problematum ad Sortem spectantium.* This Chapter gives the solutions of nine problems in Chances. The first eight of these are in the *Doctrine of Chances;* nothing of importance is added in the *Miscellanea Analytica,* except in two cases. The first of these additions is of some historical interest. Suppose we take an example of the Binomial Theorem, as $(p+q)^8$; one term will be $28p^6q^2$: then De Moivre says, page 218,

...at fortasse nesciveram hujus termini coefficientem, nimirum 28, designaturam numerum permutationum quas literæ $p, p, p, p, p, p, q, q,$ productum $p^6 q^2$ constituentes subire possint; imuò vero, hoc jam diu mihi erat exploratum, etenim ego fortasse primus omnium detexi coefficientes annexas productis Binomii, vel Multinomii cujuscunque, id denotare quotenis variationibus literæ producti positiones suas inter se permutent: sed utrum illud facile fuerit ad inveniendum, postquam lex coefficientium ex productis continuis $\frac{n}{1} \times \frac{n-1}{2} \times \frac{n-2}{3} \times \frac{n-3}{4}$ &c. jam perspecta esset, aut quisquam ante me hoc ipsum detexerit, ad rem præsentem non magni interest, cum id monere suffecerit hanc proprietatem Coefficientium a me assertam fuisse et demonstratam in *Actis Philosophicis Anno* 1697 impressis.

The second addition relates to Problem XLIX. of the *Doctrine of Chances;* some easy details relating to a maximum value are not given there which may be found in the *Miscellanea Analytica,* pages 223, 224.

332. The ninth problem in the seventh Chapter of the *Responsio...* is to find the ratio of the sum of the largest p terms in the expansion of $(1+1)^n$ to the sum of all the terms; p being an odd number and n an even number. De Moivre expresses this ratio in terms of the chances of certain events, for which chances he had already obtained formulæ. This mode of expressing the ratio is not given in the *Doctrine of Chances,* being rendered unnecessary by the application of Stirling's Theorem;

but it involves an interesting fact in approximation, and we will therefore explain it.

Suppose two players A and B of equal skill; let A have an infinite number of counters, and B have the number p. Let $\phi(n, p)$ denote the chance that B will be ruined in n games. Then the required ratio is $1 - \phi(n, p)$; this follows from the first form of solution of Problem LXV; see Art. 307. Again, suppose that each of the players starts with p counters; and let $\psi(n, p)$ then denote the chance that B will be ruined in n games; similarly if each starts with $3p$ counters let $\psi(n, 3p)$ denote the chance that B will be ruined in n games; and so on. Then De Moivre says that approximately

$$\phi(n, p) = \psi(n, p) + \psi(n, 3p),$$

and still more approximately

$$\phi(n, p) = \psi(n, p) + \psi(n, 3p) - \psi(n, 5p) + \psi(n, 7p).$$

The closeness of the approximation will depend on n being large, and p being only a moderate fraction of n.

These results follow from the formulæ given on pages 199 and 210 of the *Doctrine of Chances*... The second term of $\psi(n, p)$ is negative, and is numerically equal to the first term of $\psi(n, 3p)$, and so is cancelled; similarly the third term of $\psi(n, p)$ is cancelled by the first of $-\psi(n, 5p)$, and the fourth term of $\psi(n, p)$ by the first of $\psi(n, 7p)$. The terms which do not mutually cancel, and which we therefore neglect, involve fewer factors than that which we retain, and are thus comparatively small.

333. We now proceed to notice the *Supplement* to the *Miscellanea Analytica*. The investigations of problems in Chances had led mathematicians to consider the approximate calculation of the coefficients in the Binomial Theorem; and as we shall now see, the consequence was the discovery of one of the most striking results in mathematics. The *Supplement* commences thus:

Aliquot post diebus quam Liber qui inscribitur, *Miscellanea Analytica*, in lucem prodiisset, Doctissimus *Stirlingius* me literis admonuit Tabulam ibi a me exhibitam de summis Logarithmorum, non satis autoritatis habere ad ea firmanda quæ in speculatione niterentur, utpote cui Tabulæ subesset error perpetuus in quinta quaque figura decimali summarum: quæ cum pro humanitate sua monuisset, his subjunxit seriem celerrime convergentem, cujus ope summæ logarithmorum tot numerorum naturalium quot quis sumere voluerit obtineri possent; res autem sic exposita fuerat.

Then follows a Theorem which is not quite coincident in form with what we now usually call Stirling's Theorem, but is practically equivalent to it. De Moivre gives his own investigation of the subject, and arrives at the following result:

$$\log 2 + \log 3 + \log 4 + \ldots + \log (m - 1)$$

$$= \left(m - \frac{1}{2}\right) \log m - m + \frac{1}{12m} - \frac{1}{360m^3} + \frac{1}{1260m^5} - \frac{1}{1680m^7} + \ldots$$

$$+ 1 - \frac{1}{12} + \frac{1}{360} - \frac{1}{1260} + \frac{1}{1680} \ldots$$

With respect to the series in the last line, De Moivre says on page 9, of the Supplement to the *Miscellanea Analytica* ... quæ satis commode convergit in principio, post terminos quinque primos convergentiam amittit, quam tamen postea recuperat... The last four words involve an error, for the series is *divergent*, as we know from the nature of Bernoulli's Numbers. But De Moivre by using a result which Stirling had already obtained, arrived at the conclusion that the series $1 - \frac{1}{12} + \frac{1}{360} - \frac{1}{1260} + \ldots$ is equal to $\frac{1}{2} \log 2\pi$; and thus the theorem is deduced which we now call Stirling's Theorem. See *Miscellanea Analytica*, page 170, *Supplement*, page 10.

334. De Moivre proceeds in the *Supplement* to the *Miscellanea Analytica* to obtain an approximate value of the middle coefficient of a Binomial expansion, that is of the expression

$$\frac{(m+1)\,(m+2)\ldots 2m}{m\,(m-1)\ldots 1}.$$

He expends nearly two pages in arriving at the result, which he might have obtained immediately by putting the proposed expression in the equivalent form $\dfrac{\lfloor 2m}{\lfloor m \; \lfloor m}$.

De Moivre then gives the general theorem for the approximate summation of the series

$$\frac{1}{n^e} + \frac{1}{(n+1)^e} + \frac{1}{(n+2)^e} + \frac{1}{(n+3)^e} + \dots ;$$

We have already noticed his use of a particular case of this summation in Art. 276.

De Moivre does not demonstrate the theorem; it is of course included in the wellknown result to which Euler's name is usually attached,

$$\Sigma u_x = \int u_x \, dx - \frac{1}{2} u_x + \frac{1}{6} \cdot \frac{1}{2} \frac{du_x}{dx} - \frac{1}{30} \cdot \frac{1}{\lfloor 4} \frac{d^3 u_x}{dx^3} + \dots$$

See *Novi Comm....Petrop.* Vol. XIV. part 1, page 137; 1770.

The theorem however is also to be found in Maclaurin's *Treatise of Fluxions*, 1742, page 673.

335. We return to the *Doctrine of Chances*, to notice what is given in its pages 243—254; see Art. 324.

In these pages De Moivre begins by adverting to the theorem obtained by Stirling and himself. He deduces from this the following result: suppose n to be a very large number, then the logarithm of the ratio which a term of $\left(\frac{1}{2} + \frac{1}{2}\right)^n$, distant from the middle term by the interval l, bears to the middle term, is approximately $-\dfrac{2l^2}{n}$.

This enables him to obtain an approximate value of the sum of the l terms which immediately precede or follow the middle term. Hence he can estimate the numerical values of certain chances. For example, let $n = 3600$: then, supposing that it is an even chance for the happening or failing of an event in a single trial, De Moivre finds that the chance is ·682688 that in 3600 trials, the number of times in which the event happens, will lie between $1800 + 30$ and $1800 - 30$.

Thus by the aid of Stirling's Theorem the value of Bernoulli's Theorem is largely increased.

De Moivre adverts to the controversy between Nicolas Bernoulli and Dr Arbuthnot, respecting the inferences to be drawn from the observed fact of the nearly constant ratio of the number of births of boys to the number of births of girls; see Art. 223. De Moivre shews that Nicolas Bernoulli's remarks were not relevant to the argument really advanced by Dr Arbuthnot.

336. Thus we have seen that the principal contributions to our subject from De Moivre are his investigations respecting the Duration of Play, his Theory of Recurring Series, and his extension of the value of Bernoulli's Theorem by the aid of Stirling's Theorem. Our obligations to De Moivre would have been still greater if he had not concealed the demonstrations of the important results which we have noticed in Art. 306; but it will not be doubted that the Theory of Probability owes more to him than to any other mathematician, with the sole exception of Laplace.

CHAPTER X.

MISCELLANEOUS INVESTIGATIONS
BETWEEN THE YEARS 1700 AND 1750.

337.   THE present Chapter will contain notices of various contributions to our subject which were made between the years 1700 and 1750.

338.   The first work which claims our attention is the essay by Nicolas Bernoulli, to which we have already alluded in Art. 72 ; it is entitled *Specimina Artis conjectandi, ad quæstiones Juris applicatæ.*   This is stated to have been published at Basle in 1709; see *Gouraud,* page 36.

It is reprinted in the fourth volume of the *Act. Eruditorum...
Supplementa,* 1711, where it occupies pages 159—170.   Allusion is made to the essay in the volume which we have cited in Art. 59, pages 842, 844, 846.

339.   In this essay Nicolas Bernoulli professes to apply mathematical calculations to various questions, principally relating to the probability of human life.   He takes for a foundation some facts which his uncle James had deduced from the comparison of bills of mortality, namely that out of 100 infants born at the same time 64 are alive at the end of the sixth year, 40 at the end of the sixteenth year, and so on.   Nicolas Bernoulli considers the following questions : the time at the end of which an absent man of whom no tidings had been received might be considered as dead ; the

value of an annuity on a life ; the sum to be paid to assure to a child just born an assigned sum on his attaining a certain age ; marine assurances ; and a lottery problem. He also touches on the probability of testimony ; and on the probability of the innocence of an accused person.

The essay does not give occasion for the display of that mathematical power which its author possessed, and which we have seen was called forth in his correspondence with Montmort; but it indicates boldness, originality, and strong faith in the value and extent of the applications which might be made of the Theory of Probability.

We will take two examples from the Essay.

340. Suppose there are b men who will all die within a years, and are equally likely to die at any instant within this time : required the probable duration of the life of the last survivor. Nicolas Bernoulli really views the problem as equivalent to the following : A line of length a is measured from a fixed origin ; on this line b points are taken at random : determine the mean distance from the origin of the most distant point.

Let the line a be supposed divided into an indefinitely large number n of equal parts ; let each part be equal to c, so that $nc = a$.

Suppose that each of the b points may be at the distance c, or $2c$, or $3c$, ... up to nc ; but no two or more at exactly the same distance.

Then the whole number of cases will be the number of combinations of n things taken b at a time, say $\phi\,(n,\,b)$.

Suppose that the most distant point is at the distance xc ; then the number of ways in which this can happen is the number of ways in which the remaining $b-1$ points can be put nearer to the origin ; that is, the number of combinations of $x-1$ things, taken $b-1$ at a time, say $\phi\,(x-1,\,b-1)$.

Hence the required mean distance is

$$\frac{\Sigma\, xc\,\phi\,(x-1,\,b-1)}{\phi\,(n,\,b)},$$

where the summation extends from $x = b$ to $x = n$.

13—2

It is easily seen that the limit, when n is infinite, is $\dfrac{ncb}{b+1}$, that is $\dfrac{ab}{b+1}$.

The above is substantially the method of Nicolas Bernoulli.

341. Nicolas Bernoulli has a very curious mode of estimating the probability of innocence of an accused person. He assumes that any single evidence against the accused person is twice as likely to be false as true. Suppose we denote by u_n the probability of innocence when there are n different evidences against him; there are two chances out of three that the nth evidence is false, and then the accused prisoner is reduced to the state in which there are $n-1$ evidences against him; and there is one chance out of three that the evidence is true and his innocence therefore impossible. Thus

$$u_n = \frac{2u_{n-1} + 0}{3} = \frac{2}{3} u_{n-1}.$$

Hence $$u_n = \left(\frac{2}{3}\right)^n.$$

This is not the notation of Nicolas; but it is his method and result.

342. In the correspondence between Montmort and Nicolas Bernoulli allusion was made to a work by Barbeyrac, entitled *Traité du Jeu;* see Art. 212. I have not seen the book myself. It appears to be a dissertation to shew that religion and morality do not prohibit the use of games in general, or of games of chance in particular. It is stated that there are two editions of the work, published respectively in 1709 and 1744.

Barbeyrac is also said to have published a discourse *Sur la nature du Sort.*

See the *English Cyclopædia*, and the *Biographie Universelle*, under the head *Barbeyrac*.

343. We have next to notice a memoir by Arbuthnot to whom we have already assigned an elementary work on our subject; see Art. 79.

The memoir is entitled *An Argument for Divine Providence,*

taken from the constant Regularity observ'd in the Births of both Sexes. By Dr John Arbuthnott, *Physitian in Ordinary to Her Majesty, and Fellow of the College of Physitians and the Royal Society.*

This memoir is published in Vol. XXVII. of the *Philosophical Transactions;* it is the volume for 1710, 1711 and 1712: the memoir occupies pages 186—190.

344. The memoir begins thus :

Among innumerable Footsteps of Divine Providence to be found in the Works of Nature, there is a very remarkable one to be observed in the exact Ballance that is maintained, between the Numbers of Men and Women; for by this means it is provided, that the Species may never fail, nor perish, since every Male may have its Female, and of a proportionable Age. This Equality of Males and Females is not the Effect of Chance but Divine Providence, working for a good End, which I thus demonstrate:

345. The registers of births in London for 82 years are given; these shew that in every year more males were born than females· There is very little relating to the theory of probability in the memoir. The principal point is the following. Assume that it is an even chance whether a male or female be born; then the chance that in a given year there will be more males than females is $\frac{1}{2}$; and the chance that this will happen for 82 years in succession is $\frac{1}{2^{82}}$. This chance is so small that we may conclude that it is not an even chance whether a male or female be born.

346. The memoir attracted the attention of Nicolas Bernoulli, who in his correspondence with Montmort expressed his dissent from Arbuthnot's argument; see Art. 223. There is also a letter from Nicolas Bernoulli to Leibnitz on the subject; see page 989 of the work cited in Art. 59. De Moivre replied to Nicolas Bernoulli, as we have already intimated in Art. 335.

347. The subject is also discussed in the *Oeuvres Philosophiques et Mathématiques* of 's Gravesande, published at Amsterdam, 1774, 2 vols. 4to. The discussion occupies pages 221—248 of the second volume.

It appears from page 237, that when Nicolas Bernoulli travelled in Holland he met 'sGravesande.

In this discussion we have first a memoir by 'sGravesande. This memoir contains a brief statement of some of the elements of the theory of probability. The following result is then obtained. Assume that the chance is even for a male or female birth, and find the chance that out of 11429 births the males shall lie between 5745 and 6128. By a laborious arithmetical calculation this is found to be about $\frac{1}{4}$. Then the chance that this should happen for 82 years in succession will be $\frac{1}{4^{82}}$.

But in fact the event for which the chance is so small had happened in London. Hence it is inferred that it is not an even chance that a male or female should be born.

It appears that 'sGravesande wrote to Nicolas Bernoulli on the subject; the reply of Nicolas Bernoulli is given. This reply contains a proof of the famous theorem of James Bernoulli; the proof is substantially the same as that given by Nicolas Bernoulli to Montmort, and published by the latter in pages 389—393 of his book.

Then 'sGravesande wrote a letter giving a very clear account of his views, and, as his editor remarks, the letter seems to have impressed Nicolas Bernoulli, judging from the reply which the latter made.

Nicolas Bernoulli thus sums up the controversy:

Mr. *Arbuthnot* fait consister son argument en deux choses; 1º. en ce que, supposée une égalité de naissance entre les filles et les garçons, il y a peu de probabilité que le nombre des garçons et des filles se trouve dans des limites fort proches de l'égalité: 2º. qu'il y a peu de probabilité que le nombre des garçons surpassera un grand nombre de fois de suite le nombre des filles. C'est la première partie que je refute, et non pas la seconde.

But this does not fairly represent Arbuthnot's argument. Nicolas Bernoulli seems to have imagined, without any adequate reason, that the theorem known by his uncle's name was in some way contradicted by Arbuthnot.

348. Two memoirs on our subject are published in Vol.

XXIX. of the *Philosophical Transactions*, which is the volume for 1714, 1715, 1716 the memoirs occupy pages 133—158. They are entitled *Solutio Generalis Problematis* XV. *propositi à D. de Moivre, in tractatu de Mensura Sortis...Solutio generalis altera præcedentis Problematis, ope Combinationum et Serierum infinitarum....*

These memoirs relate to the problem which we have called *Waldegrave's;* see Art. 211.

The first memoir is by Nicolas Bernoulli; it gives substantially the same solution as he sent to Montmort, and which was printed in pages 381—387 of Montmort's work.

The second memoir is by De Moivre; it gives the solution which was reproduced in the *Doctrine of Chances.*

349. We have next to notice a work which appeared under the following title :

Christiani Hugenii Libellus de Ratiociniis in Ludo Aleæ. Or, the value of all chances in games of fortune; cards, dice, wagers, lotteries, &c. mathematically demonstrated. London : Printed by S. Keimer, for T. Woodward, near the Inner Temple-Gate in Fleet-street. 1714.

This is a translation of Huygens's treatise, by W. Browne. It is in small octavo size ; it contains a Dedication to Dr Richard Mead, an Advertisement to the Reader, and then 24 pages, which comprise the translation. The dedication commences thus :

. Honour'd Sir, When I consider the Subject of the following Papers, I can no more forbear dedicating them to Your Name, than I can refuse giving my assent to any one Proposition in these Sciences, which I have already seen clearly demonstrated. The Reason is plain, for as You have contributed the greatest Lustre and Glory to a very considerable part of the Mathematicks, by introducing them into their noblest Province, the Theory of Physick; the Publisher of any Truths of that Nature, who is desirous of seeing them come to their utmost Perfection, must of course beg Your Patronage and Application of them. By so prudent a Course as this, he may perhaps see those Propositions which it was his utmost Ambition to make capable only of directing Men in the Management of their Purses, and instructing them to what Chances and Hazards they might safely commit their Money ; turn'd some time or other to a much more glorious End, and made instrumental likewise towards the securing their Bodies from the Tricks of that too successful

Sharper, Death, and countermining the underhand Dealings of secret and overreaching Distempers.

In his Advertisement to the Reader, Browne refers to a translation of Huygens's treatise which had been made by Arbuthnot ; he also notices the labours of Montmort and De Moivre. He says further,

My Design in publishing this Edition, was to have made it as useful as possible, by an addition of a very large Appendix to it, containing a Solution of some of the most serviceable and intricate Problems I cou'd think of, and such as have not as yet, that I know of, met with a particular Consideration: But an Information I have within these few Days receiv'd, that M. Montmort's French Piece is just newly reprinted at Paris, with very considerable Additions, has made me put a Stop to the Appendix, till I can procure a Sight of what has been added anew, for fear some part of it may possibly have been honour'd with the Notice and Consideration of that ingenious Author.

I do not know whether this proposed Appendix ever appeared.

350. In the *Hist. de l'Acad.... Paris* for 1728, which was published in 1730, there is a notice respecting some results obtained by Mairan, *Sur le Jeu de Pair ou Non*. The notice occupies pages 53—57 of the volume ; it is not by Mairan himself.

Suppose a heap of counters ; a person takes a number of them at random, and asks another person to guess whether the number is *odd* or *even*. Mairan says that the number is more likely to be odd than even ; and he argues in the following way. Suppose the number in the heap to be an odd number, for example 7 ; then a person who takes from the heap may take 1, or 2, or 3, ... or 7 counters ; thus there are 7 cases, namely 4 in which he takes an odd number, and 3 in which he takes an even number. The advantage then is in favour of his having taken an odd number. If the number in the heap be an even number, then the person who takes from it is as likely to take an even number as an odd number. Thus on the whole Mairan concludes that the guess should be given for an odd number.

The modern view of this problem is different from Mairan's.

If the original heap contains n counters we should say that there are n ways of drawing one counter, $\frac{n(n-1)}{1.2}$ ways of drawing two counters, and so on. Mairan notices this view but condemns it.

Laplace treated this problem in the *Mémoires…par divers Savans…* Tome VI., Paris, 1774, and he arrives at the ordinary result, though not by the method of combinations; he refers to Mairan's result, and briefly records his dissent. The problem is solved by the method of combinations in the *Théorie…des Prob.* page 201.

In the article *Pair ou Non* of the original French *Encyclopédie*, which was published in 1765, Mairan's view is given; this article was repeated in the *Encyclopédie Méthodique*, in 1785, without any notice of Laplace's dissent.

351. On page 68 of the volume of the *Hist. de l'Acad.…Paris*, which contains Mairan's results, is the following paragraph:

M. L'Abbé Sauveur, fils de feu M. Sauveur Académicien, a fait voir une Méthode qu'il a trouvée pour déterminer au Jeu de Quadrille quelle est la probabilité de gagner sans prendre plusieurs Jeux différents, dont il a calculé une Table. On a trouvé que la matiére épineuse et délicate des Combinaisons étoit très-bien entendüe dans cet ouvrage.

352. We have next to notice a memoir by Nicole, entitled *Examen et Résolution de quelques questions sur les Jeux*.

This memoir is published in the volume for 1730 of the *Hist. de l'Acad.…Paris;* the date of publication is 1732 : the memoir occupies pages 45—56 of the part devoted to memoirs.

The problem discussed is really the Problem of Points ; the method is very laborious, and the memoir seems quite superfluous since the results had already been given in a simpler manner by Montmort and De Moivre.

One point may be noticed. Let a and b be proportional to the respective chances of A and B to win a single game; let them play for an *even* number of games, say for example 8, and let S be the sum which each stakes. Then A's advantage is

$$S\frac{a^8 + 8a^7b + 28a^6b^2 + 56a^5b^3 - 56a^3b^5 - 28a^2b^6 - 8ab^7 - b^8}{(a+b)^8}.$$

This supposes that if each wins four games, neither receives nor loses any thing. Now it is obvious that the numerator of the expression is divisible by $a + b$; thus we may simplify the expression to

$$S \frac{a^7 + 7a^6b + 21a^5b^2 + 35a^4b^3 - 35a^3b^4 - 21a^2b^5 - 7ab^6 - b^7}{(a + b)^7}.$$

This is precisely the expression we should have if the players had agreed to play *seven* games instead of eight. Nicole notices this circumstance, and is content with indicating that it is not unreasonable; we may shew without difficulty that the result is universally true. Suppose that when A and B agree to play $2n - 1$ games, p_1 is the chance that A beats B by just one game, p_2 the chance that A beats B by two or more games; and let q_1, q_2 be similar quantities with respect to B, then A's advantage is $S(p_1 + p_2 - q_1 - q_2)$. Now consider $2n$ games : A's chance of beating B by two or more games, is $p_2 + \dfrac{p_1 a}{a + b}$; B's chance of beating A by two or more games is $q_2 + \dfrac{q_1 b}{a + b}$. Hence A's advantage is

$$S\left(p_2 + \frac{p_1 a}{a + b} - q_2 - \frac{q_1 b}{a + b}\right).$$

Now we know that $\dfrac{p_1}{a} = \dfrac{q_1}{b} = \mu$ say; therefore

$$\frac{p_1 a - q_1 b}{a + b} = \frac{\mu (a^2 - b^2)}{a + b} = \mu (a - b) = p_1 - q_1.$$

Hence the advantage of A for $2n$ games is the same as for $2n - 1$ games.

353. In the same volume of the *Hist. de l'Acad....Paris*, on pages 331—344, there is another memoir by Nicole, entitled *Méthode pour déterminer le sort de tant de Joüeurs que l'on voudra, et l'avantage que les uns ont sur les autres, lorsqu'ils joüent à qui gagnera le plus de parties dans un nombre de parties déterminé.*

This is the Problem of Points in the case of any number of players, supposing that each player wants the same number of points. Nicole begins in a laborious way; but he sees that the chances of the players are represented by the terms in the expansion of a certain multinomial, and thus he is enabled to give a general rule. Suppose for example that there are three players, whose chances for a single game are a, b, c. Let them play a set of three games. Then the chance that A has of winning the whole stake is $a^3 + 3a^2(b + c)$; and similar expressions give the chances of B and C; there is also the chance $6abc$ that the three players should each win one game, and thus no one prevail over the others.

Similarly, if they play four games, A's chance of winning the whole stake is $a^4 + 4a^3(b + c) + 12a^2bc$; there is also the chance $6a^2b^2$ that A and B should share the stake between them to the exclusion of C; and so on.

But all that Nicole gives was already well known; see Montmort's page 353, and De Moivre's *Miscellanea Analytica*, page 210.

354. In the year 1733 Buffon communicated to the Academy of Sciences at Paris the solution of some problems in chances. See *Hist. de l'Acad....Paris* for 1733, pages 43—45, for a brief account of them. The solutions are given in Buffon's *Essai d'Arithmétique Morale*, and we shall notice them in speaking of that work.

355. We now return to the work entitled *Of the Laws of Chance*, the second part of which we left for examination until after an account had been given of De Moivre's works; see Arts. 78, 88.

According to the title page this second part is to be attributed to John Ham.

Although De Moivre is never named, I think the greater part of Ham's additions are taken from De Moivre.

Ham considers the game of Pharaon in his pages 53—73. This I think is all taken from De Moivre. Ham gives the same introductory problem as De Moivre; namely the problem which is XI. in De Moivre's first edition, and X. in his third edition.

In pages 74—94 we have some examples relating to the game of Ace of Hearts, or Fair Chance, and to Lotteries. Here we have frequent use made of De Moivre's results as to the number of trials in which it is an even chance that an event will happen once, or happen twice; see Art. 264.

356. There is however an addition given without demonstration, to De Moivre's results, which deserves notice.

De Moivre made the problem of finding the number of trials in which it is an even chance that an event will occur twice depend on the following equation:

$$\left(1 + \frac{1}{q}\right)^n = 2\,(1+z).$$

If we suppose q infinite this reduces to

$$z = \log 2 + \log (1 + z) \,;$$

from which De Moivre obtained $z = 1\cdot678$ approximately. But let us not suppose q infinite; put $\left(1 + \frac{1}{q}\right)^q = e^c$; so that our equation becomes

$$e^{cz} = 2\,(1+z).$$

Assume $z = 2 - y$, thus

$$e^{2c-cy} = 6 - 2y.$$

Assume $2c = \gamma + s$ where $e^\gamma = 6$.

Thus,

$$e^{s-cy} = 1 - \frac{1}{3}\, y.$$

Take the logarithms of both sides, then

$$s - cy = -\frac{1}{3}\, y - \frac{1}{18}\, y^2 - \frac{1}{81}\, y^3 - \dots ,$$

that is

$$ry - \frac{1}{18}\, y^2 - \frac{1}{81}\, y^3 - \dots = s \,;$$

where

$$r = c - \frac{1}{3}.$$

Hence by reversion of series we obtain

$$y = \frac{s}{r} + \frac{1}{18r}\left(\frac{s}{r}\right)^2 + \frac{1+2r}{162r^3}\left(\frac{s}{r}\right)^3 + \dots$$

This is Ham's formula, given as we have said without demonstration. Since we assumed

$$e^\gamma = 6,$$

we have $\gamma =$ Napierian log of $6 = 1\cdot791759$; thus

$$s = 2c - \gamma = 2c - 1\cdot791759.$$

Ham says that this series will determine the value of z in all cases when q is greater than $4\cdot1473$. This limit is doubtless obtained by making $2c - \gamma = 0$, which leads to $\left(1 + \dfrac{1}{q}\right)^q = \sqrt{6}$; and this can be solved by trial. But Ham seems to be unnecessarily scrupulous here; for if $2c$ be less than γ we shall still have $\dfrac{s}{r}$ *numerically less than unity*, so long as $\gamma - 2c$ is less than $c - \dfrac{1}{3}$, that is so long as c is greater than $\dfrac{\gamma}{3} + \dfrac{1}{9}$.

357. The work finishes with some statements of the numerical value of certain chances at Hazard and Backgammon.

358. We have next to notice a work entitled *Calcul du Jeu appellé par les François le trente-et-quarante, et que l'on nomme à Florence le trente-et-un.... Par Mr D. M.* Florence, 1739.

This is a volume in quarto. The title, notice to the reader, and preface occupy eight pages, and then the text follows on pages 1—90.

The game considered is the following : Take a common pack of cards, and reject the *eights*, the *nines*, and the *tens*, so that forty cards remain. Each of the picture cards counts for ten, and each of the other cards counts for its usual number.

The cards are turned up singly until the number formed by the sum of the values of the cards falls between 31 and 40, both inclusive. The problem is to determine the chances in favour of each of the numbers between 31 and 40 inclusive.

The problem is solved by examining all the cases which can occur, and counting up the number of ways. The operation is most laborious, and the work is perhaps the most conspicuous example of misdirected industry which the literature of Games of Chance can furnish.

The author seems to refer on page 80 to another work which I have not seen. He says, ...j'en ai déja fait la démonstration dans mon Calcul de la Loterie de Rome,...

It will be observed from our description of the game that it does not coincide with that which has been called in more recent times by the same name. See Poisson's memoir in Gergonne's *Annales de Mathématiques*, Vol. 16.

359. A treatise on the subject of Chances was published by the eminent Thomas Simpson, Professor of Mathematics at the Royal Military Academy, Woolwich. Simpson was born in 1710, and died in 1761; an account of his life and writings is prefixed to an edition of his *Select Exercises for Young Proficients in the Mathematicks*, by Charles Hutton.

Simpson's work is entitled *The Nature and Laws of Chance... The whole after a new, general, and conspicuous Manner, and illustrated with a great variety of Examples* ... 1740.

Simpson implies in his preface that his design was to produce an introduction to the subject less expensive and less abstruse than De Moivre's work; and in fact Simpson's work may be considered as an abridgement of De Moivre's. Simpson's problems are nearly all taken from De Moivre, and the mode of treatment is substantially the same. The very small amount of new matter which is contributed by a writer of such high power as Simpson shews how closely De Moivre had examined the subject so far as it was accessible to the mathematical resources of the period.

We will point out what we find new in Simpson. He divides his work into thirty Problems.

360. Simpson's Problem VI. is as follows :

There is a given Number of each of several sorts of Things, (of the same Shape and Size); as (a) of the first Sort, (b) of the second, &c. put promiscuously together; out of which a given Number (m) is to be taken, as it happens: To find the Probability that there shall come out precisely a given Number of each sort, as (p) of the first, (q) of the second, (r) of the third, &c.

The result in modern notation is a fraction of which the numerator is

$$\frac{\lfloor a}{\lfloor p \lfloor a-p} \times \frac{\lfloor b}{\lfloor q \lfloor b-q} \times \frac{\lfloor c}{\lfloor r \lfloor c-r} \times \ldots,$$

and the denominator is

$$\frac{\lfloor n}{\lfloor m \lfloor n-m},$$

where $\qquad n = a + b + c + \ldots$

This is apparently the problem which Simpson describes in his title page as "A new and comprehensive Problem of great Use in discovering the Advantage or Loss in Lotteries, Raffles, &c."

361. Simpson's Problem X. relates to the game of Bowls; see Art. 177. Simpson gives a Table containing results for the case of an indefinitely large number of players on each side, but he does not fully explain his Table; a better account of it will be found in Samuel Clark's *Laws of Chance*, pages 63—65.

362. Simpson's Problem XV. is to find in how many trials one may undertake to have an equal chance for an event to occur r times, its chance at a single trial being known. Simpson claims to have solved this problem "in a more general manner than hitherto;" but it does not seem to me that what he has added to De Moivre's result is of any importance. We will however give Simpson's addition. Suppose we require the event to happen r times, the chance for it in a single trial being $\frac{a}{a+b}$. Let $q = \frac{b}{a}$; and suppose that q is large. Then De Moivre shews that in order to have an even chance that the event shall occur r times we must make about $q\left(r - \frac{3}{10}\right)$ trials; see Art. 262. But if $q = 1$ the required number of trials is exactly $2r-1$. Simpson then proposes to take as a universal formula $q\left(r - \frac{3}{10}\right) + r - \frac{7}{10}$; this is accurate when $q = 1$, and extremely near the truth when q is large.

363. Simpson's Problem xx. is the same as De Moivre's Problem VII; it is an example of the Duration of Play: see Art. 107; Simpson's method is less artificial than that which De Moivre used, and in fact much resembles the modern method.

364. Simpson's Problem XXII. is that which we have explained in Art. 148; Simpson's method is very laborious compared with De Moivre's. Simpson however adds a useful Corollary.

By introducing or cancelling common factors we may put the result of Art. 148 in the following form:

$$\frac{(p-1)(p-2)\ldots(p-n+1)}{\lfloor n-1} - \frac{n}{1}\frac{(q-1)(q-2)\ldots(q-n+1)}{\lfloor n-1}$$
$$+ \frac{n(n-1)}{1.2}\frac{(r-1)(r-2)\ldots(r-n+1)}{\lfloor n-1} - \ldots,$$

where $q = p-f$, $r = p-2f$, $\ldots$; and the series is to continue so long as no negative factors appear.

Simpson's Corollary then assigns the chance that the sum of the numbers exhibited by the dice *shall not exceed p*. We must put successively 1, 2, 3, ... up to p for p in the preceding expression, and sum the results. This gives, by an elementary proposition respecting the summation of series, the following expression for the required chance:

$$\frac{p(p-1)\ldots(p-n+1)}{\lfloor n} - \frac{n}{1}\frac{q(q-1)\ldots(q-n+1)}{\lfloor n}$$
$$+ \frac{n(n-1)}{1.2}\frac{r(r-1)\ldots(r-n+1)}{\lfloor n} - \ldots,$$

where, as before, the series is to continue so long as no negative factor appears.

365. Simpson's Problem XXIV. is the same as De Moivre's LXXIV., namely respecting the chance of a run of p successes in n trials; see Art. 325. De Moivre gave the solution without a demonstration; Simpson gives an imperfect demonstration, for having proceeded some way he says that the "Law of Continuation is manifest."

We have shewn in effect that the solution is obtained by taking the coefficient of t^{n-p} in the expansion of

$$\frac{a^p (1 - at)}{(1-t) \{1 - t + ba^p t^{p+1}\}},$$

that is in the expansion of

$$\frac{a^p (1-at)}{(1-t)^2} \left\{ 1 - \frac{ba^p t^{p+1}}{1-t} + \left(\frac{ba^p t^{p+1}}{1-t}\right)^2 - \left(\frac{ba^p t^{p+1}}{1-t}\right)^3 + \ldots \right\}.$$

Now $\dfrac{1 - at}{(1-t)^2} = \dfrac{1}{1-t} + \dfrac{(1-a)t}{(1-t)^2} = \dfrac{1}{1-t} + \dfrac{bt}{(1-t)^2}.$

We can thus express the result as the sum of two series, which will be found to agree with the form given by Simpson.

366. Simpson's Problem XXV. is on the Duration of Play. Simpson says in his Preface respecting his Problems XXII. and XXV, that they "are two of the most intricate and remarkable in the Subject, and both solv'd by Methods entirely new." This seems quite incorrect so far as relates to Problem XXV. Simpson gives results without any demonstration; his Case I. and Case II. are taken from De Moivre, his Case III. is a particular example of his general statement which follows, and this general statement coincides with Montmort's solution; see *Montmort*, page 268, *Doctrine of Chances*, pages 193 and 211.

367. We will give the enunciation of Simpson's Problem XXVII, together with a remark which he makes relating to it in his Preface.

In a Parallelopipedon, whose Sides are to one another in the Ratio of a, b, c; To find at how many Throws any one may undertake that any given Plane, viz. ab, may arise.

The 27th is a Problem that was proposed to the Public some time ago in *Latin*, as a very difficult one, and has not (that I know of) been answered before.

We have seen the origin of this problem in Art. 87. Simpson supposes that a sphere is described round the parallelepiped, and that a radius of the sphere passes round the boundary of the given plane; he considers that the chance of the given plane being

14

uppermost in a single throw is equal to the ratio which the spherical surface bounded by the moving radius bears to the whole surface of the sphere. Thus the problem is reduced to finding the area of a certain portion of the surface of a sphere.

368. Simpson gives two examples of the Summation of Series on his pages 70—73, which he claims as new in method.

(1) Let $(a+x)^n$ be denoted by $A + Bx + Cx^2 + Dx^3 + ...$; required the sum of

$$\frac{A}{1.2...r} + \frac{Bx}{2.3...(r+1)} + \frac{Cx^2}{3.4...(r+2)} + ...$$

Integrate both sides of the identity, and determine the constant so that both sides may vanish when $x = 0$; thus

$$\frac{(a+x)^{n+1}}{n+1} - \frac{a^{n+1}}{n+1} = Ax + \frac{Bx^2}{2} + \frac{Cx^3}{3} + \frac{Dx^4}{4} + ...$$

Repeat the operation; thus

$$\frac{(a+x)^{n+2}}{(n+1)(n+2)} - \frac{a^{n+1}x}{n+1} - \frac{a^{n+2}}{(n+1)(n+2)}$$
$$= \frac{Ax^2}{1.2} + \frac{Bx^3}{2.3} + \frac{Cx^4}{3.4} + \frac{Dx^5}{4.5} + ...$$

Proceed thus for r operations, then divide both sides by x^r, and the required sum is obtained.

(2) Required the sum of $1^n + 2^n + 3^n + ... + x^n$.

Simpson's method is the same as had been already used by Nicolas Bernoulli, who ascribed it to his uncle John; see Art. 207.

369. Simpson's Problem XXIX. is as follows:

A and B, whose Chances for winning any assigned Game are in the proportion of a to b, agree to play until n stakes are won and lost, on Condition that A, at the Beginning of every Game shall set the Sum p to the Sum $p \times \frac{b}{a}$, so that they may play without Disadvantage on either Side; it is required to find the present Value of all the Winnings that may be betwixt them when the Play is ended.

The investigation presents no difficulty.

370. Simpson's Problem XXX. is as follows:

Two Gamesters, A and B, equally skilful, enter into Play together, and agree to continue the same till (n) Games are won and lost. 'Tis required to find the Probability that neither comes off a Winner of $r\sqrt{n}$ Stakes, and also the Probability that B is never a Winner of that Number of Stakes during the whole Time of the Play; r being a given, and n any very great, Number.

Simpson says in his Preface relating to his Problems XXIV. and XXX. that they

"are the same with the two new ones, added in the End of Mr De Moivre's last Edition, whose Demonstrations that learned Author was pleased to reserve to himself, and are here fully and clearly investigated...."

The same two problems are thus referred to in Simpson's title page:

Full and clear Investigations of two Problems, added at the end of Mr. De Moivre's last Edition; one of them allowed by that great Man to be the most useful on the Subject, but their Demonstrations there omitted.

Simpson is quite wrong in claiming the solution of Problem XXX, and saying that De Moivre had reserved his demonstration to himself. The investigation is that for determining the approximate value of terms near the largest in the expansion of $(a + b)^n$; it is given in the *Doctrine of Chances*, second edition, pages 233—243, third edition pages 241—251: the method of Simpson is in fact identical with De Moivre's.

371. We may remark that Simpson published a work in 1757 under the title of *Miscellaneous Tracts on some curious, and very interesting Subjects in Mechanics, Physical-Astronomy, and Speculative Mathematics;*...

In this work on pages 64—75 we have a section entitled *An Attempt to shew the Advantage arising by Taking the Mean of a Number of Observations, in Practical Astronomy.*

This is a very interesting section; the problems solved by Simpson were reproduced by Lagrange in a memoir in the fifth volume of the *Miscellanea Taurinensia*, without any allusion however to Simpson.

It will be more convenient to defer any account of the section in Simpson until we examine Lagrange's memoir, and then we will state what Simpson gave in 1757.

372. The fourth volume of the collected edition of John Bernoulli's works, which was published in 1742 has a section entitled *De Alea, sive Arte Conjectandi, Problemata quædam;* this section occupies pages 28—33 : it contains seven problems.

373. The first and second problems are simple and well-known ; they are solved completely. The third problem relates to the game of Bowls ; John Bernoulli gives, without demonstration, the result which had already been published ; see *Montmort*, page 248, and the *Doctrine of Chances*, page 117.

374. The fourth problem contains an error. John Bernoulli says that if $2n$ common dice are thrown, the number of ways in which the sum of the marks is $7n$ is

$$\frac{(7n-1)\ (7n-2)\ (7n-3) \dots (5n+1)}{1.2.3.4 \dots (2n-1)} :$$

this amounts to asserting that the expression here given is the coefficient of x^{7n} in the expansion of

$$(x + x^2 + x^3 + x^4 + x^5 + x^6)^{2n} :$$

in fact however the coefficient is a series of which the above expression is only the first term.

375. The fifth and sixth problems involve nothing new in principle ; John Bernoulli gives merely the numerical results which would require long calculation to verify. The seventh problem does not seem intelligible.

CHAPTER XI.

DANIEL BERNOULLI.

376. DANIEL BERNOULLI was the son of the John Bernoulli to whom we have often referred; Daniel was born in 1700, and died in 1782: he is the author of some important memoirs on our subject, remarkable for their boldness and originality, which we shall now proceed to examine.

377. The first memoir which we have to notice is entitled *Specimen Theoriæ Novæ de Mensura Sortis.* This memoir is contained in the *Commentarii Acad. ... Petrop.* Vol. v., which is the volume for the years 1730 and 1731; the date of publication of the volume is 1738: the memoir occupies pages 175—192.

378. This memoir contains the theory of *Moral expectation* proposed by Daniel Bernoulli, which he considered would give results more in accordance with our ordinary notions than the theory of *Mathematical expectation.* Laplace has devoted to this subject pages 432—445 of his *Théorie...des Prob.*, in which he reproduces and developes the hypothesis of Daniel Bernoulli.

379. *Mathematical expectation* is estimated by the product of the chance of obtaining a sum of money into that sum. But we cannot in practice suppose that a given sum of money is of equal importance to every man; a shilling is a matter of small moment to a person who possesses a thousand pounds, but it is of great moment to a person who only possesses a few shillings. Various hypotheses may be proposed for taking into account the

relative value of money; of these Daniel Bernoulli's has attracted most notice.

Suppose a person to possess a sum of money x, then if it receive an increment dx, Daniel Bernoulli estimates the relative value of the increment as proportional to dx directly and x inversely; that is, he takes it equal to $\dfrac{kdx}{x}$ where k is some constant. Put this equal to dy; so that

$$dy = \frac{kdx}{x};$$

therefore
$$y = k \log x + \text{constant}$$

$$= k \log \frac{x}{a} \text{ say.}$$

Laplace calls x the *fortune physique* and y the *fortune morale*. We must suppose a some positive quantity, for as Daniel Bernoulli remarks, no man is absolutely destitute unless he is dying of hunger.

Daniel Bernoulli calls y the *emolumentum*, a he calls *summa bonorum*, and $x - a$ he calls *lucrum*.

380. Suppose then that a person, starting with a for his *fortune physique*, has the chance p_1 of gaining x_1, the chance p_2 of gaining x_2, the chance p_3 of gaining x_3, and so on; and suppose the sum of these chances to be unity. Let

$$Y = kp_1 \log (a + x_1) + kp_2 \log (a + x_2) + kp_3 \log (a + x_3) + \ldots - k \log a.$$

Then Bernoulli calls Y the *emolumentum medium*, and Laplace still calls Y the *fortune morale*. Let X denote the *fortune physique* which corresponds to this *fortune morale*; then

$$Y = k \log X - k \log a.$$

Thus
$$X = (a + x_1)^{p_1} (a + x_2)^{p_2} (a + x_3)^{p_3} \ldots$$

And $X - a$ will be according to Laplace *l'accroissement de la fortune physique qui procurerait à l'individu le même avantage moral qui résulte pour lui, de son expectative.* Daniel Bernoulli calls $X - a$ the *lucrum legitime expectandum seu sors quaesita*.

381. Daniel Bernoulli in his memoir illustrates his hypothesis by drawing a curve. He does not confine himself to the case in which $y = k \log \dfrac{x}{a}$, but supposes generally $y = \phi(x)$. Thus the ordinary theory of mathematical expectation amounts to supposing that the curve becomes a straight line, or $\phi(x)$ a linear function of x.

382. After obtaining the value of X which we have given in Art. 380, the remainder of Daniel Bernoulli's memoir consists of inferences drawn from this value.

383. The first inference is that even a fair game of chance is disadvantageous. Suppose a man to start with a as his *fortune physique*, and have the chance p_1 of gaining x_1, and the chance p_2 of losing x_2. Then by Art. 380, the *fortune physique* which he may expect is

$$(a + x_1)^{p_1} (a - x_2)^{p_2} ;$$

we have to shew that this is less than a, supposing the game to be mathematically fair, so that

$$\frac{p_1}{p_2} = \frac{x_2}{x_1} .$$

Daniel Bernoulli is content with giving an arithmetical example, supposing $p_1 = p_2 = \dfrac{1}{2}$. Laplace establishes the proposition generally by the aid of the Integral Calculus. It may be proved more simply. We have

$$p_1 = \frac{x_2}{x_1 + x_2}, \quad p_2 = \frac{x_1}{x_1 + x_2} ;$$

and we have to shew that

$$\left\{ (a + x_1)^{x_2} (a - x_2)^{x_1} \right\}^{\frac{1}{x_1 + x_2}} \text{ is less than } a.$$

Now we may regard x_1 and x_2 as integers. Thus the result we require is true by virtue of the general theorem in inequalities that *the geometrical mean is less than the arithmetical mean*. For here we may suppose that there are x_2 quantities, each equal to $a + x_1$, and x_1 quantities each equal to $a - x_2$. The arithmetical mean is

$$\frac{x_2 (a + x_1) + x_1 (a - x_2)}{x_1 + x_2},$$

that is a. The geometrical mean is the quantity which we had to shew to be less than a.

384. Daniel Bernoulli proposes to determine what a man should stake at a wager, in order that the wager may not be disadvantageous to him. He takes the case in which $p_1 = p_2 = \frac{1}{2}$. Then we require that

$$(a + x_1)^{\frac{1}{2}} (a - x_2)^{\frac{1}{2}} = a.$$

This leads to

$$x_2 = \frac{ax_1}{a + x_1}.$$

Thus x_2 is less than x_1 and less than a.

385. Daniel Bernoulli now makes an application to insurances. But this application will be more readily understood if we give first a proposition from Laplace which is not in Daniel Bernoulli's memoir. Suppose that a merchant has a *fortune physique* equal to a, and that he expects the sum x to arrive by a ship. Also let p be the chance that the ship will arrive safely, and let $q = 1 - p$.

Suppose that he insures his ship on the ordinary terms of mathematical equity; then he pays qx to the insurance company, so that he has on the whole $a + x - qx$, that is $a + px$.

Suppose however that he does not insure; then his *fortune physique* is $(a + x)^p a^q$. We shall shew that $a + px$ is greater than $(a + x)^p a^q$.

Laplace establishes this by the aid of the Integral Calculus, with which however we may dispense. We have to shew that

$$(a + x)^p a^q \text{ is less than } a + px,$$

that is that

$$\left(1 + \frac{x}{a}\right)^p \text{ is less than } 1 + \frac{px}{a}.$$

Let $p = \dfrac{m}{m + n}$ where m and n are integers.

Then we know that $\left\{ \left(1 + \dfrac{x}{a}\right)^m 1^n \right\}^{\frac{1}{m+n}}$ is less than

$$\frac{m \left(1 + \dfrac{x}{a}\right) + n}{m + n}$$

by the theorem respecting the geometrical mean and the arithmetical mean which we quoted in Art. 383; and this is what we had to establish.

It follows that the merchant can afford without disadvantage to increase his payment to the insurance company beyond the sum qx. If we suppose ξ to represent the extreme additional sum, we have

$$\xi = a + px - (a + x)^p a^q.$$

386. We now return to Daniel Bernoulli. We have seen that a merchant can afford to pay more than the sum qx for insuring; but it may happen that the insurance company demand more than the merchant can afford to pay. Daniel Bernoulli proposes this question: for a given charge by the insurance company required to find the merchant's fortune, so that it may be indifferent to him whether he insures or not.

Retaining the notation of the last Article, let e be the charge of the insurance company; then we have to find a from the equation

$$a + x - e = (a + x)^p a^q.$$

Daniel Bernoulli takes for an example $x = 10000$, $e = 800$, $p = \dfrac{19}{20}$; whence by approximation $a = 5043$. Hence he infers that if the merchant's fortune is less than 5043 he ought to insure, if greater than 5043 he ought not to insure. This amounts to assuming that the equation from which a is to be found has only one positive root. It may be interesting to demonstrate this. We have to compare

$$a + x - e \text{ with } (a + x)^p a^q,$$

where a is the variable, and x is greater than e.

Let $p = \dfrac{m}{m+n}$ and $q = \dfrac{n}{m+n}$, where m and n are integers; then we have to compare

$$(a + x - e)^{m+n} \text{ with } (a + x)^m a^n.$$

When $a = 0$ the right-hand member is the less; when a is infinite the right-hand member is the greater, provided mx is greater than $(m + n)(x - e)$: we will assume that this is the case. Thus the equation

$$(a + x - e)^{m+n} = (a + x)^m a^n$$

has *one* positive root. We must examine if it has another.

Let $\log (a + x - e)^{m+n} = y, \qquad \log (a + x)^m a^n = z;$

then $\qquad \dfrac{dy}{da} = \dfrac{m+n}{a+x-e}, \qquad \dfrac{dz}{da} = \dfrac{m}{x+a} + \dfrac{n}{a}.$

Thus when a is zero $\dfrac{dz}{da}$ is greater than $\dfrac{dy}{da}$, so that z begins by increasing more rapidly than y does. If we suppose

$$\frac{dy}{da} = \frac{dz}{da}$$

we obtain $\qquad a = \dfrac{nx(x - e)}{(m+n)e - nx}.$

Now begin with $a = 0$, and let a gradually increase until we have $y = z$; then it is obvious that we have not yet reached the value of a just given. And if by increasing a we could arrive at a second value at which $y = z$, we should have passed beyond the value of a just given. Then after that value z would increase more slowly than y, and the final value of z would be less than the final value of y, which is impossible. Thus there is only one value of a which makes $y = z$, and this value is less than

$$\frac{nx(x - e)}{(m+n)e - nx}.$$

If mx is less than $(m + n)(x - e)$ the original equation has no positive root; for then we have z always increasing more rapidly than y, and yet the final value of z less than that of y; so that it is impossible that any value of a can make $y = z$.

387. Daniel Bernoulli also inquires what capital the insurance company must have so that they may safely undertake the insurance. Let y denote the least value of the capital; then y must be found from

$$(y + e)^p (y - x + e)^q = y.$$

This is merely the former equation with y in place of $a + x - e$. Thus, taking the same example as before, we have $y = 14243$.

388. Daniel Bernoulli now lays down the important principle that it is more advantageous for a person to expose his fortune to different independent risks than to expose it all to one risk. He gives this example: suppose a merchant to start with a capital of 4000, and that he expects 8000 by a ship; let $\frac{9}{10}$ be the chance of the safe arrival of the ship. The merchant's *fortune physique* is thus

$$(4000 + 8000)^{\frac{9}{10}} (4000)^{\frac{1}{10}} = 10751 \text{ approximately.}$$

But suppose him to put half of his merchandize in one ship and half in another. The chance that both ships will arrive safely is $\frac{81}{100}$; the chance that one of the two will arrive safely is $2 \times \frac{9}{10} \times \frac{1}{10}$, that is $\frac{18}{100}$; the chance that both will be lost is $\frac{1}{100}$. Hence the merchant's *fortune physique* is

$$(4000 + 8000)^{\frac{81}{100}} (4000 + 4000)^{\frac{18}{100}} (4000)^{\frac{1}{100}} = 11033$$

approximately.

Subtract the original capital 4000, and we find the expectation in the former case to be 6751, and in the latter to be 7033.

Daniel Bernoulli says that the merchant's expectation continually increases by diminishing the part of the merchandize which is intrusted to a single ship, but can never exceed 7200. This number is $\frac{9}{10}$ of 8000; so that it expresses the Mathematical expectation. The result which Daniel Bernoulli thus enunciates without demonstration is demonstrated by Laplace, *Théorie ... des Prob.*, pages 435—437; the proposition is certainly by no means easy, and it is to be wished that Daniel Bernoulli had explained how he obtained it.

389. Daniel Bernoulli now applies his theory to the problem which is known as the *Petersburg Problem*, probably from its first appearing here in the *Commentarii* of the Petersburg Academy. The problem is similar to two which Nicolas Bernoulli proposed to Montmort; see Art. 231.

A throws a coin in the air; if head appears at the first throw he is to receive a shilling from *B*, if head does not appear until the second throw he is to receive 2 shillings, if head does not appear until the third throw he is to receive 4 shillings, and so on: required the expectation of *A*.

The expectation is

$$\frac{1}{2} + \frac{2}{2^2} + \frac{4}{2^3} + \frac{8}{2^4} + \dots \textit{ in infinitum,}$$

that is

$$\frac{1}{2} + \frac{1}{2} + \frac{1}{2} + \frac{1}{2} + \dots \textit{ in infinitum.}$$

Thus *A*'s expectation is *infinite*, so that he ought to give an infinite sum to *B* to induce *B* to play with him in the manner proposed. Still no prudent man in the position of *A* would be willing to pay even a small number of shillings for the advantage to be gained.

The paradox then is that the mathematical theory is apparently directly opposed to the dictates of common sense.

390. We will now give Daniel Bernoulli's application of his theory of Moral expectation to the Petersburg Problem.

Suppose that *A* starts with the sum *a*, and is to receive 1 if head appears at the first throw, 2 if head does not appear until the second throw, and so on. *A*'s *fortune physique* is

$$(a + 1)^{\frac{1}{2}} (a + 2)^{\frac{1}{4}} (a + 4)^{\frac{1}{8}} (a + 8)^{\frac{1}{16}} \dots - a.$$

This expression is finite if *a* be finite. The value of it when $a = 0$ is easily seen to be 2. Daniel Bernoulli says that it is about 3 when $a = 10$, about $4\frac{1}{3}$ when $a = 100$, and about 6 when $a = 1000$.

Let x represent the sum which a person with the capital a might give without disadvantage for the expectation of A; then x is to be found from

$$(a+1-x)^{\frac{1}{2}} (a+2-x)^{\frac{1}{4}} (a+4-x)^{\frac{1}{8}} (a+8-x)^{\frac{1}{16}} \ldots = a.$$

Put $a - x = a'$; thus

$$(a'+1)^{\frac{1}{2}} (a'+2)^{\frac{1}{4}} (a'+4)^{\frac{1}{8}} (a'+8)^{\frac{1}{16}} \ldots - a' = x.$$

Then if a is to have any large value, from what we have already seen, x is small compared with a, so that we may put a' for a; and we have approximately

$$x = (a+1)^{\frac{1}{2}} (a+2)^{\frac{1}{4}} (a+4)^{\frac{1}{8}} (a+8)^{\frac{1}{16}} \ldots - a.$$

Laplace reproduces this part of Daniel Bernoulli's memoir with developments in pages 439—442 of the *Théorie...des Prob.*

391. Daniel Bernoulli's memoir contains a letter addressed to Nicolas Bernoulli by Cramer, in which two methods are suggested of explaining the paradox of the Petersburg Problem.

(1) Cramer considers that the value of a sum of money is not to be taken uniformly proportional to the sum; he proposes to consider all sums greater than 2^{24} as practically equal. Thus he obtains for the expectation of B

$$\frac{1}{2} + \frac{2}{2^2} + \frac{4}{2^3} + \ldots + \frac{2^{24}}{2^{25}}$$

$$+ \frac{2^{24}}{2^{26}} + \frac{2^{24}}{2^{27}} + \frac{2^{24}}{2^{28}} + \ldots.$$

The first twenty-five terms give $12\frac{1}{2}$; the remainder constitute a geometrical progression of which the sum is $\frac{1}{2}$. Thus the total is 13.

(2) Cramer suggests that the pleasure derivable from a sum of money may be taken to vary as the square root of the sum. Thus he makes the moral expectation to be

$$\frac{1}{2}\sqrt{1} + \frac{1}{4}\sqrt{2} + \frac{1}{8}\sqrt{4} + \frac{1}{16}\sqrt{8} + \ldots,$$

that is $\dfrac{1}{2-\sqrt{2}}$. This moral expectation corresponds to the sum

$$\frac{1}{(2-\sqrt{2})^2},$$ that is to 2·9 approximately; and Cramer considers this to be nearer the common notion on the subject than his former value 13.

392. It is obvious that Cramer's suppositions are entirely arbitrary, and that such suppositions might be multiplied to any extent. Montucla alludes on his page 403 to an attempt made by M. Fontaine to explain the paradox. This attempt seems to consist in limiting the game to 20 throws at most, instead of allowing it theoretically to extend to infinity. But the opponents of the mathematical theory would assert that for the game as thus understood the value of the expectation assigned by the theory is still far larger than common sense can admit.

393. The Petersburg Problem will come under our notice again as we advance with the subject. We may remark that Laplace adopts Daniel Bernoulli's view; *Théorie...des Prob.* page 439. Poisson prefers to reconcile mathematical theory with common sense by the consideration that the fortune of the person whom we represent by B is necessarily finite so that he cannot pay more than a certain sum; this in result practically coincides with the first of Cramer's two suppositions; see Poisson, *Recherches sur la Prob...* page 73; Cournot, *Exposition de la Théorie des Chances...* page 108.

394. We pass to another memoir by Daniel Bernoulli. The Academy of Sciences of Paris proposed the following question as a prize subject for 1732,

Quelle est la cause physique de l'inclinaison des Plans des Orbites des Planetes par rapport au plan de l'Équateur de la revolution du Soleil autour de son axe; Et d'où vient que les inclinaisons de ces Orbites sont differentes entre elles.

None of the memoirs sent in appeared to the judges to be worthy of the prize. The Academy then proposed the subject again for 1734, with a double prize. The prize was divided between Daniel Bernoulli and his father John Bernoulli. The memoirs of both are contained in the *Recueil des pieces qui ont remporté le prix de l'Academie Royale des Sciences*, Tom. 3, 1734.

A French translation of Daniel Bernoulli's memoir occupies
pages 95—122 of the volume; the original memoir in Latin occupies pages 125—144.

395. The portion of the memoir with which we are concerned
occurs at the beginning. Daniel Bernoulli wishes to shew that we
cannot attribute to hazard the small mutual inclinations of the
planetary orbits. He puts the calculation in three forms.

(1) He finds that the greatest mutual inclination of any two
planetary orbits is that of Mercury to the Ecliptic, which is 6° $54'$.
He imagines a zone of the breadth of 6° $54'$ on the surface of a
sphere, which would therefore contain about $\frac{1}{17}$ of the whole surface of the sphere. There being six planets altogether he takes
$\frac{1}{17^{5}}$ for the chance that the inclinations of five of the planes to one
plane shall all be less than 6° $54'$.

(2) Suppose however that all the planes intersected in a
common line. The ratio of 6° $54'$ to 90° is equal to $\frac{1}{13}$ nearly;
and he takes $\frac{1}{13^{5}}$ for the chance that each of the five inclinations
would be less than 6° $54'$.

(3) Again; take the Sun's equator as the plane of reference.
The greatest inclination of the plane of any orbit to this is 7° $30'$,
which is about $\frac{1}{12}$ of 90°; and he takes $\frac{1}{12^{6}}$ as the chance that each
of the six inclinations would be less than 7° $30'$.

396. It is difficult to see why in the first of the three pre-
ceding calculations Daniel Bernoulli took $\frac{1}{17}$ instead of $\frac{2}{17}$; that is
why he compared his zone with the surface of a sphere instead of
with the surface of a hemisphere. It would seem too that he
should rather have considered the *poles* of the orbits than the
planes of the orbits, and have found the chance that all the
other poles should lie within a given distance from one of them.

397. We shall find hereafter that D'Alembert did not admit that there was any value in Daniel Bernoulli's calculations.

Laplace proposes to find the probability that the *sum* of all the inclinations should not exceed an assigned quantity ; see *Théorie... des Prob.* page 257. The principle of Daniel Bernoulli's attempt seems more natural, because it takes more explicit account of the fact that *each* inclination is small.

398. The next memoir by Daniel Bernoulli is entitled *Essai d'une nouvelle analyse de la mortalité causée par la petite Vérole, et des avantages de l'Inoculation pour la prévenir.*

This memoir is contained in the *Hist. de l'Acad. ... Paris*, for 1760 ; the date of publication of the volume is 1766 : the memoir occupies pages 1—45 of the part devoted to memoirs.

399. The reading of the memoir commenced on April 30th, 1760, as we learn from its seventh page. Before the memoir was printed, a criticism on it appeared, which Daniel Bernoulli ascribes to a *grand mathématicien;* see his pages 4 and 18. In consequence of this, an *introduction apologétique* was written on April 16th, 1765, and now forms the first six pages of the whole.

The critic was D'Alembert; see *Montucla*, page 426, and our Chapter XIII.

400. Daniel Bernoulli's main object is to determine the mortality caused by the small-pox at various stages of age. This of course could have been determined if a long series of observations had been made; but at that time such observations had not been made. Tables of mortality had been formed, but they gave the total number of deaths at various ages without distinguishing the causes of death. Thus it required calculation to determine the result which Daniel Bernoulli was seeking.

401. Daniel Bernoulli made two assumptions : that in a year on an average 1 person out of 8 of all those who had not previously taken the disease, would be attacked by small-pox, and that 1 out of every 8 attacked would die. These assumptions he supported by appeal to observation ; but they might not be uni-

versally admitted. Since the introduction of vaccination, the memoir of Bernoulli will have no practical value; but the mathematical theory which he based on his hypotheses is of sufficient interest to be reproduced here.

402. Let x denote the age expressed in years; let ξ denote the number who survive at that age out of a given number who were born; let s denote the number of these survivors who have not had the small-pox. Assume that in a year the small-pox attacks 1 out of every n who have not had the disease, and that 1 out of every m who are attacked dies.

The number of survivors who have not had the small-pox continually diminishes; partly because the small-pox continually attacks some whom it had previously left unattacked, and partly because some persons die of other diseases without ever being attacked by the small-pox.

The number of those attacked by the small-pox during the element dx of time is by hypothesis $\dfrac{s\,dx}{n}$: because we suppose $\dfrac{s}{n}$ to be attacked in one year, and therefore $\dfrac{s\,dx}{n}$ in the element dx of a year. The number of those who die of the small-pox is by hypothesis $\dfrac{s\,dx}{mn}$; and therefore the number of those who die of other diseases is $-\,d\xi - \dfrac{s\,dx}{mn}$. But this last number must be diminished in the ratio of s to ξ, because we only want the diminution of those who have not yet had the small-pox, of whom the number is s.

Thus
$$- ds = \frac{s\,dx}{n} - \frac{s}{\xi}\left(d\xi + \frac{s\,dx}{mn}\right).$$

This equation is to be integrated. We have

$$s\,\frac{d\xi}{\xi} - ds = \frac{s\,dx}{n} - \frac{s^2\,dx}{mn\xi};$$

therefore
$$\frac{s\,d\xi - \xi\,ds}{s^2} = \frac{\xi\,dx}{ns} - \frac{dx}{mn}.$$

15

Put q for $\frac{\xi}{s}$; thus, $dq = \dfrac{mq-1}{mn}\, dx$;

therefore $\qquad n \log (mq - 1) = x + \text{constant}$;

therefore $\qquad \left(\dfrac{m\xi}{s} - 1 \right)^{n} = e^{x+C}$,

and $\qquad s = \dfrac{m\xi}{e^{\frac{x+c}{n}} + 1}$.

To determine the constant C, we observe that when $x = 0$, we have $s = \xi$; thus, finally,

$$ s = \frac{m\xi}{(m-1)\, e^{\frac{x}{n}} + 1}. $$

403. By this formula Daniel Bernoulli calculates a table on the basis of Halley's table, derived from the Breslau Observations, assuming that m and n each equal 8; Halley's table gives the values of ξ corresponding to successive integer values of x, and Daniel Bernoulli's formula then gives the values of s. The following is an extract from the table:

x	ξ	s
0	1300	1300
1	1000	896
2	855	685
3	798	571
4	760	485
5	732	416
6	710	359
7	692	311
8	680	272
9	670	237
10	661	208
11	653	182
12	646	160
13	640	140
14	634	123
15	628	108
16	622	94

Halley's table begins with 1000 at the end of the first year, and does not say to what number of births this corresponds. Daniel Bernoulli gives reasons for assuming this to be 1300, which accordingly he takes; see Art. 64.

404. On page 21 of the memoir, Daniel Bernoulli says that the following question had been asked: Of all persons alive at a given epoch what fractional part had not been attacked by the small-pox? The inquirer himself, who was D'Alembert, estimated the number at one-fourth at most. Daniel Bernoulli himself makes it about two-thirteenths. He intimates that it would be desirable to test this by observation. He adds,

Voici un autre théorème qui pourroit servir à la vérification de nos principes. Si de tous les vivans on ne prend que l'enfance et la jeunesse, jusqu'à l'âge de seize ans et demi, on trouvera le nombre de ceux qui auront eu la petite vérole à peu-près égal au nombre de ceux qui ne l'auront pas eue.

405. Daniel Bernoulli gives another interesting investigation. Required to find the number of survivors at a given age from a given number of births, supposing the small-pox altogether extinguished. Retain the notation of Article 402; and let z be the number who would have been alive at the age x if there had been no small-pox, the original number of births being supposed the same.

The whole mortality during the element dx of time being $- d\xi$, and the mortality caused by the small-pox being $\dfrac{sdx}{mn}$, we have for the mortality in the absence of small-pox $- d\xi - \dfrac{sdx}{mn}$. But this mortality arises from a population ξ; and we must multiply it by $\dfrac{z}{\xi}$ to obtain the mortality which would arise from a population z. Hence, finally,

$$- dz = - \frac{z}{\xi}\left(d\xi + \frac{sdx}{mn}\right);$$

therefore

$$\frac{dz}{z} = \frac{d\xi}{\xi} + \frac{s}{\xi}\frac{dx}{mn}.$$

Substitute for s from the result in Art. 402; then integrate, and determine the arbitrary constant by the condition that $z = \xi$ when $x = 0$. Hence we shall obtain

$$\frac{z}{\xi} = \frac{m e^{\frac{x}{n}}}{(m - 1) e^{\frac{x}{n}} + 1}.$$

Thus as x increases, the right-hand member approaches the limit $\frac{m}{m - 1}$.

406. After discussing the subject of the mortality caused by the small-pox, Daniel Bernoulli proceeds to the subject of Inoculation. He admits that there is some danger in Inoculation, but finds on the whole that it is attended with large advantages. He concluded that it would lengthen the average duration of life by about three years. This was the part of the memoir which at the time of publication would be of the greatest practical importance; but that importance happily no longer exists.

407. We shall find hereafter that D'Alembert strongly objected to the justness of Daniel Bernoulli's investigations. Laplace speaks very highly of Daniel Bernoulli; Laplace also briefly indicates the method of treating the problem respecting Inoculation, but as he does not assume m and n to be constant, he rather follows D'Alembert than Daniel Bernoulli; see *Théorie...des Prob.*, pages CXXXVII. and 413.

408. The next memoir by Daniel Bernoulli is entitled *De usu algorithmi infinitesimalis in arte conjectandi specimen.*

This memoir is contained in the *Novi Comm...Petrop.* Vol. XII, which is the volume for the years 1766 and 1767; the date of publication of the volume is 1768; the memoir occupies pages 87—98.

409. The object of the memoir is twofold. A certain problem in chances is to be solved, which is wanted in the next memoir to which we shall come; and the introduction of the Differential Calculus into the Theory of Probability is to be illustrated. The reader will see in Art. 402 that Daniel Bernoulli had already really

employed the Differential Calculus, and the present memoir contains remarks which would serve to explain the process of Art. 402; but the remarks are such as any student could easily supply for himself. We shall see the point illustrated in another memoir. See Art. 417.

410. The problem which Daniel Bernoulli solves is in its simplest form as follows : In a bag are $2n$ cards ; two of them are marked 1, two of them are marked 2, two of them are marked 3,... and so on. We draw out m cards ; required the probable number of *pairs* which remain in the bag.

We give the solution of Daniel Bernoulli with some changes of notation. Suppose that x_m pairs remain after m cards have been drawn out; let a new drawing be made. The card thus drawn out is either one of the cards of a pair, or it is not; the probabilities for these two cases are proportional to $2x_m$, and $2n - 2x_m - m$ respectively : in the former case there remain $x_m - 1$ pairs in the bag, and in the latter case there remain x_m pairs. Thus by ordinary principles

$$x_{m+1} = \frac{2x_m (x_m - 1) + (2n - 2x_m - m) x_m}{2n - m}$$

$$= \frac{2n - m - 2}{2n - m} x_m .$$

We can thus form in succession $x_1,\ x_2,\ x_3, \ldots$ As $x_0 = n$ we find that

$$x_m = \frac{(2n - m)\,(2n - m - 1)}{2\,(2n - 1)} .$$

411. The problem is extended by Daniel Bernoulli afterwards to a greater generality ; but we have given sufficient to enable the reader to understand the nature of the present memoir, and of that to which we now proceed.

412. The next memoir is entitled *De duratione media matrimoniorum, pro quacunque conjugum aetate, aliisque quaestionibus affinibus.*

This memoir is closely connected with the preceding; it follows in the same volume of the *Novi Comm...Petrop.*, and occupies pages 99—126.

413. Suppose 500 men of a given age, as for example 20 years, to marry 500 women of the same age. The tables of mortality will shew at what rate these 1000 individuals gradually diminish annually until all are dead. But these tables do not distinguish the married from the unmarried, so that we cannot learn from them the number of unbroken couples after the lapse of a given number of years. Daniel Bernoulli applies the result of Art. 410 ; the pairs of cards correspond to the married couples. From that article knowing the number of cards which remain undrawn we infer the probable number of pairs. The number of cards remaining undrawn corresponds to the number of persons remaining alive at a given age ; this is taken from the tables of mortality, and by the formula the probable number of unbroken couples is calculated. Daniel Bernoulli calculates such a table for the numbers we have supposed above.

414. Daniel Bernoulli then proceeds to the case in which the husband and wife are supposed of different ages ; this requires the extended problem to which we have referred in Art. 411. Daniel Bernoulli calculates a table for the case in which 500 men aged 40 years marry 500 women aged 20 years.

Daniel Bernoulli allows that his results must not claim implicit confidence. He has taken the same laws of mortality for both men and women, though of course he was aware that on an average women live longer than men. With respect to this fact he says, page 100, ...neque id diversæ vivendi rationi tribui potest, quia ista sequioris sexus praerogativa a primis incunabilis constantissime manifestatur atque per totam vitam in illo manet.

Daniel Bernoulli's process is criticised by Trembley in the *Mémoires de l'Acad....Berlin*, 1799, 1800.

The problem respecting the mean duration of marriages is considered by Laplace, *Théorie...des Prob.* page 415.

415. The memoir which we have noticed in Arts. 412—414 bears a close analogy to the memoir which we have noticed in Arts. 398—406. In both cases theory is employed to supply the lack of observations, in both cases the questions discussed are of the same kind, and in both cases the use of the Differential Calculus is illustrated:

416. The next memoir by Daniel Bernoulli is entitled *Disquisitiones Analyticæ de novo problemate conjecturali.*

This memoir is contained in the *Novi Comm...Petrop...*Vol. 14, 1769, *pars prior.* The date 1759 occurs by mistake in the title-page. The date of publication of the volume is 1770. The memoir occupies pages 1—25 of the part devoted to memoirs.

417. The object of the memoir is to illustrate the use of the Differential Calculus, and it is thus analogous to memoirs which we have already noticed by Daniel Bernoulli.

Suppose three urns ; in the first are n white balls, in the second n black balls, in the third n red balls. A ball is taken at random from each urn ; the ball taken from the first urn is put into the second, the ball taken from the second is put into the third, and the ball taken from the third is put into the first ; this operation is repeated for any assigned number of times : required the probable distribution of the balls at the end of these operations.

Suppose that after x operations the probable numbers of white balls in the three urns are denoted by u_x, v_x, w_x respectively. Then

$$u_{x+1} = u_x - \frac{u_x}{n} + \frac{w_x}{n}.$$

For $\dfrac{u_x}{n}$ is the probability of drawing one white ball out of the first urn, and $\dfrac{w_x}{n}$ is the probability that a white ball will be drawn from the third urn and so put into the first. Similarly

$$v_{x+1} = v_x - \frac{v_x}{n} + \frac{u_x}{n}, \qquad w_{x+1} = w_x - \frac{w_x}{n} + \frac{v_x}{n}.$$

By eliminating, using the condition $u_x + v_x + w_x = n$, we may obtain an equation in Finite Differences of the second order for u_x, namely,

$$u_{x+2} = u_{x+1}\left(2 - \frac{3}{n}\right) - u_x\left(1 - \frac{3}{n} + \frac{3}{n^2}\right) + \frac{1}{n}.$$

But the following process is more symmetrical. Put $u_{x+1} = Eu_x$, and separate the symbols in the usual way ;

thus

$$\left\{ E - \left(1 - \frac{1}{n}\right)\right\} u_x = \frac{1}{n} w_x,$$

$$\left\{ E - \left(1 - \frac{1}{n}\right)\right\} v_x = \frac{1}{n} u_x,$$

$$\left\{ E - \left(1 - \frac{1}{n}\right)\right\} w_x = \frac{1}{n} v_x,$$

therefore

$$\left\{ E - \left(1 - \frac{1}{n}\right)\right\}^3 u_x = \left(\frac{1}{n}\right)^3 u_x.$$

Therefore $u_x = A\left(1 - \frac{1}{n} + \frac{\alpha}{n}\right)^x + B\left(1 - \frac{1}{n} + \frac{\beta}{n}\right)^x + C\left(1 - \frac{1}{n} + \frac{\gamma}{n}\right)^x$,

where A, B, C are constants, and α, β, γ are the three cube roots of unity.

Then from the above equations we obtain

$$w_x = n\left\{ E - \left(1 - \frac{1}{n}\right)\right\} u_x;$$

therefore

$$w_x = \alpha A\left(1 - \frac{1}{n} + \frac{\alpha}{n}\right)^x + \beta B\left(1 - \frac{1}{n} + \frac{\beta}{n}\right)^x + \gamma C\left(1 - \frac{1}{n} + \frac{\gamma}{n}\right)^x.$$

Similarly

$$v_x = \alpha^2 A\left(1 - \frac{1}{n} + \frac{\alpha}{n}\right)^x + \beta^2 B\left(1 - \frac{1}{n} + \frac{\beta}{n}\right)^x + \gamma^2 C\left(1 - \frac{1}{n} + \frac{\gamma}{n}\right)^x.$$

The three constants A, B, C are not all arbitrary, for we require that

$$u_x + v_x + w_x = n,$$

with this condition and the facts that

$$u_0 = n, \quad v_0 = 0, \quad w_0 = 0,$$

we shall obtain $A = B = C = \frac{n}{3}$.

418. The above process will be seen to be applicable if the number of urns be any whatever, instead of being limited to three.

We need not investigate the distribution of the balls of the other colours; for it is evident from symmetry that at the end of x

operations the black balls will be probably distributed thus, u_x in the second urn, v_x in the third, and w_x in the first; similarly the red balls will be probably distributed thus, u_x in the third urn, v_x in the first, and w_x in the second.

It should be observed that the equations in Finite Differences and the solution will be the same whatever be the original distribution of the balls, supposing that there were originally n in each urn; the only difference will be in the values to be assigned to the arbitrary constants. Nor does the process require n white balls originally. Thus in fact we solve the following problem : Suppose a given number of urns, each containing n balls, m of the whole number of balls are white and the rest not white; the original distribution of the white balls is given : required their probable distribution after x operations.

419. Daniel Bernoulli does not give the investigation which we have given in Art. 417. He simply indicates the following result, which he probably obtained by induction :

$$u_x = n \left\{ \left(1 - \frac{1}{n}\right)^x + \frac{x\,(x-1)\,(x-2)}{\lfloor 3} \left(1 - \frac{1}{n}\right)^{x-3} \left(\frac{1}{n}\right)^3 \right.$$

$$\left. + \frac{x\,(x-1)\,(x-2)\,(x-3)\,(x-4)\,(x-5)}{\lfloor 6} \left(1 - \frac{1}{n}\right)^{x-6} \left(\frac{1}{n}\right)^6 + \ldots \right\},$$

together with similar expressions for v_x and w_x. These can be obtained by expanding by the Binomial Theorem the expressions we have given, using the known values of the sums of the powers of α, β, γ.

420. Now a problem involving the Differential Calculus can be framed, exactly similar to this problem of the urns. Suppose three equal vessels, the first filled with a white fluid, the second with a black fluid, and the third with a red fluid. Let there be very small tubes of equal bore, which allow fluid to pass from the first vessel into the second, from the second into the third, and from the third into the first. Suppose that the fluids have the property of mixing instantaneously and completely. Required at the end of the time t the distribution of the fluids in the vessels.

Suppose at the end of the time t the quantities of the white fluid in the three vessels to be u, v, w respectively. We obtain the following equations,

$$du = kdt\,(w - u),$$
$$dv = kdt\,(u - v),$$
$$dw = kdt\,(v - w),$$

where k is a constant.

Daniel Bernoulli integrates these equations, by an unsymmetrical and difficult process. They may be easily integrated by the modern method of separating the symbols. Put D for $\dfrac{d}{dt}$; thus

$$(D + k)\,u = kw, \quad (D + k)\,v = ku, \quad (D + k)\,w = kv,$$

therefore $$(D + k)^3\,u = k^3 u.$$

Hence $$u = e^{-kt}\{Ae^{kat} + Be^{k\beta t} + Ce^{k\gamma t}\},$$

where A, B, C are arbitrary constants, and α, β, γ are the three cube roots of unity. The values of v and w can be deduced from that of u. Let us suppose that initially $u = h$, $v = 0$, $w = 0$; we shall find that $A = B = C = \dfrac{h}{3}$, so that

$$u = \frac{h}{3}\,e^{-kt}\{e^{kat} + e^{k\beta t} + e^{k\gamma t}\}.$$

Laplace has given the result for any number of vessels in the *Théorie...des Prob.* page 303.

421. Now it is Daniel Bernoulli's object to shew, that when x and n are supposed indefinitely large in the former problem its results correspond with those of the present problem. Here indeed we do not gain any thing by this fact, because we can solve the former problem; but if the former problem had been too difficult to solve we might have substituted the latter problem for it. And thus generally Daniel Bernoulli's notion is that we may often advantageously change a problem of the former kind into one of the latter kind.

If we suppose n and x very large we can obtain by the Binomial Theorem, or by the Logarithmic Theorem,

$$\left(1 - \frac{1}{n}\right)^x = e^{-\frac{x}{n}}.$$

Hence when n and x are very large, we find that the value of u_x given in Art. 419 reduces to

$$ne^{-\frac{x}{n}}\left\{1 + \frac{1}{\underline{3}}\left(\frac{x}{n}\right)^3 + \frac{1}{\underline{6}}\left(\frac{x}{n}\right)^6 + \dots\right\}.$$

Daniel Bernoulli sums the series in the brackets by the aid of the Integral Calculus. We know however by the aid of the theorem relating to the value of the sums of the powers of α, β, γ, that this series is equal to

$$\frac{1}{3}\left\{e^{\frac{\alpha x}{n}} + e^{\frac{\beta x}{n}} + e^{\frac{\gamma x}{n}}\right\}.$$

Hence the analogy of the value of u_x, when x and n are indefinitely large, with the value of u in Art. 420 is sufficiently obvious.

Daniel Bernoulli gives some numerical applications of his general results.

Daniel Bernoulli's memoir has been criticised by Malfatti, in the *Memorie ... della Societa Italiana*, Vol. I. 1782.

422. The next memoir by Daniel Bernoulli is entitled, *Mensura Sortis ad fortuitam successionem rerum naturaliter contingentium applicata*. This memoir is in the same volume of the *Novi Comm....Petrop.* as the preceding; it occupies pages 26—45.

423. The memoir begins by noticing the near equality in the numbers of boys and girls who are born; and proposes to consider whether this is due to chance. In the present memoir only thus much is discussed: assuming that the births of a boy and of a girl are equally likely, find the probability that out of a given number of births, the boys shall not deviate from the half by more or less than a given number. The memoir gives some calculations and some numerical examples.

Daniel Bernoulli seems very strangely to be unaware that all which he effects had been done better by Stirling and De Moivre long before; see De Moivre's *Doctrine of Chances*, pages 243—254.

The following is all that Daniel Bernoulli contributes to the theory. Let m and n be large numbers; let

$$u = \frac{\lfloor 2n}{\lfloor n \lfloor n} \frac{1}{2^{2n}},$$

$$v = \frac{\lfloor 2m}{\lfloor m \lfloor m} \frac{1}{2^{2m}}.$$

He *shews* that approximately

$$\frac{u}{v} = \sqrt{\frac{4m+1}{4n+1}}.$$

He also *states* the following: in the expansion of $\left(\frac{1}{2}+\frac{1}{2}\right)^{2n}$ the μth term from the middle is approximately equal to $\dfrac{u}{e^{\frac{\mu^2}{n}}}$.

These results are included in those of Stirling and De Moivre, so that Daniel Bernoulli's memoir was useless when it appeared; see Art. 337.

424. The next memoir by Daniel Bernoulli is entitled *Dijudicatio maxime probabilis plurium observationum discrepantium atque verisimillima inductio inde formanda*. This memoir is contained in the *Acta Acad. ... Petrop.* for 1777, *pars prior;* the date of publication of the volume is 1778: the memoir occupies pages 3—23 of the part devoted to memoirs.

425. The memoir is not the first which treated of the errors of observations as a branch of the Theory of Probability, for Thomas Simpson and Lagrange had already considered the subject; see Art. 371.

Daniel Bernoulli however does not seem to have been acquainted with the researches of his predecessors.

Daniel Bernoulli says that the common method of obtaining a result from discordant observations, is to take the arithmetical mean of the result. This amounts to supposing all the observations of equal weight. Daniel Bernoulli objects to this supposition, and considers that small errors are more probable than large errors. Let e denote an error; he proposes to measure the probability of the error by $\sqrt{(r^2 - e^2)}$, where r is a constant. Then the best result from a number of observations will be that which makes the product of the probabilities of all the errors a maximum. Thus, suppose that observations have given the values a, b, c, ... for an element; denote the true value by x; then we have to find x so that the following product may be a maximum:

$$\sqrt{\{r^2 - (x-a)^2\}} \; \sqrt{\{r^2 - (x-b)^2\}} \; \sqrt{\{r^2 - (x-c)^2\}} \ldots$$

Daniel Bernoulli gives directions as to the value to be assigned to the constant r.

426. Thus Daniel Bernoulli agrees in some respects with modern theory. The chief difference is that modern theory takes for the curve of probability that defined by the equation

$$y = \sqrt{\frac{c}{\pi}} \, e^{-cx^2},$$

while Daniel Bernoulli takes a circle.

Daniel Bernoulli gives some good remarks on the subject; and he illustrates his memoir by various numerical examples, which however are of little interest, because they are not derived from real observations. It is a fatal objection to his method, even if no other existed, that as soon as the number of observations surpasses two, the equation from which the unknown quantity is to be found rises to an unmanageable degree. This objection he himself recognises.

427. Daniel Bernoulli's memoir is followed by some remarks by Euler, entitled *Observationes in praecedentem dissertationem;* these occupy pages 24—33 of the volume.

Euler considers that Daniel Bernoulli was quite arbitrary in proposing to make the product of the probabilities of the errors a maximum. Euler proposes another method, which amounts to making the sum of the fourth powers of the probabilities a maximum, that is, with the notation of Art. 425,

$$\{r^2 - (x-a)^2\}^2 + \{r^2 - (x-b)^2\}^2 + \{r^2 - (x-c)^2\}^2 + \ldots$$

is to be a maximum. Euler says it is to be a *maximum*, but

he does not discriminate between a *maximum* and a *minimum*. The equation which is obtained for determining x is a cubic, and thus it is conceivable that there may be two minima values and one maximum, or only one minimum and no maximum.

Euler seems to have objected to the wrong part of Daniel Bernoulli's method ; the particular law of probability is really the arbitrary part, the principle of making the product of the probabilities a maximum is suggested by the Theory of Probability.

Euler illustrates his method by an example derived from real observations.

CHAPTER XII.

EULER.

428. EULER was born in 1707, and died in 1783. His industry and genius have left permanent impressions in every field of mathematics; and although his contributions to the Theory of Probability relate to subjects of comparatively small importance, yet they will be found not unworthy of his own great powers and fame.

429. Euler's first memoir is entitled *Calcul de la Probabilité dans le Jeu de Rencontre*. This memoir is published in the volume for 1751 of the *Histoire de l'Acad ... Berlin;* the date of publication is 1753: the memoir occupies pages 255—270 of the volume.

430. The problem discussed is that which is called the game of Treize, by Montmort and Nicolas Bernoulli; see Art. 162. Euler proceeds in a way which is very common with him; he supposes first one card, then two cards, then three, then four, and exhibits definitely the various cases which may occur. Afterwards, by an undemonstrated inductive process, he arrives at the general law.

The results obtained by Euler had been given more briefly and simply by Nicolas Bernoulli, and published by Montmort in his page 301; so we must conclude that Euler had not read Montmort's book.

When n is infinite, the expression given in Art. 161 for the

chance that at least one card is in its right place becomes equal
to $1 - e^{-1}$, where e is the base of the Napierian logarithms; this is
noticed by Euler: see also Art. 287.

431. The next memoir by Euler is entitled *Recherches générales sur la mortalité et la multiplication du genre humain*. This
memoir is published in the volume for 1760 of the *Histoire de
l'Acad. ... Berlin;* the date of publication is 1767: the memoir
occupies pages 144—164.

432. The memoir contains some simple theorems concerning
the mortality and the increase of mankind. Suppose N infants
born at the same time; then Euler denotes by (1) N the number
of them alive at the end of one year, by (2) N the number of
them alive at the end of two years, and so on.

Then he considers some ordinary questions. For example,
a certain number of men are alive, all aged m years, how many
of them will probably be alive at the end of n years?

According to Euler's notation, $(m)\,N$ represents the number
alive aged m years out of an original number N; and $(m+n)\,N$
represents the number of those who are alive at the end of n
more years; so that $\dfrac{(m+n)}{(m)}$ is the fraction of the number
aged m years who will probably be alive at the end of n years.
Thus, if we have a number M at present aged m years, there will
probably be $\dfrac{(m+n)}{(m)}\,M$ of them alive at the end of n years.

433. Then Euler gives formulæ for annuities on a life. Suppose M persons, at present each aged m years, and that each
of them pays down the sum a, for which he is to receive x
annually as long as he lives. Let $\dfrac{1}{\lambda}$ be the present worth of the
unit of money due at the end of one year.

Then at the end of a year there will be $M\,\dfrac{(m+1)}{(m)}$ of the
persons alive, each of whom is to receive x: therefore the present
worth of the whole sum to be received is $\dfrac{x}{\lambda}\,M\,\dfrac{(m+1)}{(m)}$.

Similarly, at the end of the second year there will be $M \dfrac{(m+2)}{(m)}$ of the persons alive, each of whom is to receive x : therefore the present worth of the whole sum to be received is $\dfrac{x}{\lambda^2} M \dfrac{(m+2)}{(m)}$. And so on.

The present worth of all the sums to be received ought to be equal to Ma; hence dividing by M we get

$$a = \frac{x}{(m)} \left\{ \frac{(m+1)}{\lambda} + \frac{(m+2)}{\lambda^2} + \frac{(m+3)}{\lambda^3} + \ldots \right\}.$$

Euler gives a numerical table of the values of (1), (2), ... (95), which he says is deduced from the observations of Kerseboom.

434. Let N denote the number of infants born in one year, and rN the number born in the next year; then we may suppose that the same causes which have changed N into rN will change rN into r^2N, so that r^2N will be the number born in the year succeeding that in which rN were born. Similarly, r^3N will be born in the next succeeding year, and so on. Let us now express the number of the population at the end of 100 years.

Out of the N infants born in the present year, there will be (100) N alive; out of the rN born in the next year, there will be (99) rN alive; and so on. Thus the whole number of persons alive at the end of 100 years will be

$$Nr^{100} \left\{ 1 + \frac{(1)}{r} + \frac{(2)}{r^2} + \frac{(3)}{r^3} + \ldots \right\}.$$

Therefore the ratio of the population in the 100th year to the number of infants born in that year will be

$$1 + \frac{(1)}{r} + \frac{(2)}{r^2} + \frac{(3)}{r^3} + \ldots$$

If we assume that the ratio of the population in any year to the number of infants born in that year is constant, and we know this ratio for any year, we may equate it to the expression just given : then since (1), (2), (3), ... are known by observation, we have an equation for finding r.

435. A memoir by Euler, entitled *Sur les Rentes Viageres*, immediately follows the preceding, occupying pages 165—175 of the volume.

Its principal point is a formula for facilitating the calculation of a life annuity.

Let A_m denote the value of an annuity of one pound on the life of a person aged m years, A_{m+1} the value of an annuity of one pound on the life of a person aged $m+1$ years. Then by the preceding memoir, Art. 433,

$$A_m = \frac{1}{(m)} \left\{ \frac{((m+1)}{\lambda} + \frac{(m+2)}{\lambda^2} + \frac{(m+3)}{\lambda^3} + \ldots \right\},$$

$$A_{m+1} = \frac{1}{(m+1)} \left\{ \frac{(m+2)}{\lambda} + \frac{(m+3)}{\lambda^2} + \frac{(m+4)}{\lambda^3} + \ldots \right\};$$

therefore $\quad (m)\,\lambda\,A_m = (m+1) + (m+1)\,A_{m+1}.$

Thus when A_m has been calculated, we can calculate A_{m+1} easily.

Euler gives a table exhibiting the value of an annuity on any age from 0 to 94. But with respect to the ages 90, 91, 92, 93, 94, he says,

Mais je ne voudrois pas conseiller à un entrepreneur de se mêler avec de tels vieillards, à moins que leur nombre ne fut assez considérable; ce qui est une regle générale pour tous les établissemens fondés sur les probabilités.

Euler is of opinion that the temptations do not appear sufficient to induce many persons to buy annuities on terms which would be advantageous to the sellers. He suggests that *deferred annuities* might perhaps be more successful; for it follows from his calculations, that 350 crowns should purchase for a new born infant an annuity of 100 crowns to commence at the age of 20 years, and continue for life. He adds,

...et si l'on y vouloit employer la somme de 3500 écus, ce seroit toujours un bel établissement, que de jouir dès l'age de 20 ans d'une pension fixe de 1000 écus. Cependant il est encore douteux, s'il se trouveroit plusieurs parens qui voudroient bien faire un tel sacrifice pour le bien de leurs enfans.

436. The next memoir by Euler is entitled *Sur l'avantage du
Banquier au jeu de Pharaon.* This memoir was published in the
volume for 1764 of the *Histoire de l'Acad....Berlin;* the date of
publication is 1766 : the memoir occupies pages 144—164.

437. Euler merely solves the same problem as had been
solved by Montmort and Nicolas Bernoulli, but he makes no reference to them or any other writer. He gives a new form however
to the result which we will notice.

Consider the equation in Finite Differences,

$$u_n = \frac{m\,(m-1)}{2n\,(n-1)} + \frac{(n-m)\,(n-m-1)}{n\,(n-1)}\,u_{n-2}.$$

By successive substitution we obtain

$$u_n = \frac{m\,(m-1)\,S}{2n\,(n-1)\,(n-2)\,...\,(n-m+1)},$$

where S denotes the sum $\phi\,(n) + \phi\,(n-2) + \phi\,(n-4) + ...,$

$\phi\,(n)$ being $(n-2)\,(n-3)\,...\,(n-m+1).$

This coincides with what we have given in Art. 155, supposing
that for A we put unity.

We shall first find a convenient expression for S. We see that

$$\frac{\phi\,(n)}{\lfloor m-2} = \text{coefficient of } x^{m-2} \text{ in the expansion of } (1+x)^{n-2}.$$

Hence S is equal to $\lfloor m-2$ times the coefficient of x^{m-2} in the
expansion of

$$(1+x)^{n-2} + (1+x)^{n-4} + (1+x)^{n-6} + ...$$

Now in the game of Pharaon we have n always even; thus we
may suppose the series to be continued down to 1, and then its
sum is

$$\frac{(1+x)^n - 1}{(1+x)^2 - 1} \text{ that is } \frac{(1+x)^n - 1}{2x + x^2}.$$

Thus we require the coefficient of x^{m-1} in the expansion of

$$\frac{(1+x)^n - 1}{2 + x}.$$

This coefficient is

$$\frac{n(n-1)\ldots(n-m+2)}{2\,\underline{|m-1}} - \frac{n(n-1)\ldots(n-m+3)}{4\,\underline{|m-2}}$$

$$+ \frac{n(n-1)\ldots(n-m+4)}{8\,\underline{|m-3}} - \ldots$$

Then $S = \underline{|m-2}$ times this coefficient.

Hence with this expression for S we find that

$$u_n = \frac{1}{4}\frac{m}{n-m+1} - \frac{1}{8}\frac{m(m-1)}{(n-m+1)(n-m+2)}$$

$$+ \frac{1}{16}\frac{m(m-1)(m-2)}{(n-m+1)(n-m+2)(n-m+3)} - \ldots$$

$$+ (-1)^m \frac{1}{2^m}\frac{m(m-1)\ldots 2}{(n-m+1)\ldots(n-1)}.$$

This is the expression for the advantage of the Banker which was given by Nicolas Bernoulli, and to which we have referred in Art. 157.

Now the form which Euler gives for u_n is

$$\frac{m}{2^m}\left\{\frac{m-1}{1(n-1)} + \frac{(m-1)(m-2)(m-3)}{1.2.3(n-3)}\right.$$

$$\left.+ \frac{(m-1)(m-2)(m-3)(m-4)(m-5)}{1.2.3.4.5(n-5)} + \ldots\right\}.$$

Euler obtained this formula by trial from the cases in which $m = 2, 3, 4, \ldots 8$; but he gives no general demonstration. We will deduce it from Nicolas Bernoulli's formula.

By the theory of partial fractions we can decompose the terms in Nicolas Bernoulli's formula, and thus obtain a series of fractions having for denominators $n-1$, $n-2$, $n-3$, $\ldots n-m+1$; and the numerators will be independent of n.

We will find the numerator of the fraction whose denominator is $n - r$.

From the last term in Nicolas Bernoulli's formula we obtain

$$\frac{(-1)^{r+1}}{2^m}\frac{m(m-1)\ldots 2}{\underline{|m-1-r}\,\underline{|r-1}};$$

from the last term but one we obtain

$$\frac{(-1)^r}{2^{m-1}} \frac{m\,(m-1)\,\ldots\,3}{\lfloor m-1-r \rfloor\,\lfloor r-2 \rfloor};$$

and proceeding in this way we find for the sum

$$\frac{(-1)^{r+1}\,\lfloor m}{2^m\,\lfloor r-1\,\lfloor m-1-r} \left\{ 1 - \frac{r-1}{1\cdot 2}\,2 + \frac{(r-1)\,(r-2)}{1\cdot 2\cdot 3}\,2^2 + \ldots \right\}$$

$$= \frac{(-1)^{r+1}\,\lfloor m}{2^{m+1}\,\lfloor r\,\lfloor m-1-r} \left\{ 1 - (1-2)^r \right\}.$$

This vanishes if r be an *even* number; and is equal to

$$\frac{\lfloor m}{2^m\,\lfloor r\,\lfloor m-1-r},$$

if r be *odd*.

Thus Euler's formula follows from Nicolas Bernoulli's.

438. The next memoir by Euler is entitled *Sur la probabilité des séquences dans la Lotterie Génoise*. This memoir was published in the volume for 1765 of the *Histoire de l'Acad....Berlin*; the date of publication is 1767; the memoir occupies pages 191—230.

439. In the lottery here considered 90 tickets are numbered consecutively from 1 to 90, and 5 tickets are drawn at random. The question may be asked, what is the chance that two or more consecutive numbers should occur in the drawing? Such a result is called a *sequence*; thus, for example, if the numbers drawn are 4, 5, 6, 27, 28, there is a sequence of three and also a sequence of two. Euler considers the question generally. He supposes that there are n tickets numbered consecutively from 1 to n, and he determines the chance of a sequence, if *two* tickets are drawn, or if *three* tickets are drawn, and so on, up to the case in which *six* tickets are drawn. And having successively investigated all these cases he is able to perceive the general laws which would hold in any case. He does not formally demonstrate these laws, but their truth can be inferred from what he has previously given, by the method of induction.

440. As an example of Euler's method we will give his investigation of the case in which *three* tickets are drawn.

There are three events which may happen which may be represented as follows:

I. a, $a+1$, $a+2$, that is a sequence of three.

II. a, $a+1$, b, that is a sequence of two, the number b being neither $a+2$ nor $a-1$.

III. a, b, c, where the numbers a, b, c involve no sequence.

I. The form a, $a+1$, $a+2$. The number of such events is $n-2$. For the sequence may be (1, 2, 3), or (2, 3, 4), or (3, 4, 5), up to $(n-2, n-1, n)$.

II. The form a, $a+1$, b. In the same way as we have just shewn that the number of sequences of three, like a, $a+1$, $a+2$, is $n-2$, it follows that the number of sequences of two, like a, $a+1$, is $n-1$. Now in general b may be any number between 1 and n inclusive, except $a-1$, a, $a+1$, $a+2$; that is, b may be any number out of $n-4$ numbers. But in the case of the first sequence of two, namely 1, 2, and also of the last sequence $n-1$, n, the number of admissible values of b is $n-3$. Hence the whole number of events of the form a, $a+1$, b, is $(n-1)(n-4)+2$, that is n^2-5n+6, that is $(n-2)(n-3)$.

III. The form a, b, c. Suppose a to be any number, then b and c must be taken out of the numbers from 1 to $a-2$ inclusive, or out of the numbers from $a+2$ to n inclusive; and b and c must not be consecutive. Euler investigates the number of events which can arise. It will however be sufficient for us here to take another method which he has also given. The total number of events is the number of combinations of n things taken 3 at a time, that is $\dfrac{n(n-1)(n-2)}{1.2.3}$. The number of events of the third kind can be obtained by subtracting from the whole number the number of those of the first and second kind; it is therefore

$$\frac{n(n-1)(n-2)}{1.2.3} - (n-2)(n-3) - (n-2).$$

It will be found that this is

$$\frac{(n-2)\,(n-3)\,(n-4)}{1.2.3}.$$

The chances of the three events will be found by dividing the number of ways in which they can respectively occur by the whole number.

Thus we obtain for I, II, III, respectively

$$\frac{2.3}{n\,(n-1)}, \quad \frac{2.3\,(n-3)}{n\,(n-1)}, \quad \text{and} \quad \frac{(n-3)\,(n-4)}{n\,(n-1)}.$$

441. Euler's next memoir also relates to a lottery. This memoir is entitled *Solution d'une question tres difficile dans le Calcul des Probabilités*. It was published in the volume for 1769 of the *Histoire de l'Acad. ... Berlin;* the date of publication is 1771 : the memoir occupies pages 285—302 of the volume.

442. The first sentences give a notion of the nature of the problem.

C'est le plan d'une lotterie qui m'a fourni cette question, que je me propose de développer. Cette lotterie étoit de cinq classes, chacune de 10000 billets, parmi lesquels il y avoit 1000 prix dans chaque classe, et par conséquent 9000 blancs. Chaque billet devoit passer par toutes les cinq classes; et cette lotterie avoit cela de particulier qu'outre les prix de chaque classe on s'engageoit de payer un ducat à chacun de ceux dont les billets auroient passé par toutes les cinq classes sans rien gagner.

443. We may put it perhaps more clearly thus. A man takes the *same* ticket in 5 different lotteries, each having 1000 prizes to 9000 blanks. Besides his chance of the prizes, he is to have £1 returned to him if he gains no prize.

The question which Euler discusses is to determine the probable sum which will thus have to be paid to those who fail in obtaining prizes.

444. Euler's solution is very ingenious. Suppose k the number of classes in the lottery; let n be the number of prizes in each class, and m the number of blanks.

Suppose the tickets of the first class to have been drawn, and that the prizes have fallen on certain n tickets A, B, C...

Let the tickets of the second class be now drawn. Required the chance that the prizes will fall on the same n tickets as before. The chance is

$$\frac{1 \cdot 2 \ldots\ldots n}{(m+1)(m+2)\ldots\ldots(m+n)}.$$

And in like manner the chance that the prizes in all the classes will fall on the same tickets as in the first class, is obtained by raising the fraction just given to the power $k-1$.

Let
$$\{(m+1)(m+2)\ldots\ldots(m+n)\}^{k-1} = M,$$
and
$$\{1 \cdot 2 \ldots\ldots n\}^{k-1} = \alpha.$$

Then $\dfrac{\alpha}{M}$ is the chance that all the prizes will fall on the same n tickets. In this case there are m persons who obtain no prize, and so the managers of the lottery have to pay m ducats.

445. Now consider the case in which there are $m-1$ persons who obtain no prize at all. Here besides the n tickets A, B, C, ... which gained in the first class, one of the other tickets, of which the number is m, gains in some one or more of the remaining classes. Denote the number of ways in which this can happen by βm. Now M denotes the whole number of cases which can happen after the first class has been drawn. Moreover β *is independent of* m. This statement involves the essence of Euler's solution. The reason of the statement is, that all the cases which can occur will be produced by distributing in various ways the fresh ticket among A, B, C, ... excluding one of these to make way for it.

In like manner, in the case in which there are $m-2$ persons who obtain no prize at all, there are two tickets out of the m which failed at first that gain prizes once or oftener in the remaining classes. The number of ways in which this can occur may be denoted by $\gamma m (m-1)$, where γ *is independent of* m.

Proceeding in this way we have from the consideration that the sum of all possible cases is M

$$M = \alpha + \beta m + \gamma m (m-1) + \delta m (m-1)(m-2) + \cdots$$

Now α, β, γ, ... are *all independent of m*. Hence we may put in succession for m the values 1, 2, 3, ...; and we shall thus be able to determine β, γ ....

446. Euler enters into some detail as to the values of β, γ ...; but he then shews that it is not necessary to find their values for his object.

For he proposed to find the probable expense which will fall on the managers of the lottery. Now on the first hypothesis it is m ducats, on the second it is $m-1$ ducats, on the third it is $m-2$ ducats, and so on. Thus the probable expense is

$$\frac{1}{M}\left\{\alpha m + \beta m\,(m-1) + \gamma m\,(m-1)\,(m-2) + ...\right\},$$

$$= \frac{m}{M}\left\{\alpha + \beta\,(m-1) + \gamma\,(m-1)\,(m-2) + ...\right\}.$$

The expression in brackets is what we shall get if we change m into $m-1$ in the right-hand member of the value of M in Art. 445; the expression therefore is what M becomes when we change m into $m-1$. Thus

$$\alpha + \beta\,(m-1) + \gamma\,(m-1)\,(m-2) + ...$$
$$= \{m\,(m+1)\,...\,(m+n-1)\}^{k-1}.$$

Thus finally the probable expense is

$$m\left(\frac{m}{m+n}\right)^{k-1}.$$

Euler then confirms the truth of this simple result by general reasoning.

447. We have next to notice a memoir entitled *Éclaircissemens sur le mémoire de Mr. De La Grange, inséré dans le V⁰ volume de Mélanges de Turin, concernant la méthode de prendre le milieu entre les résultats de plusieurs observations, &c.* Présenté à l'Académie le 27 Nov. 1777. This memoir was published in the *Nova Acta Acad. ... Petrop.* Tom. 3, which contains the history of the Academy for the year 1785; the date of publication of the volume is 1788: the memoir occupies pages 289—297.

The memoir consists of explanations of part of that memoir by Lagrange to which we have alluded in Art. 371; nothing new is given. The explanations seem to have been written for the benefit of some beginner in Algebra, and would be quite unnecessary for any student unless he were very indolent or very dull.

448. The next contribution of Euler to our subject relates to a lottery; the problem is one that has successively attracted the attention of De Moivre, Mallet, Laplace, Euler and Trembley. We shall find it convenient before we give an account of Euler's solution to advert to what had been previously published by De Moivre and Laplace.

In De Moivre's *Doctrine of Chances*, Problem XXXIX. of the third edition is thus enunciated: To find the Expectation of A, when with a Die of any given number of Faces, he undertakes to fling any number of them in any given number of Casts. The problem, as we have already stated, first appeared in the *De Mensura Sortis*. See Arts. 251 and 291.

Let n be the number of faces on the die; x the number of throws, and suppose that m specified faces are to come up. Then the number of favourable cases is

$$ n^x - m\,(n-1)^x + \frac{m\,(m-1)}{1\,.\,2}\,(n-2)^x - \ldots $$

where the series consists of $m+1$ terms. The whole number of possible cases is n^x, and the required chance is obtained by dividing the number of favourable cases by the whole number of possible cases.

449. The following is De Moivre's method of investigation. First, suppose we ask in how many ways the ace can come up. The whole number of cases is n^x; the whole number of cases if the ace were expunged would be $(n-1)^x$; thus the whole number of cases in which the ace can come up is $n^x - (n-1)^x$.

Next, suppose we ask in how many ways the ace and deux can come up. If the deux were expunged, the number of ways in which the ace could come up would be $(n-1)^x - (n-2)^x$, by what we have just seen; this therefore is the number of ways in which with the given die the ace can come up *without* the deux. Subtract this number from the number of ways in which the ace can come up with or without the deux, and we have left the number of ways in which the ace can come up *with* the deux. Thus the result is

$$n^x - (n-1)^x - \{(n-1)^x - (n-2_j{}^r\} ;$$

that is, $\qquad n^x - 2(n-1)^x + (n-2_j{}^x.$

De Moivre in like manner briefly considers the case in which the ace, the deux, and the tray are to come up; he then states what the result will be when the ace, the deux, the tray, and the quatre are to come up; and finally, he enunciates verbally the general result.

De Moivre then proceeds to shew how approximate numerical values may be obtained from the formula; see Art. 292.

450. The result may be conveniently expressed in the notation of Finite Differences.

The number of ways in which m specified faces can come up is $\Delta^m (n-m)^x$; where m is of course not greater than n.

It is also obvious that if m be greater than x, the event required is impossible; and in fact we know that the expression $\Delta^m (n-m)^x$ vanishes when m is greater than x.

Suppose $n = m$; then the number of ways may be denoted by $\Delta^n 0^x$; the expression written at full is

$$n^x - n(n-1)^x + \frac{n(n-1)}{1.2}(n-2)^x - \ldots$$

451. One particular case of the general result at the end of the preceding Article is deserving of notice. If we put $x = n$, we obtain the number of ways in which all the n faces come up in n throws. The sum of the series when $x = n$ is known to be equal to the product $1.2.3\ldots n$, as may be shewn in various ways. But we may remark that this result can also be obtained by the Theory of Probability itself; for if all the n faces are to appear in n throws, there must be no repetition; and thus the

number of ways is the number of permutations of n things taken all together.

Thus we see that the sum of a certain series might be inferred indirectly by the aid of the Theory of Probability; we shall hereafter have a similar example.

452. In the *Mémoires ... par divers Savans*, Vol. VI., 1775, page 363, Laplace solves the following problem : A lottery consists of n tickets, of which r are drawn at each time; find the probability that after x drawings, all the numbers will have been drawn.

The numbers are supposed to be replaced after each drawing.

Laplace's method is substantially the same as is given in his *Théorie ... des Prob.*, page 192; but the approximate numerical calculations which occupy pages 193—201 of the latter work do not occur in the memoir.

Laplace solves the problem more generally than he enunciates it; for he finds the probability that after x drawings m specified tickets will all have been drawn, and then by putting n for m, the result for the particular case which is enunciated is obtained.

453. The most interesting point to observe is that the problem treated by Laplace is really coincident with that treated by De Moivre, and the methods of the two mathematicians are substantially the same.

In De Moivre's problem n^x is the whole number of cases; the corresponding number in Laplace's problem is $\{\phi(n, r)\}^x$, where by $\phi(n, r)$ we denote the number of combinations of n things taken r at a time. In De Moivre's problem $(n-1)^x$ is the whole number of cases that would exist if one face of the die were expunged; the corresponding number in Laplace's problem is $\{\phi(n-1, r)\}^x$. Similarly to $(n-2)^x$ in De Moivre's problem corresponds $\{\phi(n-2, r)\}^x$ in Laplace's. And so on. Hence, in Laplace's problem, the number of cases in which m specified tickets will be drawn is

$$\{\phi(n, r)\}^x - m\,\{\phi(n-1, r)\}^x + \frac{m\,(m-1)}{1\,.\,2}\,\{\phi(n-2, r)\}^x - \ldots;$$

and the probability will be found by dividing this number by the whole number of cases, that is by $\{\phi(n, r)\}^x$.

454. With the notation of Finite Differences we may denote the number of cases favourable to the drawing of m specified tickets by $\Delta^m \{\phi (n - m, r)\}^x$; and the number of cases favourable to the drawing of all the tickets by $\Delta^n \{\phi (0, r)\}^x$.

455. In the *Histoire de l'Acad. ... Paris*, 1783, Laplace gives an approximate numerical calculation, which also occurs in page 195 of the *Théorie ... des Prob*. He finds that in a lottery of 10000 tickets, in which a single ticket is drawn each time, it is an even chance that all will have been drawn in about 95767 drawings.

456. After this notice of what had been published by De Moivre and Laplace, we proceed to examine Euler's solution.

The problem appears in Euler's *Opuscula Analytica*, Vol. II., 1785. In this volume pages 331—346 are occupied with a memoir entitled *Solutio quarundam quaestionum difficiliorum in calculo probabilium*. Euler begins thus :

His quaestionibus occasionem dedit ludus passim publice institutus, quo ex nonaginta schedulis, numeris 1, 2, 3, 4,...90 signatis, statis temporibus quinae schedulae sorte extrahi solent. Hinc ergo hujusmodi quaestiones oriuntur: quanta scilicet sit probabilitas ut, postquam datus extractionum numerus fuerit peractus, vel omnes nonaginta numeri exierint, vel saltem 89, vel 88, vel pauciores. Has igitur quaestiones, utpote difficillimas, hic ex principiis calculi Probabilium jam pridem usu receptis, resolvere constitui. Neque me deterrent objectiones Illustris *D'Alembert*, qui hunc calculum suspectum reddere est conatus. Postquam enim summus Geometra studiis mathematicis valedixit, iis etiam bellum indixisse videtur, dum pleraque fundamenta solidissime stabilita evertere est aggressus. Quamvis enim hae objectiones apud ignaros maximi ponderis esse debeant, haud tamen metuendum est, inde ipsi scientiae ullum detrimentum allatum iri.

457. Euler says that he finds a certain symbol very useful in these calculations ; namely, he uses

$$\begin{bmatrix} p \\ q \end{bmatrix} \text{ for } \frac{p\,(p-1)\,......\,(p-q+1)}{1\,.\,2\,......\,q}.$$

458. Euler makes no reference to his predecessors De Moivre and Laplace. He gives the formula for the chance that all the

tickets shall be drawn. This formula corresponds with Laplace's. We have only to put $m = n$ in Art. 453.

Euler then considers the question in which $n - 1$, or $n - 2$, ... tickets *at least* are to be drawn. He discusses successively the first case and the second case briefly, and he enunciates his general result. This is the following; suppose we require that $n - \nu$ tickets at least shall be drawn, then the number of favourable cases is

$$\{\phi(n, r)\}^x - \phi(n, \nu + 1) \{\phi(n - \nu - 1, r)\}^x$$
$$+ (\nu + 1) \phi(n, \nu + 2) \{\phi(n - \nu - 2, r)\}^x$$
$$- \frac{(\nu + 1)(\nu + 2)}{1 \cdot 2} \phi(n, \nu + 3) \{\phi(n - \nu - 3, r)\}^x - \dots$$

This result constitutes the addition which Euler contributes to what had been known before.

459. Euler's method requires close attention in order to gain confidence in its accuracy; it resembles that which is employed in treatises on Algebra, to shew how many integers there are which are less than a given number and prime to it. We will give another demonstration of the result which will be found easier to follow.

The number of ways in which *exactly* m tickets are drawn is $\phi(n, m) \Delta^m \{\phi(0, r)\}^x$. For the factor $\Delta^m \{\phi(0, r)\}^x$ is, by Art. 454, the number of ways in which in a lottery of m tickets, all the tickets will appear in the course of x drawings; and $\phi(n, m)$ is the number of combinations of n things taken m at a time.

The number of ways in which $n - \nu$ tickets *at least* will appear, will therefore be given by the formula $\Sigma \phi(n, m) \Delta^m \{\phi(0, r)\}^x$, where Σ refers to m, and m is to have all values between n and $n - \nu$, both inclusive.

Thus we get

$$\Delta^n \{\phi(0, r)\}^x + n \Delta^{n-1} \{\phi(0, r)\}^x + \frac{n(n-1)}{1 \cdot 2} \Delta^{n-2} \{\phi(0, r)\}^x$$
$$+ \frac{n(n-1)(n-2)}{1 \cdot 2 \cdot 3} \Delta^{n-3} \{\phi(0, r)\}^x + \dots$$

the series extending to $\nu + 1$ terms.

We may write this for shortness thus,

$$\left\{\Delta^n + n\,\Delta^{n-1} + \frac{n\,(n-1)}{1\,.\,2}\,\Delta^{n-2} + \frac{n\,(n-1)\,(n-2)}{1\,.\,2\,.\,3}\,\Delta^{n-3} + \ldots\right\}\left\{\phi\,(0,\,r)\right\}^x.$$

Now put $E - 1$ for Δ, expand, and rearrange in powers of E; we shall thus obtain

$$\left\{E^n - \phi\,(n,\,\nu + 1)\,E^{n-\nu-1} + (\nu + 1)\,\phi\,(n,\,\nu + 2)\,E^{n-\nu-2}\right.$$

$$\left. - \frac{(\nu + 1)\,(\nu + 2)}{1\,.\,2}\,\phi\,(n,\,\nu + 3)\,E^{n-\nu-3} + \ldots\right\}\left\{\phi\,(0,\,r)\right\}^x;$$

and this coincides with Euler's result.

We shall find in fact that when we put $E - 1$ for Δ, the coefficient of E^{n-p} is

$$\frac{(-1)^p\,\lfloor n}{\lfloor p\,\lfloor n-p}\left\{1 - p + \frac{p\,(p-1)}{1\,.\,2} - \frac{p\,(n-1)\,(p-2)}{1\,.\,2\,.\,3} + \ldots\right\},$$

where the series in brackets is continued to $\nu + 1$ terms, unless p be less than $\nu + 1$ and then it is continued to $p + 1$ terms only. In the former case the sum of the series can be obtained by taking the coefficient of x^ν in the expansion of $(1 - x)^p\,(1 - x)^{-1}$, that is in the expansion of $(1 - x)^{p-1}$. In the latter case the sum would be the coefficient of x^p in the same expansion, and is therefore zero, except when p is zero and then it is unity.

460. Since r tickets are drawn each time, the greatest number of tickets which can be drawn in x drawings is xr. Thus, as Euler remarks, the expression

$$\{\phi\,(n,\,r)\}^x - n\,\{\phi\,(n-1,\,r)\}^x + \frac{n\,(n-1)}{1\,.\,2}\,\{\phi\,(n-2,\,r)\}^x - \ldots$$

must be zero if n be greater than xr; for the expression gives the number of ways in which n tickets can be drawn in r drawings. Euler also says that the case in which n is equal to xr is remarkable, for then the expression just given can be reduced to a product of factors, namely to

$$\frac{\lfloor n}{\{\,\lfloor r\,\}^x}.$$

Euler does not demonstrate this result; perhaps he deduced it from the Theory of Probability itself. For if $xr = n$, it is obvious that no ticket can be repeated, when all the tickets are drawn in r drawings. Thus the whole number of favourable cases which can occur at the first drawing must be the number of combinations of n things taken r at a time; the whole number of favourable cases which can occur at the second drawing is the number of combinations of $n - r$ things taken r at a time; and so on. Then the product of all these numbers gives the whole number of favourable cases.

This example of the summation of a series indirectly by the aid of the Theory of Probability is very curious; see also Art. 451.

461. Euler gives the following paragraph after stating his formulæ,

In his probabilitatibus aestimandis utique assumitur omnes litteras ad extrahendum aeque esse proclives, quod autem Ill. *D'Alembert* negat assumi posse. Arbitratur enim, simul ad omnes tractus jam ante peractos respici oportere; si enim quaepiam litterae nimis crebro fuerint extractae, tum eas in sequentibus tractibus rarius exituras; contrarium vero evenire si quaepiam litterae nimis raro exierint. Haec ratio, si valeret, etiam valitura esset si sequentes tractus demum post annum, vel adeo integrum saeculum, quin etiam si in alio quocunque loco instituerentur; atque ob eandem rationem etiam ratio haberi deberet omnium tractuum, qui jam olim in quibuscunque terrae locis fuerint peracti, quo certe vix quicquam absurdius excogitari potest.

462. In Euler's *Opuscula Analytica*, Vol. II., 1785, there is a memoir connected with Life Assurance. The title is *Solutio quaestionis ad calculum probabilitatis pertinentis. Quantum duo conjuges persolvere debeant, ut suis haeredibus post utriusque mortem certa argenti summa persolvatur.* The memoir occupies pages 315—330 of the volume.

Euler repeats a table which he had inserted in the Berlin Memoirs for 1760; see Art. 433. The table shews out of 1000 infants, how many will be alive at the end of any given year.

Euler supposes that in order to ensure a certain sum when both a husband and wife are dead, x is paid down and z paid annually besides, until both are dead. Ho investigates the relation which must then hold between x, z and the sum to be ensured. Thus a calculator may assign an arbitrary value to two of the three quantities and determine the third. He may suppose, for example, that the sum to bo ensured is 1000 Rubles, and that $x = 0$, and find z.

Euler does not himself calculate numerical results, but he leaves the formulæ quite ready for application, so that tables might be easily constructed.

CHAPTER XIII.

D'ALEMBERT.

463. D'ALEMBERT was born in 1717 and died in 1783. This great mathematician is known in the history of the Theory of Probability for his opposition to the opinions generally received; his high reputation in science, philosophy, and literature have secured an amount of attention for his paradoxes and errors which they would not have gained if they had proceeded from a less distinguished writer. The earliest publication of his peculiar opinions seems to be in the article *Croix ou Pile* of the *Encyclopédie ou Dictionnaire Raisonné*.... We will speak of this work simply as the *Encyclopédie*, and thus distinguish it from its successor the *Encyclopédie Méthodique*. The latter work is based on the former; the article *Croix ou Pile* is reproduced unchanged in the latter.

464. The date of the volume of the *Encyclopédie* containing the article *Croix ou Pile*, is 1754. The question proposed in the article is to find the chance of throwing *head* in the course of two throws with a coin. Let *H* stand for head, and *T* for tail. Then the common theory asserts that there are four cases equally likely, namely, *HH*, *TH*, *HT*, *TT*; the only unfavourable case is the last; therefore the required chance is $\frac{3}{4}$. D'Alembert however doubts whether this can be correct. He says that if head appears at the first throw the game is finished and therefore there is no

need of the second throw. Thus he makes only three cases, namely, II, TH, TT: therefore the chance is $\frac{2}{3}$.

Similarly in the case of three throws he makes only four cases, namely, II, TH, TTH, TTT: therefore the chance is $\frac{3}{4}$. The common theory would make eight equally likely cases, and obtain $\frac{7}{8}$ for the chance.

465. In the same article D'Alembert notices the *Petersburg Problem*. He refers to the attempts at a solution in the *Commentarii Acad....Petrop.* Vol. v, which we have noticed in Arts. 389—393 ; he adds : mais nous ne savons si on en sera satisfait ; et il y a ici quelque scandale qui mérite bien d'occuper les Algébristes. D'Alembert says we have only to see if the expectation of one player and the corresponding risk of the other really is infinite, that is to say greater than any assignable finite number. He says that a little reflexion will shew that it is, for the risk augments with the number of throws, and this number may by the conditions of the game proceed to any extent. He concludes that the fact that the game may continue for ever is one of the reasons which produce an *infinite* expectation.

D'Alembert proceeds to make some further remarks which are repeated in the second volume of his *Opuscules*, and which will come under our notice hereafter. We shall also see that in the fourth volume of his *Opuscules* D'Alembert in fact contradicts the conclusion which we have just noticed.

466. We have next to notice the article *Gageure*, of the *Encyclopédie;* the volume is dated 1757. D'Alembert says he will take this occasion to insert some very good objections to what he had given in the article *Croix ou Pile*. He says, Elles sont de M. Necker le fils, citoyen de Genève, professeur de Mathématiques en cette ville, ... nous les avons extraits d'une de ses lettres. The objections are three in number. First Necker denies that D'Alembert's three cases are *equally* likely, and justifies this denial. Secondly Necker gives a good statement of the solution on the

ordinary theory. Thirdly, he shews that D'Alembert's view is inadmissible as leading to a result which is obviously untrue: this objection is given by D'Alembert in the second volume of his *Opuscules*, and will come before us hereafter. D'Alembert after giving the objections says, Ces objections, sur-tout la derniere, méritent sans doute beaucoup d'attention. But still he does not admit that he is convinced of the soundness of the common theory.

The article *Gageure* is not reproduced in the *Encyclopédie Méthodique*.

467. D'Alembert wrote various other articles on our subject in the *Encyclopédie;* but they are unimportant. We will briefly notice them.

Absent. In this article D'Alembert alludes to the essay by Nicolas Bernoulli; see Art. 338.

Avantage. This article contains nothing remarkable.

Bassette. This article contains a calculation of the advantage of the Banker in one case, namely that given by Montmort on his page 145.

Carreau. This article gives an account of the *sorte de jeu dont M. de Buffon a donné le calcul in 1733, avant que d'être de l'Académie des Sciences;* see Art. 354.

Dé. This article shews all the throws which can be made with two dice, and also with three dice.

Loterie. This is a simple article containing ordinary remarks and examples.

Pari. This article consists of a few lines giving the ordinary rules. At the end we read: Au reste, ces règles doivent être modifiées dans certains cas, ou la probabilité de gagner est fort petite, et celle de perdre fort grande. Voyez Jeu. There is however nothing in the article *Jeu* to which this remark can apply, which is the more curious because of course *Jeu* precedes *Pari* in alphabetical order; the absurdity is reproduced in the *Encyclopédie Méthodique.*

The article *Probabilité* in the *Encyclopédie* is apparently by Diderot. It gives the ordinary view of the subject with the exception of the point which we have noticed in Art. 91.

468. In various places in his *Opuscules Mathématiques* D'Alembert gives remarks on the Theory of Probabilities. These remarks are mainly directed against the first principles of the subject which D'Alembert professes to regard as unsound. We will now examine all the places in which these remarks occur.

469. In the second volume of the *Opuscules* the first memoir is entitled *Reflexions sur le calcul des Probabilités;* it occupies pages 1—25. The date of the volume is 1761. D'Alembert begins by quoting the common rule for *expectation* in the Theory of Probability, namely that it is found by taking the product of the loss or gain which an event will produce, by the probability that this event will happen. D'Alembert says that this rule had been adopted by all analysts, but that cases exist in which the rule seems to fail.

470. The first case which D'Alembert brings forward is that of the *Petersburg Problem;* see Art. 389. By the ordinary theory *A* ought to give *B* an infinite sum for the privilege of playing with him. D'Alembert says,

Or, indépendamment de ce qu'une *somme infinie* est une chimere, il n'y a personne qui voulût donner pour jouer à ce jeu, je ne dis pas une somme infinie, mais même une somme assez modique.

471. D'Alembert notices a solution of the *Petersburg Problem* which had been communicated to him by un Géometre célébre de l'Académie des Sciences, plein de savoir et de sagacité. He means Fontaine I presume, as the solution is that which Fontaine is known to have given; see *Montucla,* page 403: in this solution the fact is considered that *B* cannot pay more than a certain sum, and this limits what *A* ought to give to induce *B* to play. D'Alembert says that this is unsatisfactory; for suppose it is agreed that the game shall only extend to a finite number of trials, say 100; then the theory indicates that *A* should give 50 crowns. D'Alembert asserts that this is too much.

The answer to D'Alembert is simple; and it is very well put in fact by Condorcet, as we shall see hereafter. The ordinary rule is entitled to be adopted, because *in the long run* it is equally fair to both parties A and B, and any other rule would be unfair to one or the other.

472. D'Alembert concludes from his remarks that when the probability of an event is very small it ought to be regarded and treated as zero. For example he says, suppose Peter plays with James on this condition ; a coin is to be tossed one hundred times, and if head appear at the last trial and not before, James shall give 2^{100} crowns to Peter. By the ordinary theory Peter ought to give to James one crown at the beginning of the game.

D'Alembert says that Peter ought not to give this crown because he will *certainly* lose, for head will appear before the hundredth trial, *certainly* though not *necessarily*.

D'Alembert's doctrine about a small probability being equivalent to zero was also maintained by Buffon.

473. D'Alembert says that we must distinguish between what is *metaphysically* possible, and what is *physically* possible. In the first class are included all those things of which the existence is not absurd ; in the second class are included only those things of which the existence is not too extraordinary to occur in the ordinary course of events. It is *metaphysically* possible to throw two sixes with two dice a hundred times running ; but it is *physically* impossible, because it never has happened and never will happen.

This is of course only saying in another way that a very small chance is to be regarded and treated as zero. D'Alembert shews however, that when we come to ask at what stage in the diminution of chance we shall consider the chance as zero, we are involved in difficulty ; and he uses this as an additional argument against the common theory.

See also Mill's *Logic*, 1862, Vol. II. page 170.

474. D'Alembert says he will propose an idea which has occurred to him, by which the ratio of probabilities may be estimated. The idea is simply to make experiments. He exemplifies it by supposing a coin to be tossed a large number of times, and the results to be observed. We shall find that this has been done at the instance of Buffon and others. It is needless to say that the advocates of the common Theory of Proba-

bility would be quite willing to accept D'Alembert's reference to experiment; for relying on the theorem of James Bernoulli, they would have no doubt that experiment would confirm their calculations. It is however curious that D'Alembert proceeds in his very next paragraph to make a remark which is quite inconsistent with his appeal to experiment. For he says that if head has arrived three times in succession, it is more likely that the next arrival will be tail than head. He says that the oftener head has arrived in succession the more likely it is that tail will arrive at the next throw. He considers that this is obvious, and that it furnishes another example of the defects of the ordinary theory. In the *Opuscules,* Vol. IV. pages 90—92, D'Alembert notices the charge of inconsistency which may be urged against him, and attempts to reply to it.

475. D'Alembert then proceeds to another example, which, as he intimates, he had already given in the *Encyclopédie,* under the titles *Croix ou Pile* and *Gageure;* see Art. 463. The question is this: required the probability of throwing a head with a coin in two trials.

D'Alembert came to the conclusion in the *Encyclopédie* that the chance ought to be $\frac{2}{3}$ instead of $\frac{3}{4}$. In the *Opuscules* however he does not insist very strongly on the correctness of the result $\frac{2}{3}$, but seems to be content with saying that the reasoning which produces $\frac{3}{4}$ is unsound.

D'Alembert urges his objections against the ordinary theory with great pertinacity; and any person who wishes to see all that a great mathematician could produce on the wrong side of a question should consult the original memoir. But we agree with every other writer on the subject in thinking that there is no real force in D'Alembert's objections.

476. The following extract will shew that D'Alembert no longer insisted on the absolute accuracy of the result $\frac{2}{3}$:

Je ne voudrois pas cependant regarder en toute rigueur les trois coups dont il s'agit, comme également possibles. Car 1°. il pourroit se faire en effet (et je suis même porté à le croire), que le cas *pile croix* ne fût pas éxactement aussi possible que le cas *croix* seul; mais le rapport des possibilités me paroît inappréciable. 2°. Il pourroit se faire encore que le coup *pile croix* fût un peu plus possible que *pile pile*, par cette seule raison que dans le dernier le même effet arrive deux fois de suite; mais le rapport des possibilités (supposé qu'elles soient inégales), n'est pas plus facile à établir dans ce second cas, que dans le premier. Ainsi il pourroit très-bien se faire que dans le cas proposé, le rapport des probabilités ne fût ni de 3 à 1, ni de 2 à 1 (comme nous l'avons supposé dans l'*Encyclopédie*) mais un incommensurable ou inappréciable, moyen entre ces deux nombres. Je crois cependant que cet incommensurable approchera plus de 2 que de 3, parce qu'encore une fois il n'y a que trois cas possibles, et non pas quatre. Je crois de même et par les mêmes raisons, que dans le cas où l'on joueroit en trois coups, le rapport de 3 à 1, que donne ma méthode, est plus près du vrai, que le rapport de 7 à 1, donné par la méthode ordinaire, et qui me paroît exorbitant.

477. D'Alembert returns to the objection which had been urged against his method, and which he noticed under the title *Gageure* in the *Encyclopédie;* see Art. 466. Let there be a die with three faces, A, B, C; then according to D'Alembert's original method in the *Encyclopédie*, the chances would always be rather against the appearance of a specified face A, however great the number of trials. Suppose n trials, then by D'Alembert's method the chance for the appearance of A is to the chance against it as $2^n - 1$ is to 2^n.

For example, suppose $n = 3$: then the favourable cases are A, BA, CA, BBA, BCA, CCA, CBA; the unfavourable cases are BBB, BBC, BCB, BCC, CBB, CBC, CCC, CCB: thus the ratio is that of 7 to 8. D'Alembert now admits that these cases are not equally likely to happen; though he believes it difficult to assign their ratio to one another.

Thus we may say that D'Alembert started with decided but erroneous opinions, and afterwards passed into a stage of general doubt and uncertainty; and the dubious honour of effecting the transformation may be attributed to Necker.

478. D'Alembert thus sums up his results, on his page 24 :

Concluons de toutes ces réfléxions; 1°. que si la régle que j'ai donnée dans l'*Encyclopédie* (faute d'en connôitre une meilleure) pour déterminer le rapport des probabilités au jeu de *croix et pile*, n'est point éxacte à la vigueur, la régle ordinaire pour déterminer ce rapport, l'est encore moins; 2°. que pour parvenir à une théorie satisfaisante du calcul des probabilités, il faudroit résoudre plusieurs Problêmes qui sont peut-être insolubles; savoir, d'assigner le vrai rapport des probabilités dans les cas qui ne sont pas également possibles, ou qui peuvent n'être pas regardés comme tels; de déterminer quand la probabilité doit être regardée comme nulle; de fixer enfin comment on doit estimer l'espérance ou l'enjeu, selon que la probabilité est plus ou moins grande.

479. The next memoir by D'Alembert which we have to notice is entitled *Sur l'application du Calcul des Probabilités à l'inoculation de la petite Vérole;* it is published in the second volume of the *Opuscules.* The memoir and the accompanying notes occupy pages 26—95 of the volume.

480. We have seen that Daniel Bernoulli had written a memoir in which he had declared himself very strongly in favour of Inoculation; see Art. 398. The present memoir is to a certain extent a criticism on that of Daniel Bernoulli. D'Alembert does not deny the advantages of Inoculation; on the contrary, he is rather in favour of it : but he thinks that the advantages and disadvantages had not been properly compared by Daniel Bernoulli, and that in consequence the former had been overestimated. The subject is happily no longer of the practical importance it was a century ago, so that we need not give a very full account of D'Alembert's memoir; we shall be content with stating some of its chief points.

481. Daniel Bernoulli had considered the subject as it related to the state, and had shewn that Inoculation was to be recommended, because it augmented the mean duration of life for the citizens. D'Alembert considers the subject as it relates to a private individual : suppose a person who has not yet been attacked by small-pox; the question for him is, whether he will be inoculated, and thus run the risk, small though it may be, of dying in the course of a few days, or whether he will take his

chance of escaping entirely from an attack of small-pox during his life, or at least of recovering if attacked.

D'Alembert thinks that the prospect held out to an individual of a gain of three or four years in the probable duration of his life, may perhaps not be considered by him to balance the immediate danger of submitting to Inoculation. The relative value of the alternatives at least may be too indefinite to be estimated; so that a person may hesitate, even if he does not altogether reject Inoculation.

482. D'Alembert lays great stress on the consideration that the additional years of life to be gained form a *remote* and not a *present* benefit; and moreover, on account of the infirmities of age, the later years of a life must be considered of far less value than the years of early manhood.

D'Alembert distinguishes between the *physical* life and the *real* life of an individual. By the former, he means life in the ordinary sense, estimated by total duration in years; by the latter, he means that portion of existence during which the individual is free from suffering, so that he may be said to enjoy life.

Again, with respect to utility to his country, D'Alembert distinguishes between the *physical* life and the *civil* life. During infancy and old age an individual is of no use to the state; he is a burden to it, for he must be supported and attended by others. During this period D'Alembert considers that the individual is a charge to the state; his value is *negative,* and becomes *positive* for the intermediate periods of his existence. The *civil* life then is measured by the excess of the productive period of existence over that which is burdensome.

Relying on considerations such as these, D'Alembert does not admit the great advantage which the advocates for Inoculation found in the fact of the prolongation of the mean duration of human life effected by the operation. He looks on the problem as far more difficult than those who had discussed it appeared to have supposed.

483. We have seen that Daniel Bernoulli assumed that the small-pox attacked every year 1 in n of those not previously

attacked, and that 1 died out of every m attacked; on these hypotheses he solved definitely the problem which he undertook. D'Alembert also gives a mathematical theory of inoculation; but he does not admit that Daniel Bernoulli's assumptions are established by observations, and as he does not replace them by others, he cannot bring out definite results like Daniel Bernoulli does. There is nothing of special interest in D'Alembert's mathematical investigation; it is rendered tedious by several figures of curves which add nothing to the clearness of the process they are supposed to illustrate.

The following is a specimen of the investigations, rejecting the encumbrance of a figure which D'Alembert gives.

Suppose a large number of infants born nearly at the same epoch; let y represent the number alive at the end of a certain time; let u represent the number who have died during this period of small-pox: let z represent the number who would have been alive if small-pox did not exist: required z in terms of y and u.

Let dz denote the decrement of z in a small time, dy the decrement of y in the same time. If we supposed the z individuals subject to small-pox, we should have

$$dz = \frac{z}{y}\, dy.$$

But we must subtract from this value of dz the decrement arising from small-pox, to which the z individuals are by hypothesis not liable: this is $\frac{z}{y}\, du$.

Thus, $$dz = \frac{z}{y}\, dy + \frac{z}{y}\, du;$$

we put $+\frac{z}{y}\, du$ and not $-\frac{z}{y}\, du$, because z and y diminish while u increases. Then

$$\frac{dz}{z} = \frac{dy}{y} + \frac{du}{y};$$

therefore $$\log z = \log y + \int \frac{du}{y};$$

therefore $$z = y c^{\int \frac{du}{y}}.$$

'ALEMBERT.

The result is not of practical use because the value of the integral $\int \dfrac{du}{y}$ is not known. D'Alembert gives several formulæ which involve this or similar unfinished integrations.

484. D'Alembert draws attention on his page 74 to the two distinct methods by which we may propose to estimate the *espérance de vivre* for a person of given age. The *mean* duration of life is the average duration in the ordinary sense of the word average; the *probable* duration is such a duration that it is an even chance whether the individual exceeds it or falls short of it. Thus, according to Halley's tables, for an infant the *mean* life is 26 years, that is to say if we take a large number N of infants the sum of the years of their lives will be $26N$; the *probable* life is 8 years, that is to say $\dfrac{N}{2}$ of the infants die under 8 years old and $\dfrac{N}{2}$ die over 8 years old.

The terms *mean* life and *probable* life which we here use have not always been appropriated in the sense we here explain; on the contrary, what we call the mean life has sometimes been called the probable life. D'Alembert does not propose to distinguish the two notions by such names as we have used. His idea is rather that each of them might fairly be called the duration of life to be expected, and that it is an objection against the Theory of Probability that it should apparently give two different results for the same problem.

485. We will illustrate the point as D'Alembert does, by means of what he calls the *curve of mortality*.

Let x denote the number of years measured from an epoch; let $\psi(x)$ denote the number of persons alive at the end of x years from birth, out of a large number born at the same time. Let $\psi(x)$ be the ordinate of a curve; then $\psi(x)$ diminishes from $x = 0$ to $x = c$, say, where c is the greatest age that persons can attain, namely about 100 years.

This curve is called the curve of mortality by D'Alembert.

The *mean* duration of life for persons of the age a years is

$$\frac{\int_a^c \psi(x)\,dx}{\psi(a)}.$$

The *probable* duration is a quantity b such that

$$\psi(b) = \frac{1}{2}\psi(a).$$

This is D'Alembert's mode. We might however use another curve or function. Let $\phi(x)$ be such that $\phi(x)\,dx$ represents the number who die in an element of time dx. Then the *mean* duration of life for persons aged a years is

$$\frac{\int_a^c (x-a)\,\phi(x)\,dx}{\int_a^c \phi(x)\,dx}.$$

The *probable* duration is a quantity b such that

$$\int_a^b \phi(x)\,dx = \int_b^c \phi(x)\,dx,$$

that is

$$\int_a^b \phi(x)\,dx = \frac{1}{2}\int_a^c \phi(x)\,dx.$$

Thus the *mean* duration is represented by the abscissa of the centre of gravity of a certain area; and the *probable* duration is represented by the abscissa corresponding to the ordinate which bisects that area.

This is the modern method of illustrating the point; see Art. 101 of the *Theory of Probability* in the *Encyclopædia Metropolitana*.

486. We may easily shew that the two methods of the preceding Article agree.

For we have $\phi(x) = -k\,\psi'(x)$, where k is some constant. Therefore

$$\frac{\int_a^c (x-a)\,\phi(x)\,dx}{\int_a^c \phi(x)\,dx} = \frac{\int_a^c (x-a)\,\psi'(x)\,dx}{\int_a^c \psi'(x)\,dx};$$

and
$$\int (x-a)\,\psi'(x)\,dx = (x-a)\,\psi(x) - \int \psi(x)\,dx,$$

therefore
$$\int_a^c (x-a)\,\psi'(x)\,dx = -\int_a^c \psi(x)\,dx;$$

and
$$\int_a^c \psi'(x)\,dx = -\psi(a).$$

Thus
$$\frac{\int_a^c (x-a)\,\phi(x)\,dx}{\int_a^c \phi(x)\,dx} = \frac{\int_a^c \psi(x)\,dx}{\psi(a)}.$$

This shews that the two methods give the same *mean* duration. In the same way it may be shewn that they give the same *probable* duration.

487. D'Alembert draws attention to an erroneous solution of the problem respecting the advantages of Inoculation, which he says was communicated to him by *un savant Géometre*. D'Alembert shews that the solution must be erroneous because it leads to untenable results in two cases to which he applies it. But he does not shew the nature of the error, or explain the principle on which the pretended solution rests; and as it is rather curious we will now consider it.

Suppose that N infants are born at the same epoch, and let a table of mortality be formed by recording how many die in each year of all diseases excluding small-pox, and also how many die of small-pox. Let the table be denoted as here; so that u_r denotes the number who die in the rth year excluding those who die of small-pox, and v_r denotes the number who die of small-pox. Then we can use the table in the following way: suppose M any other number, then if u_r die in the rth year out of N from all diseases except small-pox, $\frac{M}{N} u_r$ would die out of M; and so for any other proportion.

1	u_1	v_1
2	u_2	v_2
3	u_3	v_3
4	u_4	v_4

Now suppose small-pox eradicated from the list of human diseases; required to construct a new table of mortality from the above data. The *savant Géometre* proceeds thus. He takes the preceding table and *destroys the column* v_1, v_2, v_3, ... Then he assumes that the remaining column will shew the correct mortality for the number $N - n$ at starting, where n is the total number who died of small-pox, that is $n = v_1 + v_2 + v_3 + ...$

Thus if we start with the number M of infants $\dfrac{M}{N-n} u_r$ would die on this assumption in the rth year.

There is a certain superficial plausibility in the method, but it is easy to see that it is unsound, for it takes *too unfavourable* a view of human life after the eradication of small-pox. For let

$$u_1 + u_2 + ... u_r = U_r,$$
$$v_1 + v_2 + ... v_r = V_r;$$

then we know from the observations that at the end of r years there are $N - U_r - V_r$ survivors of the original N; of these u_{r+1} die in the next year from all diseases excluding small-pox. Thus excluding small-pox

$$\frac{u_{r+1}}{N - U_r - V_r},$$

is the ratio of those who die in the year to those who are aged r years at the beginning of the year. And this ratio will be the ratio which ought to hold in the new tables of mortality. The method of the *savant Géometre* gives instead of this ratio the *greater* ratio

$$\frac{u_{r+1}}{N - U_r - n}.$$

488. Thus we see where the *savant Géometre* was wrong, and the nature of the error. The pages in D'Alembert are 88—92; but it will require some attention to extricate the false principle really used from the account which D'Alembert gives, which is also obscured by a figure of a curve. In D'Alembert's account regard is paid to the circumstance that Inoculation is fatal to some on whom it is performed; but this is only a matter of detail: the essential principle involved is that which we have here exhibited.

489. The next publication of D'Alembert on the subject of Probabilities appears to consist of some remarks in his *Mélanges*

de Philosophie, Vol. v. I have never seen the original edition of this work ; but I have no doubt that the remarks in the *Mélanges de Philosophie* were those which are reprinted in the first volume of the collected edition of the literary and philosophical works of D'Alembert, in 5 Vols. 8vo, Paris, 1821. According to the citations of some writers on the subject I conclude that these remarks also occur in the fourth volume of the edition of the literary and philosophical works in 18 Vols. 8vo, Paris, 1805.

490. In the first volume of the edition of 1821 there are two essays, one on the general subject of Probabilities, and the other on Inoculation.

The first essay is entitled *Doutes et questions sur le Calcul des Probabilités.* These occupy pages 451—466 ; the pages being closely printed.

D'Alembert commences thus :

On se plaint assez communément que les formules des mathématiciens, appliquées aux objets de la nature, ne se trouvent que trop en défaut. Personne néanmoins n'avait encore aperçu ou cru apercevoir cet inconvénient dans *le calcul des probabilités.* J'ai osé le premier proposer des doutes sur quelques principes qui servent de base à ce calcul. De grands géomètres ont jugé ces doutes *dignes d'attention;* d'autres grands géomètres les ont trouvés *absurdes;* car pourquoi adoucirais-je les termes dont ils se sont servis ? La question est de savoir s'ils ont eu tort de les employer, et en ce cas ils auraient doublement tort. Leur décision, qu'ils n'ont pas jugé à propos de motiver, a encouragé des mathématiciens médiocres, qui se sont hâtés d'écrire sur ce sujet, et de m'attaquer sans m'entendre. Je vais tâcher de m'expliquer si clairement, que presque tous mes lecteurs seront à portée de me juger.

491. The essay which we are now considering may be described in general as consisting of the matter in the second volume of the *Opuscules* divested of mathematical formulæ and so adapted to readers less versed in mathematics. The objections against the ordinary theory are urged perhaps with somewhat less confidence ; and the particular case in which $\frac{2}{3}$ was proposed instead of $\frac{3}{4}$ as the result in an elementary question does not appear. But the other errors are all retained.

492. There is some additional matter in the essay. D'Alembert notices the calculation of Daniel Bernoulli relative to the small inclination to the ecliptic of the orbits of the planets; see Art. 394. D'Alembert considers Daniel Bernoulli's result as worthless.

D'Alembert says with respect to Daniel Bernoulli,

Ce qu'il y a de singulier, c'est que ce grand géomètre dont je parle, a trouvé *ridicules*, du moins à ce qu'on m'assure, mes raisonnemens sur le *calcul des probabilités*.

493. D'Alembert introduces an illustration which Laplace afterwards adopted. D'Alembert supposes that we see on a table the letters which form the word *Constantinopolitanensibus*, arranged in this order, or arranged in alphabetical order; and he says that although mathematically these distributions and a third case in which the letters follow at hazard are equally possible, yet a man of sense would scarcely doubt that the first or second distribution had not been produced by chance. See Laplace, *Théorie ... des Prob.* page XI.

494. D'Alembert quotes the article *Fatalité* in the *Encyclopédie*, as supporting him at least partially in one of the opinions which he maintained; namely that which we have noticed in the latter part of our Art. 474. The name of the writer of the article *Fatalité* is not given in the *Encyclopédie*.

495. The other essay which we find in the first volume of the edition of D'Alembert's literary and philosophical works of 1821, is entitled *Réflexions sur l'Inoculation;* it occupies pages 463—514.

In the course of the preface D'Alembert refers to the fourth volume of his *Opuscules*. The fourth volume of the *Opuscules* is dated 1768; in the preface to it D'Alembert refers to his *Mélanges de Philosophie*, Vol. v.

We may perhaps infer that the fifth volume of the *Mélanges ...* and the fourth volume of the *Opuscules* appeared at about the same date.

496. The essay may be said to consist of the same matter

18

as appeared on the subject in the second volume of the *Opuscules*, omitting the mathematical investigations, but expanding and illustrating all the rest.

D'Alembert's general position is that the arguments which have hitherto been brought forward for Inoculation or against it are almost all unsound. His own reflexions however lead to the conclusion that Inoculation is advantageous, and that conclusion · seems more confidently maintained in the essay than in the *Opuscules*. Some additional facts concerning the subject are referred to in the essay; they had probably been published since the second volume of the *Opuscules*.

497. D'Alembert retains the opinion he had formerly held as to the difficulty of an exact mathematical solution of the problem respecting the advantages of Inoculation. He says in summing up his remarks on this point : S'il est quelqu'un à qui la solution de ce problème soit réservée, ce ne sera sûrement pas à ceux qui la croiront facile.

498. D'Alembert insists strongly on the want of ample collections of observations on the subject. He wishes that medical men would keep lists of all the cases of small-pox which come under their notice. He says,

...ces registres, donnés au public par les Facultés de médecine ou par les particuliers, seraient certainement d'une utilité plus palpable et plus prochaine, que les recueils d'observations météorologiques publiés avec tant de soin par nos Académies depuis 70 ans, et qui pourtant, à certains égards, ne sont pas eux-mêmes sans utilité.

Combien ne serait-il pas à souhaiter que les médecins, au lieu de se quereller, de s'injurier, de se déchirer mutuellement au sujet de l'inoculation avec un acharnement théologique, au lieu de supposer ou de déguiser les faits, voulussent bien se réunir, pour faire de bonne foi toutes les expériences nécessaires sur une matière si intéressante pour la vie des hommes ?

499. We next proceed to the fourth volume of *D'Alembert's Opuscules*, in which the pages 73—105 and 283—341 are devoted to our subject. The remarks contained in these pages are presented as extracts from letters.

500. We will now take the first of the two portions, which occupies pages 73—105.

D'Alembert begins with a section *Sur le calcul des Probabilités*. This section is chiefly devoted to the *Petersburg Problem*. The chance that head will not appear before the nth throw is $\frac{1}{2^n}$ on the ordinary theory. D'Alembert proposes quite arbitrarily to change this expression into some other which will bring out a finite result for A's expectation. He suggests $\frac{1}{2^n(1+\beta n^2)}$ where β is a constant. In this case the summation which the problem requires can only be effected approximately. He also suggests $\frac{1}{2^{n+an}}$ and $\frac{1}{2^{n+a(n-1)}}$ where α is a constant.

He gives of course no reason for these suggestions, except that they lead to a finite result instead of the infinite result of the ordinary theory. But his most curious suggestion is that of replacing 2^n by $2^n\left\{1+\dfrac{B}{(K-n)^{\frac{q}{2}}}\right\}$, where B and K are constants and q an odd integer. He says,

Nous mettons le nombre pair 2 au dénominateur de l'exposant, afin que quand on est arrivé au nombre n qui donne la probabilité égale à zero, on ne trouve pas la probabilité négative, en faisant n plus grand que ce nombre, ce qui seroit choquant; car la probabilité ne sauroit jamais être au-dessous de zero. Il est vrai qu'en faisant n plus grand que le nombre dont il s'agit, elle devient imaginaire; mais cet inconvénient me paroît moindre que celui de devenir négative;...

501. D'Alembert's next section is entitled *Sur l'analyse des Jeux*.

D'Alembert first proposes une considération très-simple et très-naturelle à faire dans le calcul des jeux, et dont M. de Buffon m'a donné la premiere idée, ... This consideration we will explain when noticing a work by Buffon. D'Alembert gives it in the form which Buffon *ought* to have given it in order to do justice to his own argument. But soon after in a numerical example

18—2

D'Alembert falls back on Buffon's own statement; for he supposes that a man has 100000 crowns, and that he stakes 50000 at an equal game, and he says that this man's damage if he loses is greater than his advantage if he gains; puisque dans le premier cas, il s'appauvrira de la moitié; et que dans le second, il ne s'enrichira que du tiers.

502. If a person has the chance $\dfrac{p}{p+q}$ of gaining x and the chance $\dfrac{q}{p+q}$ of losing y, his *expectation* on the ordinary theory is $\dfrac{px-qy}{p+q}$. D'Alembert obtains this result himself on the ordinary principles; but then he thinks another result, namely $\dfrac{px-qy}{p}$, might also be obtained and defended. Let z denote the sum which a man should give for the privilege of being placed in the position stated. If he gains he receives x, so that as he paid z his balance is $x - z$. Thus $\dfrac{p(x-z)}{p+q}$ is the corresponding expectation. If he loses, as he has already paid z *he will have to pay $y - z$ additional*, so that his total loss is y, and his consequent expectation $\dfrac{-qy}{p+q}$. Then $\dfrac{p(x-z)-qy}{p+q}$ is his total expectation, which ought to be zero if z is the fair sum for him to pay. Thus $z = \dfrac{px-qy}{p}$. It is almost superfluous to observe that the words which we have printed in Italics amount to assigning a new meaning to the problem. Thus D'Alembert gives us not two discordant solutions of the *same* problem, but solutions of two different problems. See his further remarks on his page 283.

503. D'Alembert objects to the common rule of multiplying the value to be obtained by the probability of obtaining it in order to determine the expectation. He thinks that the probability is the principal element, and the value to be obtained is subordinate. He brings the following example as an objection against the ordinary theory; but his meaning is scarcely intelligible :

Qu'on propose de choisir entre 100 combinaisons, dont 99 feront gagner mille écus, et la 100ᵉ 99 mille écus; quel sera l'homme assez insensé pour préférer celle qui donnera 99 mille écus. *L'espérance* dans les deux cas n'est donc pas *réellement* la même; quoiqu'elle soit la même suivant les régles des probabilités.

504. D'Alembert appeals to the authority of Pascal, in the following words :

Un homme, dit Pascal, passeroit pour fou, s'il hésitoit à se laisser donner la mort en cas qu'avec trois dez on fît vingt fois de suite trois six, ou d'être Empereur si on y manquoit? Je pense absolument comme lui; mais pourquoi cet homme passeroit-il pour *fou*, si le cas dont il s'agit, est *physiquement* possible?

See too the edition of D'Alembert's literary and philosophical works, Paris, 1821, Vol. I. page 553, note.

505. The next section is entitled *Sur la durée de la vie.* D'Alembert draws attention to the distinction between the *mean* duration of life and the *probable* duration of life ; see Art. 484. D'Alembert seems to think it is a great objection to the Theory of Probability that there *is* this distinction.

D'Alembert's objection to the Theory of Probability is as reasonable as an objection to the Theory of Mechanics would be on the ground that the centre of gravity of an area does not necessarily fall on an assigned line which bisects the area.

D'Alembert asserts that a numerical statement of Buffon's, which Daniel Bernoulli had suspected of inaccuracy, was not really inaccurate, but that the difference between Buffon and Daniel Bernoulli arose from the distinction between what we call *mean* duration and *probable* duration of life.

506. The last section is entitled *Sur un Mémoire de M. Bernoulli concernant l'Inoculation.*

Daniel Bernoulli in the commencement of his memoir had said, il seroit à souhaiter que les critiques fussent plus réservés et plus circonspects, et sur-tout qu'ils se donnassent la peine de se mettre au fait des choses qu'ils se proposent d'avance de critiquer. The words *se mettre au fait* seem to have given great offence to

D'Alembert as he supposed they were meant for him.   He refers
to them in the *Opuscules,* Vol. IV. pages IX, 99, 100; and he
seems with ostentatious deference to speak of Daniel Bernoulli
as *ce grand Géometre;* see pages 99, 101, 315, 321, 323 of the
volume.

507.   D'Alembert objects to the hypotheses on which Daniel
Bernoulli had based his calculation; see Art. 401.   D'Alembert
brings forward another objection which is quite fallacious, and
which seems to shew that his vexation had disturbed his judgment.   Daniel Bernoulli had found that the average life of all
who die of small-pox is $6\frac{1}{12}$ years; and that if small-pox were
extinguished the average human life would be $29\frac{9}{12}$ years.   Moreover the average human life subject to small-pox is $26\frac{7}{12}$ years.
Also Daniel Bernoulli admitted that the deaths by small-pox
were $\dfrac{1}{13}$ of all the deaths.

Hence D'Alembert affirms that the following relation ought
to hold,

$$\frac{1}{13} \times 6\tfrac{1}{12} + \frac{12}{13} \times 29\tfrac{9}{12} = 26\tfrac{7}{12};$$

but the relation does not hold, for the terms on the left hand side
will give $27\frac{11}{12}$ nearly instead of $26\frac{7}{12}$.   D'Alembert here makes the
mistake which I have pointed out in Art. 487; when that Article
was written, I had not read the remarks by D'Alembert which
are now under discussion, but it appeared to me that D'Alembert
was not clear on the point, and the mistake which he now makes
confirms my suspicion.

To make the above equation correct we must remove $29\frac{9}{12}$,
and put in its place the average duration of those who die of
other diseases while small-pox still prevails; this number will be
smaller than $29\frac{9}{12}$.

508.   We pass on to the pages 283—341 of the fourth volume
of the *Opuscules.*   Here we have two sections, one *Sur le Calcul
des probabilités,* the other *Sur les Calculs relatifs à l'Inoculation.*

509.   The first section consists of little more than a repetition

of the remarks which have already been noticed. D'Alembert records the origin of his doubts in these words :

Il y a près de trente ans que j'avois formé ces doutes en lisant l'excellent livre de M. Bernoulli *de Arte conjectandi;...*

He seems to have returned to his old error respecting *Croix ou Pile* with fresh ardour ; he says,

...si les trois cas, *croix, pile et croix, pile et pile*, les seuls qui puissent arriver dans le jeu proposé, ne sont pas également possibles, ce n'est point, ce me semble, par la raison qu'on en apporte communément, que la probabilité du premier est $\frac{1}{2}$, et celle des deux autres $\frac{1}{2} \times \frac{1}{2}$ ou $\frac{1}{4}$. Plus j'y pense, et plus il me paroît que *mathématiquement* parlant, ces trois coups sont également possibles...

510. D'Alembert introduces another point in which he objects to a principle commonly received. He will not admit that it is the same thing to toss one coin m times in succession, or to toss m coins simultaneously. He says it is perhaps physically speaking more possible to have the same face occurring simultaneously an assigned number of times with m coins tossed at once, than to have the same face repeated the same assigned number of times when one coin is tossed m times. But no person will allow what D'Alembert states. We can indeed suppose circumstances in which the two cases are not quite the same ; for example if the coins used are not perfectly symmetrical, so that they have a tendency to fall on one face rather than on the other. But we should in such a case expect a *run* of resemblances rather in using one coin for m throws, than in using m coins at once. Take for a simple example $m = 2$. We should have rather more than $\frac{1}{4}$ as the chance for the former result, and only $\frac{1}{4}$ for the latter ; see Laplace, *Théorie...des Prob.* page 402.

511. D'Alembert says on his page 290, Il y a quelque temps qu'un Joueur me demanda en combien de coups consécutifs on pouvoit parier avec avantage d'amener une face donnée d'un dé.... . This is the old question proposed to Pascal by the Chevalier de

Méré. D'Alembert answered that according to the common theory in n trials, the odds would be as $6^n - 5^n$ to 5^n. Thus there would be advantage in undertaking to do it in four throws. Then D'Alembert adds, Ce Joueur me répondit que l'expérience lui avoit paru contraire à ce resultat, et qu'en jouant quatre coups de suite pour amener une face donnée, il lui étoit arrivé beaucoup plus souvent de gagner que de perdre. D'Alembert says that if this be true, the disagreement between theory and observation may arise from the fact that the former rests on a supposition which he has before stated to be false. Accordingly D'Alembert points out that on his principles the number of favourable cases in n throws instead of being $6^n - 5^n$, as by the ordinary theory, would be $1 + 5 + 5^2 + ... + 5^{n-1}$. This is precisely analogous to what we have given for a die with *three* faces in Art. 477. D'Alembert however admits that we must not regard all these cases as equally likely.

512. D'Alembert quotes testimonies in his own favour from the letters of three mathematicians to himself; see his pages 296, 297. One of these correspondents he calls, un très-profond et très-habile Analyste; another he calls, un autre Mathématicien de la plus grande réputation et la mieux méritée; and the third, un autre Ecrivain très-éclairé, qui a cultivé les Mathématiques avec succès, et qui est connu par un excellent Ouvrage de Philosophie. But this *Ecrivain très-éclairé* is a proselyte whose zeal is more conspicuous than his judgment. He says "ce que vous dites sur la probabilité est excellent et très-évident; l'ancien calcul des probabilités est ruiné... D'Alembert is obliged to add in a note, Je n'en demande pas tant, à beaucoup près; je ne prétends point *ruiner* le calcul des probabilités, je désire seulement qu'il soit éclairci et modifié.

513. D'Alembert returns to the *Petersburg Problem*. He says,

Vous dites, Monsieur, que la raison pour laquelle on trouve l'enjeu infini, c'est la supposition tacite qu'on fait que le jeu peut avoir une durée infinie, ce que n'est pas admissible, attendu que la vie des hommes ne dure qu'un temps.

D'Alembert brings forward four remarks which shew that this mode of explaining the difficulty is unsatisfactory. One of them is the following: instead of supposing that one crown is to be received for head at the first throw, two for head at the second throw, four for head at the third throw, and so on, suppose that in each case only one crown is to be received. Then, although theoretically the game may endure to infinity, yet the value of the expectation is finite. This remark may be said to contradict a conclusion at which D'Alembert arrived in his article *Croix ou Pile*, which we noticed in Art. 465.

514. The case just brought forward is interesting because D'Alembert admits that it might supply an objection to his principles. He tries to repel the objection by saying that it only leads him to suspect another principle of the ordinary theory, namely that in virtue of which the total expectation is taken to be equal to the sum of the partial expectations; see his pages 299—301.

515. D'Alembert thus sums up his objections against the ordinary theory:

Pour résumer en un mot tous mes doutes sur le calcul des probabilités, et les mettre sous les yeux des vrais Juges; voici ce que j'accorde et ce que je nie dans les raisonnemens explicites ou implicites sur lesquels ce calcul me paroît fondé.

Premier raisonnement. Le nombre des combinaisons qui amenent tel cas, est au nombre des combinaisons qui amenent tel autre cas, comme p est à q. Je conviens de cette vérité qui est purement mathématique; donc, conclut-on, la probabilité du premier cas est à celle du second comme p est à q. Voilà ce que je nie, ou du moins de quoi je doute fort; et je crois que si, par exemple, $p = q$, et que dans le second cas le même événement se trouve un très-grand nombre de fois de suite, il sera moins probable *physiquement* que le premier, quoique les probabilités mathématiques soient égales.

Second raisonnement. La probabilité $\frac{1}{m}$ est à la probabilité $\frac{1}{n}$ comme np écus est à mp écus. J'en conviens; donc $\frac{1}{m} \times mp$ écus $= \frac{1}{n} \times np$ écus; j'en conviens encore; donc l'*espérance*, ou ce qui est la même chose, le *sort* d'un Joueur qui aura la probabilité $\frac{1}{m}$ de gagner mp écus, sera égale à l'espérance, au sort d'un Joueur qui aura la probabilité $\frac{1}{n}$ de gagner np écus. Voilà ce que je nie; je dis que l'*espérance* est plus grande pour celui qui a la plus grande probabilité, quoique la somme espérée soit moindre, et qu'on ne doit pas balancer de préférer le sort d'un Joueur qui a la probabilité $\frac{1}{2}$ de gagner 1000 écus, au sort d'un Joueur qui a la probabilité $\frac{1}{2000}$ d'en gagner 1000000.

Troisième raisonnement qui n'est qu'implicite. Soit $p + q$ le nombre total des cas, p la probabilité d'un certain nombre de cas, q la probabilité des autres; la probabilité de chacun sera à la certitude totale, comme p et q sont à $p + q$. Violà ce que je nie encore; je conviens, ou plutôt j'accorde, que les probabilités de chaque cas sont comme p et q; je conviens qu'il arrivera certainement et infailliblement un des cas dont le nombre est $p + q$; mais je nie que du rapport des probabilités entr'elles, on puisse en conclure leur rapport à la certitude absolue, parce que la certitude absolue est infinie par rapport à la plus grande probabilité.

Vous me demanderez peut-être quels sont les principes qu'il faut, selon moi, substituer à ceux dont je révoque en doute l'exactitude? Ma réponse sera celle que j'ai déja faite; je n'en sais rien, et je suis même très-porté à croire que la matiere dont il s'agit, ne peut être soumise, au moins à plusieurs égards, à un calcul exact et précis, également net dans ses principes et dans ses résultats.

516. D'Alembert now returns to the calculations relating to Inoculation. He criticises very minutely the mathematical investigations of Daniel Bernoulli.

The objection which D'Alembert first urges is as follows. Let s be the number of persons alive at the commencement of the time x; then Daniel Bernoulli assumes that $\dfrac{sdx}{64}$ die from small-pox during the time dx. Therefore the whole number who die from small-pox during the $(n + 1)^{\text{th}}$ year is

$$\int_n^{n+1} \frac{sdx}{64}.$$

But this is not the same thing as $\dfrac{S}{64}$, where S denotes the number alive at the beginning of the year; for s is a variable gradually diminishing during the year from the value S with which it began. But $\dfrac{S}{64}$ is the result which Daniel Bernoulli professed to take from observation; therefore Daniel Bernoulli is inconsistent with himself. D'Alembert's objection is sound; Daniel Bernoulli would no doubt have admitted it, and have given the just reply, namely that his calculations only professed to be approximately correct, and that they were approximately correct. Moreover the error arising in taking $\displaystyle\int_{n}^{n+1} s\,dx$ and S to be equal in value becomes very small if we suppose S to be, not the value of s when $x = n$ or $n + 1$ but, the intermediate value when $x = n + \dfrac{1}{2}$; and nothing in Daniel Bernoulli's investigation forbids this supposition.

517. We have put the objection in the preceding Article as D'Alembert ought to have put it in fairness. He himself however really assumes $n = 0$, so that his attack does not strictly fall on the whole of Daniel Bernoulli's table but on its first line; see Art. 403. This does not affect the *principle* on which D'Alembert's objection rests, but taken in conjunction with the remarks in the preceding Article, it will be found to diminish the practical value of the objection considerably. See D'Alembert's pages 312—314.

518. Another objection which D'Alembert takes is also sound; see his page 315. It amounts to saying that instead of using the Differential Calculus Daniel Bernoulli ought to have used the Calculus of Finite Differences. We have seen in Art. 417 that Daniel Bernoulli proposed to solve various problems in the Theory of Probability by the use of the Differential Calculus. The reply to be made to D'Alembert's objection is that Daniel Bernoulli's investigation accomplishes what was proposed, namely an approximate solution of the problem; we shall however see hereafter in examining a memoir by Trembley that, assuming the hypotheses of Daniel Bernoulli, a solution by common algebra might be effected.

284 D'ALEMBERT.

519. D'Alembert thinks that Daniel Bernoulli might have solved the problem more simply and not less accurately. For Daniel Bernoulli made two assumptions; see Art. 401. D'Alembert says that only one is required; namely to assume some function of y for u in Art. 483. Accordingly D'Alembert suggests arbitrarily some functions, which have apparently far less to recommend them as corresponding to facts, than the assumptions of Daniel Bernoulli.

520. D'Alembert solves what he calls *un problème assez curieux*; see his page 325. He solves it on the assumptions of Daniel Bernoulli, and also on his own. We will give the former solution. Return to Art. 402 and suppose it required to determine out of the number s the number of those who will die by the small-pox. Let ω denote the number of those who do *not* die of small-pox. Hence out of this number ω during the time dx none will die of small-pox, and the number of those who die of other diseases will be, on the assumptions of Daniel Bernoulli, $\left(-d\xi - \dfrac{sdx}{mn}\right)\dfrac{\omega}{\xi}$.

Hence, $\qquad -d\omega = \left(-d\xi - \dfrac{sdx}{mn}\right)\dfrac{\omega}{\xi}$,

therefore $\qquad \dfrac{d\omega}{\omega} = \dfrac{d\xi}{\xi} + \dfrac{sdx}{\xi mn}$.

Substitute the value of s in terms of x and ξ from Art. 402, and integrate. Thus we obtain

$$\frac{\omega}{\xi} = \frac{C\epsilon^{\frac{x}{n}}}{e^{\frac{x}{n}}(m-1)+1},$$

where C is an arbitrary constant. The constant may be determined by taking a result which has been deduced from observation, namely that $\dfrac{\omega}{\xi} = \dfrac{1}{24}$ when $x = 0$.

521. D'Alembert proposes on his pages 326—328 the method which according to his view should be used to find the value of s at the time x, instead of the method of Daniel Bernoulli which

we gave in Art. 402. D'Alembert's method is too arbitrary in its hypotheses to be of any value.

522. D'Alembert proposes to develop his refutation of the *Savant Géomètre* whom we introduced in Art. 487. He shews decisively that this person was wrong; but it does not seem to me that he shews distinctly *how* he was wrong.

523. D'Alembert devotes the last ten pages of the memoir to the development of his own theory of the mode of comparing the risk of an individual if he undergoes Inoculation with his risk if he declines it. We have already given in Art. 482, a hint of D'Alembert's views; his remarks in the present memoir are ingenious and interesting, but as may be supposed, his hypotheses are too arbitrary to allow any practical value to his investigations.

524. Two remarks which he makes on the curve of mortality may be reproduced; see his page 340. It appears from Buffon's tables that the *mean* duration of life for persons aged n years is always less than $\frac{1}{2}(100 - n)$. Hence, taking 100 years as the extreme duration of human life, it will follow that the curve of mortality cannot be always *concave* to the axis of abscissæ. Also from the tables of Buffon it follows that the *probable* duration of life is almost always greater than the *mean* duration. D'Alembert applies this to shew that the *curve of mortality* cannot be always *convex* to the axis of abscissæ.

525. The fifth volume of the *Opuscules* was published in 1768. It contains two brief articles with which we are concerned.

Pages 228—231 are *Sur les Tables de mortalité.* The numerical results are given which served for the foundation of the two remarks noticed in Art. 524.

Pages 508—510 are *Sur les calculs relatifs à l'inoculation...* These remarks form an addition to the memoir in pages 283—341 of the fourth volume of the *Opuscules.* D'Alembert notices a reply which had been offered to one of his objections, and enforces the justness of his objections. Nevertheless he gives his reasons for regarding Inoculation as a useful practice.

526. The seventh and eighth volume of the *Opuscules* were published in 1780. D'Alembert says in an Advertisement prefixed to the seventh volume, "... Ce seront vraisemblablement, à peu de chose près, mes derniers Ouvrages Mathématiques, ma tête, fatiguée par quarante-cinq années de travail en ce genre, n'étant plus guère capable des profondes recherches qu'il exige." D'Alembert died in 1783. It would seem according to his biographers that he suffered more from a broken heart than an exhausted brain during the last few years of his life.

527. The seventh volume of the *Opuscules* contains a memoir *Sur le calcul des Probabilités*, which occupies pages 39—60. We shall see that D'Alembert still retained his objections to the ordinary theory. He begins thus:

Je demande pardon aux Géometres de revenir encore sur ce sujet. Mais j'avoue que plus j'y ai pensé, plus je me suis confirmé dans mes doutes sur les principes de la théorie ordinaire; je desire qu'on éclaircisse ces doutes, et que cette théorie, soit qu'on y change quelques principes, soit qu'on la conserve telle qu'elle est, soit du moins exposée désormais de maniere à ne plus laisser aucun nuage.

528. We will not delay on some repetition of the old remarks; but merely notice what is new. We find on page 42 an error which D'Alembert has not exhibited elsewhere, except in the article *Cartes* in the *Encyclopédie Méthodique*, which we shall notice hereafter. He says that taking two throws there is a chance $\frac{1}{2}$ of head at the first throw, and a chance $\frac{1}{2}$ of head at the second throw; and thus he infers that the chance that head will arrive at least once is $\frac{1}{2} + \frac{1}{2}$ or 1. He says then, Or je demande si cela est vrai, ou du moins si un pareil résultat, fondé sur de pareils principes, est bien propre à satisfaire l'esprit. The answer is that the result is false, being erroneously deduced: the error is exposed in elementary works on the subject.

529. The memoir is chiefly devoted to the *Petersburg Problem*. D'Alembert refers to the memoir in Vol. VI. of the *Mémoires...par*

divers Savans... in which Laplace had made the supposition that the coin has a greater tendency to fall on one side than the other, but it is not known on which side. Suppose that 2 crowns are to be received for head at the first trial, 4 for head at the second, 8 for head at the third, ... Then Laplace shews that if the game is to last for x trials the player ought to give to his antagonist less than x crowns if x be less than 5, and more than x crowns if x be greater than 5, and just x crowns if x be equal to 5. On the common hypothesis he would always have to give x crowns. These results of Laplace are only obtained by him as approximations; D'Alembert seems to present them as if they were exact.

530. Suppose the probability that head should fall at first to be ω and not $\frac{1}{2}$; and let the game have to extend over n trials Then if 2 crowns are to be received for head at the first trial, 4 for head at the second, and so on; the sum which the player ought to give is
$$2\omega\{1 + 2(1-\omega) + 2^2(1-\omega)^2 + \ldots + 2^{n-1}(1-\omega)^{n-1}\},$$
which we will call Ω.

D'Alembert suggests, if I understand him rightly, that if we know nothing about the value of ω we may take as a solution of the problem, for the sum which the player ought to give $\int_0^1 \Omega d\omega$. But this involves all the difficulty of the ordinary solution, for the result is infinite when n is. D'Alembert is however very obscure here; see his pages 45, 46.

He seems to say that $\int_0^1 \Omega d\omega$ will be greater than, equal to, or less than n, according as n is greater than, equal to, or less than 5. But this result is false; and the argument unintelligible or inconclusive. We may easily see by calculation that $\int_0^1 \Omega d\omega = n$ when $n = 1$; and that for any value of n from 2 to 6 inclusive $\int_0^1 \Omega d\omega$ is less than n; and that when n is 7 or any greater number $\int_0^1 \Omega d\omega$ is greater than n.

531. D'Alembert then proposes a method of solving the *Peters-burg Problem* which shall avoid the *infinite* result; this method is perfectly arbitrary. He says, if tail has arrived at the first throw, let the chance that head arrives at the next be $\dfrac{1+a}{2}$, and not $\dfrac{1}{2}$, where a is some small quantity; if tail has arrived at the first throw, and at the second, let the chance that head arrives at the next throw be $\dfrac{1+a+b}{2}$, and not $\dfrac{1}{2}$; if tail has arrived at the first throw, at the second, and at the third, let the chance that head arrives at the next throw be $\dfrac{1+a+b+c}{2}$, and not $\dfrac{1}{2}$; and so on. The quantities a, b, c, ... are supposed small positive quantities, and subjected to the limitation that their sum is less than unity, so that every chance may be less than unity.

On this supposition if the game be as it is described in Art. 389, it may be shewn that A ought to give half of the following series :

$$1$$
$$+ (1 + a)$$
$$+ (1 - a)\,(1 + a + b)$$
$$+ (1 - a)\,(1 - a - b)\,(1 + a + b + c)$$
$$+ (1 - a)\,(1 - a - b - c)\,(1 + a + b + c + d)$$
$$+ \ldots\ldots$$

It is easily shewn that this is *finite*. For

(1) Each of the factors $1 + a$, $1 + a + b$, $1 + a + b + c$, ... is less than 2.

(2) $1 - a - b$ is less than $1 - a$;

$1 - a - b - c$ is less than $1 - a - b$, and *a fortiori* less than $1 - a$;

and so on.

Thus the series excluding the first two terms is less than the Geometrical Progression

$$2\,\{1 - a + (1 - a)^2 + (1 - a)^3 + (1 - a)^4 \ldots\},$$

and is therefore finite.

This is D'Alembert's principle, only he uses it thus: he shews that all the terms beginning with

$$(1-a)(1-a-b)(1-a-b-c)(1-a-b-c-d)(1+a+b+c+d+e)$$

are less than

$$2(1-a)(1-a-b)(1-a-b-c)(1-a-b-c-d)s,$$

where s denotes the geometrical progression

$$1 + r + r^2 + r^3 + \dots,$$

r being $= 1 - a - b - c - d.$

532. Thus on his arbitrary hypotheses D'Alembert obtains a finite result instead of an infinite result. Moreover he performs what appears a work of supererogation; for he shews that the successive terms of the infinite series which he obtains form *a continually diminishing series* beginning from the second, if we suppose that $a, b, c, d, \dots$ are connected by a certain law which he gives, namely,

$$1 - a - b - c - d - e - \dots = \frac{1}{1 + (m-1)\rho},$$

where ρ is a small fraction, and $m - 1$ is the number of the quantities $a, b, c, d, e, \dots$ Again he shews that the same result holds if we merely assume that $a, b, c, d, e \dots$ form a continually diminishing series. We say that this appears to be a work of supererogation for D'Alembert, because we consider that the *infinite* result was the only supposed difficulty in the *Petersburg Problem*, and that it was sufficient to remove this without shewing that the series substituted for the ordinary series consisted of terms *continually decreasing*. But D'Alembert apparently thought differently; for after demonstrating this continual decrease he says,

En voilà assez pour faire voir que les termes de l'enjeu vont en diminuant dès le troisième coup, jusqu'au dernier. Nous avons prouvé d'ailleurs que l'enjeu total, somme de ces termes, est fini, en supposant même le nombre de coups infini. Ainsi le résultat de la solution que nous donnons ici du problème de Petersbourg, n'est pas sujet à la difficulté insoluble des solutions ordinaires.

533. We have one more contribution of D'Alembert's to our subject to notice; it contains errors which seem extraordinary,

19

even for him.   It is the article *Cartes* in the *Encyclopédie Métho-dique*.   The following problem is given,

Pierre tient huit *cartes* dans ses mains qui sont : un as, un deux, un trois, un quatre, un cinq, un six, un sept et un huit, qu'il a mêlées : Paul parie que les tirant l'une après l'autre, il les devinera à mesure qu'il les tirera.   L'on demande combien Pierre doit parier contre un . que Paul ne réussira pas dans son enterprise ?

It is correctly determined that Paul's chance is

$$\frac{1}{8} \times \frac{1}{7} \times \frac{1}{6} \times \frac{1}{5} \times \frac{1}{4} \times \frac{1}{3} \times \frac{1}{2}.$$

Then follow three problems formed on this ; the whole is absurdly false.   We give the words :

Si Paul parioit d'amener ou de deviner juste à un des sept coups seulement, son espérance seroit $\frac{1}{8} + \frac{1}{7} + \ldots + \frac{1}{2}$, et par conséquent l'enjeu de Pierre à celui de Paul, comme

$$\frac{1}{8} + \frac{1}{7} + \ldots + \frac{1}{2} \text{ à } 1 - \frac{1}{8} - \frac{1}{7} - \ldots - \frac{1}{2}.$$

Si Paul parioit d'amener juste dans les deux premiers coups seulement, son espérance seroit $\frac{1}{8} + \frac{1}{7}$, et le rapport des enjeux celui de

$$\frac{1}{8} + \frac{1}{7} \text{ à } 1 - \frac{1}{8} - \frac{1}{7}.$$

S'il parioit d'amener juste dans deux coups quelconques, son espérance seroit $\frac{1}{8 \times 7} + \frac{1}{8 \times 6} + \ldots + \frac{1}{8 \times 2} + \frac{1}{7 \times 6} + \ldots + \frac{1}{7 \times 2} + \frac{1}{6 \times 5} + \ldots$

The first question means, I suppose, that Paul undertakes to be right once in the seven cases, and wrong six times.   His chance then is

$$\frac{1}{8} \left(\frac{1}{7} + \frac{1}{6} + \frac{1}{5} + \frac{1}{4} + \frac{1}{3} + \frac{1}{2} + 1 \right).$$

For his chance of being right in the *first* case and wrong in the other six is

$$\frac{1}{8} \times \frac{6}{7} \times \frac{5}{6} \times \frac{4}{5} \times \frac{3}{4} \times \frac{2}{3} \times \frac{1}{2}, \text{ that is } \frac{1}{8 \times 7} ;$$

his chance of being right in the *second* case and wrong in all the others is

$$\frac{7}{8} \times \frac{1}{7} \times \frac{5}{6} \times \frac{4}{5} \times \frac{3}{4} \times \frac{2}{3} \times \frac{1}{2}, \text{ that is } \frac{1}{8 \times 6};$$

and so on.

If the meaning be that Paul undertakes to be right *once at least* in the seven cases, then his chance is $\frac{7}{8}$. For his chance of being wrong every time is

$$\frac{7}{8} \times \frac{6}{7} \times \frac{5}{6} \times \frac{4}{5} \times \frac{3}{4} \times \frac{2}{3} \times \frac{1}{2}; \text{ that is } \frac{1}{8};$$

therefore his chance of being right once at least is $1 - \frac{1}{8}$, that is $\frac{7}{8}$.

The second question means, I suppose, that Paul undertakes to be right in the first two cases, and wrong in the other five. His chance then is

$$\frac{1}{8} \times \frac{1}{7} \times \frac{5}{6} \times \frac{4}{5} \times \frac{3}{4} \times \frac{2}{3} \times \frac{1}{2}, \text{ that is } \frac{1}{8 \times 7 \times 6}.$$

Or it may mean that Paul undertakes to be right in the first two cases, but undertakes nothing for the other cases. Then his chance is $\frac{1}{8} \times \frac{1}{7}$.

The third question means, I suppose, that Paul undertakes to be right in two out of the seven cases and wrong in the other five cases. The chance then will be the sum of 21 terms, as 21 combinations of pairs of things can be made from 7 things. The chance that he is right in the first two cases and wrong in all the others is

$$\frac{1}{8} \times \frac{1}{7} \times \frac{5}{6} \times \frac{4}{5} \times \frac{3}{4} \times \frac{2}{3} \times \frac{1}{2}, \text{ that is } \frac{1}{8 \times 7 \times 6};$$

similarly we may find the chance that he is right in any two assigned cases and wrong in all the others. The total chance will be found to be

$$\frac{1}{8} \left\{ \frac{1}{7} \left(\frac{1}{6} + \frac{1}{5} + \frac{1}{4} + \frac{1}{3} + \frac{1}{2} + 1 \right) + \frac{1}{6} \left(\frac{1}{5} + \frac{1}{4} + \frac{1}{3} + \frac{1}{2} + 1 \right) \right.$$

$$\left. + \frac{1}{5} \left(\frac{1}{4} + \frac{1}{3} + \frac{1}{2} + 1 \right) + \frac{1}{4} \left(\frac{1}{3} + \frac{1}{2} + 1 \right) + \frac{1}{3} \left(\frac{1}{2} + 1 \right) + \frac{1}{2} \right\}.$$

19—2

Or the third question may mean that Paul undertakes to be right twice at least in the course of the seven cases, or in other words he undertakes to be right twice and undertakes nothing more. His chance is to be found by subtracting from unity his chance of being never right, and also his chance of being right only once. Thus his chance is

$$1 - \frac{1}{8} - \frac{1}{8}\left(\frac{1}{7} + \frac{1}{6} + \frac{1}{5} + \dots + 1\right).$$

534. Another problem is given unconnected with the one we have noticed, and is solved correctly.

The article in the *Encyclopédie Méthodique* is signed with the letter which denotes D'Alembert. The date of the volume is 1784, which is subsequent to D'Alembert's death; but as the work was published in parts this article may have appeared during D'Alembert's life, or the article may have been taken from his manuscripts even if published after his death. I have not found it in the original *Encyclopédie*: it is certainly not under the title *Cartes*, nor under any other which a person would naturally consult. It seems strange that such errors should have been admitted into the *Encyclopédie Méthodique*.

Some time after I read the article *Cartes* and noticed the errors in it, I found that I had been anticipated by Binet in the *Comptes Rendus* ... Vol. XIX. 1844. Binet does not exhibit any doubts as to the authorship of the article; he says that the three problems are wrong and gives the correct solution of the first.

535. We will in conclusion briefly notice some remarks which have been made respecting D'Alembert by other writers.

536. Montucla after alluding to the article *Croix ou Pile* says on his page 406,

D'Alembert ne s'est pas borné à cet exemple, il en a accumulé plusieurs autres, soit dans le quatrième volume de ses *Opuscules*, 1768, page 73, et page 283 du cinquième; il s'est aussi étayé du suffrage de divers géomètres qu'il qualifie de distingués. Condorcet a appuyé ces objections dans plusieurs articles de l'Encyclopédie méthodique ou par ordre de matières. D'un autre côté, divers autres géomètres ont entrepris de répondre aux raisonnemens de d'Alembert, et je crois qu'en particulier Daniel Bernoulli a pris la défense de la théorie ordinaire.

In this passage the word *cinquième* is wrong; it should be *quatrième*. It seems to me that there is no foundation for the statement that Condorcet supports D'Alembert's objections. Nor can I find that Daniel Bernoulli gave any defence of the ordinary theory ; he seems to have confined himself to repelling the attack made on his memoir respecting Inoculation.

537. Gouraud after referring to Daniel Bernoulli's controversy with D'Alembert says, on his page 59,

...et quant au reste des mathématiciens, ce ne fut que par le silence ou le dédain qu'il répondit aux doutes que d'Alembert s'etait permis d'émettre. Mépris injuste et malhabile où tout le monde avait à perdre et qu'une postérité moins prévenue ne devait point sanctionner.

The statement that D'Alembert's objections were received with silence and disdain, is inconsistent with the last sentence of the passage quoted from Montucla in the preceding Article. According to D'Alembert's own words which we have given in Art. 490, he was attacked by some indifferent mathematicians.

538. Laplace briefly replies to D'Alembert ; see *Théorie...des Prob.* pages VII. and X.

It has been suggested that D'Alembert saw his error respecting the game of *Croix ou Pile* before he died ; but this suggestion does not seem to be confirmed by our examination of all his writings : see *Cambridge Philosophical Transactions*, Vol. IX. page 117.

CHAPTER XIV.

BAYES.

539. THE name of Bayes is associated with one of the most important parts of our subject, namely, the method of estimating the probabilities of the causes by which an observed event may have been produced. As we shall see, Bayes commenced the investigation, and Laplace developed it and enunciated the general principle in the form which it has since retained.

540. We have to notice two memoirs which bear the following titles:

An Essay towards solving a Problem in the Doctrine of Chances. By the late Rev. Mr. Bayes, F.R.S. communicated by Mr Price in a Letter to John Canton, A.M. F.R.S. A Demonstration of the Second Rule in the Essay towards the Solution of a Problem in the Doctrine of Chances, published in the Philosophical Transactions, Vol. LIII. Communicated by the Rev. Mr. Richard Price, in a Letter to Mr. John Canton, M.A. F.R.S.

The first of these memoirs occupies pages 370—418 of Vol. LIII. of the *Philosophical Transactions;* it is the volume for 1763, and the date of publication is 1764.

The second memoir occupies pages 296—325 of Vol. LIV. of the *Philosophical Transactions;* it is the volume for 1764, and the date of publication is 1765.

541. Bayes proposes to establish the following theorem: If

an event has happened p times and failed q times, the probability
that its chance at a single trial lies between a and b is

$$\frac{\int_a^b x^p (1-x)^q \, dx}{\int_0^1 x^p (1-x)^q \, dx}.$$

Bayes does not use this notation; areas of curves, according to
the fashion of his time, occur instead of integrals. Moreover we
shall see that there is an important condition implied which we
have omitted in the above enunciation, for the sake of brevity:
we shall return to this point in Art. 552.

Bayes also gives rules for obtaining approximate values of the
areas which correspond to our integrals.

542. It will be seen from the title of the first memoir that it
was published after the death of Bayes. The Rev. Mr Richard
Price is the well known writer, whose name is famous in connexion
with politics, science and theology. He begins his letter to
Canton thus:

Dear Sir, I now send you an essay which I have found among the
papers of our deceased friend Mr Bayes, and which, in my opinion, has
great merit, and well deserves to be preserved.

543. The first memoir contains an introductory letter from
Price to Canton; the essay by Bayes follows, in which he begins
with a brief demonstration of the general laws of the Theory
of Probability, and then establishes his theorem. The enunciations are given of two rules which Bayes proposed for finding
approximate values of the areas which to him represented our
integrals; the demonstrations are not given. Price himself added
*An Appendix containing an Application of the foregoing Rules
to some particular Cases.*

The second memoir contains Bayes's demonstration of his principal rule for approximation; and some investigations by Price
which also relate to the subject of approximation.

544. Bayes begins, as we have said, with a brief demonstration of the general laws of the Theory of Probability; this part of
his essay is excessively obscure, and contrasts most unfavourably
with the treatment of the same subject by De Moivre.

Bayes gives the principle by which we must calculate the probability of a compound event.

Suppose we denote the probability of the compound event by $\frac{P}{N}$, the probability of the first event by z, and the probability of the second on the supposition of the happening of the first by $\frac{b}{N}$. Then our principle gives us $\frac{P}{N} = z \times \frac{b}{N}$, and therefore $z = \frac{P}{b}$. This result Bayes seems to present as something new and remarkable; he arrives at it by a strange process, and enunciates it as his Proposition 5 in these obscure terms:

If there be two subsequent events, the probability of the 2nd $\frac{b}{N}$ and the probability of both together $\frac{P}{N}$, and it being 1st discovered that the 2nd event has happened, from hence I guess that the 1st event has also happened, the probability I am in the right is $\frac{P}{b}$.

Price himself gives a note which shews a clearer appreciation of the proposition than Bayes had.

545. We pass on now to the remarkable part of the essay. Imagine a rectangular billiard table $ABCD$. Let a ball be rolled on it at random, and when the ball comes to rest let its perpendicular distance from AB be measured; denote this by x. Let a denote the distance between AB and CD. Then the probability that the value of x lies between two assigned values b and c is $\frac{c-b}{a}$. This we should assume as obvious; Bayes, however, demonstrates it very elaborately.

546. Suppose that a ball is rolled in the manner just explained; through the point at which it comes to rest let a line EF be drawn parallel to AB, so that the billiard table is divided into the two portions $AEFB$ and $EDCF$. A second ball is to be rolled on the table; required the probability that it will rest within the

space *AEFB*. If x denote the distance between AB and EF the required probability is $\dfrac{x}{a}$: this follows from the preceding Article.

547. Bayes now considers the following compound event : The first ball is to be rolled once, and so EF determined ; then $p+q$ trials are to be made in succession with the second ball : required the probability, before the first ball is rolled, that the distance of EF from AB will lie between b and c, and that the second ball will rest p times within the space $AEFB$, and q times without that space.

We should proceed thus in the solution : The chance that EF falls at a distance x from AB is $\dfrac{dx}{a}$; the chance that the second event then happens p times and fails q times is

$$\frac{\lfloor p+q}{\lfloor p \lfloor q} \left(\frac{x}{a}\right)^p \left(1-\frac{x}{a}\right)^q;$$

hence the chance of the occurrence of the two contingencies is

$$\frac{dx}{a} \frac{\lfloor p+q}{\lfloor p \lfloor q} \left(\frac{x}{a}\right)^p \left(1-\frac{x}{a}\right)^q.$$

Therefore the whole probability required is

$$\frac{\lfloor p+q}{a \lfloor p \lfloor q} \int_b^c \left(\frac{x}{a}\right)^p \left(1-\frac{x}{a}\right)^q dx.$$

Bayes's method of solution is of course very different from the above. With him an area takes the place of the integral, and he establishes the result by a rigorous demonstration of the *ex absurdo* kind.

548. As a corollary Bayes gives the following : The probability, before the first ball is rolled, that EF will lie between AB and CD, and that the second event will happen p times and fail q times, is found by putting the limits 0 and a instead of b and c. But it is *certain* that EF will lie between AB and CD. Hence we

have for the probability, before the first ball is thrown, that the second event will happen p times and fail q times

$$\frac{\lfloor p+q}{a \lfloor p \lfloor q} \int_0^a \left(\frac{x}{a}\right)^p \left(1-\frac{x}{a}\right)^q dx.$$

549. We now arrive at the most important point of the essay. Suppose we only know that the second event has happened p times and failed q times, and that we wish to infer from this fact the probable position of the line EF which is to us unknown. The probability that the distance of EF from AB lies between b and c is

$$\frac{\int_b^c x^p (a-x)^q \, dx}{\int_0^a x^p (a-x)^q \, dx}.$$

This depends on Bayes's Proposition 5, which we have given in our Art. 544. For let z denote the required probability; then

$z \times$ probability of second event = probability of compound event.

The probability of the compound event is given in Art. 547, and the probability of the second event in Art. 548; hence the value of z follows.

550. Bayes then proceeds to find the area of a certain curve, or as we should say to integrate a certain expression. We have

$$\int x^p (1-x)^q \, dx = \frac{x^{p+1}}{p+1} - \frac{q}{1} \frac{x^{p+2}}{p+2} + \frac{q(q-1)}{1.2} \frac{x^{p+3}}{p+3} - \cdots$$

This series may be put in another form; let u stand for $1-x$, then the series is equivalent to

$$\frac{x^{p+1} u^q}{p+1} + \frac{q}{p+1} \frac{x^{p+2} u^{q-1}}{p+2} + \frac{q(q-1)}{(p+1)(p+2)} \frac{x^{p+3} u^{q-2}}{p+3}$$

$$+ \frac{q(q-1)(q-2)}{(p+1)(p+2)(p+3)} \frac{x^{p+4} u^{q-3}}{p+4} + \cdots$$

This may be verified by putting for u its value and rearranging according to powers of x. Or if we differentiate the series with respect to x, we shall find that the terms cancel so as to leave only $x^p u^q$.

551. The general theory of the estimation of the probabilities of causes from observed events was first given by Laplace in the *Mémoires ... par divers Savans*, Vol. VI. 1774. One of Laplace's results is that if an event has happened p times and failed q times, the probability that it will happen at the next trial is

$$\frac{\int_0^1 x^{p+1}(1-x)^q\, dx}{\int_0^1 x^p (1-x)^q\, dx}.$$

Lubbock and Drinkwater think that Bayes, or perhaps rather Price, confounded the probability given by Bayes's theorem with the probability given by the result just taken from Laplace ; see *Lubbock and Drinkwater*, page 48. But it appears to me that Price understood correctly what Bayes's theorem really expressed. Price's first example is that in which $p = 1$, and $q = 0$. Price says that "there would be odds of three to one for somewhat more than an even chance that it would happen on a second trial." His demonstration is then given ; it amounts to this :

$$\frac{\int_{\frac{1}{2}}^1 x^p (1-x)^q\, dx}{\int_0^1 x^p (1-x)^q\, dx} = \frac{3}{4},$$

where $p = 1$ and $q = 0$. Thus there is a probability $\frac{3}{4}$ that the chance of the event lies between $\frac{1}{2}$ and 1, that is a probability $\frac{3}{4}$ that the event is more likely to happen than not.

552. It must be observed with respect to the result in Art. 549, that in Bayes's own problem we *know* that *a priori* any position of EF between AB and CD is equally likely ; or at least we know what amount of assumption is involved in this supposition. In the applications which have been made of Bayes's theorem, and of such results as that which we have taken from Laplace in

Art. 551, there has however often been no adequate ground for such knowledge or assumption.

553. We have already stated that Bayes gave two rules for approximating to the value of the area which corresponds to the integral. In the first memoir, Price suppressed the demonstrations to save room ; in the second memoir, Bayes's demonstration of the principal rule is given : Price himself also continues the subject. These investigations are very laborious, especially Price's.

The following are among the most definite results which Price gives. Let $n = p + q$, and suppose that neither p nor q is small ; let $h = \dfrac{\sqrt{(pq)}}{n\sqrt{(n-1)}}$. Then if an event has happened p times and failed q times, the odds are about 1 to 1 that its chance at a single trial lies between $\dfrac{p}{n} + \dfrac{h}{\sqrt{2}}$ and $\dfrac{p}{n} - \dfrac{h}{\sqrt{2}}$; the odds are about 2 to 1 that its chance at a single trial lies between $\dfrac{p}{n} + h$ and $\dfrac{p}{n} - h$; the odds are about 5 to 1 that its chance at a single trial lies between $\dfrac{p}{n} + h\sqrt{2}$ and $\dfrac{p}{n} - h\sqrt{2}$. These results may be verified by Laplace's method of approximating to the value of the definite integrals on which they depend.

554. We may observe that the curve $y = x^p (1 - x)^q$ has two points of inflexion, the ordinates of which are equidistant from the maximum ordinate ; the distance is equal to the quantity h of the preceding Article. These points of inflexion are of importance in the methods of Bayes and Price.

CHAPTER XV.

LAGRANGE.

555. LAGRANGE was born at Turin in 1736, and died at Paris in 1813. His contributions to our subject will be found to satisfy the expectations which would be formed from his great name in mathematics.

556. His first memoir, relating to the Theory of Probability, is entitled *Memoire sur l'utilité de la méthode de prendre le milieu entre les résultats de plusieurs observations; dans lequel on examine les avantages de cette méthode par le calcul des probabilités; et où l'on résoud différens problêmes relatifs à cette matière.*

This memoir is published in the fifth volume of the *Miscellanea Taurinensia*, which is for the years 1770—1773: the date of publication is not given. The memoir occupies pages 167—232 of the mathematical portion of the volume.

The memoir at the time of its appearance must have been extremely valuable and interesting, as being devoted to a most important subject; and even now it may be read with advantage.

557. The memoir is divided into the discussion of ten problems; by a mistake no problem is numbered 9, so that the last two are 10 and 11.

The first problem is as follows: it is supposed that at every observation there are a cases in which no error is made, b cases in which an error equal to 1 is made, and b cases in which an error equal to -1 is made; it is required to find the probability that in taking the mean of n observations, the result shall be exact.

In the expansion of $\{a + b\,(x + x^{-1})\}^n$ according to powers of x, find the coefficient of the term independent of x; divide this coefficient by $(a + 2b)^n$ which is the whole number of cases that can occur; we thus obtain the required probability.

Lagrange exhibits his usual skill in the management of the algebraical expansions. It is found that the probability diminishes as n increases.

558. We may notice two points of interest in the course of Lagrange's discussion of this problem. Lagrange arrives indirectly at the following relation

$$1 + n^2 + \left\{\frac{n\,(n-1)}{2}\right\}^2 + \left\{\frac{n\,(n-1)\,(n-2)}{2\,.\,3}\right\}^2 + \ldots$$

$$= \frac{1\,.\,3\,.\,5 \ldots (2n-1)}{1\,.\,2\,.\,3 \ldots n}\,2^n\,;$$

and he says it is the more remarkable because it does not seem easy to demonstrate it *a priori*.

The result is easily obtained by equating the coefficients of the term independent of x in the equivalent expressions

$$(1 + x)^n \left(1 + \frac{1}{x}\right)^n, \text{ and } \frac{(1+x)^{2n}}{x^n}.$$

This simple method seems to have escaped Lagrange's notice.

Suppose we expand $\dfrac{1}{\sqrt{1 - 2az - cz^2}}$ in powers of z; let the result be denoted by

$$1 + A_1 z + A_2 z^2 + A_3 z^3 + \ldots\,;$$

Lagrange gives as a known result a simple relation which exists between every three consecutive coefficients; namely

$$A_n = \frac{2n-1}{n}\,a\,A_{n-1} + \frac{n-1}{n}\,c\,A_{n-2}.$$

This may be established by differentiation. For thus

$$\frac{a + cz}{(1 - 2az - cz^2)^{\frac{3}{2}}} = A_1 + 2A_2z + \ldots + n A_n z^{n-1} + \ldots$$

that is

$$(a + cz) \{1 + A_1z + A_2z^2 + \ldots + A_n z^n + \ldots\}$$
$$= (1 - 2az - cz^2) \{A_1 + 2A_2z + \ldots + n A_n z^{n-1} + \ldots\};$$

then by equating coefficients the result follows.

559. In the second problem the same suppositions are made as in the first, and it is required to find the probability that the error of the mean of n observations shall not surpass $\pm \dfrac{m}{n}$.

Like the first problem this leads to interesting algebraical expansions.

We may notice here a result which is obtained. Suppose we expand $\{a + b (x + x^{-1})\}^n$ in powers of x; let the result be denoted by

$$A_0 + A_1 (x + x^{-1}) + A_2 (x^2 + x^{-2}) + A_3 (x^3 + x^{-3}) + \ldots;$$

Lagrange wishes to shew the law of connexion between the coefficients $A_0, A_1, A_2, \ldots$ This he effects by taking the logarithms of both sides of the identity and differentiating with respect to x. It may be found more easily by putting $2 \cos \theta$ for $x + x^{-1}$, and therefore $2 \cos r\theta$ for $x^r + x^{-r}$. Thus we have

$$(a + 2b \cos \theta)^n = A_0 + 2A_1 \cos \theta + 2A_2 \cos 2\theta + 2A_3 \cos 3\theta + \ldots$$

Hence, by taking logarithms and differentiating,

$$\frac{nb \sin \theta}{a + 2b \cos \theta} = \frac{A_1 \sin \theta + 2A_2 \sin 2\theta + 3A_3 \sin 3\theta + \ldots}{A_0 + 2A_1 \cos \theta + 2A_2 \cos 2\theta + \ldots}.$$

Multiply up, and arrange each side according to sines of multiples of θ; then equate the coefficients of $\sin r\theta$: thus

$$nb \{A_{r-1} - A_{r+1}\} = raA_r + b \{(r - 1) A_{r-1} + (r + 1) A_{r+1}\};$$

therefore

$$A_{r+1} = \frac{b (n - r + 1) A_{r-1} - ra A_r}{b (n + r + 1)}.$$

560. In the third problem it is supposed that there are a cases at each observation in which no error is made, b cases in which an error equal to -1 is made, and c cases in which an error equal to r is made ; the probability is required that the error of the mean of n observations shall be contained within given limits.

In the fourth problem the suppositions are the same as in the third problem ; and it is required to find the most probable error in the mean of n observations; this is a particular case of the fifth problem.

561. In the fifth problem it is supposed that every observation is subject to given errors which can each occur in a given number of cases ; thus let the errors be p, q, r, s, ..., and the numbers of cases in which they can occur be a, b, c, d, ... respectively. Then we require to find the most probable error in the mean of n observations.

In the expansion of $(ax^p + bx^q + cx^r + ...)^n$ let M be the coefficient of x^μ ; then the probability that the sum of the errors is μ, and therefore that the error in the mean is $\dfrac{\mu}{n}$ is

$$\frac{M}{(a+b+c+...)^n}.$$

Hence we have to find the value of μ for which M is greatest.

Suppose that the error p occurs α times, the error q occurs β times, the error r occurs γ times, and so on. Then

$$\alpha + \beta + \gamma + \dots\dots\dots = n,$$
$$p\alpha + q\beta + r\gamma + \dots\dots\dots = \mu.$$

It appears from common Algebra that the greatest value of μ is when

$$\frac{\alpha}{a} = \frac{\beta}{b} = \frac{\gamma}{c} = \dots\dots\dots\dots = \frac{n}{a+b+c+\dots} ;$$

so that

$$\frac{\mu}{n} = \frac{pa + qb + rc + \dots}{a+b+c+\dots}.$$

This therefore is the most probable error in the mean result.

562. With the notation of Art. 561, suppose that a, b, c, ...

are not known *à priori;* but that α, β, γ, ... are known by observation. Then in the sixth problem it is taken as evident that the most probable values of a, b, c, ... are to be determined from the results of observation by the relations

$$\frac{a}{\alpha} = \frac{b}{\beta} = \frac{c}{\gamma} = \dots ,$$

so that the value of $\frac{\mu}{n}$ of the preceding Article may be written

$$\frac{\mu}{n} = \frac{p\alpha + q\beta + r\gamma + \dots}{\alpha + \beta + \gamma + \dots}.$$

Lagrange proposes further to estimate the probability that the values of a, b, c, ... thus determined from observation do not differ from the true values by more than assigned quantities. This is an investigation of a different character from the others in the memoir; it belongs to what is usually called the theory of inverse probability, and is a difficult problem.

Lagrange finds the analytical difficulties too great to be overcome; and he is obliged to be content with a rude approximation.

563. The seventh problem is as follows. In an observation it is equally probable that the error should be any one of the following quantities $-\alpha$, $-(\alpha-1)$, ... -1, 0, 1, 2 ... β; required the probability that the error of the mean of n observations shall have an assigned value, and also the probability that it shall lie between assigned limits.

We need not delay on this problem; it really is coincident with that in De Moivre as continued by Thomas Simpson: see Arts. 148 and 364. It leads to algebraical work of the same kind as the eighth problem which we will now notice.

564. Suppose that at each observation the error must be one of the following quantities $-\alpha$, $-(\alpha-1)$, ... $0, 1, \dots \alpha$; and that the chances of these errors are proportional respectively to $1, 2, \dots \alpha+1, \alpha, \dots 2, 1$: required the probability that the error in the mean of n observations shall be equal to $\frac{\mu}{n}$.

20

We must find the coefficient of x^μ in the expansion of

$$\{x^{-a} + 2x^{-a+1} + \ldots + ax^{-1} + (a+1)x^0 + ax + \ldots + 2x^{a-1} + x^a\}^n,$$

and divide it by the value of this expression when $x = 1$, which is the whole number of cases; thus we obtain the required probability.

Now
$$1 + 2x + 3x^2 + \ldots + (a+1)x^a + \ldots + 2x^{2a-1} + x^{2a}$$

$$= (1 + x + x^2 + \ldots + x^a)^2 = \left(\frac{1 - x^{a+1}}{1 - x}\right)^2.$$

Hence finally the required probability is the coefficient of x^μ in the expansion of

$$\frac{1}{(a+1)^{2n}} \frac{x^{-na}(1 - x^{a+1})^{2n}}{(1 - x)^{2n}};$$

that is the coefficient of $x^{\mu+na}$ in the expansion of

$$\frac{1}{(a+1)^{2n}}(1 - x)^{-2n}(1 - x^{a+1})^{2n}.$$

Lagrange gives a general theorem for effecting expansions, of which this becomes an example; but it will be sufficient for our purpose to employ the Binomial Theorem. We thus obtain for the coefficient of $x^{\mu+na}$ the expression

$$\frac{1}{(a+1)^{2n}}\frac{1}{\lfloor 2n-1}\left\{\phi(na+\mu+1) - 2n\,\phi(na+\mu+1-a-1)\right.$$

$$+ \frac{2n(2n-1)}{1.2}\phi(na+\mu+1-2a-2)$$

$$\left. - \frac{2n(2n-1)(2n-2)}{1.2.3}\phi(na+\mu+1-3a-3) + \ldots\right\};$$

where $\phi(r)$ stands for the product

$$r(r+1)(r+2)\ldots(r+2n-2);$$

the series within the brackets is to continue only so long as r is positive in $\phi(r)$.

565. We can see *à priori* that the coefficient of x^μ is equal to the coefficient of $x^{-\mu}$, and therefore when we want the former we may if we please find the latter instead. Thus in the result of

Art. 564, we may if we please put $-\mu$ instead of μ, without changing the value obtained. It is obvious that this would be a gain in practical examples as it would diminish the number of terms to be calculated. This remark is not given by Lagrange.

566. We can now find the probability that the error in the mean result shall lie between assigned limits. Let us find the probability that the error in the mean result shall lie between $-\dfrac{n\alpha}{n}$ and $\dfrac{\gamma}{n}$, both inclusive. We have then to substitute in the expression of Article 564 for μ in succession the numbers

$$-n\alpha, \quad -(n\alpha-1), \quad \ldots \gamma-1, \quad \gamma,$$

and add the results. Thus we shall find that, using Σ, as is customary, to denote a summation, we have

$$\Sigma\phi\,(n\alpha+\mu+1) = \frac{1}{2n}\,\psi\,(n\alpha+\gamma+1),$$

where $\psi\,(r)$ stands for

$$r\,(r+1)\,(r+2)\,\ldots\,(r+2n-1).$$

When we proceed to sum $\phi\,(n\alpha+\mu-\alpha)$ we must remember that we have only to include the terms for which $n\alpha+\mu-\alpha$ is positive; thus we find

$$\Sigma\phi\,(n\alpha+\mu-\alpha) = \frac{1}{2n}\,\psi\,(n\alpha+\gamma-\alpha).$$

Proceeding in this way we find that the probability that the error in the mean result will lie between $-\dfrac{n\alpha}{n}$ and $\dfrac{\gamma}{n}$, both inclusive, is

$$\frac{1}{(\alpha+1)^{2n}\,\lfloor 2n}\left\{\psi\,(n\alpha+\gamma+1) - 2n\,\psi\,(n\alpha+\gamma+1-\alpha-1)\right.$$

$$+\frac{2n\,(2n-1)}{1\,.\,2}\,\psi\,(n\alpha+\gamma+1-2\alpha-2)$$

$$\left.-\frac{2n\,(2n-1)\,(2n-2)}{1\,.\,2\,.\,3}\,\psi\,(n\alpha+\gamma+1-3\alpha-3)+\ldots\right\};$$

the series within the brackets is to continue only so long as r is positive in $\psi(r)$. We will denote this by $F(\gamma)$.

The probability that the mean error will lie between β and γ, where γ is greater than β, is $F(\gamma) - F(\beta)$ if we include γ and exclude β; it is $F(\gamma - 1) - F(\beta - 1)$ if we exclude γ and include β; it is $F(\gamma) - F(\beta - 1)$ if we include both γ and β; it is $F(\gamma - 1) - F(\beta)$ if we exclude both γ and β.

It is the last of these four results which Lagrange gives.

We have deviated slightly from his method in this Article in order to obtain the result with more clearness. Our result is $F(\gamma - 1) - F(\beta)$; and the number of terms in $F(\gamma - 1)$ is determined by the law that r in $\psi(r)$ is always to be positive: the number of terms in $F(\beta)$ is to be determined in a similar manner, so that the number of terms in $F(\beta)$ is not necessarily so great as the number of terms in $F(\gamma - 1)$. Lagrange gives an incorrect law on this point. He determines the number of terms in $F(\gamma - 1)$ correctly; *and then he prolongs* $F(\beta)$ *until it has as many terms as* $F(\gamma - 1)$ *by adding fictitious terms.*

567. Let us now modify the suppositions at the beginning of Art. 564. Suppose that instead of the errors $-\alpha, -(\alpha - 1), \ldots$ we are liable to the errors $-k\alpha, -k(\alpha - 1), \ldots$ Then the investigation in Art. 564 gives the probability that the error in the mean result shall be equal to $\dfrac{\mu k}{n}$; and the investigation in Art. 566 gives the probability that the error in the mean result shall lie between $\dfrac{\beta k}{n}$ and $\dfrac{\gamma k}{n}$. Let α increase indefinitely and k diminish indefinitely, and let αk remain finite and equal to h. Let γ and β also increase indefinitely; and let $\gamma = c\alpha$ and $\beta = b\alpha$ where c and b are finite. We find in the limit that $F(\gamma) - F(\beta)$ becomes

$$\frac{1}{\lfloor 2n} \left\{ (c+n)^{2n} - 2n(c+n-1)^{2n} + \frac{2n(2n-1)}{1.2}(c+n-2)^{2n} - \ldots \right\}$$

$$-\frac{1}{\lfloor 2n} \left\{ (b+n)^{2n} - 2n(b+n-1)^{2n} + \frac{2n(2n-1)}{1.2}(b+n-2)^{2n} - \ldots \right\};$$

each series is to continue only so long as the quantities which are raised to the power $2n$ are positive.

This result expresses the probability that the error in the mean result will lie between $\dfrac{bh}{n}$ and $\dfrac{ch}{n}$ on the following hypothesis; at every trial the error may have any value between $-h$ and $+h$; positive and negative errors are equally likely; the probability of a positive error z is proportional to $h-z$, and in fact $\dfrac{(h-z)\,\delta z}{h^2}$ is the probability that the error will lie between z and $z+\delta z$.

We have followed Lagrange's guidance, and our result agrees with his, except that he takes $h=1$, and his formula involves many misprints or errors.

568. The conclusion in the preceding Article is striking. We have an exact expression for the probability that the error in the mean result will lie between assigned limits, on a very *reasonable hypothesis as to the occurrence of single errors.*

Suppose that positive errors are denoted by abscissæ measured to the right of a fixed point, and negative errors by abscissæ measured to the left of that fixed point. Let ordinates be drawn representing the probabilities of the errors denoted by the respective abscissæ. The curve which can thus be formed is called the *curve of errors* by Lagrange; and as he observes, the curve becomes an isosceles triangle in the case which we have just discussed.

569. The matter which we have noticed in Arts. 563, 564, 566, 567, 568, had all been published by Thomas Simpson, in his *Miscellaneous Tracts*, 1757; he gave also some numerical illustrations: see Art. 371.

570. The remainder of Lagrange's memoir is very curious; it is devoted to the solution and exemplification of one general problem. In Art. 567 we have obtained a result for a case in which the error at a single trial may have *any* value between fixed limits; but this result was not obtained directly: we started with the supposition that the error at a single trial must be one of a *certain specified number* of errors. In other words we started with the hypothesis of errors changing *per saltum* and passed on

to the supposition of *continuous* errors. Lagrange wishes to solve questions relative to *continuous* errors without starting with the supposition of errors changing *per saltum*.

Suppose that at every observation the error must lie between b and c; let $\phi(x)\,dx$ denote the probability that the error will lie between x and $x+dx$: required the probability that in n observations the sum of the errors will lie between assigned limits say β and γ. Now what Lagrange effects is the following. He transforms $\left\{\int_b^c \phi(x)\,a^x\,dx\right\}^n$ into $\int f(z)\,a^z\,dz$, where $f(z)$ is a known function of z which does not involve a, and the limits of the integral are known. When we say that $f(z)$ and the limits of z are known we mean that they are determined from the known function ϕ and the known limits b and c. Lagrange then *says* that the probability that the sum of the errors will lie between β and γ is $\int_\beta^\gamma f(z)\,dz$. He apparently concludes that his readers will admit this at once; he certainly does not demonstrate it. We will indicate presently the method in which it seems the demonstration must be put.

571. After this general statement we will give Lagrange's first example.

Suppose that $\phi(x)$ is constant $= K$ say; then

$$\int_b^c \phi(x)\,a^x\,dx = \frac{K(a^c - a^b)}{\log a},$$

therefore

$$\left\{\int_b^c \phi(x)\,a^x\,dx\right\}^n = \frac{K^n(a^c - a^b)^n}{(\log a)^n}.$$

Now we may suppose that a is greater than unity, and then it may be easily shewn that

$$\int_0^\infty y^{n-1} a^{-y}\,dy = \frac{\lfloor n-1}{(\log a)^n};$$

thus

$$\left\{\int_b^c \phi(x)\,a^x\,dx\right\}^n = \frac{K^n}{\lfloor n-1}(a^c - a^b)^n \int_0^\infty y^{n-1} a^{-y}\,dy.$$

Let $c - b = t$, and expand $(a^c - a^b)^n$ by the Binomial Theorem;

thus
$$\left\{ \int_b^c \phi(x)\, a^x\, dx \right\}^n$$

$$= \frac{K^n}{\lfloor n-1} \left\{ a^{nc} - na^{nc-t} + \frac{n(n-1)}{1.2} a^{nc-2t} - \ldots \right\} \int_0^\infty y^{n-1} a^{-v}\, dy.$$

Now decompose $\int_0^\infty y^{n-1} a^{-v} dy$ into its elements; and multiply them by the series within brackets. We obtain for the coefficient of a^{nc-v} the expression

$$\frac{K^n}{\lfloor n-1} \left\{ y^{n-1} - n(y-t)^{n-1} + \frac{n(n-1)}{1.2} (y - 2t)^{n-1} - \ldots \right\} dy,$$

where the series within brackets is to continue only so long as the quantities raised to the power $n-1$ are positive.

Let $nc - y = z$; then $dy = - dz$: when $y = 0$ we have $z = nc$, and when $y = \infty$ we have $z = -\infty$. Substitute $nc - z$ for y, and we obtain finally

$$\left\{ \int_b^c \phi(x)\, a^x\, dx \right\}^n = \int_{-nc}^\infty f(z)\, a^x\, dz,$$

where
$$f(z) = \frac{K^n}{\lfloor n-1} \left\{ (nc - z)^{n-1} - n(nc - z - t)^{n-1} \right.$$

$$\left. + \frac{n(n-1)}{1.2} (nc - z - 2t)^{n-1} - \ldots \right\};$$

the series within brackets being continued only so long as the quantities raised to the power $n-1$ are positive.

Lagrange then says that the probability that the sum of the errors in n observations will lie between β and γ is

$$\int_\beta^\gamma f(z)\, dz.$$

572. The result is correct, for it can be obtained in another way. We have only to carry on the investigation of the problem enunciated in Art. 563 in the same way as the problem enunciated in Art. 564 was treated in Art. 567; the result will be very similar to those in Art. 567. Lagrange thus shews that his process is verified in this example.

573. In the problem of Art. 570 it is obvious that the sum of the errors must lie between nb and nc. Hence $f(z)$ ought to vanish if z does not lie between these limits; and we can easily shew that it does.

For if z be greater than nc there is no term at all in $f(z)$, for every quantity raised to the power $n-1$ would be negative.

And if z be less than nb, then $f(z)$ vanishes by virtue of the theorem in Finite Differences which shews that the nth difference of an algebraical function of the degree $n-1$ is zero.

This remark is not given by Lagrange.

574. We will now supply what we presume would be the demonstration that Lagrange must have had in view.

Take the general problem as enunciated in Art. 570. It is not difficult to see that the following process would be suitable for our purpose. Let a be any quantity, which for convenience we may suppose greater than unity. Find the value of the expression

$$\left\{ \int \phi(x_1)\, a^{x_1}\, dx_1 \right\} \left\{ \int \phi(x_2)\, a^{x_2}\, dx_2 \right\} \ldots\ldots\ldots \left\{ \int \phi(x_n)\, a^{x_n}\, dx_n \right\},$$

where the integrations are to be taken under the following limitations; each variable is to lie between b and c, and the sum of the variables between z and $z + \delta z$. Put the result in the form $Pa^z \delta z$; then $\int_\beta^\gamma P\, dz$ is the required probability.

Now to find P we proceed in an indirect way. It follows from our method that

$$\left\{ \int_b^c \phi(x)\, a^x\, dx \right\}^n = \int_{nb}^{nc} Pa^z\, dz.$$

But Lagrange by a suitable transformation shews that

$$\left\{ \int_b^c \phi(x)\, a^x\, dx \right\}^n = \int_{z_0}^{z_1} f(z)\, a^z\, dz,$$

where z_0 and z_1 are known. Hence

$$\int_{nb}^{nc} Pa^z\, dz = \int_{z_0}^{z_1} f(z)\, a^z\, dz.$$

It will be remembered that a may be *any* quantity which

is greater than unity. We shall shew that we must then have
$P = f(z)$.

Suppose that z_0 is less than nb, and z_1 greater than nc. Then we have

$$\int_{z_0}^{nb} f(z)\, a^z\, dz + \int_{nb}^{nc} \{f(z) - P\}\, a^z\, dz + \int_{nc}^{z_1} f(z)\, a^z\, dz = 0,$$

for all values of a. Decompose each integral into elements; put $a^{\delta z} = \rho$. We have then *ultimately* a result of the following form

$$a^{z_0} \left\{ T_0 + T_1\rho + T_2\rho^2 + T_3\rho^3 + \dots in\ inf. \dots \right\} = 0,$$

where T_0, T_1,... are independent of ρ. And ρ may have any positive value we please. Hence by the ordinary method of indeterminate coefficients we conclude that

$$T_0 = 0, \quad T_1 = 0, \quad T_2 = 0, \dots$$

Thus $\qquad\qquad P = f(z)$.

The demonstration will remain the same whatever supposition be made as to the order of magnitude of the limits z_0 and z_1 compared with nb and nc.

575. Lagrange takes for another example that which we have already discussed in Art. 567, and he thus again verifies his new method by its agreement with the former.

He then takes two new examples; in one he supposes that $\phi(x) = K\sqrt{c^2 - x^2}$, the errors lying between $-c$ and c; in the other he supposes that $\phi(x) = K\cos x$, the errors lying between $-\dfrac{\pi}{2}$ and $\dfrac{\pi}{2}$.

576. We have now to notice another memoir by Lagrange which is entitled *Recherches sur les suites recurrentes dont les termes varient de plusieurs manieres différentes, ou sur l'intégration des équations linéaires aux différences finies et partielles ; et sur l'usage de ces équations dans la théorie des hazards.*

This memoir is published in the *Nouveaux Mémoires de l'Acad. ... Berlin.* The volume is for the year 1775; the date of publication is 1777. The memoir occupies pages 183—272; the application to the Theory of Chances occupies pages 240—272.

577. The memoir begins thus;

J'ai donné dans le premier Volume des Mémoires de la Société des Sciences de Turin une méthode nouvelle pour traiter la théorie des suites recurrentes, eu la faisant dépendre de l'intégration des équations linéaires aux différences finies. Je me proposois alors de pousser ces recherches plus loin et de les appliquer principalement à la solution de plusieurs problemes de la théorie des hasards; mais d'autres objets m'ayant depuis fait perdre celui là de vue, M. de la Place m'a prévenu en grand partie dans deux excellens Mémoires *sur les suites recurro-recurrentes*, et *sur l'intégration des équations différentielles finies et leur usage dans la théorie des hasards*, imprimés dans les Volumes VI et VII des Mémoires présentés à l'Académie des Sciences de Paris. Je crois cependant qu'on peut encore ajoûter quelque chose au travail de cet illustre Géometre, et traiter le même sujet d'une maniere plus directe, plus simple et surtout plus générale; c'est l'objet des Recherches que je vais donner dans ce Mémoire; on y trouvera des méthodes nouvelles pour l'intégration des équations linéaires aux différences finies et partielles, et l'application de ces méthodes à plusieurs problemes intéressans du calcul des probabilités; mais il n'est question ici que des équations dont les coëfficiens sont constants, et je réserve pour un autre Mémoire l'examen de celles qui ont des coëfficiens variables.

578. We shall not delay on the part which relates to the Integration of Equations; the methods are simple but not so good as that of Generating Functions. We proceed to the part of the memoir which relates to Chances.

579. The first problem is to find the chance of the happening of an event b times at least in a trials.

Let p denote the chance of its happening in one trial; let $y_{x,t}$ denote the probability of its happening t times in x trials; then Lagrange puts down the equation

$$y_{x,t} = p y_{x-1,\,t-1} + (1-p)\, y_{x-1,\,t}.$$

He integrates and determines the arbitrary quantities and thus arrives at the usual result.

In a Corollary he applies the same method to determine the

chance that the event shall happen *just b* times ; he starts from
the same equation and by a different determination of the arbitrary quantities arrives at the result which is well known,
namely,

$$\frac{p^b \, (1-p)^{a-b} \, \lfloor a}{\lfloor b \, \lfloor a-b}.$$

Lagrange refers to De Moivre, page 15, for one solution, and
adds : mais celle que nous venons d'en donner est non seulement
plus simple, mais elle a de plus l'avantage d'être déduite de principes directs.

But it should be observed that De Moivre solves the problem
again on his page 27; and here he indicates the modern method,
which is self-evident. See Art. 257.

It seems curious for Lagrange to speak of his method as *more
simple* than De Moivre's, seeing it involves an elaborate solution
of an equation in Finite Differences.

580. Lagrange's second problem is the following :

On suppose qu'à chaque coup il puisse arriver deux évenemens dont
les probabilités respectives soient *p* et *q*; et on demande le sort d'un
joueur qui paricroit d'amener le premier de ces évenemens *b* fois au
moins et le second *c* fois au moins, en un nombre *a* de coups.

The enunciation does not state distinctly what the suppositions
really are, namely that at every trial either the first event happens,
or the second, or neither of them ; these three cases are mutually
exclusive, so that the probability of the last at a single trial
is $1-p-q$. It is a good problem, well solved ; the solution is
presented in a more elementary shape by Trembley in a memoir
which we shall hereafter notice.

581. The third problem is the following :

Les mêmes choses étant supposées que dans le Probleme II, on demande le sort d'un joueur qui parieroit d'amener, dans un nombre de
coups indéterminé, le second des deux évenemens *b* fois avant que le
premier fût arrivé *a* fois.

Let $y_{x,t}$ be the chance of the player when he has to obtain the
second event *t* times before the first event occurs *x* times. Then

$$y_{x,t} = p y_{x-1,t} + q y_{x,t-1}.$$

This leads to

$$y_{x,t} = q^t \left\{ 1 + tp + \frac{t\,(t+1)}{2}\,p^2 + \frac{t\,(t+1)\,(t+2)}{2\,.\,3}\,p^3 + \cdots \right.$$
$$\left. + \cdots \frac{\lfloor t+x-2}{\lfloor t-1 \rfloor \lfloor x-1 \rfloor}\,p^{x-1} \right\}.$$

This result agrees with the second formula in Art. 172.

582. The fourth problem is like the third, only *three* events may now occur of which the probabilities are p, q, r respectively. In a Corollary the method is extended to four events; and in a second Corollary to any number.

To this problem Lagrange annexes the following remark :

Le Probleme dont nous venons de donner une solution très générale et très simple renferme d'une maniere générale celui qu'on nomme communément dans l'analyse des hasards le probleme des partis, et qui n'a encore été résolu complettement que pour le cas de deux joueurs.

He then refers to Montmort, to De Moivre's second edition, Problem VI, and to the memoir of Laplace.

It is very curious that Lagrange here refers to De Moivre's *second* edition, while elsewhere in the memoir he always refers to the *third* edition; for at the end of Problem VI. in the *third* edition De Moivre does give the general rule for any number of players. This he first published in his *Miscellanea Analytica*, page 210 ; and he reproduced it in his *Doctrine of Chances*. But in the second edition of the *Doctrine of Chances* the rule was not given in its natural place as part of Problem VI. but appeared as Problem LXIX.

There is however some difference between the solutions given by De Moivre and by Lagrange ; the difference is the same as that which we have noticed in Art. 175 for the case of two players. De Moivre's solution resembles the *first* of those which are given in Art. 172, and Lagrange's resembles the *second*.

It is stated by Montucla, page 397, that Lagrange intended to translate De Moivre's third edition into French.

583. Lagrange's fifth problem relates to the Duration of Play, in the case in which one player has unlimited capital ; this is De Moivre's Problem LXV: see Art. 307. Lagrange gives three solutions. Lagrange's first solution demonstrates the result given

without demonstration in De Moivre's second solution; see Art. 309. We will give Lagrange's solution as a specimen of his methods. We may remark that Laplace had preceded Lagrange in the discussion of the problem of the Duration of Play. Laplace's investigations had been published in the *Mémoires ... par Divers Savans*, Vols. VI. and VII.

Laplace did not formally make the supposition that one player had unlimited capital, but we arrive at this case by supposing that his symbol i denotes an infinite number; and we shall thus find that on page 158 of Laplace's memoir in Vol. VII. of the *Mémoires...par Divers Savans*, we have in effect a demonstration of De Moivre's result.

We proceed to Lagrange's demonstration.

584. The probability of a certain event in a single trial is p; a player bets that in a trials this event will happen at least b times oftener than it fails : determine the player's chance.

Let $y_{x,t}$ represent his chance when he has x more trials to make, and when to ensure his success the event must happen at least t times oftener than it fails. Then it is obvious that we require the value of $y_{a,b}$.

Suppose one more trial made; it is easy to obtain the following equation

$$y_{x,t} = py_{x-1,t-1} + (1-p) y_{x-1, t+1}.$$

The player gains when $t = 0$ and x has any value, and he loses when $x = 0$ and t has any value greater than zero; so that $y_{x,0} = 1$ for any value of x, and $y_{0,t} = 0$ for any value of t greater than 0.

Put q for $1 - p$, then the equation becomes

$$py_{x,t} + qy_{x,t+2} - y_{x+1,t+1} = 0.$$

To integrate this assume $y = A a^x \beta^t$; we thus obtain

$$p - a\beta + q\beta^2 = 0.$$

From this we may by Lagrange's Theorem expand β^t in powers of α; there will be two series because the quadratic equation gives two values of β for an assigned value of α. These two series are

$$\beta^t = \frac{p^t}{a^t} + \frac{tp^{t+1}q}{a^{t+2}} + \frac{t(t+3)}{1.2}\frac{p^{t+2}q^2}{a^{t+4}} + \frac{t(t+4)(t+5)}{1.2.3}\frac{p^{t+3}q^3}{a^{t+6}} + \cdots$$

$$\beta^t = \frac{a^t}{q^t} - \frac{tpa^{t-2}}{q^{t-1}} + \frac{t(t-3)}{1.2}\frac{p^2 a^{t-4}}{q^{t-2}} - \frac{t(t-4)(t-5)}{1.2.3}\frac{p^3 a^{t-6}}{q^{t-3}} + \cdots$$

If then we put in succession these values of β^t in the expression $Aa^x\beta^t$ we obtain two series in powers of α, namely,

$$Ap^t\left\{\alpha^{x-t} + tpq\alpha^{x-t-2} + \frac{t(t+3)}{1.2}p^2q^2\alpha^{x-t-4} + \cdots\right\},$$

and $$Aq^{-t}\left\{\alpha^{x+t} - tpq\alpha^{x+t-2} + \frac{t(t-3)}{1.2}p^2q^2\alpha^{x+t-4} - \cdots\right\}.$$

Either of these series then would be a solution of the equation in Finite Differences, *whatever may be the values of A and* α; so that we should also obtain a solution by the sum of any number of such series with various values of A and α.

Hence we infer that the general solution will be

$$y_{x,t} = p^t\left\{f(x-t) + tpqf(x-t-2) + \frac{t(t+3)}{1.2}p^2q^2f(x-t-4)\right.$$
$$\left. + \frac{t(t+4)(t+5)}{1.2.3}p^3q^3f(x-t-6) + \cdots\right\}$$
$$+ q^{-t}\left\{\phi(x+t) - tpq\,\phi(x+t-2) + \frac{t(t-3)}{1.2}p^2q^2\,\phi(x+t-4)\right.$$
$$\left. - \frac{t(t-4)(t-5)}{1.2.3}p^3q^3\,\phi(x+t-6) + \cdots\right\}.$$

Here $f(x)$ and $\phi(x)$ represent functions, at present arbitrary, which must be determined by aid of the known particular values of $y_{x,0}$ and $y_{0,t}$.

Lagrange says it is easy to convince ourselves, that the condition $y_{0,t} = 0$ when t has any value greater than 0 leads to the following results : *all the functions with the characteristic ϕ must be zero, and those with the characteristic f must be zero for all negative values of the quantity involved.* [Perhaps this will not appear very satisfactory; it may be observed that q^{-t} will become indefinitely great with t, and this suggests that the series which multiplies q^{-t} should be zero.]

Thus the value of $y_{x,t}$ becomes a series with a finite number of terms, namely,

$$y_{x,t} = p^t \left\{ f(x-t) + tpqf(x-t-2) + \frac{t(t+3)}{1.2} p^2 q^2 f(x-t-4) \right.$$
$$\left. + \frac{t(t+4)(t+5)}{1.2.3} p^3 q^3 f(x-t-6) + \dots \right\},$$

the series extends to $\frac{1}{2}(x-t+2)$ terms, or to $\frac{1}{2}(x-t+1)$ terms,

according as $x-t$ is even or odd.

The other condition is that $y_{x,0} = 1$, for any value of x. But if
we put $t=0$ we have $y_{x,0} = f(x)$. Hence $f(x) = 1$ for every
positive value of x. Thus we obtain

$$y_{x,t} = p^t \left\{ 1 + tpq + \frac{t(t+3)}{1.2} p^2 q^2 + \frac{t(t+4)(t+5)}{1.2.3} p^3 q^3 + \dots \right\},$$

the series is to extend to $\frac{1}{2}(x-t+2)$ terms, or to $\frac{1}{2}(x-t+1)$
terms. This coincides with the result in De Moivre's second form
of solution: see Art. 309.

585. Lagrange gives two other solutions of the problem just
considered, one of which presents the result in the same form as
De Moivre's first solution. These other two solutions by Lagrange
differ in the mode of integrating the equation of Finite Differences;
but they need not be further examined.

586. Lagrange then proceeds to the general problem of the
Duration of Play, supposing the players to start with different
capitals. He gives two solutions, one similar to that in De
Moivre's Problem LXIII, and the other similar to that in De
Moivre's Problem LXVIII. The second solution is very remarkable;
it demonstrates the results which De Moivre enunciated without
demonstration, and it puts them in a more general form, as De
Moivre limited himself to the case of *equal* capitals.

587. Lagrange's last problem coincides with that given by
Daniel Bernoulli which we have noticed in Art. 417. Lagrange
supposes that there are n urns; and in a Corollary he gives some
modifications of the problem.

588. Lagrange's memoir would not now present any novelty
to a student, or any advantage to one who is in possession of the
method of Generating Functions. But nevertheless it may be read

with ease and interest, and at the time of publication its value
must have been great. The promise held out in the introduction
that something would be added to the labours of Laplace is
abundantly fulfilled. The solution of the general problem of the
Duration of Play is conspicuously superior to that which Laplace
had given, and in fact Laplace embodied some of it subsequently
in his own work. The important pages 231—233 of the *Théorie
... des Prob.* are substantially due to this memoir of Lagrange's.

589. We may notice a memoir by Lagrange entitled *Mémoire sur une question concernant les annuités.*

This memoir is published in the volume of the *Mémoires de
l'Acad.... Berlin* for 1792 and 1793; the date of publication is
1798; the memoir occupies pages 235—246.

The memoir had been read to the Academy ten years before.

590. The question discussed is the following: A father wishes
to pay a certain sum annually during the joint continuance of his
own life and the minority of all his children, so as to ensure an
annuity to his children after his death to last until all have attained
their majority.

Lagrange denotes by $\overline{A}, \overline{B}, \overline{C}, \ldots$ the value of an annuity of
one crown for the minority of the children $A, B, C \ldots$ respectively.
Then by $\overline{AB}$ he denotes the value of an annuity of one crown
for the joint minority of two children A and B; and so on. Hence
he obtains for the value of an annuity payable as long as either
A or B is a minor,
$$\overline{A} + \overline{B} - \overline{AB}.$$

Lagrange demonstrates this; but the notation renders it almost
obviously self evident.

Similarly the value of an annuity payable as long as one of
three children A, B, C remains a minor is
$$\overline{A} + \overline{B} + \overline{C} - \overline{AB} - \overline{AC} - \overline{BC} + \overline{ABC}.$$

De Moivre however had given this result in his *Treatise of
Annuities on Lives*, and had used the same notation for an annuity
on joint lives.

Lagrange adds two tables which he calculated from his
formulæ, using the table of mortality given in the work of
Sussmilch.

CHAPTER XVI.

MISCELLANEOUS INVESTIGATIONS

BETWEEN THE YEARS 1750 AND 1780.

591. THE present Chapter will contain notices of various contributions to our subject which were made between the years 1750 and 1780.

592. We first advert to a work bearing the following title: *Piece qui a remporté le prix sur le sujet des Evenemens Fortuits, proposé par l'Academie Royale des Sciences et Belles Lettres de Berlin pour l'année* 1751. *Avec les pieces qui ont concouru.*

This work is a quarto volume of 238 pages; we notice it because the title might suggest a connexion with our subject, which we shall find does not exist.

The Academy of Berlin proposed the following subject for discussion :

Les Evenemens heureux et malheureux, ou ce que nous appellons Bonheur et Malheur dependant de la volonté ou de la permission de Dieu, de sorte que le terme de fortune est un nom sans réalité; on demande si ces Evenemens nous obligent à de certains devoirs, quels sont ces devoirs et quelle est leur étendue.

The prize was awarded to Kaestner professor of Mathematics at Leipsic ; the volume contains his dissertation and those of his competitors.

There are nine dissertations on the whole ; the prize dissertation is given both in French and Latin, and the others in French

21

or German or Latin. The subject was perhaps unpromising; the
dissertations are not remarkable for novelty or interest. One of
the best of the writers finishes with a modest avowal which might
have been used by all:

Ich mache hier den Schluss, weil ich ohnehin mit gar zu guten
Gründen fürchte, zu weitläufig gewesen zu seyn, da ich so wenig neues
artiges und scharfsinniges gesagt habe. Ich finde auch in dieser Probe,
dass mein Wille noch einmahl so gut als meine übrige Fähigkeit, ist.

593. A work entitled the *Mathematical Repository*, in three
volumes, was published by James Dodson, *Accomptant and Teacher
of the Mathematics*. The work consists of the solution of Mathematical problems. The second volume is dated 1753; pages
82—136 are occupied with problems on chances: they present
nothing that is new or important. The remainder of this volume
is devoted to annuities and kindred subjects; and so also is the
whole of the third volume, which is dated 1755.

594. Some works on Games of Chance are ascribed to Hoyle
in Watt's *Bibliotheca Britannica*. I have seen only one of them
which is entitled: *An Essay towards making the Doctrine of
Chances easy to those who understand Vulgar Arithmetick only:
to which is added, some useful tables on annuities for lives &c. &c. &c.
By Mr Hoyle...* It is not dated; but the date 1754 is given in
Watt's *Bibliotheca Britannica*.

The work is in small octavo size, with large type. The title,
preface, and dedication occupy VIII pages, and the text itself occupies 73 pages. Pages 1—62 contain rules, without demonstration,
for calculating chances in certain games; and the remainder is devoted to tables of annuities, and to Halley's Breslau table of life,
with a brief explanation of the latter. I have not tested the rules.

595. We advert in the next place to a work which is entitled *Dell' Azione del Caso nelle Invenzioni, e dell' influsso degli
Astri ne' Corpi Terrestri Dissertazioni due.*

This is a quarto volume of 220 pages, published anonymously
at Padua, 1757. It is not connected with the Theory of Probability; we notice it because the title might perhaps suggest

such connexion, especially when abbreviated, as in the Catalogues of Booksellers.

The first dissertation is on the influence of chance in inventions, and the second on the influence of the celestial bodies on men, animals, and plants. The first dissertation recognises the influence of chance in inventions, and gives various examples ; the second dissertation is intended to shew that there is no influence produced by the celestial bodies on men, animals, or plants, in the sense in which astrologers understood such influence.

The author seems to have been of a sanguine temperament ; for he obviously had hopes that the squaring of the circle would be eventually obtained ; see his pages 31, 40, 85.

On the other hand his confidence is not great in the Newtonian theory of gravitation ; he thinks it may one day follow its predecessor, the theory of vortices, into oblivion ; see his pages 45, 172.

The following is one of his arguments against Lunar influence. If there be such influence we must conceive it to arise from exhalations from the Moon, and if the matter of these exhalations be supposed of appreciable density it will obstruct the motions of the planets, so that it will be necessary from time to time to clean up the celestial paths, just as the streets of London and Paris are cleaned from dust and dirt. See his page 164.

The author is not very accurate in his statements. Take the following specimen from his page 74: Jacopo III. Re d'Inghilterra alla vista d'una spada ignuda, come riferisce il Cavaliere d'Igby, sempre era compreso d'un freddo, e ferale spavento. This of course refers to James I. Again ; we have on his page 81 : ...ciò che disse in lode d'Aristotile il Berni : *Il gran Maestro de color che sanno.* It is not often that an Italian ascribes to any inferior name the honour due to Dante.

596. We have next to notice a work by Samuel Clark entitled *The Laws of Chance : or, a Mathematical Investigation of the Probabilities arising from any proposed Circumstance of Play.* London, 1758.

This is in octavo ; there is a Preface of 2 pages, and 204 pages of text. The book may be described as a treatise based on those of De Moivre and Simpson ; the abstruse problems are

omitted, and many examples and illustrations are given in order
to render the subject accessible to persons not very far advanced
in mathematics.

The book presents nothing that is new and important. The
game of bowls seems to have been a favourite with Clark; he
devotes his pages 44—68 to problems connected with this game.
He discusses at great length the problem of finding the chance of
throwing an assigned number of points with a given number of
similar dice; see his pages 113—130. He follows Simpson, but
he also indicates De Moivre's Method; see Art. 364. Clark
begins the discussion thus:

In order to facilitate the solution of this and the following problem,
I shall lay down a lemma which was communicated to me by my ingenious friend Mr *William Payne*, teacher of mathematics.

The Lemma.

The sum of 1, 3, 6, 10, 15, 21, 28, 36, &c. continued to (n) number
of terms is equal to $\dfrac{n+2}{1} \times \dfrac{n+1}{2} \times \dfrac{n}{3}$.

It was quite unnecessary to appeal to William Payne for such
a well-known result; and in fact Clark himself had given on his
page 84 Newton's general theorem for the summation of series;
see Art. 152.

Clark discusses in his pages 139—153 the problem respecting
a *run* of events, which we have noticed in Art. 325. Clark detects
the slight mistake which occurs in De Moivre's solution; and from
the elaborate manner in which he notices the mistake we may
conclude that it gave him great trouble.

Clark is not so fortunate in another case in which he ventures
to differ with De Moivre; Clark discusses De Moivre's Problem IX.
and arrives at a different result; see Art. 269. The error is
Clark's. Taking De Moivre's notation Clark assumes that A must
either receive qG from B, or pay pL to B. This is wrong. Suppose that on the whole A wins in $q + m$ trials and loses in m trials;
then there is the required difference of q games in his favour. In
this case he receives from B the sum $(q + m) G$ and pays to him
the sum mL; thus the balance is $qG + m (G - L)$ and not qG as
Clark says.

597. We have next to notice a memoir by Mallet, entitled *Recherches sur les avantages de trois Joueurs qui font entr'eux une Poule au trictrac ou à un autre Jeu quelconque.* This memoir is published in the *Acta Helvetica...Basileœ*, Vol. v. 1762; the memoir occupies pages 230—248. The problem is that of De Moivre and Waldegrave; see Art. 211. Mallet's solution resembles that given by De Moivre in his pages 132—138. Mallet however makes some additions. In the problem as treated by De Moivre the fine exacted from each defeated player is *constant*; Mallet considers the cases in which the fines increase in arithmetical progression, or in geometrical progression. A student of De Moivre will see that the extensions given by Mallet can be treated without any difficulty by De Moivre's process, as the series which are obtained may be summed by well-known methods.

598. The same volume which contains Euler's memoir which we have noticed in Art. 438, contains also two memoirs by Beguelin on the same problem. Before we notice them it will be convenient to consider a memoir by John Bernoulli, which in fact precedes Beguelin's in date of composition but not in date of publication. This John Bernoulli was grandson of the John whom we named in Art. 194. John Bernoulli's memoir is entitled *Sur les suites ou séquences dans la loterie de Genes.* It was published in the volume for 1769 of the *Histoire de l'Acad.... Berlin;* the date of publication is 1771: the memoir occupies pages 234—253. The following note is given at the beginning:

Ce Mémoire a été lu en 1765, après le Mémoire de Mr. Euler sur cette matiere inséré dans les Mémoires de l'Académie pour cette année. Comme les Mémoires de Mr. Beguelin imprimés a la suite de celui de Mr. Euler se rapportent au mien en plusieurs endroits, et que la Loterie qui l'a occasioné est plus en vogue que jamais, je ne le supprimerai pas plus longtems. Si ma méthode ne mene pas aussi loin que celle de Mrs. Euler et Beguelin, elle a du moins, je crois, l'avantage d'être plus facile à saisir.

599. In the first paragraph of the memoir speaking of the question respecting sequences, John Bernoulli says:

Je m'en occupai donc de tems en tems jusqu'à ce que j'appris de Mr. Euler qu'il traitoit le même sujet; c'en fut assez pour me faire abandonner mon dessein, et je me réservai seulement de voir par le Mémoire de cet illustre Géometre si j'avois raisonné juste; il a eu la bonté de me le communiquer et j'ai vû que le peu que j'avois fait, étoit fondé sur des raisonnemens qui, s'ils n'étoient pas sublimes, n'etoient du moins pas faux.

600. John Bernoulli does not give an Algebraical investigation; he confines himself to the arithmetical calculation of the chances of the various kinds of sequences that can occur when there are 90 tickets and 2 or 3 or 4 or 5 are drawn. His method does not seem to possess the advantage of facility, as compared with those of Euler and Beguelin, which he himself ascribes to it.

601. There is one point of difference between John Bernoulli and Euler. John Bernoulli supposes the numbers from 1 to 90 ranged as it were in a circle; and thus he counts 90, 1 as a binary sequence; Euler does not count it as a sequence. So also John Bernoulli counts 89, 90, 1 as a ternary sequence; with Euler this would count as a binary sequence. And so on.

It might perhaps have been anticipated that from the greater symmetry of John Bernoulli's conception of a sequence, the investigations respecting sequences would be more simple than on Euler's conception; but the reverse seems to be the case on examination.

In the example of Art. 440 corresponding to Euler's results

$$ n - 2, \qquad (n-2)(n-3), \qquad \frac{(n-2)(n-3)(n-4)}{1.2.3}, $$

we shall find on John Bernoulli's conception the results

$$ n, \qquad n(n-4), \qquad \frac{n(n-4)(n-5)}{1.2.3}. $$

602. There is one Algebraical result given which we may notice. Euler had obtained the following as the chances that there would be no sequences at all in the case of n tickets; if two tickets be drawn the chance is $\frac{n-2}{n}$, if three $\frac{(n-3)(n-4)}{n(n-1)}$, if four $\frac{(n-4)(n-5)(n-6)}{n(n-1)(n-2)}$, if five $\frac{(n-5)(n-6)(n-7)(n-8)}{n(n-1)(n-2)(n-3)}$; and so the law can be easily seen. Now John Bernoulli states

that on his conception of a sequence these formulæ will hold if we change n into $n-1$. He does not demonstrate this statement, so that we cannot say how he obtained it.

It may be established by induction in the following way. Let $E(n, r)$ denote the number of ways in which we can take r tickets out of n, free from any sequence, on Euler's conception of a sequence. Let $B(n, r)$ denote the corresponding number on John Bernoulli's conception. Then we have given

$$E(n, r) = \frac{(n-r+1)(n-r)\dots(n-2r+2)}{\lfloor r}\,,$$

and we have to shew that

$$B(n, r) = \frac{n(n-r-1)\dots(n-2r+1)}{\lfloor r}\,.$$

For these must be the values of $E(n, r)$ and $B(n, r)$ in order that the appropriate chances may be obtained, by dividing by the total number of cases. Now the following relation will hold:

$$E(n, r) = B(n, r) + B(n-1, r-1) - E(n-2, r-1).$$

The truth of this relation will be seen by taking an example. Suppose n is 10, and r is 3. Now every case which occurs in the total $B(n, r)$ will occur among the total $E(n, r)$; but some which do not occur in $B(n, r)$ will occur in $E(n, r)$, and these must be added. These cases which are to be added are such as (10, 1, 3) (10, 1, 4) ...... (10, 1, 8). We must then examine by what general law we can obtain these cases. We should form all the binary combinations of the numbers $1 . 2, \dots 9$ which contain no Bernoullian sequence, and which do contain 1.

And generally we should want all the combinations $r-1$ at a time which can be made from the first $n-1$ numbers, so as to contain no Bernoullian sequence, and to contain 1 as one of the numbers. It might at first appear that $B(n-1, r-1) - B(n-2, r-1)$ would be the number of such combinations; but a little consideration will shew that it is $B(n-1, r-1) - E(n-2, r-1)$, as we have given it above.

Thus having established the relation, and found the value of $B(n, 1)$ independently we can infer in succession the values of $B(n, 2)$, $B(n, 3)$, and so on.

603. We now consider Beguelin's two memoirs. These as we
have stated are contained in the same volume as Euler's memoir
noticed in Art. 438. The memoirs are entitled *Sur les suites ou
séquences dans la lotterie de Genes;* they occupy pages 231—280
of the volume.

604. Beguelin's memoirs contain general Algebraical formulæ
coinciding with Euler's, and also similar formulæ for the results on
John Bernoulli's conception; thus the latter formulæ constitute
what is new in the memoirs.

605. We can easily give a notion of the method which
Beguelin uses. Take for example 13 letters $a, b, c, \ldots i, j, k, l, m$.
Arrange 5 files of such letters side by side, thus

$$a \quad a \quad a \quad a \quad a$$
$$b \quad b \quad b \quad b \quad b$$
$$c \quad c \quad c \quad c \quad c$$
$$\ldots\ldots\ldots\ldots\ldots$$
$$m \quad m \quad m \quad m \quad m$$

Consider first only two such files; take any letter in the first
file and associate it with any letter in the second file; we thus
get 13^2 such associations, namely aa, ab, $ac \ldots ba$, bb, bc, ...

Here we have ab and ba both occurring, and so ac and ca, and
the like. But suppose we wish to prevent such repetitions, we can
attain our end in this way. Take any letter in the first file and
associate it with those letters only in the second file, which are in the
same rank or in a lower rank. Thus the a of the first file will be
associated with any one of the 13 letters of the second file; the b of
the second file will be associated with any one of the 12 letters
in the second file beginning with b. Thus the whole number of
such associations will be $13 + 12 + \ldots + 1$; that is $\dfrac{13 \times 14}{1 \cdot 2}$.

Similarly if we take three files we shall have 13^3 associations
if we allow repetitions; but if we do not allow repetitions we
shall have $\dfrac{13 \times 14 \times 15}{1 \times 2 \times 3}$. Proceeding in this way we find that if
there are five files and we do not allow repetitions the number of
associations is $\dfrac{13 \times 14 \times 15 \times 16 \times 17}{1 \times 2 \times 3 \times 4 \times 5}$.

All this is well known, as Beguelin says, but it is introduced by him as leading the way for his further investigations.

606. Such cases as a, a, a, a, a cannot occur in the lottery because no number is there repeated. Let the second file be raised one letter, the third file two letters; and so on. Thus we have

$$
\begin{array}{ccccc}
a & b & c & d & e \\
b & c & d & e & f \\
\multicolumn{5}{c}{\dotfill} \\
i & j & k & l & m \\
j & k & l & m & \\
k & l & m & & \\
l & m & & & \\
m & & & &
\end{array}
$$

We have thus $13 - 4$ *complete* files, that is 9 complete files ; and, proceeding as before, the number of associations is found to be $\dfrac{9 \times 10 \times 11 \times 12 \times 13}{1 \times 2 \times 3 \times 4 \times 5}$; that is, the number is what we know to be the number of the combinations of 13 things taken 5 at a time.

607. Suppose now that we wish to find the number of associations in which there is no sequence at all. Raise each file two letters instead of one, so that we now have

$$
\begin{array}{ccccc}
a & c & e & g & i \\
b & d & f & h & j \\
c & e & g & i & k \\
d & f & h & j & l \\
e & g & i & k & m \\
f & h & j & l & \\
g & i & k & m & \\
h & j & l & & \\
i & k & m & & \\
j & l & & & \\
k & m & & & \\
l & & & & \\
m & & & &
\end{array}
$$

Here there are only $13-8$, that is, 5 complete files; and proceeding as in Art. 605, we find that the whole number of associations is $\dfrac{5 \times 6 \times 7 \times 8 \times 9}{1 \times 2 \times 3 \times 4 \times 5}$.

In this way we arrive in fact at the value which we quoted for $E(n, r)$ in Art. 602.

608. The method which we have here briefly exemplified is used by Beguelin in discussing all the parts of the problem. He does not however employ *letters* as we have done; he supposes a series of medals of the Roman emperors, and so instead of $a, b, c, \ldots$ he uses *Augustus, Tiberius, Caligula*, ...

609. It may be useful to state the results which are obtained when there are n tickets of which 5 are drawn.

In the following table the first column indicates the form, the second the number of cases of that form according to Euler's conception, and the third the number according to John Bernoulli's conception.

Sequence of 5,	$n-4$,	n.
Sequence of 4,	$(n-5)(n-4)$,	$n(n-6)$.
Sequence of 3 combined with a sequence of 2,	$(n-5)(n-4)$,	$n(n-6)$.
Sequence of 3, and the other numbers not in sequence,	$\dfrac{(n-6)(n-5)(n-4)}{1\cdot 2}$,	$\dfrac{n(n-7)(n-6)}{1\cdot 2}$.
Two sequences of 2,	$\dfrac{(n-6)(n-5)(n-4)}{1\cdot 2}$,	$\dfrac{n(n-7)(n-6)}{1\cdot 2}$.
Single sequence of 2,	$\dfrac{(n-7)(n-6)(n-5)(n-4)}{1\cdot 2\cdot 3}$,	$\dfrac{n(n-8)(n-7)(n-6)}{1\cdot 2\cdot 3}$.

No sequence, see Art. 602.

The chance of any assigned event is found by dividing the corresponding number by the whole number of cases, that is by the number of combinations of n things taken 5 at a time.

610. We have now to notice another memoir by Beguelin. It is entitled, *Sur l'usage du principe de la raison suffisante dans le calcul des probabilités.*

This memoir is published in the volume of the *Histoire de l'Acad....Berlin* for 1767; the date of publication is 1769 : the memoir occupies pages 382—412.

611. Beguelin begins by saying, J'ai montré dans un Mémoire précédent que la doctrine des probabilités étoit uniquement fondée sur le principe de la raison suffisante : this refers apparently to some remarks in the memoirs which we have just examined. Beguelin refers to D'Alembert in these words. Un illustre Auteur, Géometre et Philosophe à la fois, a publié depuis peu sur le Calcul des probabilités, des doutes et des questions bien dignes d'être approfondies ... Beguelin proposes to try how far metaphysical principles can assist in the Theory of Probabilities.

612. Beguelin discusses two questions. The first he says is the question :

... si les événemens simmétriques et réguliers, attribués au hazard, sont (toutes choses d'ailleurs égales) aussi probables que les événemens qui n'ont ni ordre ni régularité, et au cas qu'ils aient le même degré de probabilité, d'où vient que leur régularité nous frappe, et qu'ils nous paroissent si singuliers ?

His conclusions on this question do not seem to call for any remark.

613. His next question he considers more difficult; it is

... lorsqu'un même événement est deja arrivé une ou plusieurs fois de suite, on demande si cet événement conserve autant de probabilité pour sa future existence, que l'événement contraire qui avec une égale probabilité primitive n'est point arrivé encore.

Beguelin comes to the conclusion that the oftener an event has happened the less likely it is to happen at the next trial;

thus he adopts one of D'Alembert's errors. He considers that if the chances would have been equal according to the ordinary theory, then when an event has happened t times in succession it is $t+1$ to 1 that it will fail at the next trial.

614. Beguelin applies his notions to the *Petersburg Problem*. Suppose there are to be n trials; then instead of $\frac{n}{2}$ which the common theory gives for the expectation Beguelin arrives at

$$\frac{1}{2}+\frac{1}{2}+\frac{2}{2+1}+\frac{2^2}{2.3+1}+\frac{2^3}{\lfloor 4+1}+ \cdots + \frac{2^{n-2}}{\lfloor n-1+1}.$$

The terms of this series rapidly diminish, and the sum to infinity is about $2\frac{1}{2}$.

615. Besides the above result Beguelin gives five other solutions of the *Petersburg Problem*. His six results are not coincident, but they all give a small finite value for the expectation instead of the large or infinite value of the common theory.

616. The memoir does not appear of any value whatever; Beguelin adds nothing to the objections urged by D'Alembert against the common theory, and he is less clear and interesting. It should be added that Montucla appears to have formed a different estimate of the value of the memoir. He says, on his page 403, speaking of the *Petersburg Problem*,

Ce problème a été aussi le sujet de savantes considérations métaphysiques pour Beguelin...ce métaphysicien et analyste examine au flambeau d'une métaphysique profonde plusieurs questions sur la nature du calcul des probabilités...

617. We have next to notice a memoir which has attracted considerable attention. It is entitled *An Inquiry into the probable Parallax, and Magnitude of the fixed Stars, from the Quantity of Light which they afford us, and the particular Circumstances of their Situation, by the Rev.* John Michell, B.D., F.R.S.

This memoir was published in the *Philosophical Transactions*, Vol. LVII. Part I., which is the volume for 1767: the memoir occupies pages 234—264.

618. The part of the memoir with which we are concerned is that in which Michell, from the fact that some stars are very close together, infers the existence of design. His method will be seen from the following extract. He says, page 243,

Let us then examine what it is probable would have been the least apparent distance of any two or more stars, any where in the whole heavens, upon the supposition that they had been scattered by mere chance, as it might happen. Now it is manifest, upon this supposition, that every star being as likely to be in any one situation as another, the probability, that any one particular star should happen to be within a certain distance (as for example one degree) of any other given star, would be represented (according to the common way of computing chances) by a fraction, whose numerator would be to it's denominator, as a circle of one degree radius, to a circle, whose radius is the diameter of a great circle (this last quantity being equal to the whole surface of the sphere) that is, by the fraction $\frac{(60')^2}{(6875 \cdot 5')^2}$, or, reducing it to a decimal form, $\cdot 0000761 54$ (that is, about 1 in 13131) and the complement of this to unity, viz. $\cdot 999923846$, or the fraction $\frac{13130}{13131}$, will represent the probability that it would not be so. But, because there is the same chance for any one star to be within the distance of one degree from any given star, as for every other, multiplying this fraction into itself as many times as shall be equivalent to the whole number of stars, of not less brightness than those in question, and putting n for this number, $(\cdot999923846)^n$, or the fraction $\left(\frac{13130}{13131}\right)^n$ will represent the probability, that no one of the whole number of stars n would be within one degree from the proposed given star; and the complement of this quantity to unity will represent the probability, that there would be some one star or more, out of the whole number n, within the distance of one degree from the given star. And farther, because the same event is equally likely to happen to any one star as to any other, and therefore any one of the whole number of stars n might as well have been taken for the given star as any other, we must again repeat the last found chance n times, and consequently the number $\{(\cdot999923846)^n\}^n$, or the fraction $\left\{\left(\frac{13130}{13131}\right)^n\right\}^n$ will represent the probability, that no where, in the whole heavens, any two stars, amongst those in question, would be within the distance of one degree from each other; and the complement of this quantity to unity will represent the probability of the contrary.

619. Michell obtains the following results on his page 246,

If now we compute, according to the principles above laid down, what the probability is, that no two stars, in the whole heavens, should have been within so small a distance from each other, as the two stars β Capricorni, to which I shall suppose about 230 stars only to be equal in brightness, we shall find it to be about 80 to 1.

For an example, where more than two stars are concerned, we may take the six brightest of the Pleiades, and, supposing the whole number of those stars, which are equal in splendor to the faintest of these, to be about 1500, we shall find the odds to be near 500000 to 1, that no six stars, out of that number, scattered at random, in the whole heavens, would be within so small a distance from each other, as the Pleiades are.

Michell gives the details of the calculation in a note.

620. Laplace alludes to Michell in the *Théorie ... des Prob.*, page LXIII., and in the *Connaissance des Tems* for 1815, page 219.

621. The late Professor Forbes wrote a very interesting criticism on Michell's memoir; see the *London, Edinburgh and Dublin Philosophical Magazine*, for August 1849 and December 1850. He objects with great justice to Michell's mathematical calculations, and he also altogether distrusts the validity of the inferences drawn from these calculations.

622. Struve has given some researches on this subject in his *Catalogus Novus Stellarum Duplicium et Multiplicium ... Dorpati*, 1827, see the pages XXXVII.—XLVIII. Struve's method is very different from Michell's. Let n be the number of stars in a given area S of the celestial surface; let ϕ represent the area of a small circle of x'' radius. Then Struve takes $\dfrac{n(n-1)}{2}\dfrac{\phi}{S}$ as the chance of having a pair of the n stars within the distance x'', supposing that the stars are distributed by chance. Let S represent the surface beginning from $-15°$ of declination and extending to the north pole; let $n = 10229$, and $x = 4$: then Struve finds the above expression to become ·007814.

See also Struve's *Stellarum Duplicium et Multiplicium Men-
suræ Micrometricæ... Petrop.* 1837, page XCI., and his *Stellarum
Fixarum... Positiones Mediæ... Petrop.* 1852, page CLXXXVIII.

Sir John Herschel in his *Outlines of Astronomy*, 1849, page 565,
gives some numerical results which are attributed to Struve; but
I conclude that there is some mistake, for the results do not
appear to agree with Struve's calculations in the works above cited.

623. For a notice of some of the other subjects discussed in
Michell's memoir, see Struve's *Etudes d'Astronomie Stellaire,*
St Pétersbourg, 1847.

624. We have next to notice another memoir by John Bernoulli; it is entitled *Mémoire sur un probleme de la Doctrine du
Hazard.*

This memoir is published in the volume of the *Histoire de
l'Acad.... Berlin* for 1768; the date of publication is 1770 : the
memoir occupies pages 384—408.

The problem discussed may be thus generally enunciated.
Suppose *n* men to marry *n* women at the same time; find the
chance that when half the 2*n* people are dead all the marriages
will be dissolved; that is, find the chance that all the survivors
will be widows or widowers. John Bernoulli makes two cases;
first, when there is no limitation as to those who die; second, when
half of those who die are men and half women.

The memoir presents nothing of interest or importance ; the
formulæ are obtained by induction from particular cases, but are
not really *demonstrated.*

625. We have next to notice a memoir by Lambert, entitled *Examen d'une espece de Superstition ramenée au calcul
des probabilités.*

This memoir is published in the volume for 1771 of the
Nouveaux Mémoires... Berlin; the date of publication is 1773 :
the memoir occupies pages 411—420.

626. Lambert begins by adverting to the faith which many
people in Germany had in the predictions of the almanack makers
respecting the weather and other events. This suggests to him to

consider what is the chance that the predictions will be verified supposing the predictions to be thrown out at random.

The problem which he is thus led to discuss is really the old problem of the game of *Treize*, though Lambert does not give this name to it, or cite any preceding writers except Euler's memoir of 1751 : see Arts. 162, 280, 430.

627. We may put the problem thus : suppose n letters to be written and n corresponding envelopes to be directed ; the letters are put at random into the envelopes : required the chance that all, *or any assigned number*, of the letters are placed in the wrong envelopes.

The total number of ways in which the letters can be put into the envelopes is $\lfloor n$. There is only one way in which all can be placed in the right envelopes. There is no way in which *just one* letter is in the wrong envelope. Let us consider the number of ways in which *just two* letters are in the wrong envelopes : take a pair of letters ; this can be done in $\dfrac{n(n-1)}{1 \cdot 2}$ ways ; then find in how many ways this pair can be put in the wrong envelopes without disturbing the others : this can only be done in one way. Next consider in how many ways just three letters can be put in the wrong envelopes ; take a triad of letters ; this can be done in $\dfrac{n(n-1)(n-2)}{1 \cdot 2 \cdot 3}$ ways, and the selected triad can be put in wrong envelopes in 2 ways, as will be seen on trial.

Proceeding thus we obtain the following result,

$$\lfloor n = A_0 + A_1 n + A_2 \frac{n(n-1)}{1 \cdot 2}$$
$$+ A_3 \frac{n(n-1)(n-2)}{1 \cdot 2 \cdot 3} + \ldots + A_n \frac{\lfloor n}{\lfloor n} \ldots (1),$$

where A_r expresses the number of ways in which r letters, for which there are r appropriate envelopes, can all be placed in wrong envelopes. And

$$A_0 = 1, \quad A_1 = 0, \quad A_2 = 1, \quad A_3 = 2, \ldots$$

Now $A_0, A_1, A_2, \ldots$ *are independent of n;* thus we can determine them by putting for n in succession the values 1, 2, 3, ... in the above identity. This last remark is in fact the novelty of Lambert's memoir.

Lambert gives the general law which holds among the quantities A_1, A_2, ..., namely

$$A_r = rA_{r-1} + (-1)^r \quad\quad\quad (2).$$

He does not however demonstrate that this law holds. We have demonstrated it implicitly in the value which we have found for $\phi(n)$ in Art. 161.

We get by this law

$$A_4 = 9, \quad A_5 = 44, \quad A_6 = 265, \quad A_7 = 1854, \quad A_8 = 14833, \dots$$

We can however easily demonstrate the law independently of Art. 161.

Let $\Delta^r \lfloor 0$ stand for $\lfloor r - r \lfloor r-1 + \dfrac{r(r-1)}{1.2} \lfloor r-2 - \dots\dots,$

so that the notation is analogous to that which is commonly used in Finite Differences. Then the fundamental relation (1) suggests that

$$A_r = \Delta^r \lfloor 0 ; \quad\quad\quad (3),$$

and we can shew that this is the case by an inductive proof. For we find by trial that

$$\Delta^0 \lfloor 0 = \lfloor 0 = 1 = A_0,$$
$$\Delta^1 \lfloor 0 = 1 - 1 = 0 = A_1,$$
$$\Delta^2 \lfloor 0 = 2 - 2 + 1 = A_2;$$

and then from the fundamental relation (1) it follows that if $A_r = \Delta^r \lfloor 0$ for all values of r up to $n-1$ inclusive, then $A_n = \Delta^n \lfloor 0$. Thus (3) is established, and from (3) we can immediately shew that (2) holds.

628. We now come to another memoir by the writer whom we have noticed in Art. 597. The memoir is entitled *Sur le Calcul des Probabilités, par Mr. Mallet, Prof. d'Astronomie à Geneve.*

This memoir is published in the *Acta Helvetica ... Basileœ*, Vol. VII.; the date of publication is 1772: the memoir occupies pages 133—163.

629. The memoir consists of the discussion of two problems : the first is a problem given in the *Ars Conjectandi* of James Bernoulli ; the other relates to a lottery.

630. The problem from the *Ars Conjectandi* is that which is given on page 161 of the work ; we have given it in Art. 117. Mallet notices the fact that James Bernoulli in addition to the correct solution gave another which led to a different result and was therefore wrong, but which appeared plausible. Mallet then says,

Mr. Bernoulli s'étant contenté d'indiquer cette singularité apparente, sans en donner l'explication, j'ai crû qu'il ne seroit pas inutile d'entrer dans un plus grand détail làdessus, pour éclaircir parfaitement cette petite difficulté, on verra qu'on peut imaginer une infinité de cas semblables à celui de Mr. Bernoulli, dans la solution desquels il seroit aussi aisé d'être induit en erreur.

631. Mallet's remarks do not appear to offer any thing new or important; he is an obscure writer for want of sufficiently developing his ideas. The following illustration was suggested on reading his memoir, and may be of service to a student. Suppose we refer to the theory of duration of life. Let abscissæ measured from a fixed point denote years from a certain epoch, and the corresponding ordinates be proportional to the number of survivors out of a large number born at the certain epoch. Now suppose we wish to know whether it is more probable than not that a new born infant will live more than n years. James Bernoulli's plausible but false solution amounts to saying that the event is more probable than not, provided the *abscissa of the centre of gravity* of the area is greater than n: the true solution takes instead of the abscissa of the centre of gravity the abscissa which corresponds to the ordinate *bisecting the area* of the curve. See Art. 485.

632. We pass to Mallet's second problem which relates to a certain lottery.

The lottery is that which was called by Montmort *la lotterie de Loraine*, and which he discussed in his work ; see his pages 257—260, 313, 317, 326, 346. The following is practically the form of the lottery. The director of the lottery issues n tickets to

n persons, charging a certain sum for each ticket. He retains for himself a portion of the money which he thus receives, say a; the remainder he distributes into n prizes which will be gained by those who bought the tickets. He also offers a further inducement to secure buyers of his tickets, for he engages to return a sum, say b, to every ticket-holder who does not gain a prize. The prizes are distributed in the following manner. In a box are placed n counters numbered respectively from 1 to n. A counter is drawn, and a prize assigned to the ticket-holder whose number corresponds to the number of the counter. *The counter is then replaced in the box.* Another drawing is made and a prize assigned to the corresponding ticket-holder. *The counter is then replaced in the box.* This process is carried on until n drawings have been made ; and the prizes are then exhausted.

Hence, owing to the peculiar mode of drawing the lottery, one person might gain more than one prize, or even gain them all ; for the counter which bears his number might be drawn any number of times, or even every time.

The problem proposed is to find the advantage or disadvantage of the director of the lottery.

633. Montmort solved the problem in the following manner. Consider one of the ticket-holders. The chance that this person's number is never drawn throughout the whole process is $\left(\dfrac{n-1}{n}\right)^n$. If it is not drawn he is to receive b from the director ; so that his corresponding expectation is $b\left(\dfrac{n-1}{n}\right)^n$. A similar expectation exists for each of the ticket-holders, and the sum of these expectations is the amount by which the director's gain is diminished. Thus the director's advantage is

$$a - nb\left(\frac{n-1}{n}\right)^n.$$

In the case which Montmort notices b was equal to a, and n was 20000 ; thus the director's advantage was *negative*, that is, it was really a disadvantage. Before Montmort made a complete investigation he saw that the director's position was bad, and he suspected that there was a design to cheat the public, which actually happened.

634. Mallet makes no reference to any preceding writer on the subject; but solves the problem in a most laborious manner.. He finds the chances that the number of persons without prizes should be 1, or 2, or 3, ... up to n; then he knows the advantage of the banker corresponding to each case by multiplying the chance by the gain in that case; and by summing the results he obtains the total advantage.

635. One part of Mallet's process amounts to investigating the following problem. Suppose a die with r faces; let it be thrown s times in succession: required the chance that all the faces have appeared. The number of ways in which the desired event can happen is

$$r^s - r(r-1)^s + \frac{r(r-1)}{1.2}(r-2)^s - \frac{r(r-1)(r-2)}{1.2.3}(r-3)^s + ...$$

and the chance is obtained by dividing this number by r^s.

This is De Moivre's Problem XXXIX; it was afterwards discussed by Laplace and Euler; see Art. 448.

Mallet would have saved himself and his readers great labour if he had borrowed De Moivre's formula and demonstration. But he proceeds in a different way, which amounts to what we should now state thus: the number of ways in which the desired event can happen is the product of $\lfloor r$ by the sum of all the homogeneous products of the degree $s - r$ which can be formed of the numbers 1, 2, 3, ... r. He does not demonstrate the truth of this statement; he merely examines one very easy case, and says without offering any evidence that the other cases will be obtained by following the same method. See his page 144.

Mallet after giving the result in the manner we have just indicated proceeds to transform it; and thus he arrives at the same formula as we have quoted from De Moivre. Mallet does not demonstrate the truth of his transformation generally; he contents himself with taking some simple cases.

636. The transformation to which we have just alluded,

involves some algebraical work which we will give, since as we have intimated Mallet himself omits it.

Let there be r quantities a, b, c, ... k. Suppose x^p to be divided by $(x-a)(x-b)(x-c)...(x-k)$. The quotient will be

$$x^{p-r} + H_1 x^{p-r-1} + H_2 x^{p-r-2} + ... \text{ in infinitum,}$$

where H_r denotes the sum of all the homogeneous products of the degree r which can be formed from the quantities a, b, c, ... k. This can be easily shewn by first dividing x^p by $x-a$; then dividing the result by $x-b$, that is multiplying it by $x^{-1}\left(1-\dfrac{b}{x}\right)^{-1}$, and so on.

Again, if p be not less than r the expression

$$\frac{x^p}{(x-a)(x-b)...(x-k)}$$

will consist of an integral part and a fractional part; if p be less than r there will be no integral part. In both cases the fractional part will be

$$\frac{A}{x-a} + \frac{B}{x-b} + \frac{C}{x-c} + ... + \frac{K}{x-k},$$

where
$$A = \frac{a^p}{(a-b)(a-c)...(a-k)},$$

and similar expressions hold for B, C, ... K. Now expand each of the fractions $\dfrac{A}{x-a}$, $\dfrac{B}{x-b}$, ... according to negative powers of x; and equate the coefficient of x^{-t-1} to the coefficient in the first form which we gave for $x^p \div \{(x-a)(x-b)...(x-k)\}$. Thus

$$Aa^t + Bb^t + Cc^t + ... + Kk^t = H_{p-r+t+1}.$$

Put m for $p-r+t+1$; then $p+t = m+r-1$; thus we may express our result in the following words: the sum of the homogeneous products of the degree m, which can be formed of the r quantities a, b, c, ... k, is equal to

$$\frac{a^{m+r-1}}{(a-b)(a-c)...(a-k)} + \frac{b^{m+r-1}}{(b-a)(b-c)...(b-k)} + ...$$

This is the general theorem which Mallet enunciates, but only demonstrates in a few simple cases.

If we put 1, 2, 3, ... r respectively for a, b, c, ... k we obtain the theorem by which we pass from the formula of Mallet to that of De Moivre, namely, the sum of the homogeneous products of the degree $s - r$ which can be formed of the numbers 1, 2, ... r is equal to

$$\frac{1}{\underline{|r}} \left\{ r^s - r(r-1)^s + \frac{r(r-1)}{1.2}(r-2)^s - \frac{r(r-1)(r-2)}{1.2.3}(r-3)^s + ... \right\}.$$

The particular case in which $s = r + 1$ gives us the following result,

$$1 + 2 + 3 + ... + r$$

$$= \frac{1}{\underline{|r}} \left\{ r^{r+1} - r(r-1)^{r+1} \right.$$

$$\left. + \frac{r(r-1)}{1.2}(r-2)^{r+1} - \frac{r(r-1)(r-2)}{1.2.3}(r-3)^{r+1} + ... \right\},$$

which is a known result.

637. When Mallet has finished his laborious investigation he says, very justly, *il y a apparence que celui qui fit cette Lotterie ne s'étoit pas donné la peine de faire tous les calculs précedens.*

638. Mallet's result coincides with that which Montmort gave, and this result being so simple suggested that there might be an easier method of arriving at it. Accordingly Mallet gives another solution, in which like Montmort he investigates directly not the advantage of the director of the lottery, but the expectation of each ticket-holder. But even this solution is more laborious than Montmort's, because Mallet takes separately the case in which a ticket-holder has 1, or 2, or 3, ... , or n prizes; while in Montmort's solution there is no necessity for this.

639. Mallet gives the result of the following problem: Required the chance that in p throws with a die of n faces a specified face shall appear just m times. The chance is

$$\frac{\underline{|p}}{\underline{|m}\ \underline{|p-m}} \frac{(n-1)^{p-m}}{n^p}.$$

The formula explains itself; for the chance of throwing the specified face at each throw is $\frac{1}{n}$, and the chance of not throwing it is $\frac{n-1}{n}$. Hence by the fundamental principles of the subject the chance of having the specified face just m times in p throws is

$$\frac{\lfloor p}{\lfloor m \lfloor p-m} \left(\frac{1}{n}\right)^m \left(\frac{n-1}{n}\right)^{p-m}.$$

Since the whole number of cases in the p throws is n^p, it follows that the number of cases in which the required event can happen is

$$\frac{\lfloor p}{\lfloor m \lfloor p-m}(n-1)^{p-m};$$

and the result had been previously given by Montmort in this form: see his page 307.

640. On the whole we may say that Mallet's memoir shews the laborious industry of the writer, and his small acquaintance with preceding works on the subject.

641. William Emerson published in 1776 a volume entitled *Miscellanies, or a Miscellaneous Treatise; containing several Mathematical Subjects.*

The pages 1—48 are devoted to the Laws of Chance. These pages form an outline of the subject, illustrated by thirty-four problems. There is nothing remarkable about the work except the fact that in many cases instead of exact solutions of the problems Emerson gives only rude general reasoning which he considers may serve for approximate solution. This he himself admits; he says on his page 47,

It may be observed, that in many of these problems, to avoid more intricate methods of calculation, I have contented myself with a more lax method of calculating, by which I only approach near the truth.

See also the Scholium on his page 21.

Thus Emerson's work would be most dangerous for a beginner and quite useless for a more advanced student.

We may remark that pages 49—138 of the volume are devoted to Annuities and Insurances.

642. We have now to examine a contribution to our subject from the illustrious naturalist Buffon whose name has already occurred in Art. 354.

Buffon's *Essai d'Arithmétique Morale* appeared in 1777 in the fourth volume of the *Supplément à l'Histoire Naturelle*, where it occupies 103 quarto pages. Gouraud says on his page 54, that the *Essay* was composed about 1760.

643. The essay is divided into 35 sections.

Buffon says that there are truths of different kinds ; thus there are geometrical truths which we know by reasoning, and physical truths which we know by experience ; and there are truths which we believe on testimony.

He lays down without explanation a peculiar principle with respect to physical truths. Suppose that for n days in succession the Sun has risen, what is the probability that it will rise to-morrow ?

Buffon says it is proportional to 2^{n-1}. See his 6th section.

This is quite arbitrary; see Laplace *Théorie...des Prob.* page XIII.

644. He considers that a probability measured by so small a fraction as $\dfrac{1}{10000}$ cannot be distinguished from a zero probability. He arrives at the result thus; he finds from the tables that this fraction represents the chance that a man 56 years old will die in the course of a day, and he considers that such a man does practically consider the chance as zero. The doctrine that a very small chance is practically zero is due to D'Alembert; see Art. 472 : Buffon however is responsible for the value $\dfrac{1}{10000}$; see his 8th section.

645. Buffon speaks strongly against gambling. He says at the end of his 11th section :

Mais nous allons donner un puissant antidote centre le mal épidémique de la passion du jeu, et en même-temps quelques préservatifs contre l'illusion de cet art dangereux.

He condemns all gambling, even such as is carried on under conditions usually considered fair ; and of course still more gambling in which an advantage is ensured to one of the parties. Thus for example at a game like Pharaon, he says :

... le banquier n'est qu'un fripon avoué, et le ponte une dupe, dont on est convenu de ne se pas moquer.

See his 12th section. He finishes the section thus :

... je dis qu'en général le jeu est un pacte mal-entendu, un contrat désavantageux aux deux parties, dont l'effet est de rendre la perte toujours plus grande que le gain; et d'ôter au bien pour ajouter au mal. La démonstration en est aussi aisée qu'évidente.

646. The demonstration then follows in the 13th section.

Buffon supposes two players of equal fortune, and that each stakes half of his fortune. He says that the player who wins will increase his fortune by a third, and the player who loses will diminish his by a half ; and as a half is greater than a third there is more to fear from loss than to hope from gain. Buffon does not seem to do justice to his own argument such as it is. Let a denote the fortune of each player, and b the sum staked. Then the gain is estimated by Buffon by the fraction $\dfrac{b}{a+b}$, and the loss by $\dfrac{b}{a}$; but it would seem more natural to estimate the loss by $\dfrac{b}{a-b}$, which of course increases the excess of the loss to be feared over the gain to be hoped for.

The demonstration may be said to rest on the principle that the value of a sum of money to any person varies inversely as his whole fortune.

647. Buffon discusses at length the *Petersburg Problem* which he says was proposed to him for the first time by Cramer at Geneva in 1730. This discussion occupies sections 15 to 20 inclusive. See Art. 389.

Buffon offers four considerations by which he reduces the expectation of A from an infinite number of crowns to about five crowns only. These considerations are

(1) The fact that no more than a finite sum of money exists to pay A. Buffon finds that if head did not fall until after the

twenty-ninth throw, more money would be required to pay A than the whole kingdom of France could furnish.

(2) The doctrine of the relative value of money which we have stated at the end of the preceding Article.

(3) The fact that there would not be time during a life for playing more than a certain number of games; allowing only two minutes for each game including the time necessary for paying.

(4) The doctrine that any chance less than $\frac{1}{10000}$ is to be considered absolutely zero: see Art. 644.

Buffon cites Fontaine as having urged the first reason: see Arts. 392, 393.

648. The 18th section contains the details of an experiment made by Buffon respecting the *Petersburg Problem*. He says he played the game 2084 times by getting a child to toss a coin in the air. These 2084 games he says produced 10057 crowns. There were 1061 games which produced one crown, 494 which produced two crowns, and so on. The results are given in De Morgan's *Formal Logic*, page 185, together with those obtained by a repetition of the experiment. See also *Cambridge Philosophical Transactions*, Vol. IX. page 122.

649. The 23rd section contains some novelties.

Buffon begins by saying that up to the present time Arithmetic had been the only instrument used in estimating probabilities, but he proposes to shew that examples might be given which would require the aid of Geometry. He accordingly gives some simple problems with their results.

Suppose a large plane area divided into equal regular figures, namely squares, equilateral triangles, or regular hexagons. Let a round coin be thrown down at random; required the chance that it shall fall clear of the bounding lines of the figure, or fall on one of them, or on two of them; and so on.

These examples only need simple mensuration, and we need not delay on them; we have not verified Buffon's results.

Buffon had solved these problems at a much earlier date. We find in the *Hist. de l'Acad.... Paris* for 1733 a short account of them; they were communicated to the Academy in that year; see Art. 354.

650. Buffon then proceeds to a more difficult example which requires the aid of the Integral Calculus. A large plane area is ruled with equidistant parallel straight lines; a slender rod is thrown down: required the probability that the rod will fall across a line. Buffon solves this correctly. He then proceeds to consider what he says might have appeared more difficult, namely to determine the probability when the area is ruled with a second set of equidistant parallel straight lines, at right angles to the former and at the same distances. He merely gives the result, but it is wrong.

Laplace, without any reference to Buffon, gives the problem in the *Théorie...des Prob.*, pages 359—362.

The problem involves a compound probability; for the centre of the rod may be supposed to fall at any point within one of the figures, and the rod to take all possible positions by turning round its centre: it is sufficient to consider *one* figure. Buffon and Laplace take the two elements of the problem in the less simple order; we will take the other order.

Suppose a the distance of two consecutive straight lines of one system, b the distance of two consecutive straight lines of the other system; let $2r$ be the length of the rod and assume that $2r$ is less than a and also less than b.

Suppose the rod to have an inclination θ to the line of length a; or rather suppose that the inclination lies between θ and $\theta + d\theta$. Then in order that the rod may cross a line its centre must fall somewhere on the area

$$ab - (a - 2r\cos\theta)(b - 2r\sin\theta),$$

that is on the area

$$2r(a\sin\theta + b\cos\theta) - 4r^2\sin\theta\cos\theta.$$

Hence the whole probability of crossing the lines is

$$\frac{\int\left\{2r(a\sin\theta + b\cos\theta) - 4r^2\sin\theta\cos\theta\right\}d\theta}{\int ab\,d\theta}.$$

The limits of θ are 0 and $\frac{\pi}{2}$. Hence the result is

$$\frac{4r(a+b) - 4r^2}{\pi ab}.$$

If $a = b$ this becomes

$$\frac{8ar - 4r^2}{\pi a^2}:$$

Buffon's result expressed in our notation is

$$\frac{2(a-r)r}{\pi a^2}.$$

If we have only *one* set of parallel lines we may suppose b infinite in our general result: thus we obtain $\frac{4r}{\pi a}$.

651. By the mode of solution which we have adopted we may easily treat the case in which $2r$ is not less than a and also less than b, which Buffon and Laplace do not notice.

Let b be less than a. First suppose $2r$ to be greater than b but not greater than a. Then the limits of θ instead of being 0 and $\frac{\pi}{2}$ will be 0 and $\sin^{-1}\frac{b}{2r}$. Next suppose $2r$ to be greater than a. Then the limits of θ will be $\cos^{-1}\frac{a}{2r}$ and $\sin^{-1}\frac{b}{2r}$; this holds so long as $\cos^{-1}\frac{a}{2r}$ is less than $\sin^{-1}\frac{b}{2r}$, that is so long as $\sqrt{(4r^2 - a^2)}$ is less than b, that is so long as $2r$ is less than $\sqrt{(a^2 + b^2)}$, which is geometrically obvious.

652. Buffon gives a result for another problem of the same kind. Suppose a *cube* thrown down on the area; required the probability that it will fall across a line. With the same meaning as before for a and b, let $2r$ denote the length of a diagonal of a face of the cube. The required probability is

$$\frac{\int \{ab - (a - 2r\cos\theta)(b - 2r\cos\theta)\}\, d\theta}{\int ab\, d\theta},$$

the limits of θ being 0 and $\frac{\pi}{4}$. Thus we obtain

$$\frac{2\left(a+b\right)r\sin\frac{\pi}{4}-r^2\left(\frac{\pi}{2}+1\right)}{ab\,\frac{\pi}{4}}=\frac{4\left(a+b\right)r\sqrt{2}-r^2\left(2\pi+4\right)}{\pi ab}.$$

Buffon gives an incorrect result.

653. The remainder of Buffon's essay is devoted to subjects unconnected with the Theory of Probability. One of the subjects is the *scales of notation*: Buffon recommends the duodenary scale. Another of the subjects is the *unit of length*: Buffon recommends the length of a pendulum which beats seconds at the equator. Another of the subjects is the *quadrature of the circle*: Buffon pretends to demonstrate that this is impossible. His demonstration however is worthless, for it would equally apply to any curve, and shew that no curve could be rectified ; and this we know would be a false conclusion.

654. After the Essay we have a large collection of results connected with the duration of human life, which Buffon deduced from tables he had formerly published.

Buffon's results amount to expressing in numbers the following formula: For a person aged n years the odds are as a to b that he will live x more years.

Buffon tabulates this formula for all integral values of n up to 99, and for various values of x.

After these results follow other tables and observations connected with them. The tables include the numbers of births, marriages, and deaths, at Paris, from 1709 to 1766.

655. Some remarks on Buffon's views will be found in Condorcet's *Essai...de l'Analyse...*page LXXI., and in Dugald Stewart's *Works edited by Hamilton*, Vol. I. pages 369, 616.

656. We have next to notice some investigations by Fuss under the following titles : *Recherches sur un problème du Calcul des Probabilités par Nicolas Fuss. Supplément au mémoire sur un problème du Calcul des Probabilités...*

The *Recherches...* occupy pages 81—92 of the *Pars Posterior* of the volume for 1779 of the *Acta Acad. ...Petrop.;* the date of publication is 1783.

The *Supplément...* occupies pages 91—96 of the *Pars Posterior* of the volume for 1780 of the *Acta Acad. ... Petrop.;* the date of publication is 1784.

The problem is that considered by James Bernoulli on page 161 of the *Ars Conjectandi;* see Art. 117.

In the *Recherches ...* Fuss solves the problem ; he says he had not seen James Bernoulli's own solution but obtained his knowledge of the problem from Mallet's memoir ; see Art. 628. Fuss published his solution because his results differed from that obtained by James Bernoulli as recorded by Mallet. In the *Supplément...* Fuss says that he has since procured James Bernoulli's work, and he finds that there are two cases in the problem; his former solution agreed with James Bernoulli's solution of one of the cases, and he now adds a solution of the other case, which agrees with James Bernoulli's solution for that case.

Thus in fact Fuss would have spared his two papers if he had consulted James Bernoulli's own work at the outset. We may observe that Fuss uses the Lemma given by De Moivre on his page 39, but Fuss does not refer to any previous writer for it; see Art. 149.

CHAPTER XVII.

CONDORCET.

657. CONDORCET was born in 1743 and died in 1794. He wrote a work connected with our subject, and also a memoir. It will be convenient to examine the work first, although part of the memoir really preceded it in order of time.

658. The work is entitled *Essai sur l'application de l'analyse à la probabilité des décisions rendues à la pluralité des voix. Par M. Le Marquis de Condorcet* ... Paris 1785. This work is in quarto ; it consists of a *Discours Préliminaire* which occupies CXCI. pages, and of the *Essai* itself which occupies 304 pages.

659. The object of the Preliminary Discourse is to give the results of the mathematical investigations in a form which may be intelligible to those who are not mathematicians. It commences thus :

Un grand homme, dont je regretterai toujours les leçons, les exemples, et sur-tout l'amitié, étoit persuadé que les vérités des Sciences morales et politiques, sont susceptibles de la même certitude que celles qui forment le système des Sciences physiques, et même que les branches de ces Sciences qui, comme l'Astronomie, paroissent approcher de la certitude mathématique.

Cette opinion lui étoit chère, parce qu'elle conduit à l'espérance consolante que l'espèce humaine fera nécessairement des progrès vers le bonheur et la perfection, comme elle en a fait dans la connoissance de la vérité.

C'étoit pour lui que j'avois entrepris cet ouvrage......

The great man to whom Condorcet here refers is named in a note : it is Turgot.

Condorcet himself perished a victim of the French Revolution, and it is to be presumed that he must have renounced the faith here expressed in the necessary progress of the human race towards happiness and perfection.

660. Condorcet's *Essai* is divided into five parts.

The *Discours Préliminaire*, after briefly expounding the fundamental principles of the Theory of Probability, proceeds to give in order an account of the results obtained in the five parts of the *Essai*.

We must state at once that Condorcet's work is excessively difficult ; the difficulty does not lie in the mathematical investigations, but in the expressions which are employed to introduce these investigations and to state their results : it is in many cases almost impossible to discover what Condorcet means to say. The obscurity and self contradiction are without any parallel, so far as our experience of mathematical works extends ; some examples will be given in the course of our analysis, but no amount of examples can convey an adequate impression of the extent of the evils. We believe that the work has been very little studied, for we have not observed any recognition of the repulsive peculiarities by which it is so undesirably distinguished.

661. The Preliminary Discourse begins with a brief exposition of the fundamental principles of the Theory of Probability, in the course of which an interesting point is raised. After giving the mathematical definition of probability, Condorcet proposes to shew that it is consistent with ordinary notions ; or in other words, that the mathematical measure of probability is an accurate measure of our degree of belief. See his page VII. Unfortunately he is extremely obscure in his discussion of the point.

We shall not delay on the Preliminary Discourse, because it is little more than a statement of the results obtained in the Essay.

The Preliminary Discourse is in fact superfluous to any person who is sufficiently acquainted with Mathematics to study the Essay, and it would be scarcely intelligible to any other person.

For in general when we have no mathematical symbols to guide us in discovering Condorcet's meaning, the attempt is nearly hopeless.

We proceed then to analyse the Essay.

662. Condorcet's first part is divided into eleven sections, devoted to the examination of as many *Hypotheses;* this part occupies pages 1—136.

We will consider Condorcet's first Hypothesis.

Let there be $2q + 1$ voters who are supposed exactly alike as to judgment ; let v be the probability that a voter decides correctly, e the probability that he decides incorrectly, so that $v + e = 1$: required the probability that there will be a majority in favour of the correct decision of a question submitted to the voters. We may observe, that the letters v and e are chosen from commencing the words *vérité* and *erreur*.

The required probability is found by expanding $(v + e)^{2q+1}$ by the Binomial Theorem, and taking the terms from v^{2q+1} to that which involves $v^{q+1} e^q$, both inclusive. Two peculiarities in Condorcet's notation may here be noticed. He denotes the required probability by V^q; this is very inconvenient because this symbol has universally another meaning, namely it denotes V raised to the power q. He uses $\dfrac{n}{m}$ to denote the coefficient of $v^{n-m} e^m$ in the expansion of $(v + e)^n$; this also is very inconvenient because the symbol $\dfrac{n}{m}$ has universally another meaning, namely it denotes a fraction in which the numerator is n and the denominator is m. It is not desirable to follow Condorcet in these two innovations.

We will denote the probability required by $\phi(q)$; thus

$$\phi(q) = v^{2q+1} + (2q + 1) v^{2q} e + \frac{(2q + 1)\, 2q}{1 \cdot 2}\, v^{2q-1} e^2 + \dots$$
$$\dots + \frac{\lfloor 2q + 1}{\lfloor q + 1 \, \lfloor q}\, v^{q+1} e^q.$$

663. The expression for $\phi(q)$ is transformed by Condorcet into a shape more convenient for his purpose ; and this transformation we will now give. Let $\phi(q + 1)$ denote what $\phi(q)$

becomes when q is changed into $q + 1$, that is let $\phi(q + 1)$ denote the probability that there will be a majority in favour of a correct decision when the question is submitted to $2q + 3$ voters. Therefore

$$\phi(q+1) = v^{2q+3} + (2q+3)\, v^{2q+2}\, e + \frac{(2q+3)(2q+2)}{1 \cdot 2}\, v^{2q+1}\, e^2$$

$$+ \ldots + \frac{\lfloor 2q+3}{\lfloor q+2 \lfloor q+1}\, v^{q+2}\, e^{q+1}.$$

Since $v + e = 1$ we have

$$\phi(q) = (v + e)^2\, \phi(q).$$

Thus $\quad \phi(q+1) - \phi(q) = \phi(q+1) - (v+e)^2\, \phi(q).$

Now $\phi(q+1)$ consists of certain terms in the expansion of $(v + e)^{2q+3}$, and $\phi(q)$ consists of certain terms in the expansion of $(v + e)^{2q+1}$; so we may anticipate that in the development of $\phi(q+1) - (v + e)^2\, \phi(q)$ very few terms will remain uncancelled. In fact it will be easily found that

$$\phi(q+1) - \phi(q) = \frac{\lfloor 2q+1}{\lfloor q+1 \lfloor q}\, v^{q+2}\, e^{q+1} - \frac{\lfloor 2q+1}{\lfloor q+1 \lfloor q}\, v^{q+1}\, e^{q+2}$$

$$= \frac{\lfloor 2q+1}{\lfloor q+1 \lfloor q}\, (v - e)\, v^{q+1} e^{q+1} \quad \ldots\ldots\ldots\ldots(1).$$

Hence we deduce

$$\phi(q) = v + (v - e) \left\{ ve + \frac{3}{1}\, v^2 e^2 + \frac{5 \cdot 4}{1 \cdot 2}\, v^3 e^3 + \frac{7 \cdot 6 \cdot 5}{1 \cdot 2 \cdot 3}\, v^4 e^4 \right.$$

$$\left. \ldots + \frac{\lfloor 2q-1}{\lfloor q \lfloor q-1}\, v^q e^q \right\} \ldots\ldots\ldots(2).$$

664. The result given in equation (2) is the transformation to which we alluded. We may observe that throughout the first part of his Essay, Condorcet repeatedly uses the method of transformation just exemplified, and it also appears elsewhere in the Essay; it is in fact the chief mathematical instrument which he employs.

It will be observed that we assumed $v + e = 1$ in order to obtain equation (2). We may however obtain a result analogous to (2) which shall be *identically true*, whatever v and e may be. We have only to replace the left-hand member of (1) by

$$\phi(q+1) - (v+e)^2 \phi(q),$$

and we can then deduce

$$v^{2q+1} + (2q+1)\, v^{2q} e + \frac{(2q+1)\,2q}{1\,.\,2}\, v^{2q-1} e^2 + \cdots$$

$$\cdots + \frac{\lfloor 2q+1}{\lfloor q+1 \lfloor q}\, v^{q+1} e^q$$

$$= v(v+e)^{2q} + (v-e)\left\{ ve(v+e)^{2q-2} + \frac{3}{1}\, v^2 e^2 (v+e)^{2q-4} \right.$$

$$\left. + \frac{5\,.\,4}{1\,.\,2}\, v^3 e^3 (v+e)^{2q-6} + \cdots + \frac{\lfloor 2q-1}{\lfloor q \lfloor q-1}\, v^q e^q \right\}.$$

This is *identically true;* if we suppose $v+e=1$, we have the equation (2).

665. We resume the consideration of the equation (2).

Suppose v greater than e; then we shall find that $\phi(q)=1$ when q is infinite. For it may be shewn that the series in powers of ve which occurs in (2) arises from expanding

$$-\frac{1}{2} + \frac{1}{2}(1 - 4ve)^{-\frac{1}{2}}$$

in powers of ve as far as the term which involves $v^q e^q$. Thus when q is infinite, we have

$$\phi(q) = v + (v-e)\left\{ -\frac{1}{2} + \frac{1}{2}(1 - 4ve)^{-\frac{1}{2}} \right\}.$$

Now $1 - 4ve = (v+e)^2 - 4ve = (v-e)^2$. Therefore when q is infinite

$$\phi(q) = v + (v-e)\left\{ -\frac{1}{2} + \frac{1}{2(v-e)} \right\}$$

$$= v + (v-e)\left\{ -\frac{v-e}{2(v-e)} + \frac{v+e}{2(v-e)} \right\}$$

$$= v + e = 1.$$

The assumption that v is greater than e is introduced when we put $v-e$ for $(1 - 4ve)^{\frac{1}{2}}$.

Thus we have the following result in the Theory of Probability : if the probability of a correct decision is the same for every voter and is greater than the probability of an incorrect decision, then the probability that the decision of the majority will be correct becomes indefinitely nearly equal to unity by sufficiently in-. creasing the number of voters.

It need hardly be observed that practically the hypotheses on which the preceding conclusion rests cannot be realised, so that the result has very little value. Some important remarks on the subject will be found in Mill's *Logic*, 1862, Vol. II. pages 65, 66, where he speaks of "misapplications of the calculus of probabilities which have made it the real opprobrium of mathematics."

666. We again return to the equation (2) of Art. 663.

If we denote by $\psi(q)$ the probability that there will be a majority in favour of an incorrect decision, we can obtain the value of $\psi(q)$ from that of $\phi(q)$ by interchanging e and v.

We have also $\phi(q) + \psi(q) = 1$.

Of course if $v = e$ we have obviously $\phi(q) = \psi(q)$, *for all values* of q; the truth of this result when q is infinite is established by Condorcet in a curious way ; see his page 10.

667. We have hitherto spoken of the probability that the decision *will* be correct, that is we have supposed that the result of the voting is not yet known.

But now suppose we know that a decision *has been given* and that m voters voted for that decision and n against it, so that m is greater than n. We ask, what is the probability that the decision *is* correct? Condorcet says briefly that the number of combinations in favour of the truth is expressed by

$$\frac{\lfloor 2q+1}{\lfloor m \lfloor n} v^m e^n,$$

and the number in favour of error by

$$\frac{\lfloor 2q+1}{\lfloor m \lfloor n} e^m v^n.$$

Thus the probabilities of the correctness and incorrectness of the decision are respectively

$$\frac{v^m e^n}{v^m e^n + e^m v^n} \quad \text{and} \quad \frac{e^m v^n}{v^m e^n + e^m v^n}.$$

See his page 10.

668. The student of Condorcet's work must carefully distinguish between the probability of the correctness of a decision that has been given when we know the numbers for and against, and the probability when we do not know these numbers. Condorcet sometimes leaves it to be gathered from the context which he is considering. For example, in his Preliminary Discourse page XXIII. he begins his account of his first Hypothesis thus:

Je considère d'abord le cas le plus simple, celui où le nombre des Votans étant impair, on prononce simplement à la pluralité.

Dans ce cas, la probabilité de ne pas avoir une décision fausse, celle d'avoir une décision vraie, celle que la décision rendue est conforme à la vérité, sont les mêmes, puisqu'il ne peut y avoir de cas où il n'y ait pas de décision.

Here, although Condorcet does not say so, the words *celle que la décision rendue est conforme à la vérité* mean that we know the decision has been given, but we do *not* know the numbers for and against. For, as we have just seen, in the Essay Condorcet takes the case in which we *do* know the numbers for and against, and then the probability is not the same as that of the correctness of a decision not yet given. Thus, in short, in the Preliminary Discourse Condorcet does not say which case he takes, and he really takes the case which he does not consider in the Essay, excluding the case which he does consider in the Essay; that is, he takes the case which he might most naturally have been supposed not to have taken.

669. We will now proceed to Condorcet's second Hypothesis out of his eleven; see his page 14.

Suppose, as before, that there are $2q + 1$ voters, and that a certain plurality of votes is required in order that the decision should be valid; let $2q' + 1$ denote this plurality.

Let $\phi(q)$ denote the terms obtained from the expansion of $(v + e)^{2q+1}$, from v^{2q+1} to the term which involves $v^{q+q'+1} e^{q-q'}$, both inclusive. Let $\psi(q)$ be formed from $\phi(q)$ by interchanging e and v.

Then $\phi(q) + \psi(q)$ is the probability that there will be a valid decision, $\phi(q)$ is the probability that there will be a valid and correct decision, and $\psi(q)$ is the probability that there will be a valid and incorrect decision. Moreover $1 - \psi(q)$ is the probability that there will not be an incorrect decision, and $1 - \phi(q)$ is the probability that there will not be a correct decision.

It will be observed that here $\phi(q) + \psi(q)$ is not equal to unity. In fact $1 - \phi(q) - \psi(q)$ consists of all the terms in the expansion of $(v + e)^{2q+1}$ lying between those which involve $v^{q+q'+1} e^{q-q'}$ and $v^{q-q'} e^{q+q'+1}$ both exclusive. Thus $1 - \phi(q) - \psi(q)$ is the probability that the decision will be invalid for want of the prescribed plurality.

It is shewn by Condorcet that if v is greater than e the limit of $\phi(q)$ when q increases indefinitely is unity. See his pages 19—21.

670. Suppose we know that a valid decision has been given, but do not know the numbers for and against. Then the probability that the decision is correct is $\dfrac{\phi(q)}{\phi(q) + \psi(q)}$, and the probability that it is incorrect is $\dfrac{\psi(q)}{\phi(q) + \psi(q)}$.

Suppose we know that a valid decision has been given, and also know the numbers for and against. Then the probabilities of the correctness and incorrectness of the decision are those which have been stated in Art. 667.

671. We will now indicate what Condorcet appears to mean by the principal conditions which ought to be secured in a decision; they are:

1. That an incorrect decision shall not be given; that is $1 - \psi(q)$ must be large.

2. That a correct decision shall be given; that is $\phi(q)$ must be large.

3. That there shall be a valid decision, correct or incorrect; that is $\phi(q) + \psi(q)$ must be large.

4. That a valid decision which has been given is correct,

supposing the numbers for and against not to be known; that is
$$\frac{\phi(q)}{\phi(q)+\psi(q)}$$ must be large.

5. That a valid decision which has been given is correct, supposing the numbers for and against to be known; that is
$$\frac{v^m e^n}{v^m e^n + e^m v^n}$$ must be large, even when m and n are such as to give it the least value of which it is susceptible.

These appear to be what Condorcet means by the principal conditions, and which, in his usual fluctuating manner, he calls in various places *five* conditions, *four* conditions, and *two* conditions. See his pages XVIII, XXXI, LXIX.

672. Before leaving Condorcet's second Hypothesis we will make one remark. On his page 17 he requires the following result,

$$\frac{2^{n-1}}{\{1+\sqrt{(1-4z)}\}^{n-1}\sqrt{(1-4z)}} = 1 + \frac{n+1}{1}z + \frac{(n+3)(n+2)}{1.2}z^2 + \ldots$$
$$\ldots + \frac{\lfloor n+2r-1}{\lfloor r \lfloor n+r-1}z^r + \ldots$$

On his page 18 he gives two ingenious methods by which the result may be obtained indirectly. It may however be obtained directly in various ways. For example, take a formula which may be established by the Differential Calculus for the expansion of $\{1+\sqrt{(1-4z)}\}^{-m}$ in powers of z, and differentiate with respect to z, and put $n-2$ for m.

673. Condorcet's third Hypothesis is similar to his second; the only difference is that he here supposes $2q$ voters, and that a plurality of $2q'$ is required for a valid decision.

674. In his fourth, fifth, and sixth Hypotheses Condorcet supposes that a plurality is required which is proportional, or nearly so, to the whole number of voters. We will state the results obtained in one case. Suppose we require that at least two-thirds of the whole number of voters shall concur in order that the decision may be valid. Let n represent the whole number of voters; let $\phi(n)$ represent the probability that there will

be a valid and correct decision, and $\psi(n)$ the probability that there will be a valid and incorrect decision; let v and e have the same meaning as in Art. 662. Then, when n is infinite, if v is greater than $\frac{2}{3}$ we have $\phi(n) = 1$, if v is less than $\frac{2}{3}$ we have $\phi(n) = 0$; and similarly if e is greater than $\frac{2}{3}$, that is if v is less than $\frac{1}{3}$, we have $\psi(n) = 1$, and if e is less than $\frac{2}{3}$, that is if v is greater than $\frac{1}{3}$, we have $\psi(n) = 0$.

We shall not stop to give Condorcet's own demonstrations of these results; it will be sufficient to indicate how they may be derived from *Bernoulli's Theorem;* see Art. 123. We know from this theorem that when n is very large, the terms which are in the neighbourhood of the greatest term of the expansion of $(v+e)^n$ overbalance the rest of the terms. Now $\phi(n)$ consists of the first third of all the terms of $(v+e)^n$, and thus if v is greater than $\frac{2}{3}$ the greatest term is included within $\phi(n)$, and therefore $\phi(n) = 1$ ultimately.

The same considerations shew that when $v = \frac{2}{3}$, we have $\phi(n) = \frac{1}{2}$ ultimately.

675. Condorcet's seventh and eighth Hypotheses are thus described by himself, on his page XXXIII:

La septième hypothèse est celle où l'on renvoie la décision à un autre temps, si la pluralité exigée n'a pas lieu.

Dans la huitième hypothèse, on suppose que si l'assemblée n'a pas rendu sa première décision à la pluralité exigée, on prend une seconde fois les avis, et ainsi de suite, jusqu'à ce que l'on obtienne cette pluralité.

These two Hypotheses give rise to very brief discussions in the Essay.

676. The ninth Hypothesis relates to the decisions formed by various systems of combined tribunals. Condorcet commences it thus on his page 57:

Jusqu'ici nous avons supposé un seul Tribunal ; dans plusieurs pays cependant on fait juger la même affaire par plusieurs Tribunaux, ou plusieurs fois par le même, mais d'après une nouvelle instruction, jusqu'à ce qu'on ait obtenu un certain nombre de décisions conformes. Cette hypothèse se subdivise en plusieurs cas différens que nous allons examiner séparément. En effet, on peut exiger, 1°. l'unanimité de ces décisions ; 2°. une certaine loi de pluralité, formée ou par un nombre absolu, ou par un nombre proportionnel au nombre des décisions prises ; 3°. un certain nombre consécutif de décisions conformes. Quand la forme des Tribunaux est telle, que la décision peut être nulle, comme dans la septième hypothèse, il faut avoir égard aux décisions nulles. Enfin il faut examiner ces différens cas, en supposant le nombre de ces décisions successives, ou comme déterminé, ou comme indéfini.

677. The ninth Hypothesis extends over pages 57—86 ; it appears to have been considered of great importance by Condorcet himself. We shall give some detail respecting one very interesting case which is discussed. This case Condorcet gives on pages 73—86. Condorcet is examining the probability of the correctness of a decision which has been confirmed in succession by an assigned number of tribunals out of a series to which the question has been referred. The essential part of the discussion consists in the solution of two problems which we will now enunciate. Suppose that the probability of the happening of an event in a single trial is v, and the probability of its failing is e, required, 1st the probability that in r trials the event will happen p times in succession, 2nd the probability that in r trials the event will happen p times in succession *before it fails* p times in succession.

It is the second of these problems which Condorcet wishes to apply, but he finds it convenient to begin with the solution of the first, which is much the simpler, and which, as we have seen, in Art. 325, had engaged the attention of De Moivre.

678. We have already solved the first problem, in Art. 325, but it will be convenient to give another solution.

Let $\phi(r)$ denote the probability that in r trials the event will happen p times in succession. Then we shall have

$$\phi(r) = v^p + v^{p-1} e \phi(r-p) + v^{p-2} e \phi(r-p+1) + \ldots$$
$$\ldots + v e \phi(r-2) + e \phi(r-1) \ldots\ldots\ldots\ldots (1).$$

To shew the truth of this equation we observe that in the first p trials the following p cases may arise; the event may happen p times in succession, or it may happen $p-1$ times in succession and then fail, or it may happen $p-2$ times in succession and then fail, ..., or it may fail at the first trial. The aggregate of the probabilities arising from all these cases is $\phi(r)$. The probability from the first case is v^p. The probability from the second case is $v^{p-1} e \phi(r-p)$: for $v^{p-1} e$ is the probability that the event will happen $p-1$ times in succession, and then fail; and $\phi(r-p)$ is the probability that the event will happen p times in succession in the course of the remaining $r-p$ trials. In a similar way the term $v^{p-2} e^2 \phi(r-p+1)$ is accounted for; and so on. Thus the truth of equation (1) is established.

679. The equation (1) is an equation in Finite Differences; its solution is

$$\phi(r) = C_1 y_1^r + C_2 y_2^r + C_3 y_3^r + ... + C_p y_p^r + C .... (2).$$

Here $C_1, C_2, ... C_p$ are arbitrary constants; $y_1, y_2, ... y_p$ are the roots of the following equation in y,

$$y^p = e (v^{p-1} + v^{p-2} y + v^{p-3} y^2 + ... + y^{p-1}) .... (3);$$

and C is to be found from the equation

$$C = v^p + e(v^{p-1} + v^{p-2} + ... + v + 1) C,$$

that is $$C = v^p + e \frac{1-v^p}{1-v} C;$$

and as $e = 1-v$ we obtain $C=1$.

We proceed to examine equation (3). Put $1-v$ for e, and assume $y = \frac{v}{z}$; thus

$$\frac{v}{1-v} = z^p + z^{p-1} + ... + z$$

$$= \frac{z(1-z^p)}{1-z} .... (4).$$

We shall shew that the real roots of equation (3) are numerically less than unity, and so also are the moduli of the imaginary roots; that is, we shall shew that the real roots of equation (4) are numerically greater than v, and so also are the moduli of the imaginary roots.

We know that v is less than unity. Hence from (4) if z be real and positive it must be greater than v. For if z be less than v, then $\dfrac{z}{1-z}$ is less than $\dfrac{v}{1-v}$, and *a fortiori* $\dfrac{z(1-z^p)}{1-z}$ is less than $\dfrac{v}{1-v}$. If z be negative in (4) we must have $1-z^p$ negative, so that p must be even, and z numerically greater than unity, and therefore numerically greater than v. Thus the real roots of (4) must be numerically greater than v.

Again, we may put (4) in the form

$$v + v^2 + v^3 + \ldots = z + z^2 + \ldots + z^p \ldots\ldots\ldots\ldots(5).$$

Now suppose that z is an imaginary quantity, say

$$z = k(\cos\theta + \sqrt{-1}\sin\theta) ;$$

then if k is not greater than v, we see by aid of the theorem

$$z^n = k^n(\cos n\theta + \sqrt{-1}\sin n\theta),$$

that the real terms on the right-hand side of (5) will form an aggregate less than the left-hand side. Thus k must be greater than v.

After what we have demonstrated respecting the values of the roots of (3), it follows from (2) that when r is infinite $\phi(r) = 1$.

680. We proceed to the second problem.

Let $\phi(r)$ now denote the probability that in r trials the event will happen p times in succession before it fails p times in succession.

Let $\psi(n)$ denote the probability that the event will happen p times in succession before it fails p times in succession, *supposing that one trial has just been made in which the event failed*, and that n trials remain to be made.

Then instead of equation (1) we shall now obtain

$$\phi(r) = v^p + v^{p-1}e\psi(r-p) + v^{p-2}e\psi(r-p+1) + \ldots$$
$$\ldots + ve\psi(r-2) + e\psi(r-1) \ldots (6).$$

This equation is demonstrated in the same manner as (1) was.

We have now to shew the connexion between the functions ϕ and ψ; it is determined by the following relation;

$$\psi(n) = \phi(n) - e^{p-1}\{\phi(n-p+1) - e\psi(n-p)\} \quad \ldots\ldots\ldots(7).$$

To shew the truth of this relation we observe that $\psi(n)$ is less than $\phi(n)$ for the following reason, and for that alone. If the one failure had not taken place there might be $p-1$ failures in succession, and there would still remain some chance of the happening of the event p times in succession before its failing p times in succession; since the one failure has taken place this chance is lost. The corresponding probability is

$$e^{p-1}\{\phi(n-p+1) - e\psi(n-p)\}.$$

The meaning of the factor e^{p-1} is obvious, so that we need only explain the meaning of the other factor. And it will be seen that $\phi(n-p+1) - e\psi(n-p)$ expresses the probability of the desired result in the $n-p+1$ trials which remain to be made; for here the rejected part $e\psi(n-p)$ is that part which would coexist with failure in the first of these remaining trials, which part would of course not be available when $p-1$ failures had already taken place.

Thus we may consider that (7) is established.

In (6) change r into $r-p$; therefore

$$\phi(r-p) = v^p + v^{p-1}e\psi(r-2p) + v^{p-2}e\psi(r-2p+1) + \ldots$$
$$\ldots + ve\psi(r-p-2) + e\psi(r-p-1) \quad \ldots\ldots\ldots\ldots(8).$$

Now multiply (8) by e^p and subtract the result from (6), observing that by (7) we have

$$\psi(n) - e^p\psi(n-p) = \phi(n) - e^{p-1}\phi(n-p+1);$$

thus we obtain

$$\phi(r) - e^p\phi(r-p) = v^p - e^p v^p$$
$$+ v^{p-1}e\{\phi(r-p) - e^{p-1}\phi(r-2p+1)\}$$
$$+ v^{p-2}e\{\phi(r-p+1) - e^{p-1}\phi(r-2p+2)\}$$
$$+ \ldots$$
$$+ e\{\phi(r-1) - e^{p-1}\phi(r-p)\} \quad \ldots\ldots\ldots(9).$$

681. The equation in Finite Differences which we have just

obtained may be solved in the ordinary way; we shall not however proceed with it.

One case of interest may be noticed. Suppose r infinite; then $\phi(r-p)$, $\phi(r-2p+1)$, ... will all be equal. Thus we can obtain the probability that the event will happen p times in succession before it fails p times in succession in an indefinite number of trials. Let V denote this probability; then we have from (9),

$$V(1-e^p) = v^p(1-e^p) + eV(v^{p-1}+v^{p-2}+\ldots+v+1)$$
$$- e^p V(v^{p-1}+v^{p-2}+\ldots+v+1).$$

Hence after reduction we obtain

$$V = \frac{v^{p-1}(1-e^p)}{v^{p-1}+e^{p-1}-v^{p-1}e^{p-1}} \ldots\ldots\ldots (10).$$

682. The problems which we have thus solved are solved by Laplace, *Théorie...des Prob.* pages 247—251. In the solution we have given we have followed Condorcet's guidance, with some deviations however which we will now indicate; our remarks will serve as additional evidence of the obscurity which we attribute to Condorcet.

Our original equation (1) is given by Condorcet; his demonstration consists merely in pointing out the following identity;

$$(v+e)^r = v^p(v+e)^{r-p} + v^{p-1}e(v+e)^{r-p} + v^{p-2}e(v+e)^{r-p+1} + \ldots$$
$$\ldots + v^2e(v+e)^{r-3} + ve(v+e)^{r-2} + e(v+e)^{r-1}.$$

He arrives at an equation which coincides with (4). He shews that the real roots must be numerically greater than v; but with respect to the imaginary roots he infers that the moduli cannot be greater than unity, because if they were $\phi(r)$ would be infinite when r is infinite.

We may add that Condorcet shews that (4) has no root which is a simple imaginary quantity, that is of the form $a\sqrt{-1}$.

If in our equation (7) we substitute successively for ψ in terms of ϕ we obtain

$$\psi(r) = \phi(r) - e^{p-1}\{\phi(r-p+1)-e\phi(r-p)\}$$
$$- e^{2p-1}\{\phi(r-2p+1)-e\phi(r-2p)\}$$
$$- e^{3p-1}\{\phi(r-3p+1)-e\phi(r-3p)\}$$
$$\ldots\ldots\ldots\ldots$$

On his page 75 Condorcet gives an equivalent result without explicitly using (7); but he affords very little help in establishing it.

Let $\chi(r)$ denote what $\phi(r)$ becomes when v and e are interchanged; that is let $\chi(r)$ denote the probability that in r trials the event will fail p times in succession before it happens p times in succession.

Let E denote the value of $\chi(r)$ when r is infinite. Then we can deduce the value of E from that of V by interchanging v and e; and we shall have $V + E = 1$, as we might anticipate from the result at the end of Art. 679.

Condorcet says that we shall have

$$V = (1 + e + e^2 + \ldots + e^{p-1})\, v^p f$$
$$E = (1 + v + v^2 + \ldots + v^{p-1})\, e^p f,$$

where f is *une fonction semblable de v et de e.*

Thus it would appear that he had some way of arriving at these results less simple than that which we have employed; for in our way we assign V and E definitely.

It will be seen that

$$\frac{V}{E} = \frac{v^{p-1}}{e^{p-1}} \frac{1 - e^p}{1 - v^p},$$

and this is less than $\dfrac{v^p}{e^p}$ if v be greater than e.

We have then two results, namely

$$\frac{\phi(p)}{\chi(p)} = \frac{v^p}{e^p}, \qquad \frac{V}{E} < \frac{v^p}{e^p};$$

the first of these results is obvious and the second has just been demonstrated. From these two results Condorcet seems to draw the inference that $\dfrac{\phi(r)}{\chi(r)}$ continually diminishes as r increases; see his page 78. The statement thus made may be true but it is not demonstrated.

Condorcet says on his page 78, La probabilité en général que la décision sera en faveur de la vérité, sera exprimée par

$$\frac{v^p (1 - v)(1 - e^p)}{e^p (1 - e)(1 - v^p)}.$$

This is not true. In fact Condorcet gives $\dfrac{V}{E}$ for the probability

when he ought to give $\dfrac{V}{V+E}$, that is V.

Condorcet says on the same page, Le cas le plus favorable est celui où l'on aura d'abord p décisions consécutives, sans aucun mélange. It would be difficult from the words used by Condorcet to determine what he means; but by the aid of some symbolical expressions which follow we can restore the meaning. Hitherto he has been estimating the probability *before* the trial is made; but he now takes a different position altogether. Suppose we are told that a question has been submitted to a series of tribunals, and that at last p opinions in succession on the same side have been obtained; we are also told the opinion of every tribunal to which the question was submitted, and we wish to estimate the probability that the decision is correct. Condorcet then means to say that the highest probability will be when the first p tribunals all concurred in opinion.

Condorcet continues, S'il y a quelque mélange dans le cas de $p = 2,\,\ldots\ldots$ il est clair que le cas le plus défavorable sera celui de toutes les valeurs paires de r, où le rapport des probabilités

est $\dfrac{v^2}{e^2}\cdot\dfrac{e}{v}=\dfrac{v}{e}$. Let us examine this.

Suppose that $p = 2$. Suppose we are told that a decision has been obtained after an *odd* number of trials; then we estimate the probability of the correctness of the decision at $\dfrac{v}{v+e}$. For suppose, for example, that there were five trials. The probabilities of the correctness and of the incorrectness of the decision are proportional respectively to $evev^2$ and $veve^2$, that is to v and e. On the other hand, suppose we are told that the decision has been obtained after an *even* number of trials; then in the same way we shall find that the probabilities of the correctness and of the incorrectness of the decision are proportional respectively to v^2 and e^2. Thus the probability of the correctness of the decision is $\dfrac{v^2}{v^2+e^2}$; and this is greater than $\dfrac{v}{v+e}$, assuming that v is greater than e. Thus we see the meaning which Condorcet should have expressed, and although it is almost superfluous to attempt to correct what is nearly unintelligible, it would seem that *paires* should be changed to *impaires*.

683. Condorcet's problem may be generalised. We may ask what is the probability that in r trials the event will happen p times in succession before it fails q times in succession. In this case instead of (7) we shall have

$$\psi(n) = \phi(n) - c^{q-1}\{\phi(n-q+1) - e\psi(n-q)\};$$

instead of (9) we shall have

$$\phi(r) - c^q \phi(r-q) = v^p(1-e^q)$$
$$+ v^{p-1}e\{\phi(r-p) - c^{q-1}\phi(r-p-q+1)\}$$
$$+ v^{p-2}e\{\phi(r-p+1) - e^{q-1}\phi(r-p-q+2)\}$$
$$+ \dots$$
$$+ e\{\phi(r-1) - c^{q-1}\phi(r-q)\},$$

and instead of (10) we shall have

$$V = \frac{v^{p-1}(1-e^q)}{v^{p-1} + c^{q-1} - v^{p-1}e^{q-1}}.$$

684. We will introduce here two remarks relating to that part of Condorcet's Preliminary Discourse which bears on his ninth Hypothesis.

On page XXXVI. he says,

...c'est qu'en supposant que l'on connoisse le nombre des décisions et la pluralité de chacune, on peut avoir la somme des pluralités obtenues contre l'opinion qui l'emporte, plus grande que celle des pluralités conformes à cet avis.

This is a specimen of a kind of illogical expression which is not uncommon in Condorcet. He seems to imply that the result depends on our *knowing* something, whereas the result might happen quite independently of our knowledge. If he will begin his sentence as he does, his conclusion ought to be that we may have a certain result and *know that we have it*.

On page XXXVII. he alludes to a case which is not discussed in the Essay. Suppose that a question is submitted to a series of tribunals until a certain number of opinions in succession on the same side has been obtained, the opinions of those tribunals being disregarded in which a specified plurality did not concur. Let v be the probability of an opinion for one alternative of the question, which we will call the affirmative; let e be the probability of an opinion for the negative; and let z be the probability that the opinion will have to be disregarded for want of the requisite plurality. Thus $v + e + z = 1$. Let r be the number of opinions on the same side required, q the number of tribunals. Suppose $(v + z)^q$ to be expanded, and let all the terms be taken between v^q and v^r both inclusive; denote the aggregate by $\phi(v)$. Let $\phi(e)$ be formed from $\phi(v)$ by putting e for v. Then $\phi(v)$ is the probability that there will be a decision in the affirmative, and $\phi(e)$ is the probability that there will be a decision in the negative. But, as we have said, Condorcet does not discuss the case.

685. Hitherto Condorcet has always supposed that each voter had only two alternatives presented to him, that is the voter had a proposition and its contradictory to choose between; Condorcet now proposes to consider cases in which more than two propositions are submitted to the voters. He says on his page 86 that there will be *three* Hypotheses to examine; but he really arranges the rest of this part of his Essay under *two* Hypotheses, namely the tenth on pages 86—94, and the eleventh on pages 95—136.

686. Condorcet's tenth Hypothesis is thus given on his page XLII:

...celle où l'on suppose que les Votans peuvent non-seulement voter pour ou contre une proposition, mais aussi déclarer qu'ils ne se croient pas assez instruits pour prononcer.

The pages 89—94 seem even more than commonly obscure.

687. On his page 94 Condorcet begins his eleventh Hypothesis. Suppose that there are $6q + 1$ voters and that there are three propositions, one or other of which each voter affirms. Let v, e, i denote the probabilities that each voter will affirm these three propositions respectively, so that $v + e + i = 1$. Condorcet indicates various problems for consideration. We may for example suppose that three persons A, B, C are candidates for an office,

and that v, e, i are the probabilities that a voter will vote for A, B, C respectively. Since there are $6q + 1$ voters the three candidates cannot be bracketed, but any two of them may be bracketed. We may consider three problems.

I. Find the probability that neither B nor C stands singly at. the head.

II. Find the probability that neither B nor C is *before* A.

III. Find the probability that A stands singly at the head.

These three probabilities are in descending order of magnitude. In III. we have all the cases in which A decisively beats his two opponents. In II. we have, *in addition to the cases in* III., those in which A is bracketed with one opponent and beats the other. In I. we have, *in addition to the cases in* II., those in which A is beaten by both his opponents, who are themselves *bracketed*, so that neither of the two beats the other.

Suppose for example that $q = 1$. We may expand $(v + e + i)^7$ and pick out the terms which will constitute the solution of each of our problems.

For III. we shall have

$$v^7 + 7v^6(e + i) + 21v^5(e + i)^2 + 35v^4(e + i)^3 + 35v^3 6e^2i^2.$$

For II. we shall have in addition to these

$$35v^3(4e^3i + 4ei^3).$$

For I. we shall have in addition to the terms in II.

$$7v\,20e^3i^3.$$

These three problems Condorcet briefly considers. He denotes the probabilities respectively by W^q, W'^q, and W''^q. It will scarcely be believed that he immediately proceeds to a fourth problem in which he denotes the probability by W'^q_i, *which is nothing but the second problem over again.* Such however is the fact. His enunciations appear to be so obscure as even to have misled himself. But it will be seen on examination that his second and fourth problems are identical, and the final expressions which he gives for the probabilities agree, after allowing for some misprints.

688. It may be interesting to give Condorcet's own enunciations.

I. ...soit W^q la probabilité que ni e ni i n'obtiendront sur les deux autres opinions la pluralité,... page 95.

II. ...$W_{\prime}^q$ exprimant la probabilité que e et i n'ont pas sur v la pluralité exigée, sans qu'il soit nécessaire, pour rejeter un terme, que l'un des deux ait cette pluralité sur l'autre,... page 100.

III. ...W'^q, c'est-à-dire, la probabilité que v obtiendra sur i et e la pluralité exigée,... page 102.

IV. ...$W_{\prime}^{\prime q}$, c'est-à-dire, la probabilité que v surpassera un des deux i ou e, et pourra cependant être égal à l'autre,... page 102.

Of these enunciations I., III., and IV. present no difficulty; II. is obscure in itself and is rendered more so by the fact that we naturally suppose at first that it ought not to mean the same as IV. But, as we have said, the same meaning is to be given to II. as to IV.

Before Condorcet takes these problems individually he thus states them together on his page 95:

...nous chercherons la probabilité pour un nombre donné de Votans, ou que ni e ni i ne l'emportent sur v d'une pluralité exigée, ou que e et i l'emportent chacun sur v de cette pluralité sans l'emporter l'un sur l'autre, ou enfin que v l'emporte à la fois sur e et sur i de cette pluralité.

Thus he seems to contemplate three problems. The last clause *ou enfin...pluralité* gives the enunciation of the third problem distinctly. The clause *ou que ni...exigée* may perhaps be taken as the enunciation of the second problem. The clause *ou que... l'autre* will then be the enunciation of the first problem.

In the Preliminary Discourse the problems are stated together in the following words on page XLIV:

...qu'on cherche...ou la probabilité d'avoir la pluralité d'un avis sur les deux,..., ou la probabilité que, soit les deux autres, soit un seul des deux, n'auront pas la pluralité ;...

In these words the problems are enunciated in the order III., II., I.; and knowing what the problems are we can see that the words are not inapplicable. But if we had no other way of testing the meaning we might have felt uncertain as to what problems II. and I. were to be.

24—2

689. Condorcet does not discuss these problems with much detail. He gives some general considerations with the view of shewing how what he denotes by W^{q+1} may be derived from W^q; but he does not definitely work out his suggestions.

We will here establish some results which hold when the number of voters is infinite.

We will first shew that when q is infinite W_i^q is equal to unity, provided that v is greater than either e or i. Suppose $(v+e+i)^{6q+1}$ expanded in the form

$$(v+e)^{6q+1} + (6q+1)(v+e)^{6q}i + \frac{(6q+1)6q}{1.2}(v+e)^{6q-1}i^2 +$$

$$\dots + \frac{\lfloor 6q+1}{\lfloor 4q+1 \lfloor 2q}(v+e)^{4q+1}i^{2q} + \dots$$

Now take the last term which we have here explicitly given, and pick out from it the part which it contributes to W_i^q.

We have $(v+e)^{4q+1} = (v+e)^{4q+1}\left\{\dfrac{v}{v+e} + \dfrac{e}{v+e}\right\}^{4q+1}$.

Expand $\left\{\dfrac{v}{v+e} + \dfrac{e}{v+e}\right\}^{4q+1}$ as far as the term which involves $\left(\dfrac{v}{v+e}\right)^{2q+1}$, and denote the sum by $f\left(\dfrac{v}{v+e}, \dfrac{e}{v+e}\right)$. Then finally the part which we have to pick out is

$$\frac{\lfloor 6q+1}{\lfloor 4q+1 \lfloor 2q}(v+e)^{4q+1}i^{2q}f\left(\frac{v}{v+e}, \frac{e}{v+e}\right).$$

Now if v be greater than e, then $f\left(\dfrac{v}{v+e}, \dfrac{e}{v+e}\right)$ *is equal to unity* when q is infinite, as we have already shewn; see Art. 665.

Hence we see that when q is infinite the value of W_i^q is the limit of

$$(v+e)^{6q+1} + (6q+1)(v+e)^{6q}i + \frac{(6q+1)6q}{1.2}(v+e)^{6q-1}i^2 +$$

$$\dots + \frac{\lfloor 6q+1}{\lfloor 4q+1 \lfloor 2q}(v+e)^{4q+1}i^{2q}.$$

Now we are at liberty to suppose that i is not greater than e, and then $v+e$ is greater than $2i$; so that $v+e$ must be greater than $\frac{2}{3}$. Hence by Art. 674 the value of $W_{,}^{q}$ will be unity when q is infinite.

Let $\phi(v, ei)$ stand for $W_{,}^{q}$, where we mean by our notation to draw attention to the fact that $W_{,}^{q}$ is a *symmetrical* function of e and i. We have then the following result strictly true,

$$\phi(v, ei) + \phi(e, vi) + \phi(i, ev) = 1.$$

Now suppose q infinite. Let v be greater than e or i; then as we have just shewn $\phi(v, ei) = 1$, and therefore each of the other functions in the above equation is zero. Thus, in fact, $\phi(x, yz)$ vanishes if x be less than y or z, and is equal to unity if x be greater than *both* y and z.

Next suppose $v = e$, and i less than v or e. By what we have just seen $\phi(i, ev)$ vanishes; and $\phi(v, ei) = \phi(e, vi)$, so that each of them is $\frac{1}{2}$.

Lastly, suppose that $v = e = i$. Then

$$\phi(v, ei) = \phi(e, vi) = \phi(i, ev);$$

hence each of them is $\frac{1}{3}$.

We may readily admit that when q is infinite W^{q} and W''^{q} are each equal to $W_{,}^{q}$; thus the results which we have obtained with respect to Problem II. of Art. 687 will also apply to Problems I. and III.

Condorcet gives these results, though not clearly. He establishes them for W''^{q} without using the fundamental equation we have used. He says the same values will be obtained by examining the formula for $W_{,}^{q}$. He proceeds thus on his page 104: Si maintenant nous cherchons la valeur de W^{q}, nous trouverons que W^{q} est égal à l'unité moins la somme des valeurs de W''^{q}, où l'on auroit mis v pour e, et réciproquement v pour i, et réciproquement. The words after W''^{q} are not intelligible; but it would seem that Condorcet has in view such a fundamental equation as that we have used, put in the form

$$\phi(v, ei) = 1 - \phi(e, vi) - \phi(i, ev).$$

But such an equation will not be true except on the assumption that W'^q and W^q are equal to W_r^q ultimately; and on this assumption we have the required results at once without the five lines which Condorcet gives after the sentence we have just quoted.

690. In the course of his eleventh Hypothesis Condorcet examines the propriety of the ordinary mode of electing a person by votes out of three or more candidates. Take the following example; see his page LVIII.

Suppose A, B, C are the candidates; and that out of 60 votes 23 are given for A, 19 for B, and 18 for C. Then A is elected according to ordinary method.

But Condorcet says that this is not necessarily satisfactory. For suppose that the 23 who voted for A would all consider C better than B; and suppose that the 19 who voted for B would all consider C better than A; and suppose that of the 18 who voted for C, 16 would prefer B to A, and 2 would prefer A to B. Then on the whole Condorcet gets the following result.

The two propositions in favour of C are C is better than A, C is better than B.

The first of these has a majority of 37 to 23, and the second a majority of 41 to 19.

The two propositions in favour of B are B is better than A, B is better than C.

The first of these has a majority of 35 to 25, the second is in a minority of 19 to 41.

The two propositions in favour of A are A is better than B, A is better than C.

The first of these is in a minority of 25 to 35, and the second in a minority of 23 to 37.

Hence Condorcet concludes that C who was lowest on the poll in the ordinary way, really has the greatest testimony in his favour; and that A who was highest on the poll in the ordinary way, really has the least.

Condorcet himself shews that his own method, which has just been illustrated, will lead to difficulties sometimes. Suppose, for example, that there are 23 voters for A, 19 for B, and 18 for C. Suppose moreover that all the 23 who voted for A would have preferred B to C; and that of the 19 who voted for B, there are 17 who prefer C to A, and 2 who prefer A to C; and lastly that of the 18 who voted for C there are 10 who prefer A to B, and 8 who prefer B to A. Then on the whole, the following three propositions are affirmed:

> B is better than C, by 42 votes to 18;
>
> C is better than A, by 35 votes to 25;
>
> A is better than B, by 33 votes to 27.

Unfortunately these propositions are not consistent with each other.

Condorcet treats this subject of electing out of more than two candidates at great length, both in the Essay and in the Preliminary Discourse; and it is resumed in the fifth part of his Essay after the ample discussion which it had received in the first part. His results however appear of too little value to detain us any longer. See Laplace, *Théorie ... des Prob.* page 274.

691. The general conclusions which Condorcet draws from the first part of his work do not seem to be of great importance; they amount to little more than the very obvious principle that the voters must be enlightened men in order to ensure our confidence in their decision. We will quote his own words:

On voit donc ici que la forme la plus propre à remplir toutes les conditions exigées, est en même temps la plus simple, celle où une assemblée unique, composée d'hommes éclairés, prononce seule un jugement à une pluralité telle, qu'on ait une assurance suffisante de la vérité du jugement, même lorsque la pluralité est la moindre, et il faut de plus que le nombre des Votans soit assez grand pour avoir une grande probabilité d'obtenir une décision.

Des Votans éclairés et une forme simple, sont les moyens de réunir le plus d'avantages. Les formes compliquées ne remédient point au défaut de lumières dans les Votans, ou n'y remédient qu'imparfaitement, ou même entraînent des inconvéniens plus grands que ceux qu'on a voulu éviter. Page XLII.

... il faut, 1° dans le cas des décisions sur des questions compliquées, faire en sorte que le système des propositions simples qui les forment soit rigoureusement développé, que chaque avis possible soit bien exposé, que la voix de chaque Votant soit prise sur chacune des propositions qui forment cet avis, et non sur le résultat seul......

2°. Il faut de plus que les Votans soient éclairés, et d'autant plus
éclairés, que les questions qu'ils décident sont plus compliquées ; sans
cela on trouvera bien une forme de décision qui préservera de la crainte
d'une décision fausse, mais qui en même temps rendant toute décision
presque impossible, ne sera qu'un moyen de perpétuer les abus et les
mauvaises loix. Page LXIX.

692. We now come to Condorcet's second part, which occupies
his pages 137—175. In the first part the following three elements
were always supposed known, the number of voters, the hypothesis
of plurality, and the probability of the correctness of each voter's
vote. From these three elements various results were deduced,
the principal results being the probability that the decision will
be correct, and the probability that it will not be incorrect; these
probabilities were denoted by $\phi(q)$ and $1-\psi(q)$ in Art. 669.
Now in his second part Condorcet supposes that we know only *two*
of the three elements, and that we know one of the two results;
from these known quantities he deduces the remaining element
and the other result; this statement applies to all the cases
discussed in the second part, except to two. In those two cases
we are supposed to know the probability of the correctness of a
decision which we know has been given with the least admissible
plurality; and in one of these cases we know also the probability
of the correctness of each voter's vote, and in the other case the
hypothesis of plurality.

Condorcet himself has given three statements as to the contents of his second part; namely on pages XXII, 2, and 137; of
these only the first is accurate.

693. Before proceeding to the main design of his second part
Condorcet adverts to two subjects.

First he notices and condemns Buffon's doctrine of moral cer-
tainty; see Condorcet's pages LXXI and 138. One of his objections
is thus stated on page 138 :

Cette opinion est inexacte en elle-même, en ce qu'elle tend à con-
fondre deux choses de nature essentiellement différente, la probabilité et
la certitude : c'est précisément comme si on confondoit l'asymptote
d'une courbe avec une tangente menée à un point fort éloigné ; de telles
suppositions ne pourroient être admises dans les Sciences exactes sans en
détruire toute la précision.

Without undertaking the defence of Buffon we may remark
that the illustration given by Condorcet is not fortunate; for the
student of Geometry knows that it is highly important and useful
in many cases to regard an asymptote as a tangent at a very remote point.

Secondly, Condorcet adverts to the subject of *Mathematical
Expectation;* see his pages LXXV and 142. He intimates that
Daniel Bernoulli had first pointed out the inconveniences of the
ordinary rule and had tried to remedy them, and that D'Alembert
had afterwards attacked the rule itself; see Arts. 378, 469, 471.

694. The second part of Condorcet's Essay presents nothing
remarkable; the formulæ of the first part are now employed again,
with an interchange of given and sought quantities. Methods of
approximating to the values of certain series occupy pages 155—171.
Condorcet quotes from Euler what we now call Stirling's theorem
for the approximate calculation of $\lfloor x$; Condorcet also uses the
formula, due to Lagrange, which we now usually express symbolically thus

$$\Delta^n u_x = (e^{\frac{d}{dx}} - 1)^n u_x.$$

See also Lacroix, *Traité du Calc. Diff.* ... Vol. III. page 92.

Condorcet's investigations in these approximations are disfigured and obscured by numerous misprints. The method which
he gives on his pages 168, 169 for successive approximation to a
required numerical result seems unintelligible.

695. We now arrive at Condorcet's third part which occupies
his pages 176—241. Condorcet says on his page 176,

Nous avons suffisamment exposé l'objet de cette troisième Partie : on
a vu qu'elle devoit renfermer l'examen de deux questions différentes.
Dans la première, il s'agit de connoître, d'après l'observation, la probabilité des jugemens d'un Tribunal ou de la voix de chaque Votant ; dans
la seconde, il s'agit de déterminer le degré de probabilité nécessaire pour
qu'on puisse agir dans différentes circonstances, soit avec prudence, soit
avec justice.

Mais il est aisé de voir que l'examen de ces deux questions demande
d'abord qu'on ait établi en général les principes d'après lesquels on peut
déterminer la probabilité d'un évènement futur ou inconnu, non par la

connoissance du nombre des combinaisons possibles que donnent cet évènement, ou l'évènement opposé, mais seulement par la connoissance de l'ordre des évènemens connus ou passés de la même espèce. C'est l'objet des problèmes suivans.

696. Condorcet devotes his pages 176—212 to thirteen pre-. liminary problems, and then his pages 213—241 to the application of the problems to the main purposes of his Essay.

With respect to these preliminary problems Condorcet makes the following historical remark on his page LXXXIII,

L'idée de chercher la probabilité des évènemens futurs d'après la loi. des évènemens passés, paroît s'être présentée à Jacques Bernoulli et à Moivre, mais ils n'ont donné dans leurs ouvrages aucune méthode pour y parvenir.

M^{rs}. Bayes et Price en ont donné une dans les Transactions philosophiques, *anneés* 1764 *et* 1765, et M. de la Place est le premier qui ait traité cette question d'une manière analytique.

697. Condorcet's first problem is thus enunciated :

Soient deux évènemens seuls possibles *A* et *N*, dont on ignore la probabilité, et qu'on sache seulement que *A* est arrivé *m* fois, et *N*, *n* fois. On suppose l'un des deux évènemens arrivés, et on demande la probabilité que c'est l'évènement *A*, ou que c'est l'évènement *N*, dans l'hypothèse que la probabilité de chacun des deux évènemens est constamment la même.

We have already spoken of this problem in connexion with Bayes, see Art. 551.

Condorcet solves the problem briefly. He obtains the ordinary result that the probability in favour of *A* is,

$$\frac{\int_0^1 x^{m+1}(1-x)^n\,dx}{\int_0^1 x^m(1-x)^n\,dx},$$

and this is equal to $\frac{m+1}{m+n+2}$. Similarly the probability in favour of *N* is $\frac{n+1}{m+n+2}$.

It will of course be observed that it is only by way of abbreviation that we can speak of these results as deduced from the hypothesis that the probability of the two events is constantly the same; the real hypothesis involves much more, namely, that the probability is of unknown value, any value between zero and unity being equally likely à priori.

Similarly we have the following result. Suppose the event A has occurred m times and the event N has occurred n times; suppose that the probability of the two events is constantly the same, but of unknown value, any value between a and b being equally likely à priori; required the probability that the probability of A lies between certain limits α and β which are themselves comprised between a and b.

The required probability is

$$\frac{\int_{\alpha}^{\beta} x^m (1-x)^n \, dx}{\int_{a}^{b} x^m (1-x)^n \, dx}.$$

Laplace sometimes speaks of such a result as the *probability that the possibility* of A lies between α and β; see *Théorie...des Prob. Livre* II. *Chapitre* VI. See also De Morgan, *Theory of Probabilities*, in the *Encyclopædia Metropolitana*, Art. 77, and *Essay on Probabilities* in the *Cabinet Cyclopedia*, page 87.

698. Condorcet's second problem is thus enunciated:

On suppose dans ce Problème, que la probabilité de A et de N n'est pas la même dans tous les évènemens, mais qu'elle peut avoir pour chacun une valeur quelconque depuis zéro jusqu'à l'unité.

Condorcet's solution depends essentially on this statement. The probability of m occurrences of A, and n occurrences of N is

$$\frac{\lfloor m+n}{\lfloor m \lfloor n} \left\{ \int_0^1 x\,dx \right\}^m \left\{ \int_0^1 (1-x)\,dx \right\}^n, \text{ that is } \frac{\lfloor m+n}{\lfloor m \lfloor n} \frac{1}{2^{m+n}}.$$

The probability of having A again, after A has occurred m times and N has occurred n times, is found by changing the exponent m into $m + 1$, so that it is

$$\frac{\lfloor m+n}{\lfloor m \lfloor n} \frac{1}{2^{m+n+1}}.$$

Proceeding in this way Condorcet finally arrives at the conclusion that the probability of having A is $\frac{1}{2}$ and the probability of

having N is $\frac{1}{2}$. In fact the hypothesis leads to the same conclusion as we should obtain from the hypothesis that A and N are
always equally likely to occur.

In his first problem Condorcet assumes that the probability of
each event remains constant during the observations; in his second
problem he says that he does not assume this. But we must
observe that to abstain from assuming that an element is constant
is different from distinctly assuming that it is not constant. Condorcet, as we shall see, seems to confound these two things. His
second problem does not exclude the case of a constant probability,
for as we have remarked it is coincident with the case in which
there is a constant probability equal to $\frac{1}{2}$.

The introduction of this second problem, and of others similar
to it is peculiar to Condorcet. We shall immediately see an application which he makes of the novelty in his third problem; and we
shall not be able to commend it.

699. Condorcet's third problem is thus enunciated:

On suppose dans ce problème que l'on ignore si à chaque fois la probabilité d'avoir A ou N reste la même, ou si elle varie à chaque fois, de
manière qu'elle puisse avoir une valeur quelconque depuis zéro jusqu'à
l'unité, et l'on demande, sachant que l'on a eu m évènemens A, et n
évènemens N, quelle est la probabilité d'amener A ou N.

The following is Condorcet's solution. If the probability is
constant, then the probability of obtaining m occurrences of A
and n occurrences of N is $\dfrac{\lfloor m+n}{\lfloor m \lfloor n} \displaystyle\int_0^1 x^m (1-x)^n \, dx$, that is

$\dfrac{\lfloor m+n}{\lfloor m \lfloor n} \dfrac{\lfloor m \lfloor n}{\lfloor m+n+1}$. If the probability is not constant, then, as in
the second problem, the probability of obtaining m occurrences of A
and n occurrences of N is $\dfrac{\lfloor m+n}{\lfloor m \lfloor n} \dfrac{1}{2^{m+n}}$. Hence he infers that the

probabilities of the hypothesis are respectively $\dfrac{P}{P+Q}$ and $\dfrac{Q}{P+Q}$,

where $P = \dfrac{\lfloor m \lfloor n}{\lfloor m+n+1}$ and $Q = \dfrac{1}{2^{m+n}}$.

He continues in the usual way. If the first hypothesis be true the probability of another A is $\dfrac{m+1}{m+n+2}$; if the second hypothesis be true the probability of another A is $\dfrac{1}{2}$. Thus finally the probability in favour of A is

$$\frac{1}{P+Q}\left\{\frac{m+1}{m+n+2}P+\frac{1}{2}\,Q\right\}.$$

Similarly the probability in favour of N is

$$\frac{1}{P+Q}\left\{\frac{n+1}{m+n+2}P+\frac{1}{2}\,Q\right\}.$$

It should be noticed that in this solution it is assumed that the two hypotheses were equally probable à priori, which is a very important assumption.

700. Suppose that $m+n$ is indefinitely large; if $m=n$ it may be shewn that the ratio of P to Q is indefinitely small; this ratio obviously increases as the difference of m and n increases, and is indefinitely large when m or n vanishes. Condorcet enunciates a more general result, namely this; if we suppose $m=an$ and n infinite, the ratio of P to Q is zero if a is unity, and infinite if a is greater or less than unity. Condorcet then proceeds,

Ainsi supposons m et n donnés et inegaux ; si on continue d'observer les évènemens, et que m et n conservent la même proportion, on parviendra à une valeur de m et de n, telle qu'on aura une probabilité aussi grande qu'on voudra, que la probabilité des évènemens A et N est constante.

Par la même raison, lorsque m et n sont fort grands, leur différence, quoique très-grande en elle-même, peut être assez petite par rapport au nombre total, pour que l'on ait une très-grande probabilité que la probabilité d'avoir A ou N n'est pas constante.

The second paragraph seems quite untenable. If in a very large number of trials A and N had occurred very nearly the same number of times we should infer that there is a constant probability namely $\dfrac{1}{2}$ for A and $\dfrac{1}{2}$ for N. It is the more necessary to

record dissent because Condorcet seems to attach great importance
to his third problem, and the inferences he draws from it ; see his
pages LXXXIV, XCII, 221.

701. Condorcet's fourth problem is thus enunciated :

On suppose ici un évènement A arrivé m fois, et un évènement N ·
arrivé n fois ; que l'on sache que la probabilité inconnue d'un des évènemens soit depuis 1 jusqu'à $\frac{1}{2}$, et celle de l'autre depuis $\frac{1}{2}$ jusqu'à zéro,
et l'on demande, dans les trois hypothèses des trois problèmes précédens,

1°. la probabilité que c'est A ou N dont la probabilité est depuis 1 jusqu'à $\frac{1}{2}$;

2°. la probabilité d'avoir A ou N dans le cas d'un nouvel évènement;

3°. la probabilité d'avoir un évènement dont la probabilité soit depuis
1 jusqu'à $\frac{1}{2}$.

Condorcet uses a very repulsive notation, namely,

$$\int \frac{\frac{1}{2}}{x^m (1-x)^n \, dx} \quad \text{for} \quad \int_{\frac{1}{2}}^{1} x^m (1-x)^n \, dx.$$

The chief point in the solution of this problem is the fact to
which we have drawn attention in the latter part of Art. 697.

We may remark that Condorcet begins his solution of the
second part of his problem thus : Soit supposée maintenant la probabilité changeante à chaque évènement. He ought to say, let the
probability not be assumed constant. See Art. 698.

702. Condorcet's fifth problem is thus enunciated :

Conservant les mêmes hypothèses, on demande quelle est, dans le cas
du problème premier, la probabilité, 1°. que celle de l'évènement A n'est
pas au-dessous d'une quantité donnée ; 2°. qu'elle ne diffère de la valeur
moyenne $\frac{m}{m+n}$ que d'une quantité a ; 3°. que la probabilité d'amener A,
n'est point au-dessous d'une limite a ; 4°. qu'elle ne diffère de la probabilité moyenne $\frac{m+1}{m+n+2}$ que d'une quantité moindre que a. On
demande aussi, ces probabilités étant données, quelle est la limite a
pour laquelle elles ont lieu.

The whole solution depends on the fact to which we have
drawn attention in the latter part of Art. 697.

As is very common with Condorcet, it would be uncertain from his language what questions he proposed to consider. On examining his solution it appears that his 1 and 3 are absolutely identical, and that his 2 and 4 differ only in notation.

703. In his sixth problem Condorcet says that he proposes the same questions as in his fifth problem, taking now the hypothesis that the probability is not constant.

Here his 1 and 3 are really different, and his 2 and 4 are really different.

It seems to me that no value can be attributed to the discussions which constitute the problems from the second to the sixth inclusive of this part of Condorcet's work. See also Cournot's *Exposition de la Théorie des Chances*...page 166.

704. The seventh problem is an extension of the first. Suppose there are two events A and N, which are mutually exclusive, and that in $m + n$ trials A has happened m times, and N has happened n times: required the probability that in the next $p + q$ trials A will happen p times and N happen q times.

Suppose that x and $1 - x$ were the chances of A and N at a single trial; then the probability that in $m + n$ trials A would happen m times and N happen n times would be proportional to $x^m (1 - x)^n$. Hence, by the rule for estimating the probabilities of causes from effects, the probability that the chance of A lies between x and $x + dx$ at a single trial is

$$\frac{x^m (1-x)^n\, dx}{\int_0^1 x^m (1-x)^n\, dx}.$$

And if the chance of A at a single trial is x the probability that in $p + q$ trials A will occur p times and N occur q times is

$$\frac{\lfloor p+q}{\lfloor p \lfloor q}\, x^p (1-x)^q.$$

Hence finally the probability required in the problem is

$$\frac{\lfloor p+q}{\lfloor p \lfloor q}\, \frac{\int_0^1 x^{m+p} (1-x)^{n+q}\, dx}{\int_0^1 x^m (1-x)^n\, dx}.$$

This important result had been given in effect by Laplace in the memoir which we have cited in Art. 551; but in Laplace's memoir we must suppose the $p+q$ events to be required to happen in an *assigned* order, as the factor $\dfrac{\lfloor p+q}{\lfloor p \lfloor q}$ is omitted.

We shall see hereafter in examining a memoir by Prevost and Lhuilier that an equivalent result may also be obtained by an elementary algebraical process.

705. The remaining problems consist chiefly of deductions from the seventh, the deductions being themselves similar to the problems treated in Condorcet's first part. We will briefly illustrate this by one example. Suppose that A has occurred m times and B has occurred n times; required the probability that in the next $2q+1$ trials there will be a majority in favour of A. Let $F'(q)$ denote this probability; then

$$F'(q) = \frac{\int_0^1 x^m (1-x)^n \, \phi(q) \, dx}{\int_0^1 x^m (1-x)^n \, dx},$$

where $\phi(q)$ stands for

$$x^{2q+1} + (2q+1) x^{2q}(1-x) + \frac{(2q+1)\,2q}{1.2} x^{2q-1}(1-x)^2 +$$
$$\dots + \frac{\lfloor 2q+1}{\lfloor q \lfloor q+1} x^{q+1}(1-x)^q.$$

Hence if we use, as in Art. 663, a similar notation for the case in which q is changed into $q+1$, we have

$$F(q+1) = \frac{\int_0^1 x^m (1-x)^n \, \phi(q+1) \, dx}{\int_0^1 x^m (1-x)^n \, dx}.$$

Therefore, as in Art. 663,

$$F(q+1) - F(q) = \frac{\int_0^1 x^m (1-x)^n \left\{ \phi(q+1) - \phi(q) \right\} dx}{\int_0^1 x^m (1-x)^n \, dx},$$

where $\phi(q+1) - \phi(q) = \dfrac{\lfloor 2q+1}{\lfloor q+1 \ \lfloor q} \left\{ x^{q+2}(1-x)^{q+1} - x^{q+1}(1-x)^{q+2} \right\}.$

In this manner Condorcet deduces various formulæ similar to equation (2) of Art. 663.

We may remark that at first Condorcet does not seem to deduce his formulæ in the simplest way, namely by applying the results which he has already obtained in his first part; but he does eventually adopt this plan. Compare his pages 191 and 208.

706. Condorcet now proceeds to the application of the problems to the main purposes of his Essay. As he says in the passage we have quoted in Art. 695, there are two questions to be considered. The first question is considered in pages 213—223, and the second question in pages 223—241.

707. The first question asks for two results ; Condorcet barely notices the first, but gives all his attention to the second.

Condorcet proposes two methods of treatment for the first question ; the *premier moyen* is in pages 213—220, and the *seconde méthode* in pages 220—223. Neither method is carried out to a practical application.

708. We will give a simple illustration of what Condorcet proposes in his first method. Suppose we have a tribunal composed of a large number of truly enlightened men, and that this tribunal examines a large number of decisions of an inferior tribunal. Suppose too that we have confidence that these truly enlightened men will be absolutely correct in their estimate of the decisions of the inferior tribunal. Then we may accept from their examination the result that on the whole the inferior tribunal has recorded m votes for truth and n votes for error. We are now ready to apply the problem in Art. 704, and thus determine the probability that out of the next $2q + 1$ votes given by members of the inferior tribunal there will be a majority in favour of the truth.

This must be taken however only as a very simple case of the method proposed by Condorcet ; he himself introduces circumstances which render the method much more complex. For instance he has not complete confidence even in his truly enlightened

25

men, but takes into account the probability that they will err in their estimate of the decisions of the inferior tribunal. But there would be no advantage gained in giving a fuller investigation of Condorcet's method, especially as Condorcet seems to intimate on his page 216 that the following is the chief result :

...ce qui conduit en général à cette conclusion très-importante, que tout Tribunal dont les jugemens sont rendus à une petite pluralité, relativement au nombre total des Votans, doit inspirer peu de confiance, et que ses décisions n'ont qu'une très-petite probabilité.

Such an obvious result requires no elaborate calculation to support it.

709. In the second method of treating the first question Condorcet does not suppose any tribunal composed of truly enlightened men to review the decisions of those who are less enlightened. But he assumes that the probability of the correctness of each vote lies between $\frac{1}{2}$ and 1 ; and then he proposes to apply some of the formulæ which he obtained in the solutions of the preliminary problems. Nothing of any practical value can be extracted from this part of the book. Condorcet himself says on his page C,

Il auroit été curieux de faire à la suite des décisions de quelque Tribunal existant, l'application de ce dernier principe, mais il ne nous a été possible de nous procurer les données nécessaires pour cette application. D'ailleurs les calculs auroient été très-longs, et la nécessité d'en supprimer les résultats, s'ils avoient été trop défavorables, n'étoit pas propre à donner le courage de s'y livrer.

710. Condorcet now proceeds to the second question which we have seen in Art. 695 that he proposed to consider, namely the numerical value of the probability which ought to be obtained in various cases. This occupies pages 223—241 of the Essay ; the corresponding part of the Preliminary Discourse occupies pages CII—CXXVIII. This discussion is interesting, but not of much practical value. Condorcet notices an opinion enunciated by Buffon. Buffon says that out of 10,000 persons one will die in the course of a day ; but practically the chance of dying in the

course of a day is disregarded by mankind; so that $\dfrac{1}{10000}$ may be considered the numerical estimate of a risk which any person is willing to neglect. Condorcet objects to this on various grounds; and himself proposes a different numerical estimate. He finds from tables of mortality that the risk for a person aged 37 of a sudden death in the course of a week is $\dfrac{1}{52 \times 580}$, and that the risk for a person aged 47 is $\dfrac{1}{52 \times 480}$. He assumes that practically no person distinguishes between these risks, so that their difference is in fact disregarded. The difference between these fractions is $\dfrac{1}{144768}$, and this Condorcet proposes to take as a risk which a man would practically consider equivalent to zero in the case of his own life. See Art. 644.

711. Condorcet considers however that the risk which we may with propriety neglect will vary with the subject to which it relates. He specially considers three subjects, the establishment of a new law, the decision between claimants as to the right to a property, and the condemnation of an accused person to capital punishment. We may observe that he records the opinion that capital punishments ought to be abolished, on the ground that, however large may be the probability of the correctness of a single decision, we cannot escape having a large probability that in the course of many decisions some innocent person will be condemned. See his pages CXXVI, 241.

712. We now arrive at Condorcet's fourth part, which occupies pages 242—278. He says on his page 242,

Jusqu'ici nous n'avons considéré notre sujet que d'une manière abstraite, et les suppositions générales que nous avons faites s'éloignent trop de la réalité. Cette Partie est destinée à développer la méthode de faire entrer dans le calcul les principales données auxquelles on doit avoir égard pour que les résultats où l'on est conduit, soient applicables à la pratique.

Condorcet divides this part into six questions. In these ques-

tions he proposes to examine the modifications which the results of
the preceding parts of his book require, before they can be applied
to practice. For instance we cannot in practice suppose it true
that all the voters are of equal skill and honesty; and accordingly
one of the six questions relates to this circumstance.

But the subjects proposed for investigation are too vague to be
reduced with advantage to mathematical calculation; and accordingly we find that Condorcet's researches fall far below what
his enunciations appear to promise. For example, on page 264,
he says,

Nous examinerons ici l'influence qui peut résulter de la passion ou
de la mauvaise foi des Votans.

These words may stimulate our curiosity and excite our attention; but we are quite disappointed when we read the paragraph
which immediately follows:

Comme la probabilité n'a pu être déterminée que par l'expérience,
si l'on suit la première méthode de la *troisième Partie*, ou qu'en suivant la seconde, ou suppose que l'influence de la corruption ou de la
passion sur les jugemens ne fait pas tomber la probabilité au-dessous de
$\frac{1}{2}$, alors il est évident que cet élément est entré dans le calcul, et qu'il
n'y a par conséquent rien à corriger.

Condorcet himself admits that he has here effected very little;
he says on his page CLIV,

Ainsi l'on doit regarder sur-tout cette quatrième Partie comme un
simple essai, dans lequel on ne trouvera ni les développemens ni les
détails que l'importance du sujet pourroit exiger.

713. Condorcet himself seems to attach great importance to
his fifth question which relates to that system of forced unanimity
which is established for English juries. This question he discusses in his pages 267—276 and CXI—CLI. He believes that he
shews that the system is bad. He introduces the subject thus on
page CXL:

Les jugemens criminels en Angleterre se rendent sous cette forme:
on oblige les Jurés de rester dans le lieu d'assemblée jusqu'à ce qu'ils
soient d'accord, et on les oblige de se réunir par cette espèce de torture;
car non-seulement la faim seroit un tourment réel, mais l'ennui, la

contrainte, le mal-aise, portés à un certain point, peuvent devenir un véritable supplice. Aussi pourroit-on faire à cette forme de décision un reproche semblable à celui qu'on faisoit, avec tant de justice, à l'usage barbare et inutile de la torture, et dire qu'elle donne de l'avantage à un Juré robuste et fripon, sur le Juré intègre, mais foible.

He says that there is a class of questions to which this method of forced unanimity cannot be applied; for example, the truths of Physical Science, or such as depend on reasoning. He says on page CXLI,

Aussi, du moins dans des pays ou des siècles éclairés, n'a-t-on jamais exigé cette unanimité pour les questions dont la solution dépend du raisonnement. Personne n'hésite à recevoir comme une vérité l'opinion unanime des gens instruits, lorsque cette unanimité a été le produit lent des réflexions, du temps et des recherches : mais si l'on enfermoit les vingt plus habiles Physiciens de l'Europe jusqu'à ce qu'ils fussent convenus d'un point de doctrine, personne ne seroit tenté d'avoir la moindre confiance en cette espèce d'unanimité.

714. We shall not reproduce Condorcet's investigations on the English jury system, as they do not seem to us of any practical value. They can be easily read by a student who is interested in the subject, for they form an independent piece of reasoning, and thus do not enforce a perusal of the rest of the book.

We will make a few remarks for the use of a student who consults this part of Condorcet's book; these will occupy our next Article.

715. On page CXLI Condorcet says that we ought to distinguish three sorts of questions, and he at once states the first; as usual with him he is not careful in the subsequent pages to indicate the second and third of these questions. The second is that beginning on page CXLII, *Il y a un autre genre d'opinions....* The third is that beginning on page CLI, *On peut considérer encore....*

On his page 267 Condorcet says,

Si l'on prend l'hypothèse huitième de la première Partie, et qu'en conséquence l'on suppose que l'on prendra les voix jusqu'à ce que l'unanimité se soit réunie pour un des deux avis, nous avons vu que le calcul donnoit la même probabilité, soit que cette unanimité ait lieu immédiatement, soit qu'elle ne se forme qu'après plusieurs changemens d'avis, soit que l'on se réunisse à la majorité, soit que l'avis de la minorité finisse, par avoir tous les suffrages.

We quote this passage in order to draw attention to a practice of which Condorcet is very fond, and which causes much obscurity in his writings; the practice is that of needlessly varying the language. If we compare the words *soit que l'on se réunisse à la majorité* with those which immediately follow, we discover such a great diversity in the language that we have to ascertain whether there is a corresponding diversity in the meaning which is to be conveyed. We shall conclude on examination that there is no such diversity of meaning, and we consequently pronounce the diversity of language to be very mischievous, as it only serves to arrest and perplex the student.

It would be well in this paragraph to omit all the words *soit que l'on…suffrages;* for without these every thing is fully expressed which Condorcet had obtained in his first part.

We would indicate the first eleven lines of Condorcet's page 270 as involving so much that is arbitrary as to render all the conclusions depending on them valueless. We are not prepared to offer more reasonable suppositions than those of Condorcet, but we think that if these are the best which can be found it will be prudent to give up the attempt to apply mathematics to the question.

We may remark that what is called *Trial by Jury* would more accurately be styled *Trial by Judge and Jury.* Accordingly a most important element in such an investigation as Condorcet undertakes would be the influence which the Judge exercises over the Jury; and in considering this element we must remember that the probability is very high that the opinion of the Judge will be correct, on account of his ability and experience.

716. We now arrive at Condorcet's fifth part; which occupies the remainder of his book, that is, pages 279—304. Condorcet says on page CLVII,

L'objet de cette dernière Partie, est d'appliquer à quelques exemples les principes que nous avons développés. Il auroit été à desirer que cette application eût pu être faite d'après des données réelles, mais la difficulté de se procurer ces données, difficultés qu'un particulier ne pouvoit espérer de vaincre, a forcé de se contenter d'appliquer les principes de la théorie à de simples hypothèses, afin de montrer du moins la marche que pourroient suivre pour cette application réelle ceux à qui on auroit procuré les données qui doivent en être la base.

But it would be rather more correct to describe this part as furnishing some additions to the preceding investigations than as giving examples of them.

Four so-called examples are discussed.

717. In the first example Condorcet proposes what he thinks would be a good form of tribunal for the trial of civil cases. He suggests a court of 25 judges, to decide by majority. He adds, however, this condition; suppose the case tried is the right to a certain property, then if the majority is less than 3 the court should award compensation to the claimant against whom decision is given.

718. In the second example Condorcet proposes what he thinks would be a good form of tribunal for the trial of criminal cases. He suggests a court of 30 judges, in which a majority of at least 8 is to be required to condemn an accused person.

719. The third example relates to the mode of electing from a number of candidates to an office. This example is really a supplement to the investigation given in the first part of the Essay. Condorcet refers to the memoir on the subject by a celebrated geometer, and records his own dissent from that geometer's suggestions; the geometer alluded to is Borda. See Art. 690.

720. The fourth example relates to the probability of the accuracy of the decision of a large assembly in which the voters are not all alike. Condorcet considers the case in which the number of voters whose probability of accuracy is x, is proportional to $1-x$; and he supposes that x lies between $\frac{1}{2}$ and 1. In such a case the mean probability is

$$\frac{\int_{\frac{1}{2}}^{1} (1 - x)\, x\, dx}{\int_{\frac{1}{2}}^{1} (1 - x)\, dx},$$

which is $\frac{2}{3}$. If the value of x lies between a and 1 the mean pro-·
bability is found in the same way to be $\frac{1 + 2a}{3}$.

This example is interesting, but some parts of the investigations connected with it are very obscure.

As in other parts of his book Condorcet draws a very in-
significant inference from his difficult investigations. He says,
page 303,

On voit donc combien il est important, non-seulement que les
hommes soient éclairés, mais qu'en même temps tous ceux qui, dans
l'opinion publique, passent pour instruits ou habiles, soient exempts de
préjugés. Cette dernière condition est même la plus essentielle, puisqu'il
paroît que rien ne peut remédier aux inconvéniens qu'elle entraine.

721. Besides the *Essai* Condorcet wrote a long memoir on the
Theory of Probability, which consists of six parts, and is published
in the volumes of the *Hist. de l'Acad....Paris*, for the years 1781,
1782, 1783, and 1784.

The first and second parts appear in the volume for 1781 ;
they occupy pages 707—728. The dates of publication of the
volumes are as usual later than the dates to which the volumes
belong ; the portion of the memoir which appears in the volume
for 1781 is said to have been read on August 4th, 1784.

722. The first part of the memoir is entitled *Réflexions sur la
règle générale qui prescrit de prendre pour valeur d'un évènement
incertain, la probabilité de cet évènement, multipliée par la valeur de
l'évènement en lui-même.*

Suppose that p represents the probability that an event will
happen, and that if the event happens a person is to receive a sum
of money denoted by a ; then the general rule to which Condorcet
refers is the rule which estimates the person's advantage at the
sum pa. On this rule Condorcet makes some remarks ; and these
remarks are also given in substance in the *Essai*, in pages

142—147. The sum of the remarks is this ; Condorcet justifies the rule on the ground that it will lead to satisfactory results if a *very large number of trials* be made. Suppose for example that A and B are playing together, and that A's chance of winning a single game is p, and B's chance is q: then the rule prescribes that if A's stake be denoted by kp, then B's stake must be kq. Now we know, by Bernoulli's Theorem, that if A and B play a very large number of games, there is a very high probability that the number which A wins will bear to the number which B wins a ratio extremely near to the ratio of p to q. Thus if the stakes are adjusted according to the general rule there is a very high probability that A and B are on terms of equality as to their prospects; if any other ratio of the stakes be adopted a proportional advantage is given to one of the players.

There can be no doubt that this view of the ground on which the rule is to be justified is correct.

723. Condorcet adverts to the *Petersburg Problem*. The nature of his remarks may be anticipated. Suppose that p in the preceding Article is extremely small and q very nearly equal to unity. Then B's stake is very large indeed compared with A's. Hence it may be very imprudent for B to play with A on such terms, because B may be ruined in a few games. Still it remains true that if A and B agree to continue playing through a very long series of games no proportion of stakes can be fair except that which the general rule assigns.

724. The second part of Condorcet's memoir is entitled *Application de l'analyse à cette question: Déterminer la probabilité qu'un arrangement régulier est l'effet d'une intention de le produire.*

This question is analogous to one discussed by Daniel Bernoulli, and to one discussed by Michell; see Arts. 395 and 618.

Condorcet's investigations rest on such arbitrary hypotheses that little value can be attached to them. We will give one specimen.

Consider the following two series :
$$1, 2, 3, 4, 5, 6, 7, 8, 9, 10.$$
$$1, 3, 2, 1, 7, 13, 23, 44, 87, 167.$$

In the first series each term is equal to twice the preceding term diminished by the term which precedes that; and in the second series each term is the sum of the four which precede it. Condorcet says,

Il est clair que ces deux suites sont régulières, que tout Mathématicien qui les examinera, verra qu'elles sont toutes deux assujetties à une loi ; mais il est sensible en même temps que, si l'on arrête une de ces suites au sixième terme, par exemple, on sera plutôt porté à regarder la première, comme étant régulière, que la seconde, puisque dans la première il y aura quatre termes assujettis à une loi, tandis qu'il n'y en a que deux dans la seconde.

Pour évaluer le rapport de ces deux probabilités, nous supposerons que ces deux suites soient continuées à l'infini. Comme alors il y aura dans toutes les deux un nombre infini de termes assujettis à la loi, nous supposerons que la probabilité seroit égale ; mais nous ne connoissons qu'un certain nombre de termes assujettis à cette loi ; nous aurons donc les probabilités que l'une de ces suites sera régulière plutôt que l'autre, égales aux probabilités que ces suites étant continuées à l'infini, resteront assujetties à la même loi.

Soit donc pour une de ces suites e le nombre des termes assujettis à une loi, et e' le nombre correspondant pour une autre suite, et qu'on cherche la probabilité que pour un nombre q de termes suivans, la même loi continuera d'être observée. La première probabilité sera exprimée par $\dfrac{e+1}{e+q+1}$, la seconde par $\dfrac{e'+1}{e'+q+1}$, et le rapport de la seconde à la première par $\dfrac{(e'+1)(e+q+1)}{(e+1)(e'+q+1)}$.

Soit $q=\dfrac{1}{0}$, et e, e' des nombres finis, ce rapport devient $\dfrac{e'+1}{e+1}$. Ainsi dans l'exemple précédent, si l'on s'arrête au sixième terme, on aura $e=4$, $e'=2$, et le rapport sera $\dfrac{3}{5}$: si on s'arrête au dixième, on aura $e=8$, $e'=6$, et le rapport sera $\dfrac{7}{9}$.

Si l'on suppose que e et e' sont du même ordre que q, le même rapport devient $\dfrac{ee'+e'q}{ee'+eq}$, et si on suppose $e=q=1$, il sera $\dfrac{2e'}{1+e'}$.

We will make some remarks on this investigation.

The result, that the first probability is $\dfrac{e+1}{e+q+1}$ and the second

is $\dfrac{e'+1}{e'+q+1}$, is we presume obtained by Bayes's Theorem.

After supposing that q is infinite it is perplexing to be told that $e = q = 1$. Condorcet should have proceeded thus. Suppose $e = q$, then

$$\frac{ee' + e'q}{ee' + eq} = \frac{2e'}{e+e'} = \frac{2x}{1+x} \text{ where } x = \frac{e'}{e}.$$

The following then is the result which Condorcet considers himself to have obtained. Let us suppose we have observed in a certain series that a certain law holds during so many terms as form the fraction x of the whole series, then the comparative probability that the whole series is subject to this law is $\dfrac{2x}{1+x}$.

It is however obvious that this result has been obtained by means of several most arbitrary hypotheses.

725. The remainder of this part of Condorcet's memoir is difficult, but the meaning can be discovered by patience. There is nothing that appears self-contradictory except perhaps on page 727. In the last line Condorcet takes for the limits of a certain integration b and $1 - a + b$; it would seem that the latter limit should be $1 - a$, for otherwise his Article VII. is only a repetition of his Article VI.

726. The third part of Condorcet's memoir is entitled *Sur l'évaluation des Droits éventuels*. It is published in the *Hist. de l'Acad....Paris*, for 1782; it occupies pages 674—691.

This part commences thus:

La destruction du Gouvernement féodal a laissé subsister en Europe un grand nombre de droits éventuels, mais on peut les réduire à deux classes principales; les uns se payent lorsque les propriétés viennent à changer par vente, les autres se payent aux mutations par succession, soit directe ou collatérale, soit collatérale seulement.

Condorcet then proposes to determine the sum of money which should be paid down in order to free any property from such feudal rights over it.

727. The following paragraph appears very remarkable when
we reflect how soon the expectations it contains were falsified by
the French Revolution.

Premier Principe. Nous supposerons d'abord que l'ordre suivant
lequel les dernières mutations se sont succédées, sera indéfiniment con-
tinué.

Le motif qui nous a fait adopter ce principe, est la grande probabilité que nous avons moins de grands changemens, moins de grandes
révolutions à attendre pour l'avenir, qu'il n'y en a eu dans le passé: le
progrès des lumières en tout genre et dans toutes les parties de l'Europe,
l'esprit de modération et de paix qui y règne, l'espèce de mépris où le
Machiavelisme commence à tomber, semblent nous assurer que les guerres
et les révolutions deviendront à l'avenir moins fréquentes; ainsi le
principe que nous adoptons, en même temps qu'il rend les calculs et les
observations plus faciles, a de plus l'avantage d'être plus exact.

728. The memoir is neither important nor interesting, and it
is disfigured by the contradiction and obscurity which we have
noticed in Condorcet's Essay. Condorcet says that he will begin by
examining the case in which the event producing the right necessarily happens in a certain length of time, as for example, when
the right accrues on every succession to the property ; and then he
will consider the case in which the event does not necessarily happen, as, for example, when the right accrues on a sale of the property, or on a particular kind of succession.   He then gives three
methods for the first case, and in direct contradiction to what he
has said, it will be found that only his first method applies to the
case in which the event producing the right necessarily happens.

729. We will give the results of the second of Condorcet's
methods, though not in his manner.

Let us suppose for simplicity that the sum to be paid if
the event happens is one pound ; let c represent the present worth
of one pound due at the end of a year; let x be the probability
that the event will happen in the course of one year.   Then xc
represents the value of that part of the right which arises from the
first year, xc^2 the value of that part which arises from the second
year, xc^3 the value of that part which arises from the third year,
and so on.   Thus the value of the whole right is

$$x\,(c + c^2 + c^3 + \ldots),\ \text{that is}\ \frac{xc}{1 - c}.$$

The question now arises what is the value of x? Suppose that during $m + n$ past years the event happened m times and did not happen n times; we might reasonably take $\dfrac{m}{m + n}$ for x, so that the whole value of the right would be $\dfrac{c}{1 - c}\,\dfrac{m}{m + n}$. Condorcet however prefers to employ Bayes's Theorem, and so he makes the whole value of the right

$$\frac{\displaystyle\int_0^1 x^m\,(1 - x)^n\,\frac{xc}{1 - c}\,dx}{\displaystyle\int_0^1 x^m\,(1 - x)^n\,dx},$$

that is
$$\frac{m + 1}{m + n + 2}\,\frac{c}{1 - c}.$$

Moreover Condorcet supposes that at the present moment the event has *just happened* on which the right depends, so that he adds unity to the result and obtains for the value of the whole right

$$1 + \frac{m + 1}{m + n + 2}\,\frac{c}{1 - c}.$$

730. The investigation of the preceding Article goes over the same ground as that on page 680 of the volume which contains the memoir, but is we hope more intelligible. We proceed to make two remarks.

First. It is clear that Condorcet is quite wrong in giving this method as applicable to the first case, namely that in which the event must happen in a certain length of years. The method is quite inapplicable to such an example as he mentions, namely when the right would accrue on the next succession to the property, that is, on the death of the present holder; for the probability of such an event would not be constant from year to year for ever as this method assumes. The method would be applicable to the example of the second case in which the right is to accrue upon a sale, for that might without absurdity be supposed as likely to happen in one year as in another for ever.

Secondly. We see no advantage in applying Bayes's Theorem. Condorcet is very fond of it; and throughout this memoir as well as in his other writings on the subject indulges to excess in signs of integration. In the above example if m and n are very large numbers no practical change is made in the result by using Bayes's Theorem; if $m + n$ is a small number our knowledge of the past would be insufficient to justify any confidence in our anticipations of the future.

731. From what we have said it may be expected that when Condorcet comes to his second case he should be obscure, and this is the fact. He gives on his page 685 the modifications which his three methods now require. The second method is really unaltered, for we merely suppose that observation gives m' and n' instead of m and n. The modification of the third method seems unsound; the modification of the first method is divided into two parts, of which only the former appears intelligible.

But we leave these to students of the original memoir.

732. We may add that on pages 687—690 Condorcet gives an investigation of the total value arising from two different rights. It is difficult to see any use whatever in this investigation, as the natural method would be to calculate each separately. Some idea of the unpractical character of the result may be gathered from the fact that we have to calculate a fraction the numerator and denominator of which involve $n + n' + n'' + n''' - 2$ successive integrations. This complexity arises from an extravagant extension and abuse of Bayes's Theorem.

733. The fourth part of Condorcet's memoir is intitled *Réflexions sur la méthode de déterminer la Probabilité des évènemens futurs, d'après l'Observation des évènemens passés.* The fourth and fifth parts appeared in the *Hist. de l'Acad....Paris*, for 1783; they occupy pages 539—559. This volume was published in 1786, that is after Condorcet's *Essai* which is referred to on page 541.

734. Suppose that in $m + n$ trials an event has happened m times and failed n times; required the probability that in the next

$p + q$ trials it will happen p times and fail q times. The required probability is

$$\frac{\lfloor p + q}{\lfloor p \lfloor q} \frac{\int_0^1 x^{m+p} (1 - x)^{n+q} \, dx}{\int_0^1 x^m (1 - x)^n \, dx},$$

as we have already remarked in Art. 704.

Condorcet quotes this result; he thinks however that better formulæ may be given, and he proposes two. But these seem quite arbitrary, and we do not perceive any reason for preferring them to the usual formula. We will indicate these formulæ proposed by Condorcet.

I. Let $t = m + n + p + q$ and put

$$u = \frac{x_1 + x_2 + x_3 + \ldots + x_t}{t}:$$

then the proposed formula is

$$\frac{\lfloor p + q}{\lfloor p \lfloor q} \frac{\int\!\!\int\!\!\int \ldots u^{m+p} (1 - u)^{n+q} \, dx_1 \, dx_2 \ldots dx_t}{\int\!\!\int\!\!\int \ldots u^m (1 - u)^n \, dx_1 \, dx_2 \ldots dx}.$$

The limits of each integration are to be 0 and 1.

II. Suppose an event to have happened n times in succession, required the probability that it will happen p times more in succession.

$$\text{Let } u = x_1 \frac{x_1 + x_2}{2} \frac{x_1 + x_2 + x_3}{3} \ldots \frac{x_1 + x_2 + \ldots + x_n}{n};$$

let v be an expression similar to u but extended to $n + p$ factors; then Condorcet proposes for the required probability the formula

$$\frac{\int\!\!\int\!\!\int \ldots v \, dx_1 \, dx_2 \ldots dx_{n+p}}{\int\!\!\int\!\!\int \ldots u \, dx_1 \, dx_2 \ldots dx_n}.$$

The limits of each integration are to be 0 and 1.

Condorcet proposes some other formulæ for certain cases; they are as arbitrary as those which we have already given, and not fully intelligible ; see his pages 550—553.

735. The fifth part of Condorcet's memoir is entitled *Sur la probabilité des faits extraordinaires.*

Suppose that p is the probability of an event in itself; let t denote the probability of the truth of a certain witness. This witness asserts that the event has taken place; required the probability that the event did take place, and that it did not. The required probabilities are

$$\frac{pt}{pt + (1-p)(1-t)} \quad \text{and} \quad \frac{(1-p)(1-t)}{pt + (1-p)(1-t)}.$$

Condorcet gives these formulæ with very little explanation.

The application of these formulæ is not free from difficulty. Suppose for example a trustworthy witness asserts that one ticket of a lottery of 10000 tickets was drawn, and that the number of the ticket drawn was 297. Here if we put $p = \dfrac{1}{10000}$ we obtain such a very small value of the truth of the witness's statement that we lose our confidence in the formula. See Laplace *Théorie...des Prob.* pages 446—451. De Morgan, *Cambridge Philosophical Transactions*, Vol. IX. page 119.

736. Condorcet makes remarks on two points, namely the mode of estimating p and the mode of estimating t. He recurs to the former point in the sixth part of his memoir, and we shall give an extract which will shew the view he advocated in his fifth part, and the view which he advocated in his sixth part.

With respect to the second point Condorcet's chief remark is that the probability of a witness is not the same for all facts. If we estimate it at u for a simple fact, then we should estimate it at u^2 for a compound fact consisting of two simple facts, and so on. One witness however may be as capable of observing a compound fact consisting of two or more simple facts as another is of observing a simple fact.

737. The sixth part of Condorcet's memoir is entitled *Appli-*

cation des principes de l'article précédent à quelques questions de critique. It is published in the *Hist. de l'Acad. ... Paris* for 1784; it occupies pages 454—468.

738. In this part Condorcet begins by adverting to some remarks which he had made in his fifth part as to the mode of estimating the value of what we denoted by p in Article 735. He says,

J'ai observé en même-temps qu'il ne falloit pas dans ce cas entendre, par la probabilité propre d'un fait, le rapport du nombre des combinaisons où il a lieu, avec le nombre total des combinaisons. Par exemple, si d'un jeu de dix cartes on en a tiré une, et qu'un témoin me dise que c'est telle carte en particulier, la probabilité propre de ce fait, qu'il s'agit de comparer avec la probabilité qui naît du témoignage, n'est pas la probabilité de tirer cette carte, qui seroit $\frac{1}{10}$, mais la probabilité d'amener cette carte plutôt que telle autre carte déterminée en particulier; et comme toutes ces probabilités sont égales, la probabilité propre est ici $\frac{1}{2}$.

Cette distinction étoit nécessaire, et elle suffit .pour expliquer la contrariété d'opinions entre deux classes de philosophes. Les uns ne peuvent se persuader que les mêmes témoignages puissent produire, pour un fait extraordinaire, une probabilité égale à celle qu'ils produisent pour un fait ordinaire ; et que, par exemple, si je crois un homme de bon sens qui me dit qu'une femme est accouchée d'un garçon, je dusse le croire également s'il me disoit qu'elle est accouchée de douze.

Les autres au contraire sont convaincus que les témoignages conservent toute leur force, pour les faits extraordinaires et très-peu probables, et ils sont frappés de cette observation, que si on tire une loterie de 100000 billets, et qu'un homme, digne de foi, dise que lo numéro 256, par exemple, a eu le premier lot, personne ne doutera de son témoignage, quoiqu'il y ait 99999 à parier contre 1 que cet évènement n'est pas arrivé.

Or, au moyen de l'observation précédente, on voit que dans le second cas la probabilité propre du fait étant $\frac{1}{2}$, le témoignage conserve toute sa force, au lieu que dans le premier, cette probabilité étant très-petite, réduit presque à rien celle du témoignage.

J'ai proposé ensuite de prendre, pour la probabilité propre du fait,

26

le rapport du nombre de combinaisons qui donnent ce fait, ou un fait semblable au nombre total des combinaisons.

Ainsi, par exemple, dans le cas où on tire une carte d'un jeu de dix cartes, le nombre des combinaisons où l'on tire une carte déterminée quelconque est un ; celui des combinaisons où l'on tire une autre carte déterminée est aussi un ; donc $\frac{1}{2}$ exprimera la probabilité propre.

Si on me dit qu'on a tiré deux fois de suite la même carte, alors on trouvera qu'il n'y a que dix combinaisons qui donnent deux fois une même carte, et quatre-vingt-dix qui donnent deux cartes différentes : la probabilité propre du fait n'est donc que $\frac{1}{10}$, et celle du témoignage commence à devenir plus foible.

Mais je crois devoir abandonner cette manière de considérer la question, 1° parce qu'elle me paroît trop hypothétique ; 2° parce que souvent cette comparaison d'évènemens semblables seroit difficile à faire, ou, ce qui est encore pis, ne se feroit que d'après des suppositions arbitraires ; 3° parce qu'en l'appliquant à des exemples, elle conduit à des résultats trop éloignés de ceux que donneroit la raison commune.

J'en ai donc cherché une autre, et il m'a paru plus exact de prendre, pour probabilité propre d'un évènement, le rapport de la probabilité de cet évènement prise dans le sens ordinaire, avec la probabilité moyenne de tous les autres évènemens.

739. Thus we see that Condorcet abandons the suggestion which he made in the fifth part of his memoir and offers another. It does not seem that the new suggestion escapes any of the objections which Condorcet himself advances against the old suggestion, as will appear by the analysis we shall now give of Condorcet's examples.

740. Suppose there are ten cards and it is asserted that a specified card has been drawn twice running; we proceed to estimate the *probabilité propre* of the event. There are 9 other ways in which the same card can be drawn twice, and the ordinary probability of each drawing is $\frac{1}{100}$; there are 45 ways in which two different cards are obtained in two drawings, and the ordinary probability of each drawing is $\frac{2}{100}$. Hence the mean probability of all the other events is

$$\frac{1}{54}\left\{45 \times \frac{2}{100} + 9 \times \frac{1}{100}\right\}, \text{ that is } \frac{99}{5400}.$$

Hence according to Condorcet's own words the *probabilité propre* should be $\frac{1}{100} \div \frac{99}{5400}$, that is $\frac{54}{99}$. But he himself says that the *probabilité propre* is $\frac{54}{153}$, so that he takes $\frac{1}{100} \div \left\{\frac{99}{5400} + \frac{1}{100}\right\}$ and not $\frac{1}{100} \div \frac{99}{5400}$. That is, as is so frequently the case with Condorcet, his own words do not express his own meaning.

Again suppose that there are ten cards and it is asserted that a specified card has been drawn thrice running; we proceed to estimate the *probabilité propre* of the event. Here the mean probability of all the other events is

$$\frac{1}{219}\left\{120 \times \frac{6}{1000} + 90 \times \frac{3}{1000} + \frac{9}{1000}\right\}, \text{ that is } \frac{999}{219000}.$$

Condorcet says that the *probabilité propre* is $\frac{219}{1218}$, so that he takes $\frac{1}{1000} \div \left\{\frac{999}{219000} + \frac{1}{1000}\right\}$.

741. Condorcet now proceeds to apply these results in the following words:

Ainsi supposons, par exemple, que la probabilité du témoignage soit $\frac{99}{100}$, c'est-à-dire, que le témoin ne se trompe ou ne veuille tromper qu'une fois sur cent, on aura, d'après son témoignage, la probabilité $\frac{99}{100}$ ou $\frac{9900}{10000}$ qu'on a tiré une carte déterminée; la probabilité $\frac{9818}{10000}$ qu'on a tiré deux fois la même carte; et la probabilité $\frac{9540}{10000}$ qu'on l'a tirée trois fois.

We find some difficulties in these numbers.

Let p denote the *probabilité propre* and t the probability of the testimony; then the formula to be applied is, we presume,

$$\frac{pt}{pt + (1 - p)(1 - t)}.$$ In the first case it seems that Condorcet

supposes $p = 1$, that is he takes apparently the *probabilité propre*
to be $\frac{1}{10} \div \frac{1}{9} \left\{ 9 \times \frac{1}{10} \right\}$, which agrees indeed with his own *words*
but not with his *practice* which we have exhibited in Art. 740; if
we follow that practice we shall have $p = \frac{1}{2}$.

In the second case we have $p = \frac{54}{153}$, and with this value the
formula gives $\frac{54}{55}$ which is approximately ·9818.

In the third case we have $p = \frac{219}{1218}$, and with this value the
formula gives $\frac{803}{840}$ which however is very nearly ·9560 instead of
·9540 as Condorcet states.

742.   Condorcet's next example seems very arbitrary and obscure.   His words are,

Supposons encore que l'observation ait constaté que, sur vingt millions d'hommes, un seul ait véeu 120 ans, et que la plus longue vie
ait été de 130; qu'un homme me dise que quelqu'un vient de mourir à
120 ans, et que je cherche la probabilité propre de cet événement : je
regarderai d'abord comme un fait unique, celui de vivre plus de 130
ans, fait que je suppose n'être pas arrivé; j'aurai donc 131 faits différens, dont celui de mourir à 120 ans est un seul.   La probabilité de
celui-ci sera $\frac{1}{20000131}$; la probabilité moyenne des 130 autres sera
$\frac{20000130}{20000131 \times 130}$; donc la probabilité propre cherchée sera $\frac{130}{20000260}$,
ou environ $\frac{1}{15384}$.

743.   Condorcet's next example seems also arbitrary.   His
words are,

Cette méthode s'appliquera également aux évènemens indéterminés.
Ainsi, en continuant le même exemple, si le témoin a dit seulement
que l'on a deux fois amené la même carte, sans la nommer, alors ces dix
évènemens, ayant chacun la probabilité $\frac{1}{100}$, $\frac{1}{100}$ exprimera leur pro-

babilité moyenne; $\dfrac{2}{100}$ exprimera de même celle des 45 autres évène-

mens ayant chacun la probabilité $\dfrac{2}{100}$: ainsi la probabilité propre de

l'évènement sera $\dfrac{1}{3}$.

Condorcet himself observes that it may appear singular that the result in this case is less than that which was obtained in Art. 740; so that a man is less trustworthy when he merely says that he has seen the same card drawn twice, than when he tells us in addition what card it was that he saw drawn twice. Condorcet tries to explain this apparent singularity; but not with any obvious success.

The singularity however seems entirely to arise from Condorcet's own arbitrary choice; the rule which he himself lays down requires him to estimate *la probabilité moyenne de tous les autres évènemens*, and he estimates this mean probability differently in the two cases, and apparently without sufficient reason for the difference.

744. Condorcet's next example is as follows : We are told that a person with two dice has five times successively thrown higher than 10; find the *probabilité propre*. With two dice the number thrown may be 2, 3, ... up to 12; the respective probabilities are

$$\frac{1}{36}, \frac{2}{36}, \frac{3}{36}, \frac{4}{36}, \frac{5}{36}, \frac{6}{36}, \frac{5}{36}, \frac{4}{36}, \frac{3}{36}, \frac{2}{36}, \frac{1}{36}.$$

The whole number of events is $\dfrac{11 \times 12 \times 13 \times 14 \times 15}{\lfloor 5}$, that is 3003; and of these only 6 belong to the proposed combination. Since the probability of these 6 throws is $\dfrac{1}{12^5}$ their mean probability is $\dfrac{1}{6 \times 12^5}$. The mean probability of the other throws will be $\dfrac{11^5}{2997 \times 12^5}$. Hence the *probabilité propre* is $\dfrac{2997}{6 \times 11^5 + 2997}$.

It is obvious that all this is very arbitrary. When Condorcet says there are 6 throws belonging to the proposed combination he means that all the throws may be 12, or all 11, or four 12 and one 11, or three 12 and two 11, ... And he says the mean probability is

$\frac{1}{6 \times 12^5}$. But if we consider the different orders in which these throws can occur we may say that the whole number is 2^5 and the mean probability $\frac{1}{2^5}\left(\frac{1}{36} + \frac{2}{36}\right)^5$, that is $\frac{1}{2^5 12^5}$.

Again let us admit that there are 3003 cases in all, and that of these only 6 belong to the proposed combination. The other 2997 cases form two species, namely those in which *every* throw is below 11, and those in which *some* throws are below 11 and the others above 10; when Condorcet takes $\frac{11^5}{2997 \times 12^5}$ as the mean probability, he forgets this division of species and only considers the first species. He should take $\frac{1}{2997}\left(1 - \frac{1}{12^5}\right)$ instead of $\frac{11^5}{2997 \times 12^5}$.

745. Suppose two classes of events A and B; let the probability of an A be a and the probability of a B be b; let there be m events A and n events B. The *probabilité propre* of an assigned event of the class B will be, according to Condorcet's practice,

$$\frac{b}{\frac{ma + (n-1)b}{m+n-1} + b} \text{ that is } \frac{(m+n-1)b}{ma + (m+2n-2)b}.$$

If m and n be equal and very large this becomes $\frac{2b}{a+3b}$. If we suppose b extremely small and consequently a very nearly unity we obtain $2b$ as an approximate value.

746. Condorcet proceeds to apply his doctrine to the credibility of two statements in the History of Rome. He says,

Je vais maintenant essayer de faire à une question de critique l'application des principes que je viens d'établir. Newton paroît être le premier qui ait eu l'idée d'appliquer le calcul des probabilités à la critique des faits. Il propose, dans son ouvrage sur la chronologie, d'employer la connoissance de la durée moyenne des générations et des règnes, telle que l'expérience nous la donne, soit pour fixer d'une manière du moins approchée, des points de chronologie fort incertains, soit pour juger du plus ou du moins de confiance que méritent les différens systèmes imaginés pour concilier entr'elles des époques qui paroissent se contredire.

Condorcet names Fréret as having opposed this application of the Theory of Probability, and Voltaire as having supported it; but he gives no references.

747. According to some historians the whole duration of the reigns of the seven kings of Rome was 257 years. Condorcet proposes to examine the credibility of this statement. He assumes that in an elective monarchy we may suppose that a king at the date of his election will be between 30 years old and 60 years old. He adopts De Moivre's hypothesis respecting human mortality; this hypothesis, as Condorcet uses it, amounts to assuming that the number of people at any epoch who are y years old is $k(90-y)$, where k is some constant, and that of these k die every year.

Let n denote the greatest number of years which the youngest elected king can live, m the greatest number of years which the oldest elected king can live; then the probability that a single reign will last just r years is the coefficient of x^r in the expansion of

$$\frac{(n-m+1)\,x\,(1-x) - x^{m+1} + x^{n+2}}{(1-x)^2\,\frac{n+m}{2}\,(n-m+1)}.$$

A few words will be necessary to shew how this formula can be verified. It follows from our hypothesis that the number of persons from whom the king must be elected is

$$k\{n+(n-1)+(n-2)+\ldots+m\},$$

that is $k\frac{n+m}{2}(n-m+1)$. And if r be less than $m+1$ the number of persons who die in the rth year will be $k(n-m+1)$; if r be between $m+1$ and $n+1$, both inclusive, the number who die in the rth year will be $k(n-r+1)$; if r be greater than $n+1$ the number who die in the rth year will be zero. Now the coefficient of x^r in the expansion of

$$\frac{(n-m+1)\,x}{1-x} - \frac{x^{m+1} - x^{n+2}}{(1-x)^2}$$

will be found to be $n - m + 1$ if r is less than $m + 1$, and 0 if r is greater than $n + 1$, and in other cases to be $n - r + 1$.

748. Hence the probability that the duration of seven reigns will amount to just 257 years is the coefficient of x^{257} in the expansion of the seventh power of

$$\frac{(n - m + 1)\, x\, (1 - x) - x^{m+1} + x^{n+2}}{(1 - x)^2 \dfrac{m + n}{2}\, (n - m + 1)}.$$

Now Condorcet takes $n = 60$ and $m = 30$; and he says that the value of the required coefficient is ·000792, which we will assume he has calculated correctly.

Thus he has obtained the probability in the ordinary sense, which he denotes by P; he requires the *probabilité propre*. He considers there are 414 events possible, as the reigns may have any duration in years between 7 and 420. Thus the mean probability of all the other events is $\dfrac{1 - P}{413}$; and so the *probabilité propre* is $\dfrac{413P}{1 + 412P}$, or about $\dfrac{1}{4}$.

749. Condorcet says that other historians assign 140 years instead of 257 years for the duration of the reigns of the kings. He says the ordinary probability of this is ·008887, which we may denote by Q. He then makes the *probabilité propre* to be $\dfrac{412Q}{1 + 411Q}$, which is more than $\dfrac{1}{2}$.

He seems here to take 413, and not 414, as the whole number of events.

750. Condorcet then proceeds to compare three events, namely that of 257 years' duration, that of 140 years' duration, and what he calls *un autre évènement indéterminé quelconque qui auroit pu avoir lieu*. He makes the *probabilités propres* to be respectively

$$\frac{411P}{410\,(P + Q) + 1}, \quad \frac{411Q}{410\,(P + Q) + 1} \quad \text{and} \quad \frac{1 - P - Q}{410\,(P + Q) + 1},$$

which are approximately $\dfrac{3}{50}$, $\dfrac{37}{50}$, $\dfrac{10}{50}$.

Here again he seems to take 413 as the whole number of events.

He proceeds to combine these probabilities with probabilities arising from testimony borne to the first or second event.

751. Condorcet considers another statement which he finds in Roman History, namely that the augur Accius Nævius cut a stone with a razor. Condorcet takes $\frac{1}{1000000}$ as the ordinary probability, and then by Art. 745 makes the *probabilité propre* to be $\frac{2}{1000000}$.

752. We have spent a long space on Condorcet's memoir, on account of the reputation of the author; but we fear that the reader will conclude that we have given to it far more attention . than it deserves. It seems to us to be on the whole excessively arbitrary, altogether unpractical, and in parts very obscure.

753. We have in various places expressed so decidedly our opinion as to the obscurity and inutility of Condorcet's investigations that it will be just to notice the opinions which other writers have formed.

Gouraud devotes pages 89—104 of his work to Condorcet, and the following defects are noticed : Un style embarrassé, dénué de justesse et de coloris, une philosophie souvent obscure ou bizarre, une analyse que les meilleurs juges ont trouvée confuse. With this drawback Condorcet is praised in terms of such extravagant eulogy, that we are tempted to apply to Gouraud the reflexion which Dugald Stewart makes in reference to Voltaire, who he says "is so lavish and undistinguishing in his praise of Locke, as almost to justify a doubt whether he had ever read the book which he extols so highly." *Stewart's Works, edited by Hamilton*, Vol. I. page 220.

Galloway speaks of Condorcet's Essay as "a work of great ingenuity, and abounding with interesting remarks on subjects of the highest importance to humanity." Article *Probability* in the *Encyclopædia Britannica*.

Laplace in his brief sketch of the history of the subject does not name Condorcet; he refers however to the kind of questions which Condorcet considers and says, Tant de passions, d'intérêts divers et de circonstances compliquent les questions relatives à ces objets, qu'elles sont presque toujours insolubles. *Théorie...des Prob.* page CXXXVIII.

Poisson names Condorcet expressly; with respect to his *Preliminary Discourse*, he says, ... où sont développées avec soin les considérations propres à montrer l'utilité de ce genre de recherches. And after referring to some of Laplace's investigations Poisson adds, ... il est juste de dire que c'est à Condorcet qu'est due l'idée ingénieuse de faire dépendre la solution, du principe de Bayes, en considérant successivement la culpabilité et l'innocence de l'accusé, comme une cause inconnue du jugement prononcé, qui est alors le fait observé, duquel il s'agit de déduire la probabilité de cette cause. *Recherches sur la Prob.* ... page 2.

We have already referred to John Stuart Mill, see Art. 665. One sentence of his may perhaps not have been specially aimed at Condorcet, but it may well be so applied. Mr Mill says, " It is obvious, too, that even when the probabilities are derived from observation and experiment, a very slight improvement in the data, by better observations, or by taking into fuller consideration the special circumstances of the case, is of more use than the most elaborate application of the calculus to probabilities founded on the data in their previous state of inferiority." *Logic*, Vol. II. page 65. Condorcet seems really to have fancied that valuable results could be obtained from any data, however imperfect, by using formulæ with an adequate supply of signs of integration.

CHAPTER XVIII.

TREMBLEY.

754. WE have now to examine a series of memoirs by Trembley. He was born at Geneva in 1749, and died in 1811. The first memoir is entitled *Disquisitio Elementaris circa Calculum Probabilium.* This memoir is published in the *Commentationes Societatis Regiæ Scientiarum Gottingensis*, Vol. XII. The volume is for the years 1793 and 1794; and the date of publication is 1796. The memoir occupies pages 99—136 of the mathematical portion of the volume.

755. The memoir begins thus:

Plurimae extant hic et illie sparsae meditationes analyticae eirca caleulum Probabilium, quas hic reeensere non est animus. Quae cum plerumque quaestiones partieulares spectarent, summi Geometrae la Place et la Grange hanc theoriam generalius tractare sunt aggressi, auxilia derivantes ex intimis calculi integralium visceribus, et eximios quidem fructus inde perceperunt. Cum autem tota Probabilium theoria principiis simplicibus et obviis sit innixa, quae nihil aliud fere requirunt quam doctrinam eombinationum, et pleracque difficultates in enume- randis et distinguendis casibus versentur, e re visum est easdem quaes- tiones generaliores methodo elementari tractare, sine ullo alieno auxilio. Cujus tentaminis primum specimen hae paginae complectuntur, eontinent quippe solutiones elementares Problematum generaliorum quae vir illustrissimus la Grange soluta dedit in Commentariis Academiae Regiae Berolinensis pro anno 1775. Si haec Geometris non displicuerint, alias deinde ejusdem gencris dilucidationes, deo juvante ipsis proponam.

756. The intention expressed at the end of this paragraph was

carried into effect in a memoir in the next volume of the Gottingen *Commentationes.* The present memoir discusses nine problems,
most of which are to be found in De Moivre's *Doctrine of Chances.*
To this work Trembley accordingly often refers, and his references
obviously shew that he used the *second* edition of De Moivre's
work ; we shall change these references into the corresponding
references to the *third* edition.

In this and other memoirs Trembley proposes to give elementary investigations of theorems which had been previously treated
by more difficult methods ; but as we shall see he frequently leaves
his results really undemonstrated.

757. The first problem is, to find the chance that an event
shall happen *exactly b* times in *a* trials, the chance of its happening
in a single trial being *p.* Trembley obtains the well known result,

$$\frac{\lfloor a}{\lfloor b \; \lfloor a-b} \; p^b \, (1-p)^{a-b};$$ he uses the modern method ; see Art. 257.

758. The second problem is to find the chance that the event
shall happen *at least b* times. Trembley gives and demonstrates
independently *both* the formulæ to which we have already drawn
attention ; see Art. 172. He says, longum et taediosum foret has
formulas inter se comparare *a priori;* but as we have seen in
Art. 174 the comparison of the formulæ is not really difficult.

759. The third problem consists of an application of the second
problem to the *Problem of Points,* in the case of two players ; the
fourth problem is that of *Points* in the case of three players ; and
the fifth problem is that of *Points* in the case of four players. The
results coincide with those of De Moivre ; see Art. 267.

760. Trembley's next three problems are on the *Duration of
Play.* He begins with De Moivre's Problem LXV, which in effect
supposes one of the players to have an unlimited capital ; see
Arts. 307, 309. Trembley gives De Moivre's second mode of
solution, but his investigation is unsatisfactory ; for after having
found in succession the first six terms of the series in brackets, he
says Perspicua nunc est lex progressionis, and accordingly writes
down the general term of the series. Trembley thus leaves the
main difficulty quite untouched.

761. Trembley's seventh problem is De Moivre's Problem LXIV, and he gives a result equivalent to that on De Moivre's page 207; see Art. 306. But here again after investigating a few terms the main difficulty is left untouched with the words Perspicua nunc est lex progressionis. Trembley says, Eodem redit solutio Cel. la Grange, licet eacdem formulae non prodeant. This seems to imply that Lagrange's formulae take a different shape. Trembley probably refers to Lagrange's second solution which is the most completely worked out; see Art. 583.

Trembley adds in a Scholium that by the aid of this problem we can solve that which is LXVII. in De Moivre; finishing with these words, in secunda enim formula fieri debet $c = p - 1$, which appear to be quite erroneous.

762. Trembley's eighth problem is the second in Lagrange's memoir; see Art. 580 : the chance of one event is p and of another q, find the chance that in a given number of trials the first shall happen at least b times and the second at least c times. Trembley puts Lagrange's solution in a more elementary form, so as to avoid the Theory of Finite Differences.

763. Trembley's ninth problem is the last in Lagrange's memoir; see Art. 587. Trembley gives a good solution.

764. The next memoir is entitled *De Probabilitate Causarum ab effectibus oriunda.*

This memoir is published in the *Comm. Soc. Reg....Gott.* Vol. XIII. The volume is for the years 1795—1798; the date of publication is 1799. The memoir occupies pages 64—119 of the mathematical portion of the volume.

765. The memoir begins thus :

Hanc materiam pertractarunt eximii Geometrae, ac potissimum Cel. la Place in Commentariis Academiae Parisinensis. Cum autem in hujusce generis Problematibus solvendis sublimior et ardua analysis fuerit adhibita, easdem quaestiones methodo elementari ac idoneo usu doctrinae serierum aggredi operae pretium duxi. Qua ratione haec altera pars calculi Probabilium ad theoriam combinationum reduceretur, sicut et primam reduxi in dissertatione ad Regiam Societatem transmissa.

Primarias quaestiones hic breviter attingere conabor, methodo dilucidandae imprimis intentus.

766.   The first problem is the following.   A bag contains an infinite number of white balls and black balls in an unknown ratio ; p white balls and q black have been drawn out in $p + q$ drawings ; what is the chance that $m + n$ new drawings will give m white and n black balls ?

The known result is

$$\frac{\lfloor m + n}{\lfloor m \ \lfloor n} \frac{\int_0^1 x^{m+p} (1 - x)^{n+q} \, dx}{\int_0^1 x^p (1 - x)^q \, dx},$$

that is,

$$\frac{\lfloor m + n}{\lfloor m \ \lfloor n} \frac{\lfloor m + p \ \lfloor n + q \ \lfloor p + q + 1}{\lfloor p \ \lfloor q \ \lfloor m + p + n + q + 1}.$$

Trembley refers to the memoir which we have cited in Art. 551, where this result had been given by Laplace ; see also Art. 704.

Trembley obtains the result by ordinary Algebra ; the investigations are only approximate, the error being however inappreciable when the number of balls is infinite.

If each ball is *replaced* after being drawn we can obtain an *exact* solution of the problem by ordinary Algebra, as we shall see when we examine a memoir by Prevost and Lhuilier ; and of course if the number of the balls is supposed infinite it will be indifferent whether we replace each ball or not, so that we obtain indirectly an exact elementary demonstration of the important result which Trembley establishes approximately.

767.   We proceed to another problem discussed by Trembley. A bag is known to contain a very large number of balls which are white or black, the ratio being unknown. In $p + q$ drawings p white balls and q black have been drawn. Required the probability that the ratio of the white to the black lies between zero and an assigned fraction. This question Trembley proceeds to consider at great length ; he supposes p and q very large and obtains approximate results.

If the *assigned fraction* above referred to be denoted by

$\dfrac{p}{p+q} - \theta$, he obtains as the numerator of the required probability, approximately

$$\frac{\left(\dfrac{p}{p+q} - \theta\right)^{p+1}\left(\dfrac{q}{p+q} + \theta\right)^{q+1}}{(p+q)\,\theta}\left\{1 - \frac{pq + (p+q)^2\,\theta^2}{(p+q)^3\,\theta^2}\right\}.$$

The denominator would be $\dfrac{\lfloor p \; \lfloor q}{\lfloor p + q + 1}$.

Trembley refers to two places in which Laplace had given this result; they are the *Hist. de l'Acad....Paris* for 1778, page 270, and for 1783 page 445. In the *Théorie...des Prob.* Laplace does not reproduce the general formula; he confines himself to supposing $\dfrac{p}{p+q} - \theta = \dfrac{1}{2}$; see page 379 of the work.

Trembley's methods are laborious, and like many other attempts to bring high mathematical investigations into more elementary forms, would probably cost a student more trouble than if he were to set to work to enlarge his mathematical knowledge and then study the original methods.

768. Trembley follows Laplace in a numerical application relating to the births of boys and girls at Vitteaux in Bourgogne. Laplace first gave this in the *Hist. de l'Acad....Paris* for 1783, page 448; it is in the *Théorie...des Prob.* page 380. It appears that at Vitteaux in five years 212 girls were born to 203 boys. It is curious that Laplace gives no information in the latter work of a more recent date than he gave in the *Hist. de l'Acad....Paris* for 1783; it would have been interesting to know if the anomaly still continued in the births at Vitteaux.

769. We may observe that Laplace treats the problem of births as analogous to that of drawing black and white balls from a bag. So he arrives at this result; if we draw 212 black balls to 203 white balls out of a bag, the chance is about ·67 that the black balls in the bag are more numerous than the white. It is not very easy to express this result in words relating to births; Laplace says in the *Hist. de l'Acad....Paris*, la différence ·670198 sera la

probabilité qu'a Vitcaux, la possibilité-des naissances des filles est supérieure à celle des naissances des garçons ; in the *Théorie...des Prob.* he says, la supériorité de la facilité des naissances des filles, est donc indiquée par ces observations, avec une probabilité, égale à ·67. These phrases seem much better adapted to the idea to be expressed than Trembley's, Probabilitas numerum puellarum superaturum esse numerum puerorum erit = ·67141.

770. Trembley now takes the following problem. From a bag containing white balls and black balls in a large number but in an unknown ratio p white balls and q black have been drawn ; required the chance that if $2a$ more drawings are made the white balls shall not exceed the black. This problem leads to a series of which the sum cannot be found exactly. Trembley gives some investigations respecting the series which seem of no use, and of which he himself makes no application ; these are on his pages 103—105. On his page 106 he gives a rough approximate value of the sum. He says, Similem seriem refert Cel. la Place. This refers to the *Hist. de l'Acad....Paris* for 1778, page 280. But the word *similem* must not be taken too strictly, for Laplace's approximate result is not the same as Trembley's.

Laplace applies his result to estimate the probability that more boys than girls will be born in a given year. This is not repeated in the *Théorie...des Prob.*, but is in fact included in what is there given, pages 397—401, which first appeared in the *Hist. de l'Acad....Paris* for 1783, page 458.

771. Trembley now takes another of Laplace's problems, namely that discussed by Laplace in the *Mémoires...par divers Savans*, Vol. VI. page 633.

Two players, whose respective skills are unknown, play on the condition that he who first gains n games over his adversary shall take the whole stake ; at a certain stage when A wants f games and B wants h games they agree to leave off playing : required to know how the stake should be divided. Suppose it were given that the skill of A is x and that of B is $1 - x$. Then we know by Art. 172 that B ought to have the fraction $\phi(x)$ of the stake, where

$$\phi\left(x\right)=\left(1-x\right)^{m}\left\{1+m\,\frac{x}{1-x}+\frac{m\left(m-1\right)}{1\,.\,2}\,\frac{x^{2}}{\left(1-x\right)^{2}}\right.$$
$$+\frac{m\left(m-1\right)\left(m-2\right)}{1\,.\,2\,.\,3}\,\frac{x^{3}}{\left(1-x\right)^{3}}+\cdots$$
$$\left.+\frac{\lfloor m}{\lfloor h\,\lfloor f-1}\,\frac{x^{f-1}}{\left(1-x\right)^{f-1}}\right\},$$

where $m=f+h-1$.

Now if x represents A's skill the probability that in $2n-f-h$ games A would win $n-f$ and B would win $n-h$ is $x^{n-f}\left(1-x\right)^{n-h}$, disregarding a numerical coefficient which we do not want.

Hence if A wins $n-f$ games and B wins $n-h$, which is now the observed event, we infer that the chance that A's skill is x is

$$\frac{x^{n-f}\left(1-x\right)^{n-h}dx}{\int_{0}^{1}x^{n-f}\left(1-x\right)^{n-h}dx}.$$

Therefore the fraction of the stake to which B is entitled is

$$\frac{\int_{0}^{1}\phi\left(x\right)x^{n-f}\left(1-x\right)^{n-h}dx}{\int_{0}^{1}x^{n-f}\left(1-x\right)^{n-h}dx}.$$

All this involves only Laplace's ordinary theory. Now the following is Trembley's method. Consider $\phi\left(x\right)$; the first term is $\left(1-x\right)^{m}$; this represents the chance that B will win m games running on the supposition that his skill is $1-x$. If we do not know his skill *a priori* we must substitute instead of $\left(1-x\right)^{m}$ the chance that B will win m games running, computed from the observed fact that he has won $n-h$ games to A's $n-f$ games. This chance is, by Art. 766,

$$\frac{\lfloor n+f-1\,\lfloor 2n-f-h+1}{\lfloor n-h\,\lfloor 2n}=M\text{ say.}$$

Again consider the term $mx\left(1-x\right)^{m-1}$ in $\phi\left(x\right)$. This represents the chance that B will win $m-1$ games out of m, on the supposition that his skill is $1-x$. If we do not know his skill *a priori* we must substitute instead of this the chance that B will win

27

$m - 1$ games out of m, deduced from the observed fact that he has won $n - h$ games to A's $n - f$ games. This chance is, by Art. 766,

$$\frac{m(n-f+1)}{n+f-1}M.$$

It is needless to go farther, as the principle is clear. The final result is that the fraction of the stake to which B is entitled is

$$M\left\{1 + (f+h-1)\frac{n-f+1}{n+f-1} + \frac{(f+h-1)(f+h-2)}{1.2}\frac{n-f+1}{n+f-1}\frac{n-f+2}{n+f-2} + \cdots\right.$$

$$\left.\cdots + \frac{(f+h-1)\ldots(h+1)}{\lfloor f-1}\frac{(n-f+1)(n-f+2)\ldots(n-1)}{(n+f-1)(n+f-2)\ldots(n+1)}\right\}.$$

This process is the most interesting in Trembley's memoir. Laplace does not reproduce this problem in the *Théorie ... des Prob.*

772. Trembley gives some remarks to shew the connexion between his own methods and Laplace's. These amount in fact to illustrations of the use of the Integral Calculus in the summation of series.

For example he gives the result which we may write thus:

$$\frac{1}{p+1} - \frac{q}{1}\frac{t}{p+2} + \frac{q(q-1)}{1.2}\frac{t^2}{p+3} - \frac{q(q-1)(q-2)}{1.2.3}\frac{t^3}{p+4} + \cdots$$

$$\cdots + \frac{(-1)^q t^q}{p+q+1}$$

$$= \int_0^1 x^p (1-tx)^q \, dx = \frac{1}{t^{p+1}}\int_0^t x^p (1-x)^q \, dx.$$

773. Trembley remarks that problems in Probability consist of two parts; first the formulæ must be exhibited and then modes of approximate calculation found. He proposes to give one example from Laplace.

Observation indicates that the ratio of the number of boys born to the number of girls born is greater at London than at Paris.

Laplace says: Cette différence semble indiquer à Londres une plus grande facilité pour la naissance des garçons, il s'agit de déterminer combien cela est probable. See *Hist. de l'Acad.... Paris* for 1778, page 304, for 1783, page 449; and *Théorie ... des Prob.* page 381.

Trembley says,

Supponit Cel. la Place natos esse Parisiis intra certum tempus, p pueros q puellas, Londini autem intra aliud temporis spatium p' pueros q' puellas, et quaerit Probabilitatem, causam quae Parisiis producit pueros esse efficaciorem quam Londini. E supra dictis sequitur hanc Probabilitatem repraesentari per formulam

$$\frac{\iint x^p (1-x)^q x'^{p'} (1-x')^{q'} \, dx \, dx'}{\iint x^p (1-x)^q x'^{p'} (1-x')^{q'} \, dx \, dx'}.$$

Trembley then gives the limits of the integrations; in the numerator for x' from $x' = 0$ to $x' = x$, and then for x from $x = 0$ to $x = 1$; in the denominator both integrations are between 0 and 1.

Trembley considers the numerator. He expands $x'^p (1-x')^{q'}$ in powers of x' and integrates from $x' = 0$ to $x' = x$. Then he expands $x^p (1-x)^q$ and integrates from $x = 0$ to $x = 1$; he obtains a result which he transforms into another more convenient shape, which he might have obtained at once and saved a page if he had *not* expanded $x^p (1-x)^q$. Then he uses an algebraical theorem in order to effect another transformation; this theorem he does not demonstrate generally, but infers it from examining the first three cases of it; see his page 113.

We will demonstrate his final result, by another method. We have

$$\int_0^x x'^{p'} (1-x')^q \, dx' = x^{p'+1} \left\{ \frac{1}{p'+1} - \frac{q'}{1} \frac{x}{p'+2} + \frac{q'(q'-1)}{1 \cdot 2} \frac{x^2}{p'+3} - \cdots \right\}.$$

Multiply by $x^p (1-x)^q$ and integrate from $x = 0$ to $x = 1$; thus we obtain by the aid of known formulæ

$$\frac{\lfloor q \lfloor p+p'+1}{\lfloor p+p'+q+2} \left\{ \frac{1}{p'+1} - \frac{q'}{1} \frac{1}{p'+2} \frac{p+p'+2}{p+p'+q+3} \right.$$

$$\left. + \frac{q'(q'-1)}{1 \cdot 2} \frac{1}{p'+3} \frac{(p+p'+2)(p+p'+3)}{(p+p'+q+3)(p+p'+q+4)} - \cdots \right\}.$$

This result as we have said Trembley obtains, though he goes through more steps to reach it.

Suppose however that before effecting the integration with respect to x we use the following theorem

$$\frac{1}{p'+1} - \frac{q'}{1}\frac{x}{p'+2} + \frac{q'(q'-1)}{1.2}\frac{x^2}{p'+3} - \frac{q'(q'-1)(q'-2)}{1.2.3}\frac{x^3}{p'+4} + \cdots$$

$$= (1-x)^{q'}\left\{\frac{1}{p'+q'+1} + \frac{q'}{(p'+q'+1)(p'+q')}\frac{1}{1-x}\right.$$

$$+ \frac{q'(q'-1)}{(p'+q'+1)(p'+q')(p'+q'-1)}\frac{1}{(1-x)^2}$$

$$\left. + \frac{q'(q'-1)(q'-2)}{(p'+q'+1)(p'+q')(p'+q'-1)(p'+q'-2)}\frac{1}{(1-x)^3} + \cdots\right\}.$$

Then by integrating with respect to x, we obtain

$$\frac{\lfloor q+q' \ \lfloor p+p'+1}{\lfloor p+p'+q+q'+2}\left\{\frac{1}{p'+q'+1} + \frac{q'}{(p'+q'+1)(p'+q')}\frac{p+p'+q+q'+2}{q+q'}\right.$$

$$\left. + \frac{q'(q'-1)}{(p'+q'+1)(p'+q')(p'+q'-1)}\frac{(p+p'+q+q'+2)(p+p'+q+q'+1)}{(q+q')(q+q'-1)} + \cdots\right\}.$$

It is in fact the identity of these two results of the final integration which Trembley assumes from observing its truth when $q' = 1$, or 2, or 3.

With regard to the theorem we have given above we may remark that it may be obtained by examining the coefficient of x^r on the two sides; the identity of these coefficients may be established as an example of the theory of partial fractions.

774. Trembley then proceeds to an approximate summation of the series; his method is most laborious, and it would not repay the trouble of verification. He says at the end, Series hacc, quae similis est seriei quam refert Cel. la Place ... He gives no reference, but he probably has in view the *Hist. de l'Acad....Paris* for 1778, page 310.

775. We have next to consider a memoir entitled *Recherches sur une question relative au calcul des probabilités*. This memoir is published in the volume for 1794 and 1795 of the *Mémoires de*

l'Acad....Berlin; the date of publication is 1799 : the memoir occupies pages 69—108 of the mathematical portion of the volume. The problem discussed is that which we have noticed in Art. 448.

776. Trembley refers in the course of his memoir to what had been done by De Moivre, Laplace and Euler. He says,

L'analyse dont M. Euler fait usage dans ce Mémoire est très-ingénieuse et digne de ce grand géomètre, mais comme elle est un peu indirecte et qu'il ne seroit pas aisé de l'appliquer au problème général dont celui-ci n'est qu'un cas particulier, j'ai entrepris de traiter la chose directement d'après la doctrine des combinaisons, et de donner à la question toute l'étendue dont elle est susceptible.

777. The problem in the degree of generality which Trembley gives to it had already engaged the attention of De Moivre ; see Art. 293. De Moivre begins with the simpler case in his Problem XXXIX, and then briefly indicates how the more general question in his Problem XLI. is to be treated. Trembley takes the contrary order, beginning with the general question and then deducing the simpler case.

When he has obtained the results of his problem Trembley modifies them so as to obtain the results of the problem discussed by Laplace and Euler. This he does very briefly in the manner we have indicated in Art. 453.

778. Trembley gives a numerical example. Suppose that a lottery consists of 90 tickets, and that 5 are drawn at each time ; then he obtains ·74102 as the approximate value of the probability that all the numbers will have been drawn in 100 drawings. Euler had obtained the result ·7419 in the work which we have cited in Art. 456.

779. Trembley's memoir adds little to what had been given before. In fact the only novelty which it contains is the investigation of the probability that $n - 1$ kinds of faces *at least* should come up, or that $n - 2$ kinds of faces *at least*, or $n - 3$, and so on. The result is analogous to that which had been given by Euler and which we have quoted in Art. 458. Nor do Trembley's methods present any thing of importance ; they are in fact such as would naturally occur to a reader of De Moivre's book if he wished to

reverse the order which De Moivre has taken. Trembley does not
supply general demonstrations; he begins with a simple case, then
he proceeds to another which is a little more complex, and when
the law which governs the general result seems obvious he enun-,
ciates it, leaving to his readers to convince themselves that the law
is universally true.

780. Trembley notices the subject of the summation of a certain series which we have considered in Art. 460. Trembley says,
M. Euler remarque que dans ce cas la somme de la suite qui donne
la probabilité, peut s'exprimer par des produits. Cela peut se démontrer par le calcul intégral, par la méthode suivante qui est
fort simple. But in what follows in the memoir, there is no use of
the Integral Calculus, and the demonstration seems quite unsatisfactory. The result is verified when $x = 1$, 2, 3, or 4 and then is
assumed to be universally true. And these verifications themselves are unsatisfactory; for in each case r is put successively
equal to 1, 2, 3, 4, and the law which appears to hold is assumed
to hold universally.

Trembley also proposes to demonstrate that the sum of the
series is zero, if n be greater than rx. The demonstration however is of the same unsatisfactory character, and there is this additional defect. Trembley supposes successively that $n = r (x + 1)$,
$n = r (x + 2)$, $n = r (x + 3)$, and so on. But besides these cases n
may have any value *between* rx and $r (x + 1)$, or *between* $r(x + 1)$
and $r (x + 2)$, and so on. Thus, in fact, Trembley makes a most
imperfect examination of the possible cases.

781. Trembley deduces from his result a formula suitable for
approximate numerical calculation, for the case in which n and x
are large, and r small; his formula agrees with one given by Laplace in the *Hist. de l'Acad....Paris* 1783, as he himself observes.
Trembley obtains his formula by repeated use of an approximation
which he establishes by ordinary Algebraical expansion, namely

$$\left(1 - \frac{r}{n}\right)^x = e^{-\frac{rx}{n}} \left(1 - \frac{r^2 x}{2n^2}\right).$$

Trembley follows Laplace in the numerical example which
we have noticed in Art. 455. Trembley moreover finds that in

about 86927 drawings there is an even chance that all the tickets except *one* will have been drawn ; and he proceeds nearly to the end of the calculation for the case in which all the tickets except *two* are required to be drawn.

782. The next memoir is entitled *Recherches sur la mortalité de la petite vérole.*

This memoir is published in the *Mémoires de l'Acad....Berlin* for 1796 ; the date of publication is 1799 : the memoir occupies pages 17—38 of the mathematical portion of the volume.

783. This memoir is closely connected with one by Daniel Bernoulli ; see Art. 398. Its object may be described as twofold ; first, it solves the problem on the hypotheses of Daniel Bernoulli by common Algebra without the Integral Calculus ; secondly, it examines how far those hypotheses are verified by facts. The memoir is interesting and must have been valuable in a practical point of view at the date of publication.

784. Let m and n have the same signification as in Daniel Bernoulli's memoir ; see Art. 402 : that is, suppose that every year small-pox attacks 1 in n of those who have not had the disease, and that 1 in m of those who are attacked dies.

Let a_0 denote a given number of births, and suppose that $a_1, a_2, a_3, ...$ denote the number of those who are alive at the end of 1, 2, 3, ... years : then Trembley shews that the number of persons alive at the beginning of the xth year who have not had the small-pox is

$$\frac{a_x \left(1 - \frac{1}{n}\right)^x}{1 - \frac{1}{m} + \frac{1}{m}\left(1 - \frac{1}{n}\right)^x} .$$

For let b_x denote the number alive at the beginning of the xth year who have not had the small-pox, and b_{x+1} the number at the beginning of the $(x + 1)^{th}$ year. Then in the xth year small-pox attacks $\frac{b_x}{n}$ persons ; thus $b_x \left(1 - \frac{1}{n}\right)$ would be alive at the beginning of the next year without having had the small-pox if none of them died by other diseases. We must therefore find how many of these $b_x \left(1 - \dfrac{1}{n}\right)$ die of other diseases, and subtract. Now the total number who die of other diseases during the xth year is

$$a_x - a_{x+1} - \frac{b_x}{mn} \; ;$$

these die out of the number $a_x - \dfrac{b_x}{mn}$. Hence, by proportion, the number who die out of $b_x \left(1 - \dfrac{1}{n}\right)$ is

$$\frac{b_x \left(1 - \dfrac{1}{n}\right)}{a_x - \dfrac{b_x}{mn}} \left(a_x - a_{x+1} - \frac{b_x}{mn}\right).$$

Therefore $b_{x+1} = b_x \left(1 - \dfrac{1}{n}\right) - \dfrac{b_x \left(1 - \dfrac{1}{n}\right)}{a_x - \dfrac{b_x}{mn}} \left(a_x - a_{x+1} - \dfrac{b_x}{mn}\right)$

$$= \frac{b_x\, a_{x+1} \left(1 - \dfrac{1}{n}\right)}{a_x - \dfrac{b_x}{mn}}.$$

We can thus establish our result by induction; for we may shew in the manner just given that

$$b_1 = \frac{a_1 \left(1 - \dfrac{1}{n}\right)}{1 - \dfrac{1}{mn}},$$

and then universally that

$$b_x = \frac{a_x \left(1 - \dfrac{1}{n}\right)^x}{1 - \dfrac{1}{m} + \dfrac{1}{m} \left(1 - \dfrac{1}{n}\right)^x}.$$

785. We may put our result in the form

$$b_x = \frac{m a_x}{1 + (m-1) \left(1 - \dfrac{1}{n}\right)^{-x}}.$$

Now there is nothing to hinder us from supposing the intervals of time to be much shorter than a year; thus n may be a large number, and then

$$\left(1 - \frac{1}{n}\right)^{-x} = e^{\frac{x}{n}} \text{ nearly.}$$

The result thus agrees with that given by Daniel Bernoulli, see Art. 402 : for the intervals in his theory may be much shorter than a year.

786. Hitherto we have used Daniel Bernoulli's hypotheses; Trembley however proceeds to a more general hypothesis. He supposes that m and n are not constant, but vary from year to year; so that we may take m_x and n_x to denote their values for the xth year. There is no difficulty in working this hypothesis by Trembley's method; the results are of course more complicated than those obtained on Daniel Bernoulli's simpler hypotheses.

787. Trembley then compares the results he obtains on his general hypothesis with a table which had been furnished by observations at Berlin during the years 1758—1774. The comparison is effected by a rude process of approximation. The conclusions he arrives at are that n is very nearly constant for all ages, its value being somewhat less than 6; but m varies considerably, for it begins by being equal to 6, and mounts up to 120 at the eleventh year of age, then diminishes to 60 at the nineteenth year of age, and mounts up again to 133 at the twenty-fifth year of age, and then diminishes.

Trembley also compares the results he obtains on his general hypothesis with another table which had been furnished by observations at the Hague. It must be confessed that the values of m and n deduced from this set of observations differ very much from those deduced from the former set, especially the values of m. The observations at Berlin were nearly five times as numerous as those at the Hague, so that they deserved more confidence.

788. In the volume for 1804 of the *Mémoires de l'Acad....
Berlin*, which was published in 1807, there is a note by Trembley himself on the memoir which we have just examined. This note is entitled *Eclaircissement relatif au Mémoire sur la*

mortalité....&c.; it occurs on pages 80—82 of the mathematical portion of the volume.

Trembley corrects some misprints in the memoir, and he says :

Au reste, je dois avertir que la méthode d'approximation que j'ai donnée dans ce mémoire comme un essai, en attendant que des observations plus detaillées nous missent en état de procéder avec plus de régularité, que cette méthode, dis-je, ne vaut absolument rien, et je dois des excuses au public pour la lui avoir présentée.

He then shews how a more accurate calculation may be made ; and he says that he has found that the values of *n* instead of remaining nearly constant really varied enormously.

789. The next memoir is entitled *Essai sur la manière de trouver le terme général des séries récurrentes.*

This memoir is published in the volume for 1797 of the *Mémoires de l'Acad....Berlin;* the date of publication is 1800. The pages 97—105 of the memoir are devoted to the solution of a problem which had been solved by Laplace in Vol. VII. of the *Mémoires...par divers Savans;* Trembley refers to Laplace.

The problem is as follows : Suppose a solid having *n* equal faces numbered 1, 2, 3 ... *p*; required the probability that in the course of *n* throws the faces will appear in the order 1, 2, 3, ... *p*.

This problem is very nearly the same as that of De Moivre on the run of luck ; see Art. 325. Instead of the equation

$$u_{n+1} = u_n + (1 - u_{n-p})\, ba^r,$$

we shall now have

$$u_{n+1} = u_n + (1 - u_{n-p})\, a^v; \quad \text{and} \quad a = \frac{1}{p}.$$

Trembley solves the problem in his usual incomplete manner ; he discusses in succession the cases in which *p* = 2, 3, 4; and then he assumes that the law which holds in these cases will hold generally.

790. The next memoir is entitled *Observations sur les calculs relatifs à la durée des mariages et au nombre des époux subsistans.*

This memoir is published in the volume for 1799—1800 of the *Mémoires de l'Acad...Berlin;* the date of publication is 1803; the memoir occupies pages 110—130 of the mathematical portion of the volume.

791. The memoir refers to that of Daniel Bernoulli on the same subject which we have noticed in Art. 412. Trembley obtains results agreeing with those of Daniel Bernoulli so far as the latter was rigorous in his investigations; but Trembley urges objections against some of the results obtained by the use of the infinitesimal calculus, and which were only presented as approximate by Daniel Bernoulli.

792. As is usual with Trembley, the formulæ which occur are not demonstrated, but only obtained by induction from some simple cases. Thus he spends three pages in arriving at the result which we have given in Art. 410 from Daniel Bernoulli; he examines in succession the five most simple cases, for which $m = 1, 2, 3, 4, 5,$ and then infers the general formula by analogy.

793. For another example of his formulæ we take the following question. Suppose n men marry n women at the same time; if m out of the $2n$ die, required the chance that m marriages are dissolved.

We may take m pairs out of n in $\dfrac{\lfloor n}{\lfloor m \lfloor n - m}$ ways. In each of the m pairs only one person must die; this can happen in 2^m ways. Thus the whole number of cases favourable to the result is $\dfrac{2^m \lfloor n}{\lfloor m \lfloor n - m}$. But the whole number of cases is the whole number of ways in which m persons out of $2n$ may die; that is $\dfrac{2n}{\lfloor m \lfloor 2n - m}$. Hence the required chance is

$$\frac{2^m \lfloor n \lfloor 2n - m}{\lfloor 2n \lfloor n - m}.$$

Trembley spends two pages on this problem, and then does not *demonstrate* the result.

794. Trembley makes some applications of his formulæ to the subject of annuities for widows. He refers to a work by Karstens, entitled *Théorie von Wittwencassen*, Halle, 1784; and also names Tetens. On the other hand, he names Michelsen as a writer who

had represented the calculations of mathematicians on such subjects as destitute of foundation.

Trembley intimates his intention of continuing his investigations in another memoir, which I presume never appeared.

795. The next memoir is entitled *Observations sur la méthode de prendre les milieux entre les observations.*

This memoir is published in the volume for 1801 of the *Mémoires de l'Acad. ... Berlin;* the date of publication of the volume is 1804: the memoir occupies pages 29—58 of the mathematical portion of the volume.

796. The memoir commences thus:

La manière la plus avantageuse de prendre les milieux entre les observations a été détaillée par de grands géomètres. M. Daniel Bernoulli, M. Lambert, M. de la Place, M. de la Grange s'en sont occupés. Le dernier a donné là-dessus un très-beau mémoire dans le Tome v. des Mémoires de Turin. Il a employé pour cela le calcul intégral. Mon dessein dans ce mémoire est de montrer comment on peut parvenir aux mêmes résultats par un simple usage de la doctrine des combinaisons.

797. The preceding extract shews the object of the memoir. We observe however that although Lagrange does employ the *Integral* Calculus, yet it is only in the latter part of his memoir, on which Trembley does not touch; see Arts. 570—575. In the other portions of his memoir, Lagrange uses the *Differential* Calculus; but it was quite unnecessary for him to do so; see Art. 564.

Trembley's memoir appears to be of no value whatever. The method is laborious, obscure, and imperfect, while Lagrange's is simple, clear, and decisive. Trembley begins with De Moivre's problem, quoting from him; see Art. 149. He considers De Moivre's demonstration *indirect* and gives another. Trembley's demonstration occupies eight pages, and a reader would probably find it necessary to fill up many parts with more detail, if he were scrupulous about exactness.

After discussing De Moivre's problem in this manner, Trembley proceeds to inflict similar treatment on Lagrange's problems.

We may remark that Trembley copies a formula from Lagrange with all the misprints or errors which it involves; see Art. 567.

798. The last memoir by Trembley is entitled *Observations sur le calcul d'un Jeu de hasard.*

This memoir is published in the volume for 1802 of the *Mémoires de l'Acad. ... Berlin;* the date of publication is 1804: the memoir occupies pages 86—102 of the mathematical portion of the volume.

799. The game considered is that of *Her*, which gave rise to a dispute between Nicolas Bernoulli and others; see Art. 187. Trembley refers to the dispute.

Trembley investigates fully the chance of Paul for every case that can occur, and more briefly the chance of Peter. He states his conclusion thus :

...M. de Montmort et ses amis concluoient de là contre Nicolas Bernoulli, que ce cas étoit insoluble, car disoient-ils, si Paul sait que Pierre se tient au huit, il changera au sept, mais Pierre venant à savoir que Paul change au sept, changera au huit, ce qui fait un cercle vicieux. Mais il résulte seulement de là que chacun sera perpetuellement dans l'incertitude sur la manière de jouer de son adversaire; dès lors il conviendra à Paul de changer au sept dans un coup donné, mais il ne pourroit suivre constamment ce système plusieurs coups de suite. Il conviendra de même à Pierre de changer au huit dans un coup donné, sans pouvoir le faire plusieurs coups de suite, ce qui s'accorde avec les conclusions de M. Nicolas Bernoulli contre celles de M. de Montmort.

800. It is hardly correct to say that the conclusion here obtained agrees with that of Nicolas Bernoulli against that of Montmort. The opponents of Nicolas Bernoulli seem only to have asserted that it was impossible to say on which rule Paul should *uniformly act,* and this Trembley allows.

801. In Trembley's investigation of the chance of Peter, he considers this chance at the epoch *before Paul has made his choice whether he will exchange or not.* But this is of little value for Peter himself; Peter would want to know how to act under certain circumstances, and before he acted he would know whether Paul retained the card he obtained at first or compelled an ex-

change. Hence Trembley's investigation of Peter's chance differs from the method which we have exemplified in Art. 189.

802. Trembley makes an attempt to solve the problem of *Her* for three players; but his solution is quite unsound. Suppose there are three players, Paul, James, and Peter. Trembley considers that the chances of Paul and James are in the proportion of the chance of the first and second players when there are only two players; and he denotes these chances by x and y. He takes x to y as 8496 to 8079; but these numbers are of no consequence for our purpose. He supposes that the chances of James and Peter are also in the same proportion. This would not be *quite* accurate, because when James is estimating his chance with respect to Peter he would have some knowledge of Paul's card; whereas in the case of Paul and James, the former had no knowledge of any other card than his own to guide him in retaining or exchanging.

But this is only a minute point. Trembley's error is in the next step. He considers that $\dfrac{x}{x+y}$ is the chance that Paul will beat James, and that $\dfrac{y}{x+y}$ is the chance that Peter will beat James; he infers that $\dfrac{xy}{(x+y)^2}$ is the chance that both Paul and Peter will beat James, so that James will be thrown out at the first trial. This is false: the game is so constructed that the players are *nearly* on the same footing, so that $\dfrac{1}{3}$ is very nearly the chance that a given player will be excluded at the first trial. Trembley's solution would give $\dfrac{1}{4}$ as the chance that James will be excluded if $x=y$; whereas $\dfrac{1}{3}$ should then be the value.

The error arises from the fact that $\dfrac{x}{x+y}$ and $\dfrac{y}{x+y}$ do not here represent *independent* chances; of course if Paul has a higher card than James, this alone affords presumption that James will rather have a card inferior to that of Peter than superior. This error at the beginning vitiates Trembley's solution.

803. As a subsidiary part of his solution Trembley gives a tedious numerical investigation which might be easily spared. He wishes to shew that supposing James to have a higher card than both Peter and Paul, it is an even chance whether Peter or Paul is excluded. He might have proceeded thus, which will be easily intelligible to a person who reads the description of the game in *Montmort*, pages 278, 279 :

Let n denote the number of James's card.

I. Suppose $n-r$ and $n-s$ the other two cards; where r and s are positive integers and different. Then either Paul or Peter may have the lower of the two $n-r$ and $n-s$; that is, there are as many cases favourable to one as the other.

II. Peter's card may also be n; then Paul's must be 1, or 2, or 3, ... or $n-1$. Here are $n-1$ cases favourable to Peter.

III. Peter and Paul may both have a card with the same mark $n-r$; this will give $n-1$ cases favourable to Paul.

Thus II. and III. balance.

CHAPTER XIX.

MISCELLANEOUS INVESTIGATIONS

Between the Years 1780 and 1800.

804. THE present Chapter will contain notices of various contributions to our subject which were made between the years 1750 and 1780.

805. We have first to mention two memoirs by Prevost, entitled, *Sur les principes de la Théorie des gains fortuits.*

The first memoir is in the volume for 1780 of the *Nouveaux Mémoires...Berlin;* the date of publication is 1782: the memoir occupies pages 430—472. The second memoir is in the volume for 1781; the date of publication is 1783: the memoir occupies pages 463—472. Prevost professes to criticise the account of the elementary principles of the subject given by James Bernoulli, Huygens, and De Moivre. It does not seem that the memoirs present anything of value or importance; see Art. 103.

806. We have next to notice a memoir by Borda, entitled *Mémoire sur les Elections au Scrutin.*

This is in the *Hist....de l'Acad....Paris* for 1781; the date of publication is 1784: the memoir occupies pages 657—665.

This memoir is not connected with Probability, but we notice it because the subject is considered at great length by Condorcet, who refers to Borda's view; see Art. 719.

Borda observes that the ordinary mode of election is liable to error. Suppose, for example, that there are 21 voters, out of whom 8 vote for A, 7 for B, and 6 for C; then A is elected. But it is possible that the 7 who voted for B and the 6 who voted for C may agree in considering A as the worst of the three candidates, although they differ about the merits of B and C. In such a case there are 8 voters for A and 13 against him out of the 21 voters; and so Borda considers that A ought not to be elected. In fact in this case if there were only A and B as candidates, or only A and C as candidates, A would lose; he gains because he is opposed by two men who are both better than himself.

Borda suggests that each voter should arrange the candidates in what he thinks the order of merit. Then in collecting the results we may assign to a candidate a marks for each lowest place, $a + b$ marks for each next place, $a + 2b$ marks for each next place, and so on if there are more than three candidates. Suppose for example that there are three candidates, and that one of them is first in the lists of 6 voters, second in the lists of 10 voters, and third in the lists of 5 voters; then his aggregate merit is expressed by $6(a + 2b) + 10(a + b) + 5a$, that is by $21a + 22b$. It is indifferent what proportion we establish between a and b, because in the aggregate merit of each candidate the coefficient of a will be the whole number of voters.

Condorcet objects to Borda's method, and he gives the following example. Let there be three candidates, A, B, and C; and suppose 81 voters. Suppose that the order ABC is adopted by 30 voters, the order ACB by 1, the order CAB by 10, the order BAC by 29, the order BCA by 10, and the order CBA by 1. In this case B is to be elected on Borda's method, for his aggregate merit is expressed by $81a + 109b$, while that of A is expressed by $81a + 101b$, and that of C by $81a + 33b$. Condorcet decides that A ought to be elected; for the proposition A is better than B is affirmed by $30 + 1 + 10$ voters, while the proposition B is better than A is affirmed by $29 + 10 + 1$ voters, so that A has the advantage over B in the ratio of 41 to 40.

Thus suppose a voter to adopt the order ABC; then Condorcet considers him to affirm with equal emphasis the three propositions A is better than B, B is better than C, A is better than C; but

Borda considers him to affirm the first two with equal emphasis, and the last with double emphasis. See Condorcet's *Discours Préliminaire*, page CLXXVII, Laplace, *Théorie...des Prob.* page 274.

807. We have next to notice a memoir by Malfatti, entitled *Esame Critico di un Problema di probabilità del Sig. Daniele Bernoulli, e soluzione d'un altro Problema analogo al Bernulliano. Del Sig. Gio: Francesco Malfatti Professore di Matematica nell' Università di Ferrara.*

This memoir is published in the *Memorie di Matematica e Fisica della Società Italiana*, Tomo I. 1782; the memoir occupies pages 768—824. The problem is that which we have noticed in Art. 416. Malfatti considers the solution of the problem about the balls to be erroneous, and that this problem is essentially different from that about the fluids which Daniel Bernoulli used to illustrate the former; see Art. 420. Malfatti restricts himself to the case of two urns.

Malfatti in fact says that the problem ought to be solved by an exact comparison of the numbers of the various cases which can arise, and not by the use of such equations as we have given in Art. 417, which are only probably true; this of course is quite correct, but it does not invalidate Daniel Bernoulli's process for its own object.

Let us take a single case. Suppose that originally there are two white balls in A and two black balls in B; required the probable state of the urn A after x of Daniel Bernoulli's operations have been performed. Let u_x denote the probability that there are two black balls in A; v_x the probability that there is one black ball and one white one, and therefore $1 - u_x - v_x$ the probability that there are two white balls.

808. We will first give a Lemma of Malfatti's. Suppose there are $n - p$ white balls in A, and therefore p black balls; then there are $n - p$ black balls in B and p white balls. Let one of Daniel Bernoulli's operations be performed, and let us find the number of cases in which each possible event can happen. There are n^2 cases altogether, for any ball can be taken from A and any ball from B. Now there are three possible events; for after the operation A may contain $n - p + 1$ white balls, or $n - p$, or $n - p - 1$.

For the first event a black ball must be taken from A and a white ball from B; the number of cases is p^2. For the second event a black ball must be taken from A and a black one from B, or else a white one from A and a white one from B; the number of cases is $2p\,(n-p)$. For the third event a white ball must be taken from A and a black ball from B; the number of cases is $(n-p)^2$.

It is obvious that

$$n^2 = p^2 + 2p\,(n-p) + (n-p)^2$$

as should be the case.

809. Now returning to the problem in Art. 807 it will be easy to form the following equations:

$$u_{z+1} = \frac{1}{4}\,v_z,$$

$$v_{z+1} = u_z + \frac{1}{2}v_z + 1 - u_z - v_z.$$

Integrating these equations and determining the constant by the condition that $v_1 = 1$, we obtain

$$v_z = \frac{2}{3}\left\{1 - \frac{(-1)^z}{2^z}\right\}, \qquad u_z = \frac{1}{6}\left\{1 + \frac{(-1)^z}{2^{z-1}}\right\}.$$

Daniel Bernoulli's general result for the probable number of white balls in A after x trials if there were n originally would be

$$\frac{n}{2}\left\{1 - \left(\frac{n-2}{n}\right)^x\right\}.$$

Thus supposing x is infinite Daniel Bernoulli finds that the probable number is $\frac{n}{2}$. This is not inconsistent with our result; for we have when x is infinite $v_z = \frac{2}{3}$, $u_z = \frac{1}{6}$, and therefore $1 - v_z - u_z = \frac{1}{6}$, so that the case of one white ball and one black ball is the most probable.

810. Malfatti advances an objection against Daniel Bernoulli's result which seems of no weight. Daniel Bernoulli obtains as

we see $\frac{n}{2}$ for the probable number of white balls in A after an infinite number of operations. Now Malfatti makes Daniel Bernoulli's statement imply conversely that it will require an infinite number of trials before the result $\frac{n}{2}$ will probably be reached. But Daniel Bernoulli himself does not state or imply this converse, so that Malfatti is merely criticising a misapprehension of his own.

811. Malfatti himself gives a result equivalent to our value of u_x in Art. 809; he does not obtain it in the way we use, but by induction founded on examination of successive cases, and not demonstrated generally.

812. The problem which Malfatti proposes to solve and which he considers analogous to Daniel Bernoulli's is the following. Let r be zero or any given integer not greater than n: required to determine the probability that in x operations the event will never occur of having just $n-r$ white balls in A. This he treats in a most laborious way; he supposes $r = 2, 3, 4, 5$ in succession, and obtains the results. He extracts by inspection certain laws from these results which he assumes will hold for all the other values of r between 6 and n inclusive. The cases $r = 0$, and $r = 1$, require special treatment.

Thus the results are not *demonstrated*, though perhaps little doubt of their exactness would remain in the mind of a student. The patience and acuteness which must have been required to extract the laws will secure high admiration for Malfatti.

813. We will give one specimen of the results which Malfatti obtains, though we shall adopt an exact method instead of his induction from particular cases.

Required the probability that in x trials the number $n-2$ of white balls will never occur in A. Let $\phi(x, n)$ represent the whole number of favourable cases in x trials which end with n white balls in A; let $\phi(x, n-1)$ be the whole number of favourable cases which end with $n-1$ white balls in A. There is no other class of favourable cases; by favourable cases we mean cases of non-occurrence of $n-2$ white balls.

By aid of the Lemma in Art. 808 the following equations are immediately established,

$$\phi(x+1,\, n) = \phi(x,\, n-1),$$
$$\phi(x+1,\, n-1) = n^2\phi(x,\, n) + 2(n-1)\,\phi(x,\, n-1).$$

By aid of the first the second becomes

$$\phi(x+1,\, n-1) = n^2\phi(x-1,\, n-1) + 2(n-1)\,\phi(x,\, n-1).$$

Thus denoting $\phi(x,\, n-1)$ by u_x we have

$$u_{x+1} = n^2 u_{x-1} + 2(n-1)\,u_x.$$

This shews that u_x is of the form $A\alpha^x + B\beta^x$ where α and β are the roots of the quadratic

$$z^2 - 2(n-1)z - n^2 = 0.$$

From the first of the above equations we see that $\phi(x+1,\, n)$ is of the same *form* as $\phi(x,\, n-1)$; thus finally we have

$$\phi(x,\, n) + \phi(x,\, n-1) = a\alpha^x + b\beta^x,$$

where a and b are constants. The required probability is found by dividing by the whole number of cases, that is by n^{2x}. Thus we obtain

$$\frac{a\alpha^x + b\beta^x}{n^{2x}}.$$

We must determine the constants a and b by special examination of the first and second operations. After the first operation we must have $n-1$ white balls and one black ball in A; all the cases are favourable; this will give

$$a\alpha + b\beta = n^2.$$

Similarly we get

$$a\alpha^2 + b\beta^2 = n^2\{1 + 2(n-1)\};$$

for the second operation must either give n white balls in A, or $n-1$, or $n-2$; and the first and second cases are favourable.

Thus a and b become known, and the problem is completely solved.

814. We will briefly indicate the steps for the solution of the problem in which we require the probability that $n-3$ white balls shall never occur in A.

Let $\phi(x, n)$, $\phi(x, n-1)$, $\phi(x, n-2)$ represent the number of favourable cases in x trials, where the final number of white balls in A is n, $n-1$, $n-2$, respectively.

Then we have the following equations

$$\phi(x+1, n) = \phi(x, n-1),$$

$$\phi(x+1, n-1) = n^2\phi(x, n) + 2(n-1)\phi(x, n-1) + 4\phi(x, n-2),$$

$$\phi(x+1, n-2) = (n-1)^2\phi(x, n-1) + 4(n-2)\phi(x, n-2).$$

If we denote $\phi(x, n-2)$ by u_x we shall arrive by elimination at the equation

$$u_{x+3} - (6n-10)u_{x+2} + (3n^2 - 16n + 12)u_{x+1} + 4n^2(n-2)u_x = 0.$$

Then it will be seen that $\phi(x, n-1)$ and $\phi(x, n)$ will be expressions of the same form as $\phi(x, n-2)$. Thus the whole number of favourable cases will be $a\alpha^x + b\beta^x + c\gamma^x$, where a, b, c are arbitrary constants, and α, β, γ are the roots of

$$z^3 - (6n-10)z^2 + (3n^2 - 16n + 12)z + 4n^2(n-2) = 0.$$

815. A work on our subject was published by Bicquilley, entitled *Du Calcul des Probabilités. Par C. F. de Bicquilley, Garde-du-Corps du Roi.* 1783.

This work is of small octavo size, and contains a preface of three pages, the *Privilége du Roi*, and a table of contents; then 164 pages of text with a plate.

According to the Catalogues of Booksellers there is a second edition published in 1805 which I have not seen.

816. The author's object is stated in the following sentence from the Preface:

La théorie des Probabilités ébauchée par des Géomètres célèbres m'a paru susceptible d'être approfondée, et de faire partie de l'enseignement élémentaire : j'ai pensé qu'un traité ne seroit point indigne d'être offert au public, qui pourroit enrichir de nouvelles vérités cette matière intéressante, et la mettre à la portée du plus grand nombre des lecteurs.

The choice of matter seems rather unsuitable for an elementary work on the Theory of Probability.

817. Pages 1—15 contain the definitions and fundamental principles. Pages 15—25 contain an account of Figurate numbers. Pages 26—39 contain various theorems which we should now describe as examples of the Theory of Combinations. Pages 40—80 contain a number of theorems which amount to little more than easy developments of one fundamental theorem, namely that which we have given in Art. 281, supposing $p = 0$.

818. Pages 81—110 may be said to amount to the following theorem and its consequences: if the chance of an event at a single trial be p the chance that it will occur m times and fail n times in $m+n$ trials is $\dfrac{\lfloor m+n}{\lfloor m \lfloor n} p^m (1-p)^n$.

Here we may notice one problem which is of interest. Suppose that at every trial we must have either an event P alone, or an event Q alone, or both P and Q, or neither P nor Q. Let p denote the chance of P alone, q the chance of Q alone, t the chance of both P and Q: then $1 - p - q - t$ is the chance of neither P nor Q; we will denote this by u. Various problems may then be proposed; Bicquilley considers the following: required the chance that in μ trials P will happen exactly m times, and Q exactly n times.

I. Suppose P and Q do not happen together in any case. Then we have P happening m times, Q happening n times, and neither P nor Q happening $\mu - m - n$ times. The corresponding chance is

$$\frac{\lfloor \mu}{\lfloor m \lfloor n \lfloor \mu - m - n} p^m q^n u^{\mu - m - n}.$$

II. Suppose that P and Q happen together once. Then we have also P happening $m-1$ times, Q happening $n-1$ times, and neither P nor Q happening $\mu - m - n + 1$ times. The corresponding chance is

$$\frac{\lfloor \mu}{\lfloor m-1 \lfloor n-1 \lfloor \mu - m - n + 1} p^{m-1} q^{n-1} t u^{\mu - m - n + 1}.$$

III.　Suppose that P and Q happen together *twice*. The corresponding chance is

$$\frac{\lfloor \mu}{\lfloor 2 \; \lfloor m-2 \; \lfloor n-2 \; \lfloor \mu-m-n+2} \, p^{m-2} q^{n-2} t^2 w^{\mu-m-n+2}.$$

And so on.

819.　As another example of the kind of problem noticed in the preceding Article, we may require the chance that in μ trials P and Q shall each happen at least once. The required chance is

$$1-(1-p-t)^{\mu}-(1-q-t)^{\mu}+(1-p-q-t)^{\mu}.$$

See also *Algebra*, Chapter LVI.

820.　Pages 111—133 contain the solution of some examples. Two of them are borrowed from Buffon, namely those which we have noticed in Art. 649, and in the beginning of Art. 650.

One of Bicquilley's examples may be given. Suppose p and q to denote respectively the chances of the happening and failing of an event in a single trial. A player lays a wager of a to b that the event will happen; if the event does not happen he repeats the wager, making the stakes ra to rb; if the event fails again he repeats the wager, making the stakes r^2a to r^2b; and so on. If the player is allowed to do this for a series of n games, required his advantage or disadvantage.

The player's disadvantage is

$$(qa-pb)\{1+qr+q^2r^2+\ldots+q^{n-1}r^{n-1}\}.$$

This is easily shewn. For $qa-pb$ is obviously the player's disadvantage at the first trial. Suppose the event fails at the first trial, of which the chance is q; then the wager is renewed; and the disadvantage for that trial is $qar-pbr$. Similarly q^2 is the chance that the event will fail twice in succession; then the wager is renewed, and the disadvantage is qar^2-pbr^2. And so on. If then qa is greater than pb the disadvantage is positive and increases with the number of games.

Bicquilley takes the particular case in which $a=1$, and $r=\dfrac{b+1}{b}$; his solution is less simple than that which we have given. The object of the problem is to shew to a gambler, by an example, that if a wager is really unfavourable to him he suffers still more by increasing his stake while the same proportion is maintained between his stake and that of his adversary.

821. Pages 134—149 relate to the evaluation of probability from experience or observation. If an event has happened m times and failed n times the book directs us to take $\dfrac{m}{m+n}$ as its chance in a single trial.

822. Pages 150—164 relate to the evaluation of probability from testimony. Bicquilley adopts the method which we have exhibited in Art. 91. Another of his peculiarities is the following. Suppose from our own experience, independent of testimony, we assign the probability P to an event, and suppose that a witness whose probability is p offers his evidence to the event, Bicquilley takes for the resulting probability $P + (1 - P)\,Pp$, and not as we might have expected from him $P + (1 - P)\,p$. He says that the reliance which we place on a witness is proportional to our own previous estimate of the probability of the event to which he testifies.

823. We will now notice the matter bearing on our subject which is contained in the *Encyclopédie Méthodique;* the mathematical portion of this work forms three quarto volumes which are dated respectively 1784, 1785, 1789.

Absent. This article is partly due to Condorcet: he applies the Theory of Probability to determine when a man has been absent long enough to justify the division of his property among his heirs, and also to determine the portions which ought to be assigned to the different claimants.

Assurances. This article contains nothing remarkable.

Probabilité. The article from the original *Encyclopédie* is repeated : see Art. 467. This is followed by another article under the same title, which professes to give the general principles of the subject. The article has not Condorcet's signature formally attached to it ; but its last sentence shews that he was the author. It may be described as an outline of Condorcet's own writings on the subject, but from its brevity it would be far less intelligible than even those writings.

Substitutions. Condorcet maintains that a State has the authority to change the laws of succession to property; but when such changes are made the rights which existed under the old laws should be valued and compensation made for them. In this article Condorcet professes to estimate the amount of compensation. The formulæ however are printed in such an obscure and repulsive manner that it would be very difficult to determine whether they are correct; and certainly the attempt to examine them would be a waste of time and labour.

824. It should be observed that in the *Encyclopédie Méthodique* various threats are uttered which are never carried into execution. Thus in the article *Assurances* we are referred to *Évènemens* and to *Société;* and in the article *Probabilité* we are referred to *Vérité* and to *Votans.* Any person who is acquainted with Condorcet's writings will consider it fortunate that no articles are to be found under the titles here named.

825. The only important article connected with our subject in the *Encyclopédie Méthodique* is that under the title *Milieu,* which we will now proceed to notice. The article is by John Bernoulli, the same person, we presume, whom we have noticed in Arts. 598 and 624.

The article gives an account of two memoirs which it asserts had not then been printed. The article says:

Le premier mémoire dont je me propose de donner l'extrait, est un petit écrit latin de M. Daniel Bernoulli, qu'il me communiqua, en 1769, et qu'il gardoit depuis long-tems parmi ses manuscrits dans le dessein sans doute de l'étendre davantage. Il a pour titre : *Dijudicatio maximæ probabilis plurium observationum discrepantium ; atque verisimillima inductio inde formanda.*

The title is the same as that of the memoir which we have noticed in Art. 424; but this article *Milieu* gives an account of the memoir which does not correspond with what we find in the *Acta Acad....Petrop.,* so we conclude that Daniel Bernoulli modified his memoir before publishing it.

The following is the method given in the article *Milieu*. Let the numerical results of discordant observations be set off as abscissæ from a fixed point; draw ordinates to represent the probabilities of the various observations; trace a curve through the extremities of these ordinates and take the abscissa of the centre of gravity of the area of the curve as the correct value of the element sought. The probabilities are to be represented by the ordinates of a certain semi-ellipse or semicircle. The article says that to determine analytically the centre of the semicircle would be very difficult, because we arrive at an equation which is almost unmanageable; accordingly a method of approximation is proposed. First take for the centre the point corresponding to the mean of all the observations, and determine the centre of gravity of the area corresponding to the observations; take this point as a new centre of a semicircle, and repeat the operation; and so on, until the centre of gravity obtained corresponds with the centre of the respective semicircle. The magnitude of the radius of the semicircle must be assigned arbitrarily by the calculator.

This is ingenious, but of course there is no evidence that we thus obtain a result which is specially trustworthy.

The other memoir which is noticed in this article *Milieu* is that by Lagrange, published in the *Miscellanea Taurinensia;* see Art. 556. It is strange that the memoirs by Daniel Bernoulli and Lagrange should be asserted to be unprinted in 1785, when Daniel Bernoulli had published a memoir with the same title in the *Acta Acad....Petrop.* for 1777, and Lagrange's memoir was published in the *Miscellanea Taurinensia* for 1770—1773. The date of publication of the last volume is not given, but that it was prior to 1777 we may infer from a memoir by Euler; see Art. 447.

826. We will now notice the portions of the *Encyclopédie Méthodique* which relate to games of chance. The three volumes which we have mentioned in Art. 817 contain articles on various games; they do not give mathematical investigations, with a slight exception in the case of *Bassette:* see Art. 467. The commencement of the article *Breland* is amusing: *il se joue à tant de*

*personnes que l'on veut: mais il n'est beau, c'est-à-dire, très-ruineux,
qu'à trois ou cinq.*

There is however a distinct work on games, entitled *Dictionnaire des Jeux, faisant suite au Tome III. des Mathématiques.*
1792. The *Avertissement* begins thus : Comme il y a, dit Mon-
tesquieu, une infinité de choses sages qui sont menées d'une
manière très-folle, il y a aussi des folies qui sont conduites d'une
manière très-sage. The work contains 316 pages of text and
16 plates. There are no mathematical investigations, but in three
cases the numerical values of the chances are given. One of these
cases is the game of *Trente et quarante ;* but the results given are
inaccurate, as Poisson shewed in the memoir which we have cited
in Art. 358. The other two cases in which the results are given
are the games *Krabs* and *Passe-dix.*

The copy of the *Encyclopédie Méthodique* which belongs to the
Cambridge University Library includes another work on games
which is wanting in other copies that I have examined. This is
entitled *Dictionnaire des Jeux Mathématiques....*An. VII. The
advertisement states that after the publication of the Dictionary
of Games in 1792 many of the subscribers requested that this
treatise should be enlarged and made more complete. The present Dictionary is divided into two parts ; first, the *Dictionnaire
des Jeux Mathématiques,* which occupies 212 pages ; secondly, a
Dictionnaire de Jeux familiers, which is unfinished, for it extends
only from *A* to *Grammairien,* occupying 80 pages.

The *Dictionnaire des Jeux Mathématiques* does not contain
any thing new or important in the calculation of chances. The
investigations which are given are chiefly taken from Montmort,
in some cases with a reference to him, but more often without.
Under the title *Joueur* we have the names of some writers on the
subject, and we find a very faint commendation of Montmort to
whose work the Dictionary is largely indebted :

Plusieurs auteurs se sont exercés sur l'analyse des jeux ; on en a un
traité élémentaire de Huygens ; on en a un plus profond de Moivre ;
on a des morceaux très-savans de Bernoulli sur cette matière. Il y a
un analyse des jeux de hasard par Montmaur, qui n'est pas sans mérite.

The game of *Draughts* obtains 16 pages, and the game of *Chess*

73 pages. Under the title *Cartes (jeu de)* we have the problem which we noticed in Art. 533, omitting however the part which is false.

Under the title *Whisk ou Wisth* we have 8 pages, beginning thus :

Jeu de cartes mi-parti de hasard et de science. Il a été inventé par les Anglais, et continue depuis long tems d'être en vogue dans la Grand-Bretagne.

C'est de tous les jeux de cartes le plus judicieux dans ses principes, le plus convenable à la société, le plus difficile, le plus intéressant, le plus piquant, et celui qui est combiné avec le plus d'art.

The article quotes some of the results obtained by De Moivre in his calculations of the chances of this game : it also refers to Hoyle's work, which it says was translated into French in 1770.

With respect to the *Dictionnaire de Jeux familiers* we need only say that it comprises descriptions of the most trifling games which serve for the amusement of children ; it begins with *J'aime mon amant par A*, and it includes *Colin-Maillard*.

827. We next advert to a memoir by D'Anieres, entitled *Réflexions sur les Jeux de hazard.*

This memoir is published in the volume of the *Nouveaux Mémoires de l'Acad....Berlin* for 1784 ; the date of publication is 1786 ; the memoir occupies pages 391—398 of the volume.

The memoir is not mathematical ; it alludes to the fact that games of hazard are prohibited by governments, and shews that there are different kinds of such games, namely, those in which a man may ruin his fortune, and those which cannot produce more than a trifling loss in any case.

There is a memoir by the same author, entitled *Sur les Paris,* in the volume of the *Nouveaux Mémoires de l'Acad....Berlin* for 1786 ; the date of publication is 1788 : the memoir occupies pages 273—278 of the volume.

This memoir is intended as a supplement to the former by the same author, and is also quite unconnected with the mathematical Theory of Probability.

828. We have now to notice a curious work, entitled *On the*

Principles of translating Algebraic quantities into probable rela-
tions and annuities, &c.   By E. Waring, M.D. Lucasian Professor
of Mathematics at Cambridge, and Fellow of the Royal Societies
of London, Bononia and Gottingen.   Cambridge, Printed by J. Arch-
deacon, Printer to the University; For J. Nicholson, Bookseller, in
Cambridge. 1792.

This is an octavo pamphlet.   Besides the leaf on which the
title is printed there are 59 pages of text, and then a page with
a few *corrigenda.*   The work is excessively scarce; for the use
of a copy I am indebted to the authorities of Queens' College,
Cambridge.

829.   The author and the printer seem to have combined their
efforts in order to render the work as obscure and repulsive as
possible; and they have attained a fair measure of success.   The
title is singularly inaccurate; it is absurd to pretend to translate
algebraical quantities into probable relations or into annuities.
What Waring means is that algebraical identities may be translated so as to afford propositions in the Theory of Probabilities or
in the Theory of Annuities.

830.   Waring begins with a Lemma.   He proposes to sum the
series

$$1 + 2^{z-1}r + 3^{z-1}r^2 + 4^{z-1}r^3 + 5^{z-1}r^4 + \dots \text{ in infinitum.}$$

The sum will be

$$\frac{A + Br + Cr^2 + Dr^3 + \dots + r^{z-2}}{(1-r)^z}.$$

The coefficients A, B, $C \dots$ are independent of r; they must
be determined by multiplying up and equating coefficients.   Thus

$$A = 1,$$
$$B = 2^{z-1} - z,$$
$$C = 3^{z-1} - z2^{z-1} + \frac{z(z-1)}{2},$$
$$D = 4^{z-1} - z3^{z-1} + \frac{z(z-1)}{2} 2^{z-1} - \frac{z(z-1)(z-2)}{2.3}.$$

Proceeding in this way we shall find that in the numerator of
the fraction which represents the sum the last term is r^{z-2}; that

is there is no power of r higher than this power, and the coefficient of this power is unity. Waring refers to another work by himself for the demonstration; the student will see that it may be deduced from the elementary theorem in Finite Differences respecting the value of $\Delta^n x^m$, when n is not less than m.

Waring does not apply his Lemma until he comes to the part of the work which relates to Annuities, which forms his pages 27—59.

831. Waring now proceeds to his propositions in the Theory of Probabilities; one of his examples will suffice to indicate his method.

It is identically true that $\dfrac{a}{N}\dfrac{N-a}{N}=\dfrac{a}{N}-\dfrac{a^2}{N^2}$. Suppose $\dfrac{a}{N}$ to represent the chance of the happening of an assigned event in one trial, and therefore $\dfrac{N-a}{N}$ the chance of its failing: then the identity shews that the chance of the happening of the event in the first trial and its failing in the second trial is equal to the difference between the chance of the happening of the event once and the chance of its happening twice in succession.

832. There is nothing of any importance in the work respecting the Theory of Probability until we come to page 19. Here Waring says,

Let the chances of the events A and B happening be respectively $\dfrac{a}{a+b}$ and $\dfrac{b}{a+b}$; then the chance of the event A happening r times more than B in r trials will be $\dfrac{a^r}{(a+b)^r}$;

in $r+2$ trials will be

$$\frac{a^r}{(a+b)^r}\left\{1+r\,\frac{ab}{(a+b)^2}\right\};$$

in $r+4$ trials will be

$$\frac{a^r}{(a+b)^r}\left\{1+r\,\frac{ab}{(a+b)^2}+\frac{r\,(r+3)}{2}\,\frac{a^2b^2}{(a+b)^4}\right\},$$

and in general it will be

$$\frac{a^r}{(a+b)^r}\left\{1 + r\,\frac{ab}{(a+b)^2} + \frac{r\,(r+3)}{2}\,\frac{a^2b^2}{(a+b)^4} + \frac{r\,(r+4)\,(r+5)}{\underline{|3}}\,\frac{a^3b^3}{(a+b)^6}\right.$$

$$+ \ldots\ldots + \frac{r\,(r+l+1)\,(r+l+2)\ldots(r+2l-1)}{\underline{|l}}\,\frac{a^l b^l}{(a+b)^{2l}} + \ldots\ldots\left.\text{in infinitum}\right\}.$$

This may be deduced from the subsequent arithmetical theorem, viz.

$$\frac{2m\,(2m-1)\,(2m-2)\ldots(2m-s)}{\underline{|s+1}} + r\,\frac{(2m-2)(2m-3)\ldots(2m-s-1)}{\underline{|s}}$$

$$+ \frac{r\,(r+3)}{2}\,\frac{(2m-4)\,(2m-5)\ldots(2m-s-2)}{\underline{|s-1}}$$

$$+ \frac{r\,(r+4)\,(r+5)}{\underline{|3}}\,\frac{(2m-6)\ldots(2m-s-3)}{\underline{|s-2}}$$

$$+ \ldots$$

$$+ \frac{r\,(r+s+2)\,(r+s+3)\ldots(r+2s+1)}{\underline{|s+1}},$$

$$= \frac{(r+2m)\,(r+2m-1)\,\ldots\,(r+2m-s)}{\underline{|s+1}}.$$

Waring's words, "*A happening r times more than B*" are scarcely adequate to convey his meaning. We see from the formula he gives that he really means to take the problem of the Duration of Play in the case where B has a capital r and A has unlimited capital. See Art. 309.

Waring gives no hint as to the demonstration of his *arithmetical theorem*. We may demonstrate it thus : take the formula in Art. 584, suppose $a = 1 + z$, $p = 1$, $q = z$; we shall find that

$$\beta = \frac{1 + z - (1 - z)}{2z} = 1.$$

Thus we get

$$1 = \frac{1}{(1+z)^t} + t\,\frac{z}{(1+z)^{t+2}} + \frac{t\,(t+3)}{2}\,\frac{z^2}{(1+z)^{t+4}}$$

$$+ \frac{t\,(t+4)\,(t+5)}{\underline{|3}}\,\frac{z^3}{(1+z)^{t+6}}$$

$$+ \frac{t\,(t+5)\,(t+6)\,(t+7)}{\underline{|4}}\,\frac{z^4}{(1+z)^{t+8}} + \ldots.$$

Multiply both sides by $(1+z)^{2n+t}$: thus

$$(1+z)^{2n+t} = (1+z)^{2n} + tz(1+z)^{2n-2} + \frac{t(t+3)}{2} z^2 (1+z)^{2n-4}$$

$$+ \frac{t(t+4)(t+5)}{\lfloor 3} z^3 (1+z)^{2n-6} + \dots$$

If we expand the various powers of $1+z$ and equate the coefficients of z^t we shall obtain the *arithmetical theorem* with t in place of r.

But it is not obvious how Waring intended to deduce the theorem on the Duration of Play from this *arithmetical theorem*. If we put $\frac{b}{a}$ for z we obtain

$$(a+b)^{2n+t} = a^t(a+b)^{2n} + ta^t(a+b)^{2n-2} ab + \frac{t(t+3)}{2} a^t(a+b)^{2n-4} a^2 b^2$$

$$+ \frac{t(t+4)(t+5)}{\lfloor 3} a^t(a+b)^{2n-6} a^3 b^3 + \dots$$

and it was perhaps from this result that Waring considered that the theorem on the Duration of Play might be deduced; but it seems difficult to render the process rigidly strict.

833. Waring gives another problem on the Duration of Play; see his page 20.

If it be required to find the chance of A's succeeding n times as oft as B's precisely: in $n+1$ trials it will be found

$$(n+1) \frac{a^n b}{(a+b)^{n+1}} = P;$$

in $2n+2$ trials it will be found

$$P + n(n+1) \frac{a^{2n} b^2}{(a+b)^{2n+2}} = Q;$$

in $3n+3$ it will be

$$Q + \frac{n(n+1)(3n+1)}{2} \frac{a^{3n} b^3}{(a+b)^{3n+3}}.$$

Waring does not give the investigation; as usual with him until we make the investigation we do not feel quite certain of the meaning of his problem.

The first of his three examples is obvious.

29

In the second example we observe that the event *may* occur in the first $n + 1$ trials, and the chance of this is P; or the event may have failed in the first $n + 1$ trials and yet may occur if we proceed to $n + 1$ more trials. This second case may occur in the following ways : B may happen twice in the first $n + 1$ trials, or twice in the second $n + 1$ trials; while A happens in the remaining $2n$ trials. Thus we obtain

$$2 \frac{(n+1)\,n}{2} \frac{a^{2n}b^2}{(a+b)^{2n+2}},$$

which must be added to P to give the chance in the second example.

In the third example we observe that the event may occur in the first $2n + 2$ trials, and the chance of this is Q; or the event may have failed in the first $2n + 2$ trials, and yet may occur if we proceed to $n + 1$ more trials. This second case may occur in the following ways :

B may happen three times in the *first* $n + 1$ trials, or three times in the *second* $n + 1$ trials, or *three* times in the last $n + 1$ trials; while A happens in the remaining $3n$ trials.

Or B may happen twice in the first $n + 1$ trials and once in the second $n + 1$ trials, or once in the second $n + 1$ trials and twice in the third $n + 1$ trials; while A happens in the remaining $3n$ trials.

Thus we obtain

$$\left\{ 3 \frac{(n+1)\,n\,(n-1)}{\lfloor 3} + 2 \frac{(n+1)^2\,n}{2} \right\} \frac{a^{3n}b^3}{(a+b)^{3n+3}},$$

which must be added to Q to give the chance in the third example.

834. The following specimen may be given of Waring's imperfect enunciations; see his page 21 :

Let a, b, c, d, &c. be the respective chances of the happening of a, β, γ, δ, &c.: in one trial, and

$$(ax^a + bx^\beta + cx^\gamma + dx^\delta + \&c.)^n = a^n x^{na} + \ldots + N x^\pi + \&c.;$$

then will N be the chance of the happening of π in n trials.

Nothing is said as to what π means. The student will see that the only meaning which can be given to the enunciation is to suppose that a, b, c, d, ... are the chances that the numbers σ, β, γ, δ, ... respectively will occur in one trial; and then N is the chance that in n trials the sum of the numbers will be π.

835. Waring gives on his page 22 the theorem which we now sometimes call by the name of Vandermonde. The theorem is that

$$(a+b)(a+b-1)\dots(a+b-n+1)$$
$$= a(a-1)\dots(a-n+1)$$
$$+ na(a-1)\dots(a-n+2)b$$
$$+ \frac{n(n-1)}{1.2}a(a-1)\dots(a-n+3)b(b-1)$$
$$+ \frac{n(n-1)(n-2)}{1.2.3}a(a-1)\dots(a-n+4)b(b-1)(b-2)$$
$$+ \dots\dots\dots$$
$$+ b(b-1)\dots(b-n+1).$$

From this he deduces a corollary which we will give in our own notation. Let $\phi(x, y)$ denote the sum of the products that can be made from the numbers 1. 2, 3, ... x, taken y together. Then will

$$\frac{\lfloor s}{\lfloor r \rfloor\lfloor s-r}\,\phi(n-1, n-s)$$
$$= \frac{\lfloor n}{\lfloor r \rfloor\lfloor n-r}\,\phi(n-r-1, n-s)$$
$$+ \frac{\lfloor n}{\lfloor r+1 \rfloor\lfloor n-r-1}\,\phi(n-r-2, n-s-1)\,\phi(r, 1)$$
$$+ \frac{\lfloor n}{\lfloor r+2 \rfloor\lfloor n-r-2}\,\phi(n-r-3, n-s-2)\,\phi(r+1, 2)$$
$$+ \frac{\lfloor n}{\lfloor r+3 \rfloor\lfloor n-r-3}\,\phi(n-r-4, n-s-3)\,\phi(r+2, 3)$$
$$+ \dots\dots\dots$$

It must be observed that s is to be less than n, and r less than s; and the terms on the right-hand side are to continue until we arrive at a term of the form $\phi(x, 0)$, and this must be replaced by unity.

This result is obtained by equating the coefficients of the term $a^{s-r}b^r$ in the two members of Vandermonde's identity.

The result is enunciated and printed so badly in Waring's work that some difficulty arose in settling what the result was and how it had been obtained.

836. I do not enter on that part of Waring's work which relates to annuities. I am informed by Professor De Morgan that the late Francis Baily mentions in a letter the following as the interesting parts of the work :—the series $S - mS' + \dfrac{m(m+1)}{2} S'' - ....$, the Problem III, and the observations on assurances payable *immediately* at death.

837. Another work by Waring requires a short notice; it is entitled *An essay on the principles of human knowledge. Cambridge* 1794. This is an octavo volume; it contains the title-leaf, then 240 pages, then 3 pages of *Addenda*, and a page containing *Corrigenda*.

838. This work contains on pages 35—40 a few common theorems of probability; the first two pages of the Addenda briefly notice the problem discussed by De Moivre and others about a series of letters being in their proper places; see Art. 281, and De Moivre Prob. XXXV. Waring remarks that if the number of letters is infinite the chance that they will occur all in their right places is infinitesimal. He gives page 49 of his work as that on which this remark bears, but it would seem that 49 is a misprint for 41.

839. Two extracts may be given from this book.

I know that some mathematicians of the first class have endeavoured to demonstrate the degree of probability of an event's happening n times from its having happened m preceding times; and consequently that such an event will probably take place; but, alas, the problem far exceeds the extent of human understanding: who can determine the time when the sun will probably cease to run its present course ? Page 35.

...I have myself wrote on most subjects in pure mathematics, and in these books inserted nearly all the inventions of the moderns with which I was acquainted.

In my prefaces I have given an history of the inventions of the different writers, and ascribed them to their respective authors ; and likewise some account of my own. To every one of these sciences I have been able to make some additions, and in the whole, if I am not mistaken in enumerating them, somewhere between three and four hundred new propositions of one kind or other, considerably more than have been given by any English writer ; and in novelty and difficulty not inferior ; I wish I could subjoin in utility : many more might have been added, but I never could hear of any reader in England out of Cambridge, who took the pains to read and understand what I have written. Page 115.

Waring proceeds to console himself under this neglect in England by the honour conferred on him by D'Alembert, Euler and Le Grange.

Dugald Stewart makes a remark relating to Waring; see his *Works edited by Hamilton*, Vol. IV. page 218.

840. A memoir by Ancillon, entitled *Doutes sur les bases du calcul des probabilités*, was published in the volume for 1794 and 1795 of the *Mémoires de l'Acad....Berlin;* the memoir occupies pages 3—32 of the part of the volume which is devoted to speculative philosophy.

The memoir contains no mathematical investigations; its object is to throw doubts on the possibility of constructing a Theory of Probability, and it is of very little value. The author seems to have determined that no Theory of Probability *could* be constructed without giving any attention to the Theory which *had* been constructed. He names Moses Mendelsohn and Garve as having already examined the question of the admissibility of such a Theory.

841. There are three memoirs written by Prevost and Lhuilier in conjunction and published in the volume for 1796 of the *Mémoires de l'Acad....Berlin.* The date of publication is 1799.

842. The first memoir is entitled *Sur les Probabilités;* it was read Nov. 12, 1795. It occupies pages 117—142 of the mathematical portion of the volume.

843. The memoir is devoted to the following problem. An urn contains m balls some of which are white and the rest black, but the number of each is unknown. Suppose that p white balls and q black balls have been drawn and *not* replaced; required the probability that out of the next $r + s$ drawings r shall give white balls and s black balls.

The possible hypotheses as to the original state of the urn are, that there were q black balls, or $q + 1$ black balls, or $q + 2, \ldots$ or $m - p$. Now form the probability of these various hypotheses according to the usual principles. Let

$$P_n = (m - q - n + 1)(m - q - n) \ldots\ldots \text{ to } p \text{ factors,}$$

$$Q_n = (q + n - 1)(q + n - 2) \ldots\ldots\ldots \text{ to } q \text{ factors;}$$

then the probability of the nth hypothesis is

$$\frac{P_n Q_n}{\Sigma},$$

where Σ denotes the sum of all such products as $P_n Q_n$. Now if this hypothesis were certainly true the chance of drawing r white balls and s black balls in the next $r + s$ drawings would be

$$\frac{R_n S_n}{\underline{|r}\ \underline{|s}\ N},$$

where

$$R_n = (m - q - p - n + 1)(m - q - p - n)\ldots\ldots\ldots \text{ to } r \text{ factors,}$$

$$S_n = (n - 1)(n - 2)\ldots\ldots\ldots\ldots\ldots\ldots\ldots \text{ to } s \text{ factors,}$$

$N =$ number of combinations of $m - p - q$ things $r + s$ at a time.

Thus the whole required probability is the sum of all the terms of which the type is

$$\frac{P_n Q_n R_n S_n}{\Sigma\ \underline{|r}\ \underline{|s}\ N}.$$

We have first to find Σ. The method of induction is adopted in the original memoir; we may however readily obtain Σ by the aid of the binomial theorem : see *Algebra*, Chapter L. Thus we shall find

$$\Sigma = \frac{\underline{|p}\ \underline{|q}}{\underline{|p + q + 1}} \frac{\underline{|m + 1}}{\underline{|m - p - q}}.$$

Now $P_n R_n$ differs from P_n only in having $p+r$ instead of p; and $Q_n S_n$ differs from Q_n only in having $q+s$ instead of q. Therefore the sum of all the terms of the form $P_n Q_n R_n S_n$ is

$$\frac{\lfloor p+r \;\lfloor q+s}{\lfloor p+q+r+s+1}\;\frac{\lfloor m+1}{\lfloor m-p-q-r-s}.$$

And $N = \dfrac{\lfloor m-p-q}{\lfloor r+s \;\lfloor m-p-q-r-s}.$

Thus finally the required probability is

$$\frac{\lfloor r+s}{\lfloor r \;\lfloor s}\;\frac{\lfloor p+r \;\lfloor q+s}{\lfloor p \;\lfloor q}\;\frac{\lfloor p+q+1}{\lfloor p+q+r+s+1}.$$

844. Let us suppose that r and s vary while their sum $r+s$ remains constant; then we can apply the preceding general result to $r+s+1$ different cases; namely the case in which *all* the $r+s$ drawings are to give white balls, or all but one, or all but two, and so on, down to the case in which none are white. The sum of these probabilities *ought to be unity*, which is a test of the accuracy of the result. This verification is given in the original memoir, by the aid of a theorem which is proved by induction. No new theorem however is required, for we have only to apply again the formula by which we found Σ in the preceding Article. The variable part of the result of the preceding Article is

$$\frac{\lfloor p+r \;\lfloor q+s}{\lfloor r \;\lfloor s},$$

that is the product of the following two expressions,

$(r+1)(r+2)\ldots\ldots p$ factors,

$(s+1)(s+2)\ldots\ldots q$ factors.

The sum of such products then is to be found supposing $r+s$ constant; and this is

$$\frac{\lfloor p \;\lfloor q}{\lfloor p+q+1}\;\frac{\lfloor p+q+r+s+1}{\lfloor r+s}.$$

Hence the required result, unity, is obtained by multiplying this expression by the constant part of the result in the preceding Article.

This result had been noticed by Condorcet; see page 189 of the *Essai... de l'Analyse...*

845. Out of the $r + s + 1$ cases considered in the preceding Article, suppose we ask which has the greatest probability? This question is answered in the memoir approximately thus. A quantity when approaching its maximum value varies slowly; thus we have to find when the result at the end of Article 843 remains nearly unchanged if we put $r - 1$ for r and $s + 1$ for s. This leads to

$$\frac{p+r}{r} = \frac{q+s+1}{s+1}, \text{ nearly};$$

therefore $\dfrac{p}{r} = \dfrac{q}{s+1}$ nearly.

Thus if r and s are large we have $\dfrac{r}{s} = \dfrac{p}{q}$ nearly.

846. It will be observed that the expression at the end of Art. 843 is independent of m the number of balls originally contained in the urn; the memoir notices this and draws attention to the fact that this is not the case if each ball *is replaced in the urn* after it has been drawn. It is stated that another memoir will be given, which will consider this form of the problem when the number of balls is supposed infinite; but it does not seem that this intention was carried into effect.

847. It will be instructive to make the comparison between the two problems which we may presume would have formed the substance of the projected memoir. Suppose that p white balls have been drawn and q black balls, and not replaced; and suppose the whole number of balls to be infinite: then by Art. 704 the probability that the next $r + s$ drawings will give r white balls and s black balls is

$$\frac{\lfloor r+s}{\lfloor r \, \lfloor s}\frac{\displaystyle\int_0^1 x^{p+r}(1-x)^{q+s}\,dx}{\displaystyle\int_0^1 x^p(1-x)^q\,dx};$$

and on effecting the integration we obtain the same result as in

Art. 843. The coincidence of the results obtained on the two different hypotheses is remarkable.

848. Suppose that $r = 1$ and $s = 0$ in the result of Art. 843; we thus obtain

$$\frac{p+1}{p+q+2}.$$

Again suppose $r = 2$ and $s = 0$; we thus obtain

$$\frac{(p+1)(p+2)}{(p+q+2)(p+q+3)}.$$

The factor $\frac{p+1}{p+q+2}$ is, as we have just seen, the probability of drawing another white ball after drawing p white balls and q black balls; the factor $\frac{p+2}{p+q+3}$ expresses in like manner the probability of drawing another white ball after drawing $p+1$ white balls and q black balls: thus the formula makes the probability of drawing two white balls in succession equal to the product of the probability of drawing the first into the probability of drawing the second, as should be the case. This property of the formula holds generally.

849. The memoir which we have now examined contains the first discussion of the problem to which it relates, namely, the problem in which the balls are *not* replaced. A particular case of the problem is considered by Bishop Terrot in the *Transactions of the Royal Society of Edinburgh*, Vol. XX.

850. The other two memoirs to which we have referred in Art. 841 are less distinctly mathematical, and they are accordingly printed in the portion of the volume which is devoted to speculative philosophy. The second memoir occupies pages 3—24, and the third memoir pages 25—41. A note relating to a passage of the third memoir, by the authors of the memoir, is given in the volume for 1797 of the *Mémoires de l'Acad....Berlin*, page 152.

851. The second memoir is entitled *Sur l'art d'estimer la probabilité des causes par les effets*. It consists of two sections. The first section discusses the general principle by which the

probabilities of causes are estimated. The principle is quoted as given by Laplace in the *Mémoires...par divers Savans*, Vol. VI.: Si un événement peut être produit par un nombre n de causes différentes, les probabilités de l'existence de ces causes prises de l'événement, sont entre elles comme les probabilités de l'événement prises de ces causes. The memoir considers it useful and necessary to demonstrate this principle; and accordingly deduces it from a simple hypothesis on which it is conceived that the whole subject rests. Some remarks made by Condorcet are criticised; and it is asserted that our persuasion of the constancy of the laws of nature is not of the same kind as that which is represented by a fraction in the Theory of Probability. See Dugald Stewart's *Works edited by Hamilton*, Vol. I. pages 421, 616.

The second section of the memoir applies Laplace's principle to some easy examples of the following kind. A die has a certain number of faces; the markings on these faces are not known, but it is observed that out of $p+q$ throws p have given ace and q not-ace. Find the probability that there is a certain number of faces marked ace. Also find the probability that in $p'+q'$ more throws there will be p' aces and q' not-aces.

It is shewn that the result in the last case is

$$\frac{\Sigma\, m^{p+p'}\,(n-m)^{q+q'}}{n^{p'+q'}\,\Sigma\, m^p\,(n-m)^q},$$

where Σ denotes a summation taken with respect to m from $m=1$ to $m=n$; and n is the whole number of faces. This is the result if the aces and not-aces are to come in a prescribed order; if they are not we must multiply by $\dfrac{\lfloor p'+q'}{\lfloor p' \lfloor q'}$.

The memoir *states* without demonstration what the approximate result is when n is supposed very great; namely, for the case in which the order is prescribed,

$$\frac{\lfloor q+q'}{\lfloor q}\,\frac{\lfloor p+p'}{\lfloor p}\,\frac{\lfloor p+q+1}{\lfloor p+q+p'+q'+1}.$$

852. The third memoir is entitled *Remarques sur l'utilité et l'étendue du principe par lequel on estime la probabilité des causes.* This memoir also relates to the principle which we have quoted

in Art. 851 from Laplace. The memoir is divided into four sections.

853. The first section is on the *utility* of the principle. It is asserted that before the epoch when this principle was laid down many errors had occurred in the writers on Probability.

The following paragraph is given :

Dans l'appréciation de la valeur du témoignage de deux témoins simultanés, il paroit que, jusqu'à LAMBERT, on n'a point usé d'un autre artifice, que de prendre le complément de la formule employée pour le témoignage successif. On suivoit à cet égard la trace de l'appréciation des argumens conspirans, telle que l'avoit faite Jac. Bernoulli. Si l'on avoit connu la vraie méthode de l'estimation des causes, on n'auroit pas manqué d'examiner avant tout si ce cas s'y rapportoit ; et l'on auroit vu que l'accord entre les témoins est un événement postérieur à la cause quelconque qui a déterminé les dépositions : en sorte qu'il s'agit ici d'estimer la cause par l'effet. On seroit ainsi retombé tout naturellement et sans effort dans la méthode que Lambert a trouvée par un effet de cette sagacité rare qui caractérisoit son génie.

854. The authors of the memoir illustrate this section by quoting from a French translation, published in Paris in 1786, of a work by Haygarth on the small-pox. Haygarth obtained from a mathematical friend the following remark. Assuming that out of twenty persons exposed to the contagion of the small-pox only one escapes, then, however violent the small-pox may be in a town if an infant has not taken the disease we may infer that it is 19 to 1 that he has not been exposed to the contagion ; if *two* in a family have escaped the probability that both have not been exposed to the contagion is more than 400 to 1 ; if *three* it is more than 8000 to 1.

With respect to this statement the memoir says that M. de la Roche the French translator has shewn that it is wrong by a judicious discussion. The end of the translator's note is quoted ; the chief part of this quotation is the following sentence :

Si l'on a observé que sur vingt personnes qui pontent à une table de pharaon il y en a dix-neuf qui se ruinent, on ne pourra pas en déduire qu'il y a un à parier contre dix-neuf que tout homme dont la fortune n'est pas dérangée, n'a pas pouté au pharaon, ni qu'il y ait dix-neuf à parier contre un, que cet homme est un joueur.

This would be absurd, M. de la Roche says, and he asserts that the reasoning given by Haygarth's friend is equally absurd. We may remark that there must be some mistake in this note; he has put 19 to 1 for 1 to 19, and vice versâ. And it is difficult to see how Prevost and Lhuilier can commend this note; for M. de la Roche argues that the reasoning of Haygarth's friend is entirely absurd, while they only find it slightly inaccurate. For Prevost and Lhuilier proceed to calculate the chances according to Laplace's principle; and they find them to be $\frac{20}{21}$, $\frac{400}{401}$, $\frac{8000}{8001}$, which, as they say, are nearly the same as the results obtained by Haygarth's friend.

855. The second section is on the *extent* of the principle. The memoir asserts that we have a conviction of the constancy of the laws of nature, and that we rely on this constancy in our application of the Theory of Probability; and thus we reason in a vicious circle if we pretend to apply the principle to questions respecting the constancy of such laws.

856. The third section is devoted to the comparison of some results of the Theory of Probability with common sense notions. In the formula at the end of Art. 843 suppose $s = 0$; the formula reduces to

$$\frac{(p+1)(p+2)\dots(p+r)}{(p+q+2)(p+q+3)\dots(p+q+r+1)},$$

it is this result of which particular cases are considered in the third section. The cases are such as according to the memoir lead to conclusions coincident with the notions of common sense; in one case however this is not immediately obvious, and the memoir says, Ceci donne l'explication d'une espèce de paradoxe remarqué (sans l'expliquer) par M. De La Place; and a reference is given to *Ecoles normales, 6ième cahier*. We will give this case. Nothing is known *à priori* respecting a certain die; it is observed on trial that in five throws ace occurs twice and not-ace three times; find the probability that the next four throws will all give ace. Here

$p=2$, $q=3$, $r=4$; the above result becomes $\dfrac{3 \cdot 4 \cdot 5 \cdot 6}{7 \cdot 8 \cdot 9 \cdot 10}$, that is $\dfrac{1}{14}$.

If we knew à priori that the die had as many faces ace as not-ace we should have $\dfrac{1}{2^4}$, that is $\dfrac{1}{16}$, for the required chance. The paradox is that $\dfrac{1}{14}$ is greater than $\dfrac{1}{16}$; while the fact that we have had only two aces out of five throws suggests that we ought to have a smaller chance for obtaining four consecutive aces, than we should have if we knew that the die had the same number of faces ace as not-ace. We need not give the explanation of the paradox, as it will be found in connexion with a similar example in Laplace, *Théorie...des Prob.* page CVI.

857. The fourth section gives some mathematical developments. The following is the substance. Suppose n dice, each having r faces; and let the number of faces which are marked ace be m', m'', m''', ... respectively. If a die is taken at random, the probability of throwing ace is

$$\frac{m' + m'' + m''' + \dots}{nr}.$$

If an ace has been thrown the probability of throwing ace again on a second trial with the same die is

$$\frac{m'^2 + m''^2 + m'''^2 + \dots}{r\,(m' + m'' + m''' + \dots)}.$$

The first probability is the greater; for $(m' + m'' + m''' + \dots)^2$ is greater than $n\,(m'^2 + m''^2 + m'''^2 + \dots)$. The memoir demonstrates this simple inequality.

858. Prevost and Lhuilier are also the authors of a memoir entitled *Mémoire sur l'application du Calcul des probabilités à la valeur du témoignage.*

This memoir is published in the volume for 1797 of the *Mémoires de l'Acad....Berlin;* the date of publication is 1800 : the memoir occupies pages 120—151 of the portion of the volume devoted to speculative philosophy.

The memoir begins thus :

Le but de ce mémoire est plutôt de reconnoitre l'état actuel de cette théorie, que d'y rien ajouter de nouveau.

The memoir first notices the criticism given in Lambert's *Organon* of James Bernoulli's formula which we have already given in Art. 122.

It then passes on to the theory of *concurrent* testimony now commonly received. Suppose a witness to speak truth m times and falsehood n times out of $m + n$ times; let m' and n' have similar meanings for a second witness. Then if they agree in an assertion the probability of its truth is $\dfrac{mm'}{mm' + nn'}$.

The ordinary theory of *traditional* testimony is also given. Using the same notation as before if one witness reports a statement from the report of another the probability of its truth is

$$\frac{mm' + nn'}{(m + m')(n + n')};$$

for the statement is true if they both tell the truth or if they both tell a falsehood. If there be two witnesses in succession each of whom reverses the statement he ought to give, the result is true; that is a double falsehood gives a truth. It is stated that this consequence was first indicated in 1794 by Prevost.

The hypothesis of Craig is noticed; see Art. 91.

The only new point in the memoir is an hypothesis which is proposed relating to *traditional* testimony, and which is admitted to be arbitrary, but of which the consequences are examined. The hypothesis is that *no testimony founded on falsehood can give the truth*. The meaning of this hypothesis is best seen by an example: suppose the two witnesses precisely alike, then instead of taking $\dfrac{m^2 + n^2}{(m + n)^2}$ as the probability of the truth in the case above considered we should take $\dfrac{m^2}{(m + n)^2}$; that is we reject the term n^2 in the numerator which arises from the agreement of the witnesses in a falsehood.

Thus we take $\dfrac{m^2}{(m + n)^2}$ and $\dfrac{2mn + n^2}{(m + n)^2}$ to represent respectively the probabilities of the truth and falsehood of the statement on which the witnesses agree.

Suppose now that there is a second pair of witnesses independent of the former, of the same character, and that the same

statement is also affirmed by this pair. Then the memoir combines
the two pairs by the ordinary rule for concurrent testimony, and so
takes for the probability arising from the two ¯pairs

$$\frac{m^4}{m^4 + (2nm + n^x)^2}.$$

Then the question is asked for what ratio of m to n this expression is equal to $\frac{m}{m+n}$, so that the force of the two pairs of witnesses may be equal to that of a single witness. The approximate
value of $\frac{m}{n}$ is said to be 4·864 so that $\frac{m}{m+n}$ is about $\frac{5}{6}$.

859. In Vol. VII. of the Transactions of the Royal Irish
Academy there is a memoir by the Rev. Matthew Young, D.D.
S.F.T.C.D. and M.R.I.A., entitled *On the force of Testimony in establishing Facts contrary to Analogy.* The date of publication of
the volume is 1800; the memoir was read February 3rd, 1798: it
occupies pages 79—118 of the volume.

The memoir is rather metaphysical than mathematical. Dr
Young may be said to adopt the modern method of estimating the
force of the testimony of concurrent witnesses; in this method,
supposing the witnesses of equal credibility, we obtain a formula
coinciding with that in Art. 667. Dr Young condemns as erroneous
the method which we noticed in Art. 91; he calls it "Dr Halley's
mode," but gives no authority for this designation. Dr Young
criticises two rules given by Waring on the subject; in the first of
the two cases however it would not be difficult to explain and
defend Waring's rule.

CHAPTER XX.

LAPLACE.

860. LAPLACE was born in 1749, and died in 1827. He wrote elaborate memoirs on our subject, which he afterwards embodied in his great work the *Théorie analytique des Probabilités,* and on the whole the Theory of Probability is more indebted to him than to any other mathematician. We shall give in the first place a brief account of Laplace's memoirs, and then consider more fully the work in which they are reproduced.

861. Two memoirs by Laplace on our subject are contained in the *Mémoires...par divers Savans,* Vol. VI. 1774. A brief notice of the memoirs is given in pages 17—19 of the preface to the volume which concludes thus :

Ces deux Mémoires de M. de la Place, ont été choisis parmi un très-grand nombre qu'il a présentés depuis trois ans, à l'Académie, où il remplit actuellement une place de Géomètre. Cette Compagnie qui s'est empressée de récompenser ses travaux et ses talens, n'avoit encore vu personne aussi jeune, lui présenter en si peu de temps, tant de Mémoires importans, et sur des matières si diverses et si difficiles.

862. The first memoir is entitled *Mémoire sur les suites récurro-récurrentes et sur leurs usages dans la théorie des hasards.* It occupies pages 353—371 of the volume.

A recurring series is connected with the solution of an equation in Finite Differences where there is *one* independent variable ; see Art. 318. A recurro-recurrent series is similarly connected with the solution of an equation in Finite Differences where there are *two* independent variables. Laplace here first introduces the term and the subject itself; we shall not give any account of his investigations, but confine ourselves to the part of his memoir which relates to the Theory of Probability.

863. Laplace considers three problems in our subject. The first is the problem of the Duration of Play, supposing two players of unequal skill and unequal capital; Laplace, however, rather shews how the problem may be solved than actually solves it. He begins with the case of equal skill and equal capital, and then passes on to the case of unequal skill. He proceeds so far as to obtain an equation in Finite Differences with one independent variable which would present no difficulty in solving. He does not actually discuss the case of unequal capital, but intimates that there will be no obstacle except the length of the process.

The problem is solved completely in the *Théorie...des Prob.* pages 225—238 ; see Art. 588.

864. The next problem is that connected with a lottery which appears in the *Théorie...des Prob.* pages 191—201. The mode of solution is nearly the same in the two places, but it is easier to follow in the *Théorie...des Prob.* The memoir does not contain any of the approximate calculation which forms a large part of the discussion in the *Théorie...des Prob.* We have already given the history of the problem; see Arts. 448, 775.

865. The third problem is the following : Out of a heap of counters a number is taken at random ; find the chances that this number will be odd or even respectively. Laplace obtains what we should now call the ordinary results ; his method however is more elaborate than is necessary, for he uses Finite Differences : in the *Théorie...des Prob.* page 201, he gives a more simple solution. We have already spoken of the problem in Art. 350.

866. The next memoir is entitled *Mémoire sur la Probabilité des causes par les évènemens;* it occupies pages 621—656 of the volume cited in Art. 861.

The memoir commences thus :

La Théorie des hasards est une des parties les plus curieuses et les plus délicates de l'analyse, par la finesse des combinaisons qu'elle exige et par la difficulté de les soumettre au calcul ; celui qui paroît l'avoir traitée avec le plus de succès est M. Moivre, dans un excellent Ouvrage qui a pour titre, *Theory of Chances;* nous devons à cet habile Géomètre les premières recherches que l'on ait faites sur l'intégration des équations différencielles aux différences finies ; ...

867. Laplace then refers to Lagrange's researches on the theory of equations in Finite Differences, and also to two of his own memoirs, namely that which we have just examined, and one which was about to appear in the volume of the Academy for 1773. But his present object, he says, is very different, and is thus stated :

...je me propose de déterminer la probabilité des causes par les évènemens, matière neuve à bien des égards et qui mérite d'autant plus d'être cultivée que c'est principalement sous ce point de vue que la science des hasards peut être utile dans la vie civile.

868. This memoir is remarkable in the history of the subject, as being the first which distinctly enunciated the principle for estimating the probabilities of the causes by which an observed event may have been produced. Bayes must have had a notion of the principle, and Laplace refers to him in the *Théorie...des Prob.* page CXXXVII. though Bayes is not named in the memoir. See Arts. 539, 696.

869. Laplace states the general principle which he assumes in the following words :

Si un évènement peut être produit par un nombre n de causes différentes, les probabilités de l'existence de ces causes prises de l'évènement, sont entre elles comme les probabilités de l'évènement prises de ces causes, et la probabilité de l'existence de chacune d'elles, est égale à la probabilité de l'évènement prise de cette cause, divisée par la somme de toutes les probabilités de l'évènement prises de chacune de ces causes.

870. Laplace first takes the standard problem in this part of our subject : Suppose that an urn contains an infinite number of white tickets and black tickets in an unknown ratio ; $p + q$ tickets

are drawn of which p are white and q are black : required the probability of drawing m white tickets and n black tickets in the next $m + n$ drawings.

Laplace gives for the required probability

$$\frac{\int_0^1 x^{r+m} (1-x)^{q+n}\, dx}{\int_0^1 x^p (1-x)^q\, dx},$$

so that of course the m white tickets and n black tickets are supposed to be drawn in an assigned order ; see Arts. 704, 766, 843. Laplace effects the integration, and approximates by the aid of a formula which he takes from Euler, and which we usually call Stirling's Theorem.

The problem here considered is not explicitly reproduced in the *Théorie...des Prob.*, though it is involved in the Chapter which forms pages 363—401.

871. After discussing this problem Laplace says,

La solution de ce Problème donne une méthode directe pour déterminer la probabilité des évènemens futurs d'après ceux qui sont déja arrivés ; mais cette matière étant fort étendue, je me bornerai ici à donner une démonstration assez singulière du théorème suivant.

On peut supposer les nombres p *et* q *tellement grands, qu'il devienne aussi approchant que l'on voudra de la certitude, que le rapport du nombre de billets blancs au nombre total des billets renfermés dans l'urne, est compris entre les deux limites* $\frac{p}{p+q} - \omega$, *et* $\frac{p}{p+q} + \omega$, ω *pouvant être supposé moindre qu'aucune grandeur donnée.*

The probability of the ratio lying between the specified limits is

$$\frac{\int x^r (1-x)^q\, dx}{\int_0^1 x^p (1-x)^q\, dx},$$

where the integral in the numerator is to be taken between the limits $\frac{p}{p+q} - \omega$ and $\frac{p}{p+q} + \omega$. Laplace by a rude process of approximation arrives at the conclusion that this probability does not differ much from unity.

872. Laplace proceeds to the Problem of Points. He quotes the second formula which we have given in Art. 172; he says that it is now demonstrated in several works. He also refers to his own memoir in the volume of the Academy for 1773; he adds the following statement:

... on y trouvera pareillement une solution générale du Problème des partis dans le cas de trois ou d'un plus grand nombre de joueurs, problème qui n'a encore été résolu par personne, que je sache, bien que les Géomètres qui ont travaillé sur ces matières en aient desiré la solution.

Laplace is wrong in this statement, for De Moivre had solved the problem; see Art. 582.

873. Let x denote the skill of the player A, and $1 - x$ the skill of the player B; suppose that A wants f games in order to win the match, and that B wants h games: then, if they agree to leave off and divide the stakes, the share of B will be a certain quantity which we may denote by $\phi(x, f, h)$. Suppose *the skill of each player unknown;* let n be the whole number of games which A or B ought to win in order to entitle him to the stake. Then Laplace says that it follows from the general principle which we have given in Art. 869, that the share of B is

$$\frac{\int_0^1 x^{n-f}(1-x)^{n-h}\,\phi(x, f, h)\,dx}{\int_0^1 x^{n-f}(1-x)^{n-h}\,dx}.$$

The formula depends on the fact that A must already have won $n - f$ games, and B have won $n - h$ games. See Art. 771.

874. Laplace now proceeds to the question of the mean to be taken of the results of observations. He introduces the subject thus:

On peut, au moyen de la Théorie précédente, parvenir à la solution du Problème qui consiste à déterminer le milieu que l'on doit prendre entre plusieurs observations données d'un même phénomène. Il y a deux ans que j'en donnai une à l'Académie, à la suite du Mémoire *sur les Séries récurrorécurrentes*, imprimé dans ce volume ; mais le peu d'usage dont elle pouvoit être, me la fit supprimer lors de l'impression. J'ai appris depuis par le Journal astronomique de M. Jean Bernoulli, que M^r. Daniel Bernoulli et la Grange se sont occupés du même problème dans deux Mémoires manuscrits qui ne sont point venus à ma connoissance. Cette annonce jointe à l'utilité de la matière, a réveillé mes idées sur cet objet ; et quoique je ne doute point que ces deux illustres Géomètres ne l'aient traité beaucoup plus heureusement que moi, je vais cependant exposer ici les réflexions qu'il m'a fait nâitre, persuadé que les différentes manières dont on peut l'envisager produiront une méthode moins hypothétique et plus sûre pour déterminer le milieu que l'on doit prendre entre plusieurs observations.

875. Laplace then enunciates his problem thus :

Déterminer le milieu que l'on doit prendre entre trois observations données d'un même phénomène.

Laplace supposes positive and negative errors to be equally likely, and he takes for the probability that an error lies between x and $x + dx$ the expression $\frac{m}{2} e^{-mx} dx$; for this he offers some reasons, which however are very slight. He restricts himself as his enunciation states, to *three* observations. Thus the investigation cannot be said to have any practical value.

876. Laplace says that by the *mean* which ought to be taken of several observations, two things may be understood. We may understand such a value that it is equally likely that the true value is above or below it ; this he says we may call the *milieu de probabilité*. Or we may understand such a value that the sum of the errors, each multiplied by its probability, is a minimum ; this he says we may call the *milieu d'erreur*, or the *milieu astronomique*, as being that which astronomers ought to adopt. The errors are here supposed to be all taken positively.

It might have been expected from Laplace's words that these two notions of a *mean* value would lead to different results ; he shews however that they lead to the same result. In both cases the mean value corresponds to the point at which the ordinate to a certain curve of probability bisects the area of the curve. See *Théorie...des Prob.* page 335.

Laplace does not notice another sense of the word *mean*, namely an *average* of all the values; in this case the mean would correspond to the abscissa of the centre of gravity of the area of a certain curve. See Art. 485.

877. Laplace now proceeds to the subject which is considered in Chapter VII. of the *Théorie...des Prob.*, namely the influence produced by the want of perfect symmetry in coins or dice on the chances of repetitions of events. The present memoir and the Chapter in the *Théorie...des Prob.* give different illustrations of the subject.

The first case in the memoir is that of the *Petersburg Problem*, though Laplace does not give it any name. Suppose the chance for head to be $\dfrac{1+\varpi}{2}$, and therefore the chance for tail to be $\dfrac{1-\varpi}{2}$; suppose there are to be x trials, and that 2 crowns are to be received if head appears at the first trial, 4 crowns if head does not appear until the second trial, and so on. Then the expectation is

$$(1+\varpi)\left\{1+(1-\varpi)+(1-\varpi)^2+...+(1-\varpi)^{x-1}\right\}.$$

If the chance for head is $\dfrac{1-\varpi}{2}$, and therefore the chance for tail is $\dfrac{1+\varpi}{2}$, we must change the sign of ϖ in the expression for the expectation. If we do not know which is the more likely to appear, head or tail, we may take half the sum of the two expressions for the expectation. This gives

$$1+\frac{1-\varpi^2}{2\varpi}\left\{(1+\varpi)^{x-1}-(1-\varpi)^{x-1}\right\}.$$

If we expand, and reject powers of ϖ higher than ϖ^2, we obtain

$$x+\left\{\frac{(x-1)(x-2)(x-3)}{1\cdot2\cdot3}-(x-1)\right\}\varpi^2.$$

If we suppose that ϖ may have any value between 0 and c we may multiply the last expression by $d\varpi$ and integrate from 0 to c. See Art. 529.

878. As another example Laplace considers the following question. A undertakes to throw a given face with a common die in n throws: required his chance.

If the die be perfectly symmetrical the chance is $1 - \left(\dfrac{5}{6}\right)^n$; but if the die be not perfectly symmetrical this result must be modified. Laplace gives the investigation: the principle is the same as in another example which Laplace also gives, and to which we will confine ourselves. Instead of a common die with *six* faces we will suppose a triangular prism which can only fall on one of its *three* rectangular faces: required the probability that in n throws it will fall on an assigned face. Let the chance of its falling on the three faces be $\dfrac{1+\varpi}{3}$, $\dfrac{1+\varpi'}{3}$ and $\dfrac{1+\varpi''}{3}$ respectively, so that

$$\varpi + \varpi' + \varpi'' = 0.$$

Then if we are quite ignorant which of the three chances belongs to the assigned face, we must suppose in succession that each of them does, and take one-third of the sum of the results. Thus we obtain one-third of the following sum,

$$\left\{1 - \left(\frac{2-\varpi}{3}\right)^n\right\} + \left\{1 - \left(\frac{2-\varpi'}{3}\right)^n\right\} + \left\{1 - \left(\frac{2-\varpi''}{3}\right)^n\right\},$$

that is $1 - \dfrac{1}{3}\left\{\left(\dfrac{2-\varpi}{3}\right)^n + \left(\dfrac{2-\varpi'}{3}\right)^n + \left(\dfrac{2-\varpi''}{3}\right)^n\right\}$.

If we reject powers of ϖ, ϖ', and ϖ'' beyond the square we get approximately

$$1 - \frac{2^n}{3^n} - \frac{n(n-1)}{1.2} \cdot \frac{2^{n-2}}{3^{n+1}} (\varpi^2 + \varpi'^2 + \varpi''^2).$$

Suppose we know nothing about ϖ, ϖ', and ϖ'', except that each must lie between $-c$ and $+c$; we wish to find what we may call the average value of $\varpi^2 + \varpi'^2 + \varpi''^2$.

We may suppose that we require the mean value of $x^2 + y^2 + z^2$, subject to the conditions that $x + y + z = 0$, and that x, y, and z must each lie between $-c$ and $+c$.

The result is

$$\frac{2\int_0^c \int_{-c}^{c-x} \{x^2 + y^2 + (x+y)^2\}\, dx\, dy}{2\int_0^c \int_{-c}^{c-x} dx\, dy}.$$

Laplace works out this result, giving the reasons for the steps briefly. Geometrical considerations will furnish the result very readily. We may consider $x + y + z = 0$ to be the equation to a plane, and we have to take all points in this plane lying within a certain regular hexagon. The projection of this hexagon on the plane of (x, y) will be a hexagon, four of whose sides are equal to c, and the other two sides to $c\sqrt{2}$. The result of the integration is $\frac{5}{6} c^2$. Thus the chance is

$$1 - \frac{2^n}{3^n} - \frac{n(n-1)}{1.2} \frac{2^{n-3}}{3^{n+2}} 5c^2.$$

879. It easily follows from Laplace's process that if we suppose a coin to be not perfectly symmetrical, but do not know whether it is more likely to give head or tail, then the chance of two heads in two throws or the chance of two tails in two throws is rather more than $\frac{1}{4}$: it is in fact equal to such an expression as

$$\frac{1}{2} \left\{ \left(\frac{1-\varpi}{2}\right)^2 + \left(\frac{1+\varpi}{2}\right)^2 \right\}$$

instead of being equal to $\frac{1}{2} \times \frac{1}{2}$. Laplace after adverting to this case says,

Cette aberration de la Théorie ordinaire, qui n'a encore été observée par personne, que je sache, m'a paru digne de l'attention des Géomètres, et il me semble que l'on ne peut trop y avoir égard, lorsqu'on applique le calcul des probabilités, aux différens objets de la vie civile.

880. Scarcely any of the present memoir is reproduced by Laplace in his *Théorie...des Prob.* Nearly all that we have noticed in our account of the memoir up to Art. 876 inclusive is

indeed superseded by Laplace's later researches; but what we
have given from Art. 877 inclusive might have appeared in
Chapter VII. of the *Théorie...des Prob.*

881. Laplace's next memoir on our subject is in the *Mémoires
...par divers Savans...*1773; the date of publication is 1776. The
memoir is entitled *Recherches sur l'intégration des Equations différentielles aux différences finies, et sur leur usage dans la théorie
des hasards,* &c.

The portion on the theory of chances occupies pages 113—163.
Laplace begins with some general observations. He refers to the
subject which he had already discussed, which we have noticed
in Art. 877. He says that the advantage arising from the want
of symmetry is on the side of the player who bets that head
will not arrive in two throws : this follows from Art. 879 ; for to
bet that head will not arrive in two throws is to bet that both
throws will give tail.

882. The first problem he solves is that of *odd* and *even; see*
Art. 865.

The next problem is an example of Compound Interest, and
has nothing connected with probability.

The next problem is as follows. A solid has p equal faces,
which are numbered 1, 2,...p : required the probability that in
the course of n throws the faces will occur in the order 1, 2,...p.

This problem is nearly the same as that about a *run of events*
which we have reproduced from De Moivre in Art. 325 : instead
of the equation there given we have

$$u_{n+1} = u_n + (1 - u_{n+1-p})\, a^p, \text{ where } a = \frac{1}{p}.$$

883. The next problem is thus enunciated:

Je suppose un nombre n de joueurs (1), (2), (3), ... (n), jouant de
cette manière ; (1) joue avec (2), et s'il gagne il gagne la partie ; s'il ne
perd ni gagne, il continue de jouer avec (2), jusqu'à ce que l'un des
deux gagne. Que si (1) perd, (2) joue avec (3) ; s'il le gagne, il gagne la
partie ; s'il ne perd ni gagne, il continue de jouer avec (3) ; mais s'il
perd, (3) joue avec (4), et ainsi de suite jusqu'à ce que l'un des joueurs
ait vaincu celui qui le suit ; c'est-à-dire que (1) soit vainqueur de (2),

ou (2) de (3), ou (3) de (4), ... ou $(n-1)$ de (n), ou (n) de (1). De plus, la probabilité d'un quelconque des joueurs, pour gagner l'autre $= \frac{1}{3}$, et celle de ne gagner ni perdre $= \frac{1}{3}$. Cela posé, il faut déterminer la probabilité que l'un de ces joueurs gagnera la partie au coup x.

This problem is rather difficult; it is not reproduced in the *Théorie...des Prob.* The following is the general result: Let v_x denote the chance that any assigned player will win the match at the xth trial; then

$$v_x - \frac{n}{3} v_{x-1} + \frac{n(n-1)}{1.2} \frac{1}{3^2} v_{x-2} - \frac{n(n-1)(n-2)}{1.2.3} \frac{1}{3^3} v_{x-3} + \cdots$$
$$= \frac{1}{3^n} v_{x-n}.$$

884. Laplace next takes the Problem of Points in the case of two players, and then the same problem in the case of three players; see Art. 872. Laplace solves the problem by Finite Differences. At the beginning of the volume which contains the memoir some errata are corrected, and there is also another solution indicated of the Problem of Points for three players; this solution depends on the expansion of a multinomial expression, and is in fact identical with that which had been given by De Moivre.

Laplace's next problem may be considered an extension of the Problem of Points; it is reproduced in the *Théorie...des Prob.* page 214, beginning with the words *Concevons encore.*

885. The next two problems are on the Duration of Play; in the first case the capitals being equal, and in the second case unequal; see Art. 863. The solutions are carried further than in the former memoir, but they are still much inferior to those which were subsequently given in the *Théorie...des Prob.*

886. The next problem is an extension of the problem of Duration of Play with equal capitals.

It is supposed that at every game there is the chance p for A, the chance q for B, and the chance r that neither wins; each player has m crowns originally, and the loser in any game gives a crown to the winner: required the probability that the play will be finished in x games. This problem is not reproduced in the *Théorie...des Prob.*

887. The present memoir may be regarded as a collection of examples in the theory of Finite Differences; the methods exemplified have however since been superseded by that of Generating Functions, which again may be considered to have now given way to the Calculus of Operations. The problems involve only questions in *direct* probability; none of them involve what are called questions in *inverse* probability, that is, questions respecting the probability of causes as deduced from observed events.

888. In the same volume as the memoir we have just analysed there is a memoir by Laplace entitled, *Mémoire sur l'inclinaison moyenne des orbites des comètes; sur la figure de la Terre, et sur les Fonctions.* The part of the memoir devoted to the mean inclination of the orbits of comets occupies pages 503—524 of the volume.

In these pages Laplace discusses the problem which was started by Daniel Bernoulli; see Art. 395. Laplace's result agrees with that which he afterwards obtained in the *Théorie...des Prob.* pages 253—260, but the method is quite different; both methods are extremely laborious.

Laplace gives a numerical example; he finds that supposing 12 comets or planets the chance is ·339 that the mean inclination of the planes of the orbits to a fixed plane will lie between $45° - 7\frac{1}{2}°$ and $45°$, and of course the chance is the same that the mean inclination will lie between $45°$ and $45° + 7\frac{1}{2}°$.

889. The volume with which we have been engaged in Articles 881—888 is remarkable in connexion with Physical Astronomy. Historians of this subject usually record its triumphs, but omit its temporary failures. In the present volume Lagrange affects to shew that the secular acceleration of the Moon's motion cannot be explained by the ordinary theory of gravitation; and Laplace affects to shew that the inequalities in the motions of Jupiter and Saturn cannot be attributed to the mutual action of these planets: see pages 47, 213 of the volume. Laplace lived to correct both his rival's error and his own, by two of his greatest contributions to Physical Astronomy.

890. Laplace's next memoir on our subject is entitled *Mémoire sur les Probabilités;* it is contained in the volume for 1778 of the *Histoire de l'Acad....Paris:* the date of publication of the volume is 1781. The memoir occupies pages 227—332.

In the notice of the memoir which is given in the introductory part of the volume the names of Bayes and Price are mentioned. Laplace does not allude to them in the memoir. See Art. 540.

891. Laplace begins with remarks, similar to those which we have already noticed, respecting the chances connected with the tossing of a coin which is not quite symmetrical; see Arts. 877, 881. He solves the simple problem of Duration of Play in the way we have given in Art. 107. Thus let p denote A's skill, and $1 - p$ denote B's skill. Suppose A to start with m stakes, and B to start with $n - m$ stakes: then A's chance of winning all B's stakes is

$$\frac{p^{n-m}\{p^m - (1 - p)^m\}}{p^n - (1 - p)^n}.$$

Laplace puts for p in succession $\frac{1}{2}(1 + a)$ and $\frac{1}{2}(1 - a)$, and takes half the sum. Thus he obtains for A's chance

$$\frac{\frac{1}{2}\left\{(1 + a)^{n-m} + (1 - a)^{n-m}\right\}\left\{(1 + a)^m - (1 - a)^m\right\}}{(1 + a)^n - (1 - a)^n},$$

which he transforms into

$$\frac{1}{2} - \frac{1}{2}(1 - a^2)^m \frac{(1 + a)^{n-2m} - (1 - a)^{n-2m}}{(1 + a)^n - (1 - a)^n}.$$

The expression for A's chance becomes $\frac{m}{n}$ when a vanishes; Laplace proposes to shew that the expression increases as a increases, if $2m$ be less than n. The factor $(1 - a^2)^m$ obviously diminishes as a increases. Laplace says that if $2m$ is less than n it is clear that the fraction

$$\frac{(1 + a)^{n-2m} - (1 - a)^{n-2m}}{(1 + a)^n - (1 - a)^n}$$

also diminishes as α increases. We will demonstrate this.

Put r for $n - 2m$, and denote the fraction by u; then

$$\frac{1}{u}\frac{du}{d\alpha} = r\,\frac{(1+\alpha)^{r-1}+(1-\alpha)^{r-1}}{(1+\alpha)^r-(1-\alpha)^r} - n\,\frac{(1+\alpha)^{n-1}+(1-\alpha)^{n-1}}{(1+\alpha)^n-(1-\alpha)^n}\,.$$

Thus

$$(1-\alpha)\frac{1}{u}\frac{du}{d\alpha} = \frac{r\,(z^{r-1}+1)}{z^r-1} - \frac{n\,(z^{n-1}+1)}{z^n-1}\,,$$

where $z = \dfrac{1+\alpha}{1-\alpha}$. We have to shew that this expression is negative; this we shall do by shewing that $\dfrac{r\,(z^{r-1}+1)}{z^r-1}$ increases as successive integral values are ascribed to r. We have

$$\frac{(r+1)\,(z^r+1)}{z^{r+1}-1} - \frac{r\,(z^{r-1}+1)}{z^r-1}$$

$$= \frac{(r+1)\,(z^{2r}-1) - r\,(z^{r+1}-1)\,(z^{r-1}+1)}{(z^{r+1}-1)\,(z^r-1)}\,;$$

thus we must shew that $z^{2r}-1$ is greater than $r\,(z^{r+1}-z^{r-1})$.

Expand by the exponential theorem; then we find we have to shew that

$$(2r)^p \text{ is greater than } r\left\{(r+1)^p - (r-1)^p\right\},$$

where p is any positive integer; that is, we must shew that

$$2^{p-1}\,r^{p-1} \text{ is greater than } pr^{p-1} + \frac{p\,(p-1)\,(p-2)}{1\,.\,2\,.\,3}\,r^{p-3} + \dots$$

But this is obvious, for r is supposed greater than unity, and the two members would be equal if all the exponents of r on the right hand side of the inequality were $p - 1$.

We observe that r must be supposed not less than 2; if $r = 1$ we have $z^{2r}-1 = r\,(z^{r+1}-z^{r-1})$.

We have assumed that r and n are *integers*, and this limitation is necessary. For return to the expression

$$\frac{(1+\alpha)^r-(1-\alpha)^r}{(1+\alpha)^n-(1-\alpha)^n}\,,$$

and put for a in succession 0 and 1; then we have to compare $\dfrac{r}{n}$ with $\dfrac{2^r}{2^n}$; that is, we have to compare $\dfrac{r}{2^r}$ with $\dfrac{n}{2^n}$. Now consider $\dfrac{x}{2^x}$; the differential coefficient with respect to x is $\dfrac{1 - x \log 2}{2^x}$; so that $\dfrac{x}{2^x}$ increases as x changes from 0 to $\dfrac{1}{\log 2}$, and then diminishes.

Laplace treats the same question in the *Théorie...des Prob.* page 406; there also the difficulty is dismissed with the words *il est facile de voir*. In the memoir prefixed to the fourth volume of Bowditch's Translation of the *Mécanique Céleste*, page 62, we read:

Dr Bowditch himself was accustomed to remark, "Whenever I meet in La Place with the words 'Thus it plainly appears' I am sure that hours, and perhaps days of hard study will alone enable me to discover *how* it plainly appears."

892. The pages 240—258 of the memoir contain the important but difficult investigation which is reproduced in the *Théorie...des Prob.* pages 262—272. Laplace gives in the memoir a reference to those investigations by Lagrange which we have noticed in Art. 570; the reference however is omitted in the *Théorie...des Prob.*

893. Laplace now proceeds to the subject which he had considered in a former memoir, namely, the probability of causes as deduced from events; see Art. 868. Laplace repeats the general principle which he had already enunciated in his former memoir; see Art. 869. He then takes the problem which we have noticed in Art. 870, enunciating it however with respect to the births of boys and girls, instead of the drawings of white and black balls. See Art. 770.

894. Laplace is now led to consider the approximate evaluation of definite integrals, and he gives the method which is reproduced almost identically in pages 88—90 of the *Théorie...des Prob.* He applies it to the example $\int x^p \, (1-x)^q \, dx$, and thus demonstrates the theorem he had already given; see Art. 871: the present demonstration is much superior to the former.

895. There is one proposition given here which is not reproduced in the *Théorie...des Prob.*, but which is worthy of notice.

Suppose we require the value of $\int y dx$ where $y = x^p (1-x)^q$,
the integral being taken between assigned limits.

Put $p = \dfrac{1}{\alpha}$ and $q = \dfrac{\mu}{\alpha}$; and let

$$z = \frac{1}{\alpha} y \frac{dx}{dy}.$$

Then, by integrating by parts,

$$\int y dx = \int \alpha z \, dy = \alpha y z - \alpha \int y \, dz \dots\dots\dots(1),$$

$$\int y dz = \alpha \int z \frac{dz}{dx} \, dy = \alpha y z \frac{dz}{dx} - \alpha \int y \frac{d}{dx} \left(z \frac{dz}{dx} \right) dx \, ;$$

so that

$$\int y dx = \alpha y z - \alpha^2 y z \frac{dz}{dx} + \alpha^2 \int y \frac{d}{dx} \left(z \frac{dz}{dx} \right) dx \dots\dots\dots(2).$$

Now y vanishes with x. Laplace shews that the value of
$\int y dx$ when the lower limit is zero and the upper limit is any
value of x less than $\dfrac{1}{1+\mu}$, is less than $\alpha y z$ and is greater than
$\alpha y z - \alpha^2 y z \dfrac{dz}{dx}$; so that we can test the closeness of the approximation. This proposition depends on the following considera-
tions: $\dfrac{dz}{dx}$ is positive so long as x is less than $\dfrac{1}{1+\mu}$, and therefore $\int y dx$ is less than $\alpha y z$ by (1); and $\dfrac{d}{dx} \left(z \dfrac{dz}{dx} \right)$ is also positive,
so that $\int y dx$ is greater than $\alpha y z - \alpha^2 y z \dfrac{dz}{dx}$ by (2). For we have

$$z = \frac{x(1-x)}{1-(1+\mu)x},$$

and this can be put in the form

$$z = -\frac{\mu}{(1+\mu)^2} + \frac{x}{1+\mu} + \frac{\mu}{(1+\mu)^2\{1-(1+\mu)\,x\}}.$$

Hence we see that z and $\frac{dz}{dx}$ both increase with x so long as x is less than $\frac{1}{1+\mu}$: this establishes the required proposition.

See also Art. 767.

896. Laplace then takes the following problem. In 26 years it was observed in Paris that 251527 boys were born and 241945 girls: required the probability that the possibility of the birth of a boy is greater than $\frac{1}{2}$. The probability is found to differ from unity by less than a fraction having for its numerator 1·1521 and for its denominator the seventh power of a million.

This problem is reproduced in the *Théorie...des Prob.* pages 377—380, the data being the numbers of births during 40 years instead of during 26 years.

897. Taking the same data as in the preceding Article, Laplace investigates the probability that in a given year the number of boys born shall not exceed the number of girls born. He finds the probability to be a little less than $\frac{1}{259}$. The result of a similar calculation from data furnished by observations in London is a little less than $\frac{1}{12416}$. In pages 397—401 of the *Théorie...des Prob.* we have a more difficult problem, namely to find the probability that during a century the annual births of boys shall never be less than that of girls. The treatment of the simpler problem in the memoir differs from that of the more difficult problem in the *Théorie...des Prob.* In the memoir Laplace obtains an equation in Finite Differences

$$y_m = z_m \Delta y_m;$$

hence he deduces

$$\Sigma y_m = \text{constant} + y_m z_{m-1}\Big\{1 - \Delta z_{m-2} + \Delta\,(z_{m-2}\Delta z_{m-3})$$
$$- \Delta\,[z_{m-2}\Delta\,(z_{m-3}\Delta z_{m-4})] + \ldots\Big\},$$

which as he says is analogous to the corresponding theorem in
the Integral Calculus given in Art. 895; and, as in that Article, he
shews that in the problem he is discussing the exact result lies
between two approximate results. See also Art. 770.

898. The memoir contains on page 287 a brief indication of a
problem which is elaborately treated in pages 369—376 of the
Théorie...des Prob.

899. Laplace now developes another form of his method of
approximation to the value of definite integrals. Suppose we
require $\int y\,dx$; let Y be the maximum value of y within the
range of the integration. Assume $y = Ye^{-t^2}$, and thus change
$\int y\,dx$ into an integral with respect to t. The investigation is
reproduced in the *Théorie...des Prob.* pages 101—103.

Laplace determines the value of $\int_0^\infty e^{-t^2}dt$. He does this by
taking the double integral $\int_0^\infty\int_0^\infty e^{-s(1+u^2)}ds\,du$, and equating the
results which are obtained by considering the integrations in
different orders.

900. Laplace also considers the case in which instead of assuming $y = Ye^{-t^2}$, we may assume $y = Ye^{-t^4}$. Something similar is
given in the *Théorie...des Prob.* pages 93—95.

Some formulæ occur in the memoir which are not reproduced
in the *Théorie...des Prob.*, and which are quite wrong: we will
point out the error. Laplace says on pages 298, 299 of the
memoir :

Considérons présentement la double intégrale $\iint \dfrac{dx\,dz}{(1-z^2-x^4)^{\frac{3}{4}}}$, prise
depuis $x=0$ jusqu'à $x=1$, et depuis $z=0$ jusqu'à $z=1$; en faisant
$\dfrac{x}{(1-z^2)^{\frac{1}{4}}} = x'$, elle se changera dans celle-ci $\int\dfrac{dz}{\sqrt{(1-z^2)}}\int\dfrac{dx'}{(1-x'^4)^{\frac{3}{4}}}$, ces
intégrales étant prises depuis $x'=0$ et $z=0$, jusqu'à $x'=1$ et $z=1$, ...

31

Then, as $\int_0^1 \frac{dz}{\sqrt{(1-z^2)}} = \frac{\pi}{2}$, Laplace infers that

$$\int_0^1\int_0^1 \frac{dx\,dz}{(1-z^2-x^4)^{\frac{3}{4}}} = \frac{\pi}{2}\int_0^1 \frac{dx'}{(1-x'^4)^{\frac{3}{4}}}.$$

But this is wrong; for the limits of x' are 0 and $\frac{1}{(1-z^2)^{\frac{1}{4}}}$, and not 0 and 1, as Laplace says; and so the process fails.

Laplace makes the same mistake again immediately afterwards; he puts $\frac{z}{\sqrt{(1-x^4)}} = z'$, and thus deduces

$$\int_0^1\int_0^1 \frac{dx\,dz}{(1-z^2-x^4)^{\frac{3}{4}}} = \int_0^1 \frac{dx}{(1-x^4)^{\frac{1}{4}}}\int_0^1 \frac{dz'}{(1-z'^2)^{\frac{3}{4}}}.$$

But the upper limit for z' should be $\frac{1}{\sqrt{(1-x^4)}}$, and not 1 as Laplace assumes; and so the process fails.

901. Laplace applies his method to evaluate approximately $\int_0^1 x^p (1-x)^q\,dx$; and he finds an opportunity for demonstrating Stirling's Theorem. See Art. 333.

902. Laplace discusses in pages 304—313 of the memoir the following problem. Observation shews that the ratio of the number of births of boys to that of girls is sensibly greater at London than at Paris; this seems to indicate a greater facility for the birth of a boy at London than at Paris: required to determine the amount of probability. See Art. 773.

Let u be the probability of the birth of a boy at Paris, p the number of births of boys observed there, and q the number of births of girls; let $u-x$ be the possibility of the birth of a boy at London, p' the number of births of boys observed there, and q' the number of births of girls. If P denote the probability that the birth of a boy is less possible at London than at Paris, we have

$$P = \frac{\iint u^p (1-u)^q (u-x)^{p'} (1-u+x)^{q'}\,du\,dx}{\iint u^p (1-u)^q (u-x)^{p'} (1-u+x)^{q'}\,du\,dx}.$$

Laplace says that the integral in the numerator is to be taken from $u = 0$ to $u = x$, and from $x = 0$ to $x = 1$, and that the integral in the denominator is to be taken for all possible values of x and u. Thus putting $u - x = s$ the denominator becomes

$$\int_0^1 \int_0^1 u^p (1 - u)^q s^{p'} (1 - s)^{q'} \, du \, ds.$$

Laplace's *statement* of the limits for the numerator is wrong; we should integrate for x from 0 to u, and then for u from 0 to 1. There is also another mistake. Laplace has the equation

$$\frac{p}{X} - \frac{q}{1 - X} + \frac{p'}{X - x} - \frac{q'}{1 - X + x} = 0.$$

He finds correctly that when $x = 0$ this gives

$$X = \frac{p + p'}{p + p' + q + q'}.$$

He says that when $x = 1$ it gives $X = 1$, which is wrong.

Laplace however really uses the right limits of integration in his work. His solution is very obscure; it is put in a much clearer form in a subsequent memoir which we shall presently notice; see Art. 909. He uses the following values,

$$p = 251527, \quad q = 241945,$$
$$p' = 737629, \quad q' = 698958,$$

and he obtains in the present memoir

$$P = \frac{1}{410458};$$

he obtains in the subsequent memoir

$$P = \frac{1}{410178}.$$

The problem is also solved in the *Théorie... des Prob.* pages 381—384; the method there is different and free from the mistakes which occur in the memoir. Laplace there uses values of p and q derived from longer observations, namely

$$p = 393386, \quad q = 377555;$$

he retains the same values of p' and q' as before, and he obtains

$$P = \frac{1}{328269}.$$

It will be seen that the new values of p and q make $\frac{p}{q}$ a little larger than the old values; hence it is natural that P should be increased.

903. Laplace gives in the memoir some important investigations on the probability of future events as deduced from observed events; these are reproduced in the *Théorie...des Prob.* pages 394—396.

904. Laplace devotes the last ten pages of his memoir to the theory of errors; he says that after his memoir in the sixth volume of the *Mémoires...par divers Savans* the subject had been considered by Lagrange, Daniel Bernoulli and Euler. Since, however, their principles differed from his own he is induced to resume the investigation, and to present his results in such a manner as to leave no doubt of their exactness. Accordingly he gives, with some extension, the same theory as before; see Art. 874. The theory does not seem, however, to have any great value.

905. The present memoir deserves to be regarded as very important in the history of the subject. The method of approximation to the values of definite integrals, which is here expounded, must be esteemed a great contribution to mathematics in general and to our special department in particular. The applications made to the problems respecting births shew the power of the method and its peculiar value in the theory of probability.

906. Laplace's next memoir on our subject is entitled *Mémoire sur les Suites;* it is published in the volume for 1779 of the *Histoire de l'Acad...Paris;* the date of publication is 1782. The memoir occupies pages 207—309 of the volume.

This memoir contains the theory of Generating Functions. With the exception of pages 269—286 the whole memoir is reproduced almost identically in the *Théorie...des Prob.;* it forms pages 9—80 of the work. The pages which are not reproduced relate to the solution of partial differential equations of the second order, and have no connexion with our subject.

The formulæ which occur at the top of pages 18 and 19 of the *Théorie...des Prob.* are stated in the memoir to agree with those which had been given in Newton's *Methodus differentialis;* this reference is omitted in the *Théorie...des Prob.*

907. Laplace's next memoir on our subject is entitled *Sur les approximations des Formules qui sont fonctions de très-grands nombres;* it is published in the volume for 1782 of the *Histoire de l'Acad...Paris:* the date of publication is 1785. The memoir occupies pages 1—88 of the volume.

Laplace refers at the commencement to the evaluation of the middle coefficient of a binomial raised to a high power by the aid of Stirling's Theorem ; Laplace considers this to be one of the most ingenious discoveries which had been made in the theory of Series. His object in the memoir is to effect similar transformations for other functions involving large numbers, in order that it might be practicable to calculate the numerical values of such functions.

The memoir is reproduced without any important change in the *Théorie...des Prob.*, in which it occupies pages 88—174. See Arts. 894, 899.

A mistake occurs at the beginning of page 29 of the memoir, and extends its influence to the end of page 30. Suppose that a function of two independent variables, θ and θ', is to be expanded in powers of these variables: we may denote the terms of the second degree by $M\theta^2 + 2N\theta\theta' + P\theta'^2$: Laplace's mistake amounts to omitting the term $2N\theta\theta'$. The mistake does not occur in the corresponding passage on page 108 of the *Théorie...des Prob.*

908. Laplace's next memoir is the continuation of the preceding; it is entitled, *Suite du Mémoire sur les approximations des Formules qui sont fonctions de très-grands Nombres;* it is published in the volume for 1783 of the *Histoire de l'Acad...Paris:* the date of publication is 1786. The memoir occupies pages 423—467 of the volume.

909. Laplace gives here some matter which is reproduced in the *Théorie...des Prob.* pages 363—365, 394—396. Pages 440—444

of the memoir are not reproduced in the *Théorie...des Prob.;* they depend partly on those pages of the memoir of 1782 which are erroneous, as we saw in Art. 907.

Laplace in this memoir applies his formulæ of approximation to the solution of questions in probability. See Arts. 767, 769. He takes the problem which we have noticed in Art. 896, and arrives at a result practically coincident with the former. He takes the problem which we have noticed in Art. 902, gives a much better investigation, and arrives at a result practically coincident with the former. He solves the problem about the births during a century to which we have referred in Art. 897, using the smaller values of p and q which we have given in Art. 902; he finds the required probability to be ·664. In the *Théorie...des Prob.* page 401 he uses the larger values of p and q which we have given in Art. 902, and obtains for the required probability ·782.

910. This memoir also contains a calculation respecting a lottery which is reproduced in the *Théorie...des Prob.* page 195. See Arts. 455, 864.

Laplace suggests on page 433 of the memoir that it would be useful to form a table of the value of $\int e^{-t} dt$ for successive limits of the integration : such a table we now possess.

911. In the same volume there is another memoir by Laplace which is entitled, *Sur les naissances, les mariages et les morts à Paris....* This memoir occupies pages 693—702 of the volume.

The following problem is solved. Suppose we know for a large country like France the number of births in a year; and suppose that for a certain district we know both the population and the number of births. If we assume that the ratio of the population to the number of births in a year is the same for the whole country as it is for the district, we can determine the population of the whole country. Laplace investigates the probability that the error in the result will not exceed an assigned amount. He concludes from his result that the district ought to contain not less than a million of people in order to obtain a sufficient accuracy in the number of the population of France.

The problem is reproduced in the *Théorie...des Prob.* pages 391—394. The necessary observations were made by the French government at Laplace's request; the population of the district selected was a little more than two millions. The solutions of the problem in the memoir and in the *Théorie...des Prob.* are substantially the same.

912. In the *Leçons de Mathématiques données à l'école normale, en* 1795, *par M. Laplace*, we have one *leçon* devoted to the subject of probabilities. The *leçons* are given in the *Journal de l'Ecole Polytechnique*, viie *et* viiie *cahiers*, 1812; but we may infer from page 164 that there had been an earlier publication. The *leçon* on probabilities occupies pages 140—172. It is a popular statement of some of the results which had been obtained in the subject, and was expanded by Laplace into the *Introduction* which appeared with the second edition of the *Théorie...des Prob.*, as he himself states at the beginning of the *Introduction*.

913. With the exception of the unimportant matter noticed in the preceding Article, Laplace seems to have left the Theory of Probability untouched for more than twenty-five years. His attention was probably fully engaged in embodying his own researches and those of other astronomers in his *Mécanique Céleste*, the first four volumes of which appeared between 1798 and 1805.

914. Laplace's next memoir connected with the Theory of Probability is entitled *Mémoire sur les approximations des formules qui sont fonctions de très-grands nombres, et sur leur application aux probabilités*. This memoir is published in the *Mémoires...de l'Institut* for 1809; the date of publication is 1810; the memoir occupies pages 353—415 of the volume, and a supplement occupies pages 559—565.

915. The first subject which is discussed is the problem relating to the inclination of the orbits of the planets and comets which is given in the *Théorie...des Prob.* pages 253—261; see also Art. 888. The mode of discussion is nearly the same. There is however some difference in the process relating to the *planets*, for in the memoir Laplace takes *two* right angles as the extreme

angle instead of *one* right angle which he takes in the *Théorie...
des Prob.* Laplace's words are, on page 362 of the memoir:

Si l'on fait varier les inclinaisons depuis zéro jusqu'à la demi-cir-
conférence, on fait disparoître la considération des mouvemens rétrogrades ; car le mouvement direct se change en rétrograde, quand l'inclinaison surpasse un angle droit.

Laplace obtains in the memoir the same numerical result as on
page 258 of the *Théorie...des Prob.;* but in the latter place the
fact of the motions being all in the same direction is expressly
used, while in the former place Laplace implies that this fact still
remains to be considered.

The calculation for the comets, which follows some investigations noticed in the next Article, does not materially differ from
the corresponding calculation in the *Théorie...des Prob.;* 97 is
taken as the number of comets in the memoir, and 100 in the
Théorie...des Prob.

916. Laplace gives an investigation the object of which is
the approximate calculation of a formula which occurs in the
solution of the problem noticed in the preceding Article. The
formula is the series for $\Delta^n s^i$, so far as the terms consist of
positive quantities raised to the power which i denotes. A large
part of the memoir bears on this subject, which is also treated
very fully in the *Théorie...des Prob.* pages 165—171, 475—482.
This memoir contains much that is not reproduced in the
Théorie...des Prob., being in fact superseded by better methods.

We may remark that Laplace gives two methods for finding the
value of $\int_0^\infty t^i e^{-ct^2} \cos bt\, dt$, but he does not notice the simplest
method, which would be to differentiate $\int_0^\infty e^{-ct^2} \cos bt\, dt$ four times
with respect to b, or twice with respect to c; see pages 368—370
of the memoir.

917. In pages 383—389 of the memoir we have an important
investigation resembling that given in pages 329—332 of the
Théorie...des Prob., which amounts to finding the probability that
a linear function of a large number of errors shall have a certain

value, the law of facility of a single error being any whatever.

Pages 390—397 of the memoir are spent in demonstrating the formula marked (*q*) which occurs at the top of page 170 of the *Théorie...des Prob.* The remaining pages of the memoir amount to demonstrating the formula marked (*p*) on page 168 of the *Théorie...des Prob.*, which is again discussed in pages 475—482 of the *Théorie...des Prob.* The methods of the memoir are very laborious and inferior to those of the *Théorie...des Prob.*

918. The supplement to the memoir consists of the matter which is reproduced in pages 333—335 and 340—342 of the *Théorie...des Prob.* In his supplement Laplace refers to his memoir of 1778; see Art. 904: the reference is not preserved in the *Théorie...des Prob.* He names Daniel Bernoulli, Euler, and Gauss; in the corresponding passage on page 335 of the *Théorie...des Prob.*, he simply says, *des géomètres célèbres.*

919. Laplace's next memoir is entitled, *Mémoire sur les Intégrales Définies, et leur application aux Probabilités, et spécialement à la recherche du milieu qu'il faut choisir entre les résultats des observations.* This memoir is published in the *Mémoires... de l'Institut* for 1810 ; the date of publication is 1811 : the memoir occupies pages 279—347 of the volume.

920. Laplace refers to his former memoirs on Generating Functions and on Approximations; he speaks of the approaching publication of his work on Probabilities. In his former memoirs he had obtained the values of some definite integrals by the passage from real to imaginary values; but he implies that such a method should be considered one of invention rather than of demonstration. Laplace says that Poisson had demonstrated several of these results in the *Bulletin de la Société Philomatique* for March 1811 ; Laplace now proposes to give direct investigations.

921. The first investigation is that which is reproduced in pages 482—484 of the *Théorie...des Prob.* Then follow those which are reproduced in pages 97—99 of the *Théorie...des Prob.* Next we have the problem of the Duration of Play, when the

players are of equal skill and one of them has an infinite capital; there is an approximate calculation which is reproduced in pages 235—238 of the *Théorie...des Prob.* Next we have the problem about balls and the long dissertation on some integrals which we find reproduced in pages 287—298 of the *Théorie...des Prob.* Lastly we have the theory of errors substantially coincident with so much of the same theory as we find in pages 314—328 and 340—342 of the *Théorie...des Prob.*

922. A theorem may be taken from page 327 of the memoir, which is not reproduced in the *Théorie...des Prob.*

To shew that if $\psi(x)$ always decreases as x increases between 0 and 1 we shall have

$$\int_0^1 \psi(x)\, dx \quad \text{greater than} \quad 3\int_0^1 x^2 \psi(x)\, dx.$$

It is sufficient to shew that

$$x^2 \int_0^x \psi(x)\, dx \text{ is greater than } 3\int_0^x x^2 \psi(x)\, dx,$$

or that $\quad 2x \int_0^x \psi(x)\, dx$ is greater than $2x^2 \psi(x)$,

or that $\quad \int_0^x \psi(x)\, dx$ is greater than $x\psi(x)$,

or that $\quad \psi(x)$ is greater than $\psi(x) + x\dfrac{d\psi(x)}{dx}$;

but this is obviously true, for $\dfrac{d\psi(x)}{dx}$ is negative.

The result stated on page 321 of the *Théorie...des Prob.*, that under a certain condition $\dfrac{k''}{k}$ is less than $\dfrac{1}{6}$, is an example of this theorem.

923. In the *Connaissance des Tems* for 1813, which is dated July 1811, there is an article by Laplace on pages 213—223, entitled, *Du milieu qu'il faut choisir entre les résultats d'un grand nombre d'observations.* The article contains the matter which is reproduced in pages 322—329 of the *Théorie...des Prob.* Laplace speaks of his work as soon about to appear.

924. In the *Connaissance des Tems* for 1815, which is dated November 1812, there is an article on pages 215—221 relating to Laplace's *Théorie...des Prob*. The article begins with an extract from the work itself, containing Laplace's account of its object and contents. After this follow some remarks on what is known as Laplace's nebular hypothesis respecting the formation of the solar system. Reference is made to the inference drawn by Michell from the group of the Pleiades ; see Art. 619.

925. In the *Connaissance des Tems* for 1816, which is dated November 1813, there is an article by Laplace, on pages 213—220, entitled, *Sur les Comètes*.

Out of a hundred comets which had been observed not one had been ascertained to move in an hyperbola; Laplace proposes to shew by the Theory of Probability that this result might have been expected, for the probability is very great that a comet would move either in an ellipse or parabola or in an hyperbola of so great a transverse axis that it would be undistinguishable from a parabola.

The solution of the problem proposed is very difficult, from the deficiency of verbal explanation. We will indicate the steps.

Laplace supposes that r denotes the radius of the sphere of the sun's activity, so that r represents a very great length, which may be a hundred thousand times as large as the radius of the earth's orbit. Let V denote the velocity of the comet at the instant when it enters the sphere of the sun's activity, so that r is the comet's radius vector at that instant. Let a be the semi-axis major of the orbit which the comet proceeds to describe, e its excentricity, D its perihelion distance, ϖ the angle which the direction of V makes with the radius r. Take the mass of the sun for the unit of mass, and the mean distance of the sun from the earth as the unit of distance; then we have the well-known formulæ;

$$\frac{1}{a} = \frac{2}{r} - V^2,$$

$$r V \sin \varpi = \sqrt{a (1 - e^2)},$$

$$D = a (1 - e).$$

From these equations by eliminating a and e we have

$$\sin^2 \varpi = \frac{2D - \dfrac{2D^2}{r} + D^2 V^2}{r^2 V^2},$$

and from this we deduce

$$1 - \cos \varpi = 1 - \frac{\sqrt{\left(1 - \dfrac{D}{r}\right)}}{rV} \sqrt{\left\{ r^2 V^2 \left(1 + \dfrac{D}{r}\right) - 2D \right\}}.$$

Now if we suppose that when the comet enters the sphere of the sun's activity all directions of motion which tend inwards are equally probable, we find that the chance that the direction will make an angle with the radius vector lying between zero and ϖ is $1 - \cos \varpi$. The values of the perihelion distance which correspond to these limiting directions are 0 and D. Laplace then proceeds thus:

...en supposant donc toutes les valeurs de D également possibles, on a pour la probabilité que la distance périhélie sera comprise entre zéro et D,

$$1 - \frac{\sqrt{\left(1 - \dfrac{D}{r}\right)}}{rV} \sqrt{\left\{ r^2 V^2 \left(1 + \dfrac{D}{r}\right) - 2D \right\}}.$$

Il faut multiplier cette valeur par dV; en l'intégrant ensuite dans des limites déterminées, et divisant l'intégrale par la plus grande valeur de V, valeur que nous désignerons par U; on aura la probabilité que la valeur de V sera comprise dans ces limites. Cela posé, la plus petite valeur de V est celle qui rend nulle la quantité renfermée sous le radical précédent; ce qui donne

$$rV = \frac{\sqrt{2D}}{\sqrt{\left(1 + \dfrac{D}{r}\right)}}.$$

It would seem that the above extract is neither clear nor correct; not clear for the real question is left uncertain; not correct in what relates to U. We will proceed in the ordinary way, and not as Laplace does. Let $\psi(V)$ stand for

$$1 - \frac{\sqrt{\left(1 - \dfrac{D}{r}\right)}}{rV} \sqrt{\left\{ r^2 V^2 \left(1 + \dfrac{D}{r}\right) - 2D \right\}};$$

then we have found that supposing all directions of projection
equally probable, if a comet starts with the velocity V the chance
is $\psi(V)$ that its perihelion distance will lie between 0 and D.
Now suppose we assume as a fact that the perihelion distance
does lie between 0 and D, but that we do not know the initial
velocity: required the probability that such initial velocity lies
between assigned limits. This is a question in inverse probability;
and the answer is that the chance is

$$\frac{\int \psi(V)\,dV}{\int \psi(V)\,dV},$$

where the integral in the numerator is to be taken between the
assigned limits; and the integral in the denominator between the
extreme admissible values of V.

Laplace finds the value of $\int \psi(V)dV$; for this purpose he
assumes

$$\sqrt{\left\{r^2 V^2\left(1+\frac{D}{r}\right)-2D\right\}}=rV\sqrt{\left(1+\frac{D}{r}\right)}-z.$$

For the assigned limits of V he takes $\dfrac{\sqrt{2D}}{r\sqrt{\left(1+\frac{D}{r}\right)}}$ and $\dfrac{i}{\sqrt{r}}$.

The value of $\int \psi(V)\,dV$ between these limits he finds to be approximately

$$\frac{(\pi-2)\sqrt{2D}}{2r}-\frac{D}{ir\sqrt{r}};$$

the other terms involve higher powers of r in the denominator,
and so are neglected.

The above expression is the numerator of the chance which
we require. For the denominator we may suppose that the upper
limit of the velocity is infinite, so that i will now be infinite.
Hence we have for the required chance

$$\left\{\frac{(\pi-2)\sqrt{2D}}{2r}-\frac{D}{ir\sqrt{r}}\right\} \div \frac{(\pi-2)\sqrt{2D}}{2r},$$

that is,

$$1 - \frac{\sqrt{2D}}{i\,(\pi - 2)\,\sqrt{r}}.$$

If for example we supposed $i^2 = 2$, we should have the extreme velocity which would allow the orbit to be an ellipse.

In the equation $\dfrac{1}{a} = \dfrac{2}{r} - V^2$ suppose $a = -100$; then

$$V^2 = \frac{r + 200}{100r}; \quad \text{thus} \quad i^2 = \frac{r + 200}{100}.$$

If we use this value of i we obtain the chance that the orbit shall be either an ellipse or a parabola or an hyperbola with transverse axis greater than a hundred times the radius of the earth's orbit. The chance that the orbit is an hyperbola with a smaller transverse axis will be

$$\frac{\sqrt{2D}}{i\,(\pi - 2)\,\sqrt{r}}.$$

Laplace obtains this result by his process.

Laplace supposes $D = 2$, $r = 100000$; and the value of i to be that just given: he finds the chance to be about $\dfrac{1}{5714}$.

Laplace then says that his analysis supposes that all values of D between 0 and 2 are equally probable for such comets as can be perceived; but observation shews that the comets for which the perihelion distance is greater than 1 are far less numerous than those for which it lies between 0 and 1. He proceeds to consider how this will modify his result.

926. In the *Connaissance des Tems* for 1818, which is dated 1815, there are two articles by Laplace on pages 361—381; the first is entitled, *Sur l'application du Calcul des Probabilités à la Philosophie naturelle;* the second is entitled, *Sur le Calcul des Probabilités, appliqué à la Philosophie naturelle.* The matter is reproduced in the first Supplement to the *Théorie...des Prob.* pages 1—25, except two pages, namely, 376, 377: these contain an application of the formulæ of probability to determine from observations the length of a seconds' pendulum.

927. In the *Connaissance des Tems* for 1820, which is dated 1818, there is an article by Laplace on pages 422—440, entitled, *Application du Calcul des Probabilités, aux opérations géodésiques:* it is reproduced in the second Supplement to the *Théorie...des Prob.* pages 1—25.

928. In the *Connaissance des Tems* for 1822, which is dated 1820, there is an article by Laplace on pages 346—348, entitled, *Application du Calcul des Probabilités aux opérations géodésiques de la méridienne de France:* it is reproduced in the third Supplement to the *Théorie...des Prob.* pages 1—7.

929. We have now to speak of the great work of Laplace which is entitled, *Théorie analytique des Probabilités.* This was published in 1812, in quarto. There is a dedication to Napoléon-le-Grand ; then follow 445 pages of text, and afterwards a table of contents which occupies pages 446—464 : on another page a few errata are noticed.

The second edition is dated 1814, and the third edition is dated 1820.

The second edition contains an introduction of CVI. pages ; then the text paged from 3 to 484 inclusive ; then a table of contents which occupies pages 485—506 : then two pages of errata are given.

The pages 9—444 of the first edition *were not reprinted* for the second or third edition ; a few pages were cancelled and replaced, apparently on account of errata.

The third edition has an introduction of CXLII. pages ; and then the remainder as in the second edition. There are, however, four supplements to the work which appeared subsequently to the · first edition. The exact dates of issue of these supplements do not seem to be given ; but the first and second supplements were probably published between 1812 and 1820, the third in 1820, and the fourth after 1820. Copies of the third edition generally have the first three supplements, but not the fourth.

930. Since the bulk of the text of Laplace's work *was not reprinted* for the editions which appeared during his life time,

a reference to the page of the work will in general suffice for
any of these editions : accordingly we shall adopt this mode of
reference.

An edition of the works of Laplace was published in France
at the national expense. The seventh volume consists of the
Théorie...des Prob.; it is dated 1847. This volume is a reprint of
the third edition. The title, advertisement, introduction, and
table of contents occupy CXCV. pages; the text occupies 532
pages, and the four supplements occupy pages 533—691.

It will be found that in the text a page n of the editions published by Laplace himself will correspond nearly to the page $n + \dfrac{n}{10}$
of the national edition : thus our references will be easily available
for the national edition. We do not think that the national
edition is so good as it ought to have been ; we found, for example,
that in the second supplement the misprints of the original were
generally reproduced.

931. We shall now proceed to analyse the work: We take the
third edition, and we shall notice the places in which the introduction differs from the introduction to the second edition.

The dedication was not continued after the first edition, so that
it may be interesting to reproduce it here.

A Napoléon-le-Grand. Sire, La bienveillance avec laquelle Votre
Majesté a daigné accueillir l'hommage de mon Traité de Mécanique
Céleste, m'a inspiré le desir de Lui dédier cet Ouvrage sur le Calcul des
Probabilités. Ce calcul délicat s'étend aux questions les plus importantes de la vie, qui ne sont en effet, pour la plupart, que des problèmes
de probabilité. Il doit, sous ce rapport, intéresser Votre Majesté dont
le génie sait si bien apprécier et si dignement encourager tout ce qui
peut contribuer au progrès des lumières, et de la prospérité publique.
J'ose La supplier d'agréer ce nouvel hommage dicté par la plus vive
reconnaissance, et par les sentimens profonds d'admiration et de respect,
avec lesquels je suis, Sire, de Votre Majesté, Le très-humble et très-
obéissant serviteur et fidèle sujet, Laplace.

Laplace has been censured for suppressing this dedication after
the fall of Napoleon ; I do not concur in this censure. The dedication appears to me to be mere adulation ; and it would have

been almost a satire to have repeated it when the tyrant of Europe
had become the mock sovereign of Elba or the exile of St Helena :
the fault was in the original publication, and not in the final sup-
pression.

932. We have said that some pages of the original impression
were cancelled, and others substituted; the following are the pages :
25, 26, 27, 28, 37, 38, 147, 148, 303, 304, 359, 360, 391, 392 ; we
note them because a student of the first edition will find some
embarrassing errata in them.

933. The introduction to the *Théorie...des Prob.* was published separately in octavo under the title of *Essai philosophique
sur les Probabilités;* we shall however refer to the introduction
by the pages of the third edition of the *Théorie...des Prob.*

934. On pages I—XVI. of the introduction we have some general remarks on Probability, and a statement of the first principles
of the mathematical theory ; the language is simple and the
illustrations are clear, but there is hardly enough space allotted to
the subject to constitute a good elementary exposition for beginners.

935. On pages XVI—XXXVII. we have a section entitled *Des
méthodes analytiques du Calcul des Probabilités;* it is principally
devoted to an account of the Theory of Generating Functions, the
account being given in words with a very sparing use of symbols.
This section may be regarded as a complete waste of space ; it
would not be intelligible to a reader unless he were able to master
the mathematical theory delivered in its appropriate symbolical
language, and in that case the section would be entirely superfluous.

This section differs in the two editions ; Laplace probably
thought he improved in his treatment of the difficult task he had
undertaken, namely to explain abstruse mathematical processes in
ordinary language. We will notice two of the changes. Laplace
gives on pages XXIII. and XXIV. some account of De Moivre's
treatment of Recurring Series; this account is transferred from page
CI. of the second edition of the introduction : a student however

32

who wished to understand the treatment would have to consult the original work, namely De Moivre's *Miscellanea Analytica*, pages 28—33. Also some slight historical reference to Wallis and others is introduced on pages XXXV—XXXVII.; this is merely an abridgement of the pages 3—8 of the *Théorie...des Prob.*

936. We have next some brief remarks on games, and then some reference to the unknown inequalities which may exist in chances supposed to be equal, such as would arise from a want of symmetry in a coin or die; see Arts. 877, 881, 891.

937. We have next a section on the laws of probability, which result from an indefinite multiplication of events; that is the section is devoted to the consideration of James Bernoulli's theorem and its consequences. Some reflexions here seem aimed at the fallen emperor to whom the first edition of the work was dedicated; we give two sentences from page XLIII.

Voyez au contraire, dans quel abime de malheurs, les peuples ont été souvent précipités par l'ambition et par la perfidie de leurs chefs. Toutes les fois qu'une grande puissance enivrée de l'amour des conquêtes, aspire à la domination universelle; le sentiment de l'indépendance produit entre les nations menacées, une coalition dont elle devient presque toujours la victime.

The section under consideration occurs in the second edition, but it occupies a different position there, Laplace having made some changes in the arrangement of the matter in the third edition.

We may notice at the end of this section an example of the absurdity of attempting to force mathematical expressions into unmathematical language. Laplace gives a description of a certain probability in these words:

La théorie des fonctions génératrices donne une expression très simple de cette probabilité, que l'on obtient en intégrant le produit de la différentielle de la quantité dont le résultat déduit d'un grand nombre d'observations s'écarte de la vérité, par une constante moindre que l'unité, dépendante de la nature du problème, et élevée à une puissance dont l'exposant est le rapport du carré de cet écart, au nombre des observations. L'intégrale prise entre des limites données, et divisée par la même intégrale étendue à l'infini positif et négatif, exprimera la probabilité que l'écart de la vérité, est compris entre ces limites.

A student familiar with the *Théorie...des Prob.* itself might not find it easy to say what formula Laplace has in view ; it must be that which is given on page 309 and elsewhere, namely

$$\sqrt{\frac{k}{k''\pi}}\int dr\, e^{-\frac{kr^2}{4K'}}.$$

Other examples of the same absurdity will be found on page LI. of the introduction, and on page 5 of the first supplement.

938. A section occupies pages XLIX—LXX. entitled *Application du Calcul des Probabilités, à la Philosophie naturelle.* The principle which is here brought forward is simple ; we will take one example which is discussed in the *Théorie...des Prob.* If a large number of observations be taken of the height of a barometer at nine in the morning and at four in the afternoon, it is found that the average in the former case is higher than in the latter ; are we to ascribe this to chance or to a constant cause? The theory of probabilities shews that if the number of observations be large enough the existence of a constant cause is very strongly indicated. Laplace intimates that in this way he had been induced to undertake some of his researches in Physical Astronomy, because the theory of probabilities shewed irresistibly that there were constant causes in operation.

Thus the section contains in reality a short summary of Laplace's contributions to Physical Astronomy ; and it is a memorable record of the triumphs of mathematical science and human genius. The list comprises—the explanation of the irregularity in the motion of the moon arising from the spheroidal figure of the earth—the secular equation of the moon—the long inequalities of Jupiter and Saturn—the laws connecting the motions of the satellites of Jupiter—the theory of the tides. See Gouraud, page 115 ; he adds to the list—the temperature of the earth shewn to be constant for two thousand years : it does not appear that Laplace himself here notices this result.

939. In the second edition of the *Théorie...des Prob.*

Laplace did not include the secular acceleration of the moon and the theory of the tides in the list of his labours suggested by the Theory of Probability. Also pages LI—LVI. of the introduction seem to have been introduced into the third edition, and taken from the first supplement.

Laplace does not give references in his *Théorie...des Prob.*, so we cannot say whether he published all the calculations respecting probability which he intimates that he made; they would however, we may presume, be of the same kind as that relating to the barometer which is given in page 350 of the *Théorie...des Prob.*, and so would involve no novelty of principle.

Laplace alludes on page LIV. to some calculations relating to the masses of Jupiter and Saturn; the calculations are given in the first supplement. Laplace arrived at the result that it was 1000000 to 1 that the error in the estimation of the mass of Jupiter could not exceed $\frac{1}{100}$ of the whole mass. Nevertheless it has since been recognised that the error was as large as $\frac{1}{50}$; see Poisson, *Recherches sur la Prob...*, page 316.

940. Laplace devotes a page to the *Application du Calcul des Probabilités aux Sciences morales;* he makes here some interesting remarks on the opposing tendencies to change and to conservatism.

941. The next section is entitled, *De la Probabilité des témoignages;* this section occupies pages LXXI—LXXXII: it is an arithmetical reproduction of some of the algebraical investigations of Chapter XI. of the *Théorie...des Prob.* One of Laplace's discussions has been criticised by John Stuart Mill in his *Logic;* see Vol. II. page 172 of the fifth edition. The subject is that to which we have alluded in Art. 735. Laplace makes some observations on miracles, and notices with disapprobation the language of Racine, Pascal and Locke. He examines with some detail a famous argument by Pascal which he introduces thus:

Ici se présente naturellement la discussion d'un argument fameux de Pascal, que Craig, mathématicien anglais, a reproduit sous une forme géométrique. Des témoins attestent qu'ils tiennent de la Divinité même, qu'en se conformaut à telle chose, on jouira, nou pas d'une ou de deux, mais d'une infinité de vies heureuses. Quelque faible que soit la probabilité des témoignages, pourvu qu'elle ne soit pas infiniment petite; il est clair que l'avantage de ceux qui se conforment à la chose prescrite, est infui, puisqu'il est le produit de cette probabilité par un bien infini; on ne doit donc point balancer à se procurer cet avantage.

See also the *Athenæum* for Jan. 14th, 1865, page 55.

942. The next section is entitled, *Des choix et des décisions des assemblées;* it occupies four pages: results are stated respecting voting on subjects and for candidates which are obtained at the end of Chapter II. of the *Théorie...des Prob.*

The next section is entitled, *De la probabilité des Jugemens des tribunaux;* it occupies five pages: results are stated which are obtained in the first supplement to the *Théorie...des Prob.* This section is nearly all new in the third edition of the *Théorie...des Prob.*

The next section is entitled, *Des Tables de mortalité, et des durées moyennes de la vie, des mariages et des associations quelconques;* it occupies six pages : results are stated which are obtained in Chapter VIII. of the *Théorie...des Prob.*

The next section is entitled, *Des bénéfices des établissemens qui dépendent de la probabilité des évènemens;* it occupies five pages. This section relates to insurances: results are given which are obtained in Chapter IX. of the *Théorie...des Prob.*

943. The next section is entitled, *Des illusions dans l'estimation des Probabilités;* this important section occupies pages CII—CXXVIII: in the second edition of the *Théorie...des Prob.* the corresponding section occupied little more than seven pages.

The illusions which Laplace notices are of various kinds. One of the principal amounts to imagining that past events influence future events when they are really unconnected. This is illustrated from the example of lotteries, and by some remarks on page CIV. relating to the birth of a son, which are new in the third edition. Another illusion is the notion of a kind of fatality which gamblers often adopt.

Laplace considers that one of the great advantages of the

theory of probabilities is that it teaches us to mistrust our first impressions; this is illustrated by the example which we have noticed in Art. 856, and by the case of the Chevalier de Méré: see Art. 10. Laplace makes on his page CVIII. some remarks respecting the excess of the births of boys over the births of girls; these remarks are new in the third edition.

Laplace places in the list of illusions an application of the Theory of Probability to the summation of series, which was made by Leibnitz and Daniel Bernoulli. They estimated the infinite series

$$1 - 1 + 1 - 1 + \ldots$$

as equal to $\frac{1}{2}$; because if we take an even number of terms we obtain 0, and if we take an odd number of terms we obtain 1, and they assumed it to be equally probable that an infinite number of terms is odd or even. See Dugald Stewart's *Works edited by Hamilton*, Vol. IV. page 204.

Laplace makes some remarks on the apparent verification which occasionally happens of predictions or of dreams; and justly remarks that persons who attach importance to such coincidences generally lose sight of the number of cases in which such anticipations of the future are falsified by the event. He says,

Ainsi, le philosophe de l'antiquité, auquel on montrait dans un temple, pour exalter la puissance du dieu qu'on y adorait, les *ex voto* de tous ceux qui après l'avoir invoqué, s'étaient sauvés du naufrage, fit une remarque conforme au calcul des probabilités, en observant qu'il ne voyait point inscrits, les noms de ceux qui, malgré cette invocation, avaient péri.

944. A long discussion on what Laplace calls *Psychologie* occupies pages CXIII—CXXVIII of the present section. There is much about the *sensorium*, and from the close of the discussion it would appear that Laplace fancied all mental phenomena ought to be explained by applying the laws of Dynamics to the vibrations of the sensorium. Indeed we are told on page CXXIV. that faith is a modification of the sensorium, and an extract from Pascal is used in a manner that its author would scarcely have approved.

945. The next section is entitled, *Des divers moyens d'approcher de la certitude;* it occupies six pages. Laplace says,

L'induction, l'analogie, des hypothèses fondées sur les faits et rectifiées sans cesse par de nouvelles observations, un tact heureux donné par la nature et fortifié par des comparaisons nombreuses de ses indications avec l'expérience; tels sont les principaux moyens de parvenir à la vérité.

A paragraph beginning on page CXXIX, with the words *Nous jugeons* is new in the third edition, and so are the last four lines of page CXXXII. Laplace cites Bacon as having made a strange abuse of induction to demonstrate the immobility of the earth. Laplace says of Bacon,

Il a donné pour la recherche de la vérité, le précepte et non l'exemple. Mais en insistant avec toute la force de la raison et de l'éloquence, sur la nécessité d'abandonner les subtilités insignifiantes de l'école, pour se livrer aux observations et aux expériences, et en indiquant la vraie méthode de s'élever aux causes générales des phénomènes; ce grand philosophe a contribué aux progrès immenses que l'esprit humain a faits dans le beau siècle où il a terminé sa carrière.

Some of Laplace's remarks on *Analogy* are quoted with approbation by Dugald Stewart; see his *Works edited by Hamilton,* Vol. IV. page 290.

946. The last section of the introduction is entitled, *Notice historique sur le Calcul des Probabilités;* this is brief but very good. The passage extending from the middle of page CXXXIX. to the end of page CXLI. is new in the third edition; it relates principally to Laplace's development in his first supplement of his theory of errors. Laplace closes this passage with a reference to the humble origin of the subject he had so much advanced; he says it is remarkable that a science which began with the consideration of games should have raised itself to the most important objects of human knowledge.

A brief sketch of the plan of the *Théorie...des Prob.,* which appeared on the last page of the introduction in the second edition, is not repeated in the third edition.

947. The words in which at the end of the introduction La-

place sums up the claims of the Theory of Probability well deserve to be reproduced here:

.   On voit par cet Essai, que la théorie des probabilités n'est au fond, que le bon sens réduit au calcul : elle fait apprécier avec exactitude, ce que les esprits justes sentent par une sorte d'instinct, sans qu'ils puissent souvent s'en rendre compte. Si l'on considère les méthodes analytiques auxquelles cette théorie a donné naissance, la vérité des principes qui lui servent de base, la logique fine et délicate qu'exige leur emploi dans la solution des problèmes, les établissemens d'utilité publique qui s'appuient sur elle, et l'extension qu'elle a reçue et qu'elle peut recevoir encore, par son application aux questions les plus importantes de la Philosophie naturelle et des sciences morales; si l'on observe ensuite, que dans les choses mêmes qui ne peuvent être soumises au calcul, elle donne les aperçus les plus sûrs qui puissent nous guider dans nos jugemens, et qu'elle apprend à se garantir des illusions qui souvent nous égarent; on verra qu'il n'est point de science plus digne de nos méditations, et qu'il soit plus utile de faire entrer dans le système de l'instruction publique.

948.   We now leave the introduction and pass to the *Théorie... des Prob.* itself. Laplace divides this into two books. *Livre* I. is entitled *Du Calcul des Fonctions Génératrices:* this occupies pages 1—177; *Livre* II. is entitled *Théorie générale des Probabilités;* this occupies pages 179—461. Then follow *Additions* on pages 462—484.

949.   The title which Laplace gives to his *Livre* I. does not adequately indicate its contents. The subject of generating functions, strictly so called, forms only the first part of the book ; the second part is devoted to the consideration of the approximate calculation of various expressions which occur in the Theory of Probability.

950.   The first part of *Livre* I. is almost a reprint of the memoir of 1779 in which it originally appeared ; see Art. 906. This part begins with a few introductory remarks on pages 3—8 ; these pages 3—8 of the third edition do not quite agree with the pages 1—8 of the first edition, but there is nothing of consequence peculiar to the first edition. Laplace draws attention to the importance of notation in mathematics; and he illustrates the point by the

advantage of the notation for denoting powers, which leads him to speak of Descartes and Wallis.

Laplace points out that Leibnitz made a remarkable use of the notation of powers as applied to differentials; this use we might describe in modern terms as an example of the separation of the symbols of operation and quantity. Lagrange followed up this analogy of powers and differentials; his memoir inserted in the volume for 1772 of the memoirs of the Academy of Berlin is characterised by Laplace as one of the finest applications ever made of the method of inductions.

951. The first Chapter of the first part of *Livre* I. is entitled *Des Fonctions génératrices, à une variable;* it occupies pages 9—49.

The method of generating functions has lost much of its value since the cultivation of the Calculus of Operations by Professor Boole and others; partly on this account, and partly because the method is sufficiently illustrated in works on the Theory of Finite Differences, we shall not explain it here.

Pages 39—49 contain various formulæ of what we now call the Calculus of Operations; these formulæ cannot be said to be *demonstrated* by Laplace; he is content to rely mainly on analogy. Lagrange had led the way here; see the preceding Article.

One of the formulæ may be reproduced; see Laplace's page 41. If we write Taylor's theorem symbolically we obtain

$$\Delta y_x = \left(e^{h\frac{d}{dx}} - 1\right)y_x,$$

where Δ indicates the difference in y_x arising from a difference h in x. Then

$$\Delta^n y_x = \left(e^{h\frac{d}{dx}} - 1\right)^n y_x.$$

Laplace transforms this into the following result,

$$\Delta^n y_x = \left(e^{\frac{h}{2}\frac{d}{dx}} - e^{-\frac{h}{2}\frac{d}{dx}}\right)^n y_{x+\frac{nh}{2}}.$$

The following is his method:

$$\left(e^{h\frac{d}{dx}} - 1\right)^n y_x = e^{\frac{nh}{2}\frac{d}{dx}}\left(e^{\frac{h}{2}\frac{d}{dx}} - e^{-\frac{h}{2}\frac{d}{dx}}\right)^n y_x.$$

Now let $k \left(\dfrac{d}{dx}\right)^r$ denote any term arising from the development of

$$\left(e^{\frac{h}{2}\frac{d}{dx}} - e^{-\frac{h}{2}\frac{d}{dx}}\right)^n.$$

Then
$$k \left(\dfrac{d}{dx}\right)^r e^{\frac{nh}{2}\frac{d}{dx}} y_x = k \left(\dfrac{d}{dx}\right)^r y_{x+\frac{nh}{2}};$$

and the term on the right hand may be supposed to have arisen from the development of $\left(e^{\frac{h}{2}\frac{d}{dx}} - e^{-\frac{h}{2}\frac{d}{dx}}\right)^n y_{x+\frac{nh}{2}}$. Thus the formula is considered to be established.

We ought to observe that Laplace does not express the formula quite in the way which we adopt. His mode of writing Taylor's Theorem is

$$\Delta y_x = e^{h\frac{dy_x}{dx}} - 1,$$

and then he would write

$$\Delta^n y_x = \left(e^{h\frac{dy_x}{dx}} - 1\right)^n.$$

He gives verbal directions as to the way in which the symbols are to be treated, which of course make his formulæ really identical with those which we express somewhat differently. We may notice that Laplace uses c for the base of the Napierian logarithms, which we denote by e.

If in the formula we put $h = 1$ and change x into $x - \dfrac{n}{2}$ we obtain

$$\Delta^n y_{x-\frac{n}{2}} = \left(e^{\frac{1}{2}\frac{d}{dx}} - e^{-\frac{1}{2}\frac{d}{dx}}\right)^n y_x,$$

which Laplace obtains on his page 45 by another process.

952. The second Chapter of the first part of *Livre* I. is entitled *Des fonctions génératrices à deux variables:* it occupies pages 50—87.

Laplace applies the theory of generating functions to solve equations in Finite Differences with two independent variables. He gives on his pages 63—65 a strange process for integrating the following equation in Finite Differences,

$$z_{x+1,\,y+1} - a z_{x,\,y+1} - b z_{x+1,\,y} - c z_{x,\,y} = 0.$$

We might suppose that $z_{x,y}$ is the coefficient of $t^x \tau^y$ in the expansion of a function of t and τ; then it would easily follow that this function must be of the form

$$\frac{\phi(t) + \psi(\tau)}{\tau t \left(\dfrac{1}{\tau t} - \dfrac{a}{\tau} - \dfrac{b}{t} - c\right)},$$

where $\phi(t)$ is an arbitrary function of t, and $\psi(\tau)$ an arbitrary function of τ.

Laplace, however, proceeds thus. He puts

$$\frac{1}{\tau t} - \frac{a}{\tau} - \frac{b}{t} - c = 0,$$

and he calls this the *équation génératrice* of the given equation in Finite Differences. He takes u to denote the function of t and τ which when expanded in powers of t and τ has $z_{x,y}$ for the coefficient of $t^x \tau^y$. Then in the expansion of $\dfrac{u}{t^x \tau^y}$ the coefficient of $t^0 \tau^0$ will be $z_{x,y}$.

Laplace then transforms $\dfrac{u}{t^x \tau^y}$ thus. By the *équation génératrice* we have

$$\frac{1}{t} = \frac{c + \dfrac{a}{\tau}}{\dfrac{1}{\tau} - b};$$

therefore,

$$\frac{u}{t^x \tau^y} = \frac{u \left(\dfrac{1}{\tau} - \dfrac{1}{b} + \dfrac{1}{b}\right)^y \left[c + ab + a \left(\dfrac{1}{\tau} - b\right)\right]^x}{\left(\dfrac{1}{\tau} - b\right)^x}.$$

Develope the second member according to powers of $\dfrac{1}{\tau} - b$; thus

$$\frac{u}{t^x \tau^y} = u \left\{ \left(\frac{1}{\tau} - b\right)^y + yb \left(\frac{1}{\tau} - b\right)^{y-1} + \frac{y(y-1)}{1 \cdot 2} b^2 \left(\frac{1}{\tau} - b\right)^{y-2} + \dots \right\}$$

$$\times \left\{ a^x + \frac{x(c+ab)\,a^{x-1}}{\dfrac{1}{\tau} - b} + \frac{x(x-1)}{1 \cdot 2} (c+ab)^2 \frac{a^{x-2}}{\left(\dfrac{1}{\tau} - b\right)^2} + \dots \right\}.$$

Multiply the two series together. Let

$$V = a^x,$$

$$V_1 = yba^x + x\,(c + ab)\,a^{x-1},$$

$$V_2 = \frac{y\,(y-1)}{1\cdot 2}\,b^2 a^x + yxb\,(c + ab)\,a^{x-1} + \frac{x\,(x-1)}{1\cdot 2}\,(c + ab)^2\,a^{x-2},$$

$$V_3 = \frac{y\,(y-1)\,(y-2)}{1\cdot 2\cdot 3}\,b^3 a^x + \dots$$

$$\dots\dots\dots\dots$$

Then

$$\frac{u}{t^x \tau^y} = u\left\{ V\left(\frac{1}{\tau} - b\right)^y + V_1 \left(\frac{1}{\tau} - b\right)^{y-1} + \dots + V_y \right.$$

$$\left. + \frac{V_{y+1}}{\dfrac{1}{\tau} - b} + \frac{V_{y+2}}{\left(\dfrac{1}{\tau} - b\right)^2} + \dots + \frac{V_{y+x}}{\left(\dfrac{1}{\tau} - b\right)^x} \right\}.$$

But the equation

$$\frac{1}{t\tau} - \frac{a}{\tau} - \frac{b}{t} - c = 0$$

gives

$$\frac{1}{\dfrac{1}{\tau} - b} = \frac{\dfrac{1}{t} - a}{c + ab};$$

therefore

$$\frac{u}{t^x \tau^y} = u\left\{ V\left(\frac{1}{\tau} - b\right)^y + V_1 \left(\frac{1}{\tau} - b\right)^{y-1} + \dots + V_y \right.$$

$$\left. + \frac{V_{y+1}}{c + ab}\left(\frac{1}{t} - a\right) + \frac{V_{y+2}}{(c + ab)^2}\left(\frac{1}{t} - a\right)^2 + \dots + \frac{V_{y+x}}{(c + ab)^x}\left(\frac{1}{t} - a\right)^x \right\}.$$

Now we pass from the generating functions to the coefficients, and we pick out the coefficients of $t^0 \tau^0$ on both sides. This gives $z_{x,y}$ on the left-hand side, and on the right-hand side a series which we shall now proceed to express.

Let Δ apply to x, and indicate a Finite Difference produced by the change of x into $x + 1$; and let δ similarly apply to y, and indicate a Finite Difference produced by the change of y into $y + 1$.

Now $\left(\frac{1}{\tau}-b\right)^r = b^r\left(\frac{1}{b\tau}-1\right)^r$; hence in $u\left(\frac{1}{\tau}-b\right)^r$ the coefficient

of $t^0\tau^0$ will be $b^r\delta^r\left(\frac{z_{0,y}}{b^y}\right)$, provided we suppose that y is made zero

after the operation denoted by δ^r has been performed on $\frac{z_{0,y}}{b^y}$.

Similarly in $u\left(\frac{1}{t}-a\right)^r$ the coefficient of $t^0\tau^0$ will be $a^r\Delta^r\left(\frac{z_{x,0}}{a^x}\right)$,

provided we suppose that x is made zero after the operation de-

noted by Δ^r has been performed on $\frac{z_{x,0}}{a^x}$.

In this way we obtain

$$z_{x,y} = Vb^y\delta^y\left(\frac{z_{0,y}}{b^y}\right) + V_1 b^{y-1}\delta^{y-1}\left(\frac{z_{0,y}}{b^y}\right) + \dots + V_y z_{0,0}$$

$$+\frac{a}{c+ab}V_{y+1}\Delta\left(\frac{z_{x,0}}{a^x}\right) + \frac{a^2}{(c+ab)^2}V_{y+2}\Delta^2\left(\frac{z_{x,0}}{a^x}\right)$$

$$+ \dots + \frac{a^x}{(c+ab)^x}V_{y+x}\Delta^x\left(\frac{z_{x,0}}{a^x}\right):$$

Thus we see that in order to obtain $z_{x,y}$ we must know $z_{0,1}, z_{0,2},\dots$ up to $z_{0,y}$, and we must know $z_{1,0}, z_{2,0},\dots$ up to $z_{x,0}$.

Now we have to observe that this process as given by Laplace cannot be said to be demonstrative or even intelligible. His method of *connecting the two independent variables* by the *équation génératrice* without explanation is most strange.

But the student who is acquainted with the modern methods of the Calculus of Operations will be able to translate Laplace's process into a more familiar language.

Let E denote the change of x into $x+1$, and F the change of y into $y+1$: then the fundamental equation we have to integrate will be written

$$(EF - aF - bE - c)\, z_{x,y} = 0,$$

or for abbreviation

$$EF - aF - bE - c = 0.$$

Then $E^x F^y$ will be expanded in the way Laplace expands $\frac{1}{t^x\tau^y}$ and his result obtained from $E^x F^y z_{0,0}$. Thus we rely on the foundations on which the Calculus of Operations is based.

We may notice that we have changed Laplace's notation in order to avoid the dashes which are difficult in printing. Laplace uses x' where we use y, t' where we use τ, and $'\Delta$ where we use δ.

953. Laplace takes another equation in Finite Differences. The equation we will denote thus

$$\Delta^n z_{x, y} + \frac{a}{a}\Delta^{n-1}\delta z_{x, y} + \frac{b}{a^2}\Delta^{n-2}\delta^2 z_{x, y} + \ldots = 0.$$

Here Δ belongs to x of which the difference is unity; and δ belongs to y of which the difference is a.

Laplace says that the *équation génératrice* is

$$\left(\frac{1}{t}-1\right)^n + \frac{a}{a}\left(\frac{1}{t}-1\right)^{n-1}\left(\frac{1}{\tau^a}-1\right) + \frac{b}{a^2}\left(\frac{1}{t}-1\right)^{n-2}\left(\frac{1}{\tau^a}-1\right)^2 + \ldots = 0.$$

He supposes that this equation is solved, and thus decomposed into the following n equations:

$$\frac{1}{t}-1 = \frac{q}{a}\left(1 - \frac{1}{\tau^a}\right),$$

$$\frac{1}{t}-1 = \frac{q_1}{a}\left(1 - \frac{1}{\tau^a}\right),$$

$$\frac{1}{t}-1 = \frac{q_2}{a}\left(1 - \frac{1}{\tau^a}\right),$$

$$\ldots\ldots\ldots$$

where $q, q_1, q_2, \ldots$ are the n roots of the equation

$$\zeta^n - a\zeta^{n-1} + b\zeta^{n-2} - \ldots = 0.$$

Then, using the first root

$$\frac{u}{t^x \tau^y} = \frac{u}{\tau^y}\left(1 + \frac{q}{a} - \frac{q}{a\tau^a}\right)^x$$

$$= \frac{u}{\tau^y}(-1)^x\left\{\frac{q^x}{a^x \tau^{ax}} - x\frac{q^{x-1}}{a^{x-1}}\left(1 + \frac{q}{a}\right)\frac{1}{\tau^{a(x-1)}} + \ldots\right\}.$$

Then passing from the generating functions to the coefficients, that is equating the coefficients of $t^x \tau^0$, we obtain

$$z_{x, y} = (-1)^x\left\{\frac{q^x}{a^x}z_{0, y+ax} - x\frac{q^{x-1}}{a^{x-1}}\left(1 + \frac{q}{a}\right)z_{0, y+a(x-1)} + \ldots\right\}.$$

The second member may be put in the form

$$\left(1+\frac{a}{q}\right)^{x+\frac{y}{a}}\left(-\frac{q}{a}\right)^{x}\delta^{x}\left\{\left(\frac{q}{a+q}\right)^{\frac{y}{a}}z_{0,\,v}\right\}.$$

Denote the quantity $\left(\frac{q}{a+q}\right)^{\frac{y}{a}}z_{0,\,v}$ by the arbitrary function $\phi\,(y)$. Thus

$$z_{x,\,y}=\left(1+\frac{a}{q}\right)^{x+\frac{y}{a}}\left(-\frac{q}{a}\right)^{x}\delta^{x}\phi\,(y).$$

This value of $z_{x,\,y}$ will then satisfy the equation in Finite Differences.

Each of the n roots $q,\,q_{1},\,q_{2},\,\ldots$ gives rise to a similar expression; and the sum of the n particular values thus obtained for $z_{x,\,y}$ will furnish the general value, involving n arbitrary functions.

The student will as before be able to translate this process into the language of the Calculus of Operations.

Laplace continues thus: Suppose a indefinitely small, and equal to dy. Then

$$\left(1+\frac{dy}{q}\right)^{x+\frac{y}{dy}}=e^{\frac{y}{q}},$$

as we may see by taking logarithms. Thus we shall obtain

$$z_{x,\,y}=e^{\frac{y}{q}}(-q)^{x}\frac{d^{x}\phi\,(y)}{dy^{x}}+e^{\frac{y}{q_{1}}}(-q_{1})^{x}\frac{d^{x}\phi_{1}\,(y)}{dy^{x}}+\cdots$$

This is the complete integral of the equation

$$\Delta^{n}z_{x,\,y}+a\,\Delta^{n-1}\left(\frac{dz_{x,\,y}}{dy}\right)+b\,\Delta^{n-2}\left(\frac{d^{2}z_{x,\,y}}{dy^{2}}\right)+\ldots=0.$$

Laplace next gives some formulæ of what we now call the Calculus of Operations, in the case of two independent variables; see his pages 68—70.

954. In his pages 70—80 Laplace offers some remarks on the transition from the finite to the indefinitely small; his object is to shew that the process will furnish rigorous demonstrations. He illustrates by referring to the problem of vibrating strings, and this leads him to notice a famous question, namely that of the admissibility of discontinuous functions in the solution of partial dif-

ferential equations; he concludes that such functions are admissible under certain conditions. Professor Boole regards the *argument* as unsound; see his *Finite Differences*, Chapter x.

955. Laplace closes the Chapter with some general considerations respecting generating functions. The only point to which we need draw attention is that there is an important error in page 82; Laplace gives an incomplete form as the solution of an equation in Finite Differences; the complete form will be found on page 5 of the fourth supplement. We shall see the influence of the error hereafter in Arts. 974, 980, 984.

956. We now arrive at the second part of *Livre* I., this is nearly a reprint of the memoir for 1782; the method of approximation had however been already given in the memoir for 1778. See Arts. 894, 899, 907, 921.

The first chapter of the second part of *Livre* I. is entitled *De l'intégration par approximation, des différentielles qui renferment des facteurs élevés à de grandes puissances;* this Chapter occupies pages 88—109.

957. The method of approximation which Laplace gives is of great value: we will explain it. Suppose we require the value of $\int y\,dx$ taken between two values of x which include a value for which y is a maximum. Assume $y = Ye^{-t^2}$, where Y denotes this maximum value of y. Then

$$\int y\,dx = Y\int e^{-t^2}\frac{dx}{dt}\,dt.$$

Let $y = \phi(x)$; suppose a the value of x which makes y have the value Y: assume $x = a + \theta$.

Thus $\qquad\qquad \phi(a + \theta) = Ye^{-t^2};$

therefore $\qquad\qquad t^2 = \log\dfrac{Y}{\phi(a+\theta)}.$

From this equation we may expand t in a series of ascending powers of θ, and then by reversion of series we may obtain θ in a series of ascending powers of t. Suppose that thus we have

$$\theta = B_1 t + B_2 t^2 + B_3 t^3 + \cdots;$$

then
$$\frac{dx}{dt} = \frac{d\theta}{dt} = B_1 + 2B_2t + 3B_3t^2 + \dots;$$

$$\int y\,dx = Y\int e^{-t^2}(B_1 + 2B_2t + 3B_3t^2 + \dots)\,dt.$$

Such is the method of Laplace. It will be practically advantageous in the cases where $B_1, B_2, B_3, \dots$ form a rapidly converging series; and it is to such cases that we shall have to apply it, when we give some examples of it from Laplace's next Chapter. In these examples there will be no difficulty in calculating the terms $B_1, B_2, B_3, \dots$, so far as we shall require them. An investigation of the general values of these coefficients as far as B_5 inclusive will be found in De Morgan's *Differential and Integral Calculus*, page 602.

If we suppose that the limits of x are such as to make the corresponding values of y zero, the limits of t will be $-\infty$ and $+\infty$.
Now if r be odd $\int_{-\infty}^{\infty} e^{-t^2} t^r\,dt$ vanishes, and if r be even it is equal to

$$\frac{(r-1)(r-3)\dots 3.1}{2^{\frac{r}{2}}}\sqrt{\pi}.$$

Thus we have
$$\int y\,dx = Y\sqrt{\pi}\left\{B_1 + \frac{3}{2}B_3 + \frac{5.3}{2^2}B_5 + \dots\right\}.$$

Besides the transformation $y = Ye^{-t^2}$ Laplace also takes cases in which the exponent of e instead of being $-t^2$ has other values. Thus on his page 88 the exponent is $-t$, and on his page 93 it is $-t^{2i}$; in the first of these cases Y is not supposed to be a maximum value of y.

958. Some definite integrals are given on pages 95—101, in connexion with which it may be useful to supply a few references.
The formula marked (T) on page 95 occurs in Laplace's memoir of 1782, page 17.
$$\int_0^{\infty} \cos rx\, e^{-a^2x^2}\,dx = \frac{\sqrt{\pi}}{2a}e^{-\frac{r^2}{4a^2}};$$

this was given by Laplace in the *Mémoires...de l'Institut* for 1810, page 290; see also *Tables d'Intégrales Définies*, 1858, by D. Bierens de Haan, page 376.

$$\int_0^\infty \frac{\sin rx}{x}\,dx = \frac{\pi}{2}\,;$$

see D. Bierens de Haan, page 268.

$$\int_0^\infty \frac{\cos ax}{1+x^2}\,dx = \frac{\pi}{2}\,e^{-a}, \quad \int_0^\infty \frac{x\sin ax}{1+x^2}\,dx = \frac{\pi}{2}\,e^{-a},$$

where a is supposed positive; these seem due to Laplace; see D. Bierens de Haan, page 282, *Théorie...des Prob.*, pages 99—134. We may remark that these two results, together with

$$\int_0^\infty \frac{\sin ax}{1+x^2}\frac{dx}{x} = \frac{\pi}{2}\,(1 - e^{-a}),$$

are referred by D. F. Gregory, in his *Examples of the... Differential and Integral Calculus*, to Laplace's memoir of 1782; but they are not explicitly given there: with respect to the last result see D. Bierens de Haan, page 293.

959. Since the integral $\int e^{-t^2}\,dt$ occurs in the expressions of Art. 957, Laplace is led to make some observations on modes of approximating to the value of this integral. He gives the following series which present no difficulty:

$$\int_0^\tau e^{-t^2}\,dt = \tau - \frac{\tau^3}{3} + \frac{1}{\lfloor 2}\frac{\tau^5}{5} - \frac{1}{\lfloor 3}\frac{\tau^7}{7} + \dots\,;$$

$$\int_0^\tau e^{-t^2}\,dt = \tau e^{-\tau^2}\left(1 + \frac{2\tau^2}{1.3} + \frac{(2\tau^2)^2}{1.3.5} + \frac{(2\tau^2)^3}{1.3.5.7} + \dots\right);$$

$$\int_\tau^\infty e^{-t^2}\,dt = \frac{e^{-\tau^2}}{2\tau}\left(1 - \frac{1}{2\tau^2} + \frac{1.3}{2^2\tau^4} - \frac{1.3.5}{2^3\tau^6} + \dots\right).$$

In the memoir of 1782 the second of these three expressions does not occur.

Laplace also gives a development of $\int_\tau^\infty e^{-t^2}\,dt$ into the form of a continued fraction, which he takes from his *Mécanique Céleste, Livre* x. See also De Morgan's *Differential and Integral Calculus*, page 591, for this and some similar developments.

960. Laplace extends the method of approximation given in Art. 957 to the case of double integrals. The following is substantially his process. Suppose we require $\iint y\,dx\,dx'$ taken between such limits of x and x' as make y vanish. Let Y denote the maximum value of y, and suppose that a and a' are the corresponding values of x and x'. Assume

$$y = Ye^{-t^2-t'^2},$$
$$x = a + \theta, \quad x' = a' + \theta'.$$

Substitute these values of x and x' in the function $\log \dfrac{Y}{y}$ and expand it in powers of θ and θ'; then since Y is by hypothesis the maximum value of y the coefficients of θ and θ' will vanish in this expansion: hence we may write the result thus

$$M\theta^2 + 2N\theta\theta' + P\theta'^2 = t^2 + t'^2,$$

that is $\qquad M\left(\theta + \dfrac{N}{M}\theta'\right)^2 + \left(P - \dfrac{N^2}{M}\right)\theta'^2 = t^2 + t'^2.$

Since we have made only one assumption respecting the independent variables t and t' we are at liberty to make another; we will assume

$$\theta\sqrt{M} + \frac{\theta'N}{\sqrt{M}} = t,$$

and therefore $\qquad \theta'\sqrt{\left(P - \dfrac{N^2}{M}\right)} = t'.$

Now by the ordinary theory for the transformation of double integrals we have

$$\iint y\,dx\,dx' = \iint \frac{Ye^{-t^2-t'^2}\,dt\,dt'}{D},$$

where D stands for $\qquad \dfrac{dt}{d\theta}\dfrac{dt'}{d\theta'} - \dfrac{dt}{d\theta'}\dfrac{dt'}{d\theta}.$

Thus far the process is exact. For an approximation we may suppose M, N, P to be functions of a and a' only; then we have

$$M = -\frac{1}{2Y}\frac{d^2Y}{da^2}, \quad N = -\frac{1}{2Y}\frac{d^2Y}{da\,da'}, \quad P = -\frac{1}{2Y}\frac{d^2Y}{da'^2}.$$

Then we shall find that

$$D = \sqrt{(PM - N^2)} = \frac{1}{2Y} \sqrt{\left\{ \frac{d^2 Y}{da^2} \frac{d^2 Y}{da'^2} - \left(\frac{d^2 Y}{da\, da'} \right)^2 \right\}}.$$

And the limits of t and t' will be $-\infty$ and $+\infty$; thus finally we have approximately

$$\iint y\, dx\, dx' = \frac{2\pi Y^2}{\sqrt{\left\{ \dfrac{d^2 Y}{da^2} \dfrac{d^2 Y}{da'^2} - \left(\dfrac{d^2 Y}{da\, da'} \right)^2 \right\}}}.$$

See Art. 907.

961. The second Chapter of the second part of *Livre* I. is entitled *De l'intégration par approximation, des équations linéaires aux différences finies et infiniment petites:* this Chapter occupies pages 110—125.

This Chapter exemplifies the process of solving linear differential equations by the aid of definite integrals. Laplace seems to be the first who drew attention to this subject: it is now fully discussed in works on differential equations. See Boole's *Differential Equations.*

962. The third Chapter of the second part of *Livre* I. is entitled *Application des méthodes précédentes, à l'approximation de diverses fonctions de très-grands nombres:* this Chapter occupies pages 126—177.

The first example is the following. Suppose we have to integrate the equation in Finite Differences,

$$y_{s+1} = (s + 1)\, y_s.$$

Assume $y_s = \int x^s \phi\, dx$, where ϕ is a function of x at present undetermined, and the limits of the integration are also undetermined.

Let δy stand for x^s; then $\dfrac{d\delta y}{dx} = sx^{s-1}$. Hence the proposed equation becomes

$$0 = \int \phi\, dx \left\{ (1 - x)\, \delta y + x\, \frac{d\delta y}{dx} \right\};$$

that is, by integrating by parts,

$$0 = [x\,\delta y\,\phi] + \int \left\{(1-x)\,\phi - \frac{d}{dx}(x\phi)\right\} \delta y\,dx.$$

Where by $[x\,\delta y\,\phi]$ we mean that $x\,\delta y\,\phi$ is to be taken between limits.

Assume ϕ such that

$$(1-x)\,\phi - \frac{d}{dx}(x\phi) = 0,$$

and take the limits of integration such that $[x\,\delta y\,\phi] = 0$; then our proposed equation is satisfied.

From $(1-x)\,\phi - \frac{d}{dx}(x\phi) = 0$, we obtain

$$\phi = Ae^{-x},$$

where A is a constant. Then $x\,\delta y\,\phi$ will vanish when $x = 0$ and also when $x = \infty$. Thus, finally

$$y = A \int_0^\infty x^s e^{-x}\,dx.$$

Now we proceed to put this integral in the form of a series. The maximum value of $x^s e^{-x}$ is easily found to be that which corresponds to $x = s$. Assume, according to Art. 957,

$$x^s e^{-x} = s^s e^{-s} e^{-t^2},$$

and put $x = s + \theta$; thus

$$\left(1 + \frac{\theta}{s}\right)^s e^{-\theta} = e^{-t^2}.$$

Take the logarithms of both sides; thus

$$t^2 = -s \log\left(1 + \frac{\theta}{s}\right) + \theta$$

$$= \frac{\theta^2}{2s} - \frac{\theta^3}{3s^2} + \frac{\theta^4}{4s^3} - \dots$$

Hence by reversion of series we get

$$\theta = t\sqrt{2s} + \frac{2}{3}t^2 + \frac{t^3}{9\sqrt{2s}} + \dots;$$

therefore $\qquad dx = d\theta = dt\sqrt{2s}\left\{1 + \dfrac{4t}{3\sqrt{2s}} + \dfrac{t^2}{6s} + ...\right\}.$

The limits of t corresponding to the limits 0 and ∞ of x will be $-\infty$ and $+\infty$. Therefore

$$\int_0^\infty x^s e^{-x}\, dx = s^s e^{-s} \int_{-\infty}^\infty e^{-t^2}\sqrt{2s}\left\{1 + \frac{4t}{3\sqrt{2s}} + \frac{t^2}{6s} + ...\right\} dt.$$

By integration we obtain

$$y_s = As^{s+\frac{1}{2}} e^{-s}\sqrt{2\pi}\left\{1 + \frac{1}{12s} + ...\right\}.$$

Laplace says we may determine the value of the factor

$$1 + \frac{1}{12s} + ...$$

very simply thus.

Denote it by $1 + \dfrac{B}{s} + \dfrac{C}{s^2} + ...$ so that

$$y_s = As^{s+\frac{1}{2}} e^{-s}\sqrt{2\pi}\left\{1 + \frac{B}{s} + \frac{C}{s^2} + ...\right\}$$

Substitute this value in the equation

$$y_{s+1} = (s+1)\, y_s,$$

thus

$$\left(1 + \frac{1}{s}\right)^{s+\frac{1}{2}} e^{-1}\left\{1 + \frac{B}{s+1} + \frac{C}{(s+1)^2} + ...\right\} = 1 + \frac{B}{s} + \frac{C}{s^2} + ...;$$

therefore

$$\left(1 + \frac{B}{s} + \frac{C}{s^2} + ...\right)\left\{e^{1 - \left(s+\frac{1}{2}\right)\,\log\left(1+\frac{1}{s}\right)} - 1\right\}$$

$$= -\frac{B}{s^2} + \frac{B-2C}{s^3} + ...$$

And

$$1 - \left(s + \frac{1}{2}\right)\log\left(1 + \frac{1}{s}\right) = 1 - \left(s + \frac{1}{2}\right)\left(\frac{1}{s} - \frac{1}{2s^2} + \frac{1}{3s^3} - ...\right)$$

$$= -\frac{1}{12s^2} + \frac{1}{12s^3} - ...$$

Thus

$$\left(1 + \frac{B}{s} + \frac{C}{s^2} + ...\right)\left\{-\frac{1}{12s^2} + \frac{1}{12s^3} - ...\right\} = -\frac{B}{s^2} + \frac{B-2C}{s^3} - ...$$

Hence, equating coefficients,

$$B = \frac{1}{12}, \quad C = \frac{1}{288}, \quad \ldots\ldots$$

The value of A in the expression for y_s must be determined by some particular value of y_s. Suppose that when $s = \mu$ we have $y_s = Y$.

Then
$$Y = A \int_0^\infty x^\mu e^{-x} dx ;$$

thus
$$A = \frac{Y}{\displaystyle\int_0^\infty x^\mu e^{-x} dx}.$$

Hence
$$y_s = \frac{Y s^{s+\frac{1}{2}} e^{-s} \sqrt{2\pi}}{\displaystyle\int_0^\infty x^\mu e^{-x} dx} \left\{ 1 + \frac{1}{12s} + \frac{1}{288 s^2} + \ldots \right\}.$$

The original equation can be very easily integrated; and we obtain

$$y_s = Y (\mu + 1)(\mu + 2) \ldots s.$$

Hence, by equating the two values of y_s,

$$(\mu + 1)(\mu + 2) \ldots s = \frac{s^{s+\frac{1}{2}} e^{-s} \sqrt{2\pi} \left\{ 1 + \dfrac{1}{12s} + \dfrac{1}{288 s^2} + \ldots \right\}}{\displaystyle\int_0^\infty x^\mu e^{-x} dx}.$$

It will be observed that $s - \mu$ is assumed to be a positive integer, but there is nothing to require that s itself should be an integer.

963. One remark must be made on the process which we have just given. Let $\phi(s)$ denote

$$1 + \frac{1}{12s} + \frac{1}{288 s^2} + \ldots ;$$

then
$$1 - \frac{1}{12s} + \frac{1}{288 s^2} - \ldots$$

will be denoted by $\phi(-s)$.

Now Laplace does not shew that

$$\phi(s)\, \phi(-s) = 1,$$

although he assumes the truth of this on his page 134. It may be shewn by adopting the usual mode of proving Stirling's Theorem. For by using Euler's theorem for summation, given in Art. 334, it will appear that

$$1 . 2 \ldots s = s^{s+\frac{1}{2}} e^{-s} \sqrt{2\pi} \, e^{\psi(s)},$$

where

$$\psi(s) = \frac{B_1}{2s} - \frac{B_3}{3 . 4s^3} + \frac{B_5}{5 . 6s^5} - \ldots,$$

the coefficients being the well-known *numbers of Bernoulli.*

Thus

$$\psi(s) + \psi(-s) = 0 ;$$

therefore

$$e^{\psi(s)} \times e^{\psi(-s)} = e^0 = 1,$$

that is

$$\phi(s) \, \phi(-s) = 1.$$

964. Laplace, after investigating a formula sometimes deduces another from it by passing from real to imaginary quantities. This method cannot be considered demonstrative; and indeed Laplace himself admits that it may be employed to discover new formulæ, but that the results thus obtained should be confirmed by direct demonstration. See his pages 87 and 471; also Art. 920.

Thus as a specimen of his results we may quote one which he gives on his page 134.

Let

$$Q = \cos \varpi \, \frac{(\mu + \varpi \sqrt{-1})^\mu + (\mu - \varpi \sqrt{-1})^\mu}{(\mu^2 + \varpi^2)^\mu}$$

$$+ \sqrt{-1} \sin \varpi \, \frac{(\mu - \varpi \sqrt{-1})^\mu - (\mu + \varpi \sqrt{-1})^\mu}{(\mu^2 + \varpi^2)^\mu} ;$$

then

$$\int_0^\infty Q d\varpi = \frac{2\mu\pi \, e^{-\mu}}{\int_0^\infty x^\mu e^{-x} \, dx} .$$

A memoir by Cauchy on Definite Integrals is published in the *Journal de l'École Polytechnique, 28ᵉ Cahier;* this memoir was presented to the Academy of Sciences, Jan. 2nd, 1815, but not printed until 1841. The memoir discusses very fully the results given by Laplace in the Chapter we are now considering. Cauchy says, page 148,

... je suis parvenu à quelques résultats nouveaux, ainsi qu'à la démonstration directe de plusieurs formules, que M. Laplace a déduites du passage du réel à l'imaginaire, dans le 3me chapitre du *Calcul des Probabilités*, et qu'il vient de confirmer par des méthodes rigoureuses dans quelques additions faites à cet ouvrage.

The *additions* to which Cauchy refers occupy pages 464—484 of the *Théorie...des Prob.*, and first appeared in the second edition, which is dated 1814.

965. An important application which Laplace makes of his method of approximation is to evaluate the coefficients of the terms in the expansion of a high power of a certain polynomial.

Let the polynomial consist of $2n+1$ terms and be denoted by

$$\frac{1}{a^n} + \frac{1}{a^{n-1}} + \frac{1}{a^{n-2}} + \ldots + \frac{1}{a} + 1 + a + \ldots + a^{n-2} + a^{n-1} + a^n;$$

and suppose the polynomial raised to the power s.

First, let it be required to find the coefficient of the term independent of a.

Substitute $e^{\theta\sqrt{-1}}$ for a; then we require the term which is independent of θ when

$$\left\{1 + 2\cos\theta + 2\cos 2\theta + \ldots + 2\cos n\theta\right\}^s$$

is expanded and arranged according to cosines of multiples of θ. This term will be found by integrating the above expression with respect to θ from 0 to π, and dividing by π. Sum the series of cosines by the usual formula; then the required term

$$= \frac{1}{\pi}\int_0^\pi \left\{\frac{\sin\frac{2n+1}{2}\theta}{\sin\frac{1}{2}\theta}\right\}^s d\theta$$

$$= \frac{2}{\pi}\int_0^{\frac{1}{2}\pi} \left(\frac{\sin m\phi}{\sin\phi}\right)^s d\phi,$$

where $\phi = \frac{1}{2}\theta$, and $m = 2n+1$.

Now the expression $\left(\frac{\sin m\phi}{\sin\phi}\right)^s$ vanishes when

$$\phi = \frac{\pi}{m} \text{ or } \frac{2\pi}{m} \text{ or } \frac{3\pi}{m} \ldots;$$

and between each of these values it will be found that the expression is numerically a maximum, and it is also a maximum when $\phi = 0$. Thus we may calculate by Art. 957 the value of the integral $\int \left(\frac{\sin m\phi}{\sin \phi}\right)^s d\phi$ when the limits are consecutive multiples of $\frac{\pi}{m}$.

The equation which determines the maxima values of $\frac{\sin m\phi}{\sin \phi}$ is

$$\frac{m \cos m\phi \sin \phi - \cos \phi \sin m\phi}{\sin^2 \phi} = 0.$$

It will be found that this is satisfied when $\phi = 0$; the situation of the other values of ϕ will be more easily discovered by putting the equation in the form

$$\tan m\phi - m \tan \phi = 0 :$$

now we see that the next solution will lie between $m\phi = \frac{5\pi}{4}$ and $m\phi = \frac{3\pi}{2}$, and then the next between $m\phi = \frac{9\pi}{4}$ and $m\phi = \frac{5\pi}{2}$, and so on.

We proceed then to find

$$\int_0^{\frac{\pi}{m}} \left(\frac{\sin m\phi}{\sin \phi}\right)^s d\phi.$$

The maximum value of the function which is to be integrated occurs when $\phi = 0$, and is therefore m^s; assume

$$\left(\frac{\sin m\phi}{\sin \phi}\right)^s = m^s e^{-t^2},$$

therefore
$$\left(\frac{m\phi - \frac{1}{6} m^3\phi^3 + \dots}{\phi - \frac{1}{6} \phi^3 + \dots}\right)^s = m^s e^{-t^2};$$

take logarithms, thus we obtain

$$t^2 = \frac{s}{6} (m^2 - 1) \phi^2 + \dots$$

Therefore approximately

$$\frac{d\phi}{dt} = \frac{\sqrt{6}}{\sqrt{\{s\,(m^2-1)\}}},$$

and

$$\int \left(\frac{\sin m\phi}{\sin \phi}\right)^s d\phi = \frac{m^s\,\sqrt{6}}{\sqrt{\{s\,(m^2-1)\}}} \int e^{-t^2}\, dt.$$

The limits of t will be 0 and ∞. Hence approximately

$$\frac{2}{\pi} \int_0^{\frac{\pi}{m}} \left(\frac{\sin m\phi}{\sin \phi}\right)^s d\phi = \frac{2}{\pi}\, \frac{m^s\,\sqrt{6}}{\sqrt{\{s\,(m^2-1)\}}} \int_0^\infty e^{-t^2}\, dt$$

$$= \frac{m^s\,\sqrt{6}}{\sqrt{\{s\pi\,(m^2-1)\}}} = \frac{(2n+1)^s\,\sqrt{3}}{\sqrt{\{n\,(n+1)\,2s\pi\}}}.$$

Laplace next considers the value of the integral with respect to ϕ between the limits $\frac{\pi}{m}$ and $\frac{2\pi}{m}$, and then the value between the limits $\frac{2\pi}{m}$ and $\frac{3\pi}{m}$, and so on; he shews that when s is a very large number these definite integrals diminish rapidly, and may be neglected in comparison with the value obtained for the limits 0 and $\frac{\pi}{m}$. This result depends on the fact that the successive numerical maxima values of $\frac{\sin m\phi}{\sin \phi}$ diminish rapidly; as we shall now shew. At a numerical maximum we have

$$\frac{\sin m\phi}{\sin \phi} = \frac{m\cos m\phi}{\cos \phi} = \frac{m}{\cos\phi\,\sqrt{(1+m^2\tan^2\phi)}} = \frac{m}{\sqrt{(\cos^2\phi+m^2\sin^2\phi)}} :$$

this is less than $\frac{1}{\sin\phi}$, that is less than $\frac{\phi}{\sin\phi}\cdot\frac{1}{\phi}$, and therefore a fortiori less than $\frac{\pi}{2}\frac{1}{\phi}$, that is less than $\frac{\pi}{2}\frac{m}{m\phi}$.

Hence at the second maximum $\frac{\sin m\phi}{\sin \phi}$ is less than $\dfrac{\dfrac{\pi}{2}\dfrac{m}{5}}{\dfrac{5}{4}\pi}$,

that is less than $\frac{2m}{5}$, and therefore the ratio of the second numerical maximum value of $\left(\dfrac{\sin m\phi}{\sin \phi}\right)^s$ to the first is less than $\left(\dfrac{2}{5}\right)^s$. Similarly the ratio of the third numerical maximum value to the first is less than $\left(\dfrac{2}{9}\right)^s$. And so on.

Next suppose that we require the coefficient of a^l in the expansion of

$$\left\{\frac{1}{a^n} + \frac{1}{a^{n-1}} + \frac{1}{a^{n-2}} + \dots + \frac{1}{a} + 1 + a + \dots + a^{n-2} + a^{n-1} + a^n\right\}^s.$$

The coefficient of a^r in this expansion will be the same as the coefficient of a^{-r}; denote the coefficient of a^r by A_r. Put $e^{\theta\sqrt{-1}}$ for a and suppose the expression to be arranged according to cosines of the multiples of θ; then $2A_r \cos r\theta$ will be the term corresponding to $A_r(a^r + a^{-r})$. If we multiply the expression by $\cos l\theta$, and integrate between the limits 0 and π, all the terms will vanish except that for which r is equal to l; so that the integral reduces to $2A_l \displaystyle\int_0^\pi \cos^2 l\theta\, d\theta$. Hence

$$A_l = \frac{1}{\pi}\int_0^\pi \left\{\frac{\sin\dfrac{2n+1}{2}\theta}{\sin\dfrac{1}{2}\theta}\right\}^s \cos l\theta\, d\theta.$$

We put, as before, $m = 2n+1$, and $\phi = \dfrac{1}{2}\theta$; thus we have

$$A_l = \frac{2}{\pi}\int_0^{\frac{1}{2}\pi}\left(\frac{\sin m\phi}{\sin \phi}\right)^s \cos 2l\phi\, d\phi.$$

As before assume

$$\left(\frac{\sin m\phi}{\sin \phi}\right)^s = m^s e^{-t^2},$$

then $\qquad\qquad \phi = \dfrac{t\sqrt{6}}{\sqrt{\{s\,(m^2-1)\}}}$, approximately.

Hence the integral becomes

$$\frac{2}{\pi}\,\frac{m^s\sqrt{6}}{\sqrt{\{s\,(m^2-1)\}}}\int e^{-t^2}\cos\frac{2lt\sqrt{6}}{\sqrt{\{s\,(m^2-1)\}}}\, dt.$$

As before we take 0 and ∞ for the limits of t, and thus neglect all that part of the integral with respect to ϕ which is not included between the limits 0 and $\dfrac{\pi}{m}$. Hence by Art. 958 we have finally

$$\frac{2}{\pi}\,\frac{m^s\sqrt{6}}{\sqrt{\{s\,(m^2-1)\}}}\,\frac{\sqrt{\pi}}{2}\,e^{-\frac{6l^2}{s(m^2-1)}},\ \text{ or }\ \frac{(2n+1)^s\sqrt{3}}{\sqrt{\{n\,(n+1)\,2s\pi\}}}\,e^{-\frac{8l^2}{2n(n+1)s}}.$$

Suppose now that we require the *sum* of the coefficients, from that of a^{-l} to that of a^l both inclusive; we must find the sum of

$$2A_l+2A_{l-1}+2A_{l-2}+\ldots+2A_1+A_0:$$

this is best effected by the aid of Euler's Theorem; see Art. 334. We have approximately

$$\Sigma_0^{l-1}u_x=\int_0^l u_x dx-\frac{1}{2}u_x+\frac{1}{2}u_0;$$

therefore
$$\Sigma_0^l u_x=\int_0^l u_x dx+\frac{1}{2}u_x+\frac{1}{2}u_0;$$

therefore
$$2\Sigma_0^l u_x-u_0=2\int_0^l u_x dx+u_x.$$

Hence the required result is

$$\frac{(2n+1)^s\sqrt{6}}{\sqrt{\{n\,(n+1)\,s\pi\}}}\left\{\int_0^l e^{-\frac{3l^2}{2n(n+1)s}}\,dl+\frac{1}{2}e^{-\frac{3l^2}{2n(n+1)s}}\right\}.$$

We may observe that Laplace demonstrates Euler's Theorem in the manner which is now usual in elementary works, that is by the aid of the Calculus of Operations.

966. Laplace gives on his page 158 the formula

$$\frac{\displaystyle\int_0^\infty x^{i-1}e^{-sx}\,dx}{\displaystyle\int_0^\infty x^{i-1}e^{-x}\,dx}=\frac{1}{s^i},$$

He demonstrates this in his own way; it is sufficient to observe that it may be obtained by putting x' for sx in the integral in the numerator of the left-hand side.

Hence he deduces

$$\Delta^n \frac{1}{s^i} = \frac{\int_0^\infty x^{i-1} e^{-sx} (e^{-x} - 1)^n \, dx}{\int_0^\infty x^{i-1} e^{-x} \, dx}.$$

Laplace calculates the approximate value of this expression, supposing i very large. He *assumes* that the result which he obtains will hold when the sign of i is changed; so that he obtains an approximate expression for $\Delta^n s^i$; see page 159 of his work. He gives a *demonstration* in the *additions;* see page 474 of the *Théorie...des Prob.* The demonstration involves much use of the symbol $\sqrt{(-1)}$. Cauchy gives a demonstration on page 247 of the memoir cited in Art. 964. Laplace gives another formula for $\Delta^n s^i$ on his page 163; he arrives at it by the aid of integrals with imaginary limits, and then confirms his result by a demonstration.

967. Laplace says, on his page 165, that in the theory of chances we often require to consider in the expression for $\Delta^n s^i$ only those terms in which the quantity raised to the power i is *positive;* and accordingly he proceeds to give suitable approximate formulæ for such cases. Then he passes on to consider specially the approximate value of the expression

$$(n + r\sqrt{n})^\mu - n(n + r\sqrt{n} - 2)^\mu + \frac{n(n-1)}{1 \cdot 2}(n + r\sqrt{n} - 4)^\mu - \ldots,$$

where the series is to extend only so long as the quantities raised to the power μ are positive, and μ is an integer a little greater or a little less than n. See Arts. 916, 917.

The methods are of the kind already noticed; that is they are not demonstrative, but rest on a free use of the symbol $\sqrt{(-1)}$.

A point should be noticed with respect to Laplace's page 171. He has to establish a certain formula; but the whole difficulty of the process is passed over with the words *déterminant convenablement la constante arbitraire.* Laplace's formula is established by Cauchy; see page 240 of the memoir cited in Art. 964.

968. In conclusion we may observe that this Chapter contains many important results, but it is to be regretted that the demon-

strations are very imperfect. The memoir of Cauchy to which we have referred, is very laborious and difficult, so that this portion of the *Théorie...des Prob.* remains in an unsatisfactory state.

969. We now arrive at *Livre* II, which is entitled *Théorie Générale des Probabilités.*

It will be understood that when we speak of any Chapter in Laplace's work without further specification, we always mean a Chapter of *Livre* II.

The first Chapter is entitled *Principes généraux de cette Théorie.* This occupies pages 179—188; it gives a brief statement, with exemplification, of the first principles of the subject.

970. The second Chapter is entitled *De la Probabilité des événemens composés d'événemens simples dont les possibilités respectives sont données.* This occupies pages 189—274; it contains the solution of several problems in direct probability; we will notice them in order.

971. The first problem is one connected with a lottery; see Arts. 291, 448, 455, 775, 864, 910.

The present discussion adds to what Laplace had formerly given an approximate calculation. The French lottery was composed of 90 numbers, 5 of which were drawn at a time. Laplace shews that it is about an even chance that in 86 drawings all the numbers will appear. This approximate calculation is an example of the formula for $\Delta^n s^i$ given by Laplace on page 159 of his work; see Art. 966.

We may remark that Laplace also makes use of a rougher approximation originally given by De Moivre; see Art. 292.

972. On his page 201 Laplace takes the problem of *odd* and *even;* see Arts. 350, 865, 882.

Laplace adds the following problem. Suppose that an urn contains x white balls, and the same number of black balls; an even number of balls is to be drawn out: required the probability that as many white balls as black balls will be drawn out.

The whole number of cases is found to be $2^{2x-1} - 1$, and the whole number of favourable cases to be $\dfrac{\underline{|2x}}{\underline{|x}\,\underline{|x}} - 1$; the required probability therefore is the latter number divided by the former.

973. The next problem is the Problem of Points. Laplace treats this very fully under its various modifications; the discussion occupies his pages 203—217. See Arts. 872, 884.

We will exhibit in substance, Laplace's mode of investigation. Two players A and B want respectively x and y points of winning a set of games; their chances of winning a single game are p and q respectively, where the sum of p and q is unity; the stake is to belong to the player who first makes up his set: determine the probabilities in favour of each player.

Let $\phi(x, y)$ denote A's probability. Then his chance of winning the next game is p, and if he wins it his probability becomes $\phi(x-1, y)$; and q is his chance of losing this game, and if he loses it his probability becomes $\phi(x, y-1)$: thus

$$\phi(x, y) = p\,\phi(x-1, y) + q\,\phi(x, y-1) \,\ldots\ldots\ldots (1).$$

Suppose that $\phi(x, y)$ is the coefficient of $t^x \tau^y$ in the development according to powers of t and τ of a certain function u of these variables. From (1) we shall obtain

$$u - \Sigma\,\phi(x, 0)\,t^x - \Sigma\,\phi(0, y)\,\tau^y + \phi(0, 0)$$
$$= u\,(pt + q\tau) - pt\,\Sigma\,\phi(x, 0)\,t^x - q\tau\,\Sigma\,\phi(0, y)\,\tau^y \ldots\ldots(2),$$

where $\Sigma\,\phi(x, 0)\,t^x$ denotes a summation with respect to x from $x = 0$ inclusive to $x = \infty$; and $\Sigma\,\phi(0, y)\,\tau^y$ denotes a summation with respect to y from $y = 0$ inclusive to $y = \infty$. In order to shew that (2) is true we have to observe two facts.

First, the coefficient of any such term as $t^m \tau^n$, where neither m nor n is less than unity, is the same on both sides of (2) by virtue of (1).

Secondly, on the left-hand side of (2) such terms as $t^m \tau^n$, where m or n is less than unity, cancel each other; and so also do such terms on the right-hand side of (2).

Thus (2) is fully established. From (2) we obtain

$$u = \frac{(1 - pt)\,\Sigma\,\phi(x, 0)\,t^x + (1 - q\tau)\,\Sigma\,\phi(0, y)\,\tau^y - \phi(0, 0)}{1 - pt - q\tau};$$

we may write this result thus,

$$u = \frac{F(t) + f(\tau)}{1 - pt - q\tau} \quad\quad\quad (3),$$

where $F(t)$ and $f(\tau)$ are functions of t and τ respectively, which are at present undetermined. By supposing that the term in $f(\tau)$ which is independent of τ is included in $F(t)$, we may write the result thus,

$$u = \frac{\chi(t) + \tau\psi(\tau)}{1 - pt - q\tau} \quad\quad\quad (4).$$

Thus either (3) or (4) may be taken as the general solution of the equation (1) in Finite Differences; and this general solution involves two arbitrary functions which must be determined by special considerations. We proceed to determine these functions in the present case, taking the form (4) which will be the most convenient.

Now A loses if B first makes up his set, so that $\phi(x, 0) = 0$ for every value of x from unity upwards, and $\phi(0, 0)$ does not occur, that is it may also be considered zero. But from (4) it follows that $\phi(x, 0)$ is the coefficient of t^x in the development of $\frac{\chi(t)}{1 - pt}$; therefore $\chi(t) = 0$.

Again, A wins if he first makes up his set, so that $\phi(0, y) = 1$ for every value of y from unity upwards. But from (4) it follows that $\phi(0, y)$ is the coefficient of τ^y in the development of $\frac{\tau\psi(\tau)}{1 - q\tau}$, so that

$$\frac{\tau\psi(\tau)}{1 - q\tau} = \frac{\tau}{1 - \tau};$$

therefore $\quad\quad \tau\psi(\tau) = \frac{\tau(1 - q\tau)}{1 - \tau}.$

Thus finally

$$u = \frac{\tau(1 - q\tau)}{(1 - \tau)(1 - pt - q\tau)}.$$

Now $\phi(x, y)$ is the coefficient of $t^x\tau^y$ in the development of u. First expand according to powers of t; thus we obtain for the coefficient of t^x the expression $\dfrac{p^x\tau}{(1-\tau)(1-q\tau)^x}$. Then expand this expression according to powers of τ, and we finally obtain for the coefficient of $t^x\tau^y$

$$p^x\left\{1 + xq + \frac{x(x+1)}{1.2}q^2 + \ldots + \frac{x(x+1)\ldots(x+y-2)}{\lfloor y-1}q^{y-1}\right\}.$$

This is therefore the probability in favour of A; and that in favour of B may be obtained by interchanging p with q and x with y.

The result is identical with the second of the two formulæ which we have given in Art. 172.

974. The investigation just given is in substance Laplace's; he takes the particular case in which $p = \frac{1}{2}$ and $q = \frac{1}{2}$; but this makes no difference in principle. But there is one important difference. At the stage where we have

$$u = \frac{F(t) + f(\tau)}{1 - pt - q\tau},$$

Laplace puts

$$u = \frac{f(\tau)}{1 - pt - q\tau}.$$

This is an error, it arises from a false formula given by Laplace on his page 82; see Art. 955. Laplace's error amounts to neglecting the considerations involved in the second of the facts on which equation (2) of the preceding Article depends: this kind of neglect has been not uncommon with those who have used or expounded the method of Generating Functions.

975. We will continue the discussion of the Problem of Points, and suppose that there are more than two players. Let the first player want x_1 points, the second x_2 points, the third x_3 points, and so on. Let their respective chances of winning a single game be $p_1, p_2, p_3, \ldots$ Let $\phi(x_1, x_2, x_3, \ldots)$ denote the probability in favour of the first player. Then, as in Art. 973, we obtain the equation

$$\phi(x_1, x_2, x_3, \ldots) = p_1\phi(x_1-1, x_2, x_3, \ldots) + p_2\phi(x_1, x_2-1, x_3, \ldots)$$
$$+ p_3\phi(x_1, x_2, x_3-1, \ldots) + \ldots\ldots\ldots\ldots(1).$$

Suppose that $\phi(x_1, x_2, x_3, ...)$ is the coefficient of $t_1^{x_1} t_2^{x_2} t_3^{x_3} ...$ in the development of a function u of these variables. Laplace then proceeds thus. From (1) he passes to

$$u = u(p_1 t_1 + p_2 t_2 + p_3 t_3 + ...) \qquad\qquad (2),$$

and then he deduces

$$1 = p_1 t_1 + p_2 t_2 + p_3 t_3 + ...\qquad\qquad (3).$$

Hence

$$\frac{1}{t_1} = \frac{p_1}{1 - p_2 t_2 - p_3 t_3 - ...};$$

therefore

$$\frac{u}{t_1^{x_1}} = \frac{u p_1^{x_1}}{(1 - p_2 t_2 - p_3 t_3 - ...)^{x_1}}$$

$$= u p_1^{x_1} \left\{ 1 + x_1(p_2 t_2 + p_3 t_3 + ...) \right.$$

$$+ \frac{x_1(x_1 + 1)}{1 . 2}(p_2 t_2 + p_3 t_3 + ...)^2$$

$$+ \frac{x_1(x_1 + 1)(x_1 + 2)}{1 . 2 . 3}(p_2 t_2 + p_3 t_3 + ...)^3$$

$$\left. + ... \right\}.$$

Now the coefficient of $t_1^0 t_2^{x_2} t_3^{x_3} ...$ in $\frac{u}{t_1^{x_1}}$ is $\phi(x_1, x_2, x_3, ...)$.

Let $k u p_1^{x_1} t_2^m t_3^n ...$ denote any term of the right-hand member of the last equation. Then the coefficient of $t_1^0 t_2^{x_2} t_3^{x_3} ...$ in this term will be $k p_1^{x_1} \phi(0, x_2 - m, x_3 - n, ...)$. But $\phi(0, x_2 - m, x_3 - n, ...)$ is equal to unity, for if the first player wants no points he is entitled to the stake. Moreover we must reject all the values of $\phi(0, x_2 - m, x_3 - n, ...)$ in which m is equal to or greater than x_2, in which n is equal to or greater than x_3, and so on; for these terms in fact do not exist, that is must be considered to be zero. Hence finally

$$\phi(x_1, x_2, x_3, ...) = p_1^{x_1} \left\{ 1 + x_1(p_2 + p_3 + ...) \right.$$

$$+ \frac{x_1(x_1 + 1)}{1 . 2}(p_2 + p_3 + ...)^2$$

$$+ \frac{x_1(x_1 + 1)(x_1 + 2)}{1 . 2 . 3}(p_2 + p_3 + ...)^3$$

$$\left. + ... \right\},$$

34—2

provided we reject all terms in which the power of p_2 surpasses $x_2 - 1$, in which the power of p_3 surpasses $x_3 - 1$, and so on.

Now on this process of Laplace's we remark:

First, the equation (2) is not true; as in Art. 973 we ought to allow for terms in which one or more of the variables x_1, x_2, x_3, ... is zero; and therefore additional terms ought to be placed in each member of equation (2) of the present Article, like those in equation (2) of Article 973.

Secondly, Laplace's treatment of his equation (3) is unintelligible, as we have already remarked in a similar case; see Art. 952. By making use of the Calculus of Operations we might however translate Laplace's process into another free from objection.

976. At this stage we shall find it convenient to introduce an account of the fourth Supplement to the *Théorie...des Probabilités*. This supplement contains 28 pages. Laplace begins with a few remarks on Generating Functions; he gives the correct formula for the solution of an equation in Finite Differences for which he had formerly given an incorrect formula: see Art. 955. He does not refer to the *Théorie...des Prob.* nor take any notice of the discrepancy of the two formulæ. He says, on page 4 of the Supplement,

Un des principaux avantages de cette manière d'intégrer les équations aux différences partielles, consiste en ce que l'analyse algébrique fournissant divers moyens pour développer les fonctions, on peut choisir celui qui convient le mieux à la question proposée. La solution des problèmes suivans, par le Comte de Laplace, mon fils, et les considérations qu'il y a jointes, répandront un nouveau jour sur le calcul des fonctions génératrices.

We have therefore to ascribe all the rest of the fourth Supplement to Laplace's son.

977. The main part of the fourth Supplement consists of the solution of problems which may be considered as generalisations of the Problem of Points. There are three of these problems; we will enunciate them.

I. A player A draws a ball from an urn containing white balls and black balls; his chance of drawing a white ball is p, and his chance of drawing a black ball is q: after the ball has been drawn it is replaced. Then a second player B draws a ball from a second urn containing white balls and black balls; his chance of drawing a white ball is p', and his chance of drawing a black ball is q': after the ball has been drawn it is replaced. The two players continue thus to draw alternately a ball, each from his own urn, and to replace the ball after it has been drawn. If a player draws a white ball he counts a point; if he draws a black ball he counts nothing. Suppose that A wants x points, and B wants x' points to complete an assigned set, required the probabilities in favour of each player.

II. Suppose A draws from an urn in which there are balls of three kinds; for a ball of the first kind he counts two points, for a ball of the second kind he counts one point, and for a ball of the third kind he counts no point: let his chances be p, p_1, and q for the three cases.

Similarly let B draw from a second urn containing similar balls; let p', p_1', and q' be his chances for the three cases. Then, as before, we require the probabilities for each player of his making up an assigned set of points before his adversary makes up an assigned set.

III. An urn contains a known number of black balls, and a known number of white balls; a ball is drawn and not replaced; then another ball, and so on: required the probability that a given number of white balls will be drawn before another given number of black balls.

These three problems are solved by the method of Generating Functions used carefully and accurately; that is, the terms which are required to make the equations true are given, and not omitted. See Art. 974. After the problems are solved generally particular cases are deduced.

The student of the fourth Supplement will have to bear in mind that in the first problem $p + q = 1$ and $p' + q' = 1$, and in the second problem $p + p_1 + q = 1$, $p' + p_1' + q' = 1$.

978. After the solutions of these problems we have a few pages headed *Remarque sur les fonctions génératrices;* and this is the part of the fourth Supplement with which we are chiefly interested. It is here observed that in a case like that of our Art. 975, the equation (2) is not an accurate deduction from equation (1) ; for additional terms ought to be added to each side, in the manner of our Art. 973.

There is however a mistake at the top of page 24 of the fourth Supplement : instead of adding a function of t, *two* functions must be added, one of t and the other of t'.

The fourth Supplement then proceeds thus, on its page 24 :

Faute d'avoir égard à ces fonctions, on peut tomber dans des erreurs graves, en se servant de ce moyen pour intégrer les équations aux différences partielles. Par cette même raison, la marche suivie dans la solution des problèmes des n°s 8 et 10 du second livre de la Théorie analytique des Probabilités n'est nullement rigoureuse, et semble impliquer contradiction, en ce qu'elle établit une liaison entre les variables qui sont et doivent être toujours indépendantes. Sans entrer dans les considérations particulières qui ont pu la faire réussir ici, et qu'il est aisé de saisir, nous allons faire voir que la méthode d'intégration exposée au commencement de ce Supplément s'applique également à ces questions, et les résout avec non moins de simplicité.

The problem referred to as contained in No. 8 of the *Théorie...des Prob.* is that which we have given in Art. 975 ; the problem referred to as contained in No. 10 of the *Théorie...des Prob.* is that which we shall notice in Art. 980. The fourth Supplement gives solutions of these problems by the accurate use of Generating Functions, in the manner of our Art. 973.

Thus as Laplace himself attached the fourth Supplement to his work, we may conclude that he admitted the solutions in question to be unsound. We consider that they are unsound, and in fact unintelligible, as they are presented by Laplace ; but on the other hand, we believe that they may be readily translated into the language of the Calculus of Operations, and thus become clear and satisfactory. See Art. 952.

979. We return from the fourth Supplement to the *Théorie...des Prob.* itself. Laplace's next problem is that which

is connected with the game which is called *Treize* or *Rencontre;* see Arts. 162, 280, 286, 430, 626.

Laplace devotes his pages 217—225 to this problem ; he gives the solution, and then applies his method of approximation in order to obtain numerical results when very high numbers are involved.

980. Laplace takes next on his pages 225—238 the problem of the Duration of Play. The results were enunciated by De Moivre and demonstrated by Lagrange ; Laplace has made great use of Lagrange's memoir on the subject ; see Arts. 311, 583, 588, 863, 885, 921. We may observe that before Laplace gives his analytical solution he says, Ce problème peut être résolu avec facilité par le procédé suivant qui est en quelque sorte, mécanique ; the process which he gives is due to De Moivre ; it occurs on page 203 of the *Doctrine of Chances.* See also Art. 303. In the course of the investigation, Laplace gives a process of the kind we have already noticed, which is criticised in the fourth Supplement ; see Art. 978.

981. Laplace takes next on his pages 238—247 the problem which we have called Waldegrave's problem ; see Arts. 210, 249, 295, 348.

There are $n+1$ players C_1, C_2, ... C_{n+1}. First C_1 and C_2 play together ; the loser deposits a shilling in a common stock, and the winner plays with C_3 ; the loser again deposits a shilling, and the winner plays with C_4 ; the process is continued until some one player has beaten in succession all the rest, the turn of C_1 coming on again after that of C_{n+1}. The winner is to take all the money in the common stock.

Laplace determines the probability that the play will terminate precisely at the xth game, and also the probability that it will terminate at or before the xth game. He also determines the probability that the rth player will win the money precisely at the xth game ; that is to say, he exhibits a complex algebraical function of a variable t which must be expanded in powers of x and the coefficient of t^x taken. He then deduces a general expression for the advantage of the rth player.

The part of the solution which is new in Laplace's discussion is that which determines the probability that the rth player will win the money *precisely at the xth game;* Nicolas Bernoulli had confined himself to the probability which each player has of winning the money *on the whole.*

982. We will give, after Laplace, the investigation of the probability that the play will terminate precisely at the xth game.

Let z_x denote this probability. In order that the play may terminate at the xth game, the player who enters into play at the $(x-n+1)^{th}$ game must win this game and the $n-1$ following games.

Suppose that the winner of the money starts with a player who has won only one game; let P denote the probability of this event; then $\dfrac{P}{2^n}$ will be the corresponding probability that the play will terminate at the xth game. But the probability that the play will terminate at the $(x-1)^{th}$ game, that is z_{x-1}, is equal to $\dfrac{P}{2^{n-1}}$. For it is necessary to this end that a player who has already won one game just before the $(x-n+1)^{th}$ game should win this game and the $n-2$ following games; and the probabilities of these component events being respectively P and $\dfrac{1}{2^{n-1}}$, the probability of the compound event is $\dfrac{P}{2^{n-1}}$. Thus

$$\frac{P}{2^n} = \frac{1}{2} z_{x-1};$$

and therefore $\dfrac{1}{2} z_{x-1}$ is the probability that the play will terminate at the xth game, relative to this case.

Next suppose that the winner of the money starts with a player who has won two games; let P' denote the probability of this event; then $\dfrac{P'}{2^n}$ will be the corresponding probability that the play will terminate at the xth game. And $\dfrac{P'}{2^{n-2}} = z_{x-2}$: for in order that the play should terminate at the $(x-2)^{th}$ game it is necessary that a player who has already won two games just before the $(x-n+1)^{th}$

game should win this game and the $n-2$ following games. Thus

$$\frac{P'}{2^n} = \frac{1}{2^2} z_{x-2};$$

and therefore $\frac{1}{2^2} z_{x-2}$ is the probability that the play will terminate at the xth game relative to this case.

By proceeding thus, and collecting all the partial probabilities we obtain

$$z_x = \frac{1}{2} z_{x-1} + \frac{1}{2^2} z_{x-2} + \frac{1}{2^3} z_{x-3} + \dots + \frac{1}{2^{n-1}} z_{x-n+1} \dots \dots \dots (1).$$

Suppose that z_x is the coefficient of t^x in the expansion according to powers of t of a certain function u of this variable. Then from (1) we have, as in Art. 937,

$$u = \frac{F(t)}{1 - \frac{1}{2}t - \frac{1}{2^2}t^2 - \frac{1}{2^3}t^3 - \dots - \frac{1}{2^{n-1}}t^{n-1}},$$

where $F(t)$ is a function of t which is at present undetermined.

Now if (1) were true for $x = n$ as well as for higher values of n, the function $F(t)$ would be of the degree $n-1$. But (1) does not hold when $x = n$, for in forming (1) the player who wins the money was supposed to start against an opponent who had won one game at least; so that in (1) we cannot suppose x to be less than $n+1$. Hence the function $F(t)$ will be of the degree n, and we may put

$$u = \frac{a_0 + a_1 t + a_2 t^2 + \dots + a_n t^n}{1 - \frac{1}{2}t - \frac{1}{2^2}t^2 - \frac{1}{2^3}t^3 - \dots - \frac{1}{2^{n-1}}t^{n-1}}.$$

Now the play cannot terminate before the nth game, and the probability of its terminating at the nth game is $\frac{1}{2^{n-1}}$; therefore a_x vanishes for values of x less than n, and $a_n = \frac{1}{2^{n-1}}$. Thus

$$u = \frac{1}{2^{n-1}} \frac{t^n}{1 - \frac{1}{2}t - \frac{1}{2^2}t^2 - \frac{1}{2^3}t^3 - \dots - \frac{1}{2^{n-1}}t^{n-1}}$$

$$= \frac{1}{2^n} \frac{t^n(2-t)}{1 - t + \frac{1}{2^n}t^n}.$$

The coefficient of t^x in the expansion of u in powers of t gives the probability that the play will terminate *at* the xth game.

The probability that the play will terminate *at or before* the xth game will be the sum of the coefficients of t^x and of the inferior powers of t in the expansion of u, which will be equal to the coefficient of t^x in the expansion of $\dfrac{u}{1-t}$; that is, it will be the coefficient of t^x in the expansion of

$$\frac{1}{2^n} \frac{t^n (2 - t)}{(1 - t)\left(1 - t + \dfrac{1}{2^n} t^n\right)} .$$

This expression is equal to

$$\frac{1}{2^n} \frac{t^n (2 - t)}{(1 - t)^2} \left\{ 1 - \frac{t^n}{2^n (1 - t)} + \frac{t^{2n}}{2^{2n} (1 - t)^2} - \frac{t^{3n}}{2^{3n} (1 - t)^3} + \dots \right\}.$$

The rth term of this development is

$$\frac{(-1)^{r-1}}{2^{rn}} \frac{(2 - t) t^{rn}}{(1 - t)^{r+1}} ,$$

that is

$$(-1)^{r-1} \left\{ \frac{1}{2^{rn-1}} \frac{t^{rn}}{(1 - t)^{r+1}} - \frac{1}{2^{rn}} \frac{t^{rn+1}}{(1 - t)^{r+1}} \right\}.$$

The expansion in powers of t of this r^{th} term may now be readily effected; the coefficient of t^x will be

$$(-1)^{r-1} \cdot \left\{ \frac{1}{2^{rn-1}} \frac{\lfloor x + r - rn}{\lfloor x - rn \lfloor r} - \frac{1}{2^{rn}} \frac{\lfloor x + r - rn - 1}{\lfloor x - rn - 1 \lfloor r} \right\} ,$$

that is

$$\frac{(-1)^{r-1}}{2^{rn}} \frac{\lfloor x + r - rn - 1}{\lfloor x - rn \lfloor r} (x - rn + 2r).$$

The final result is that the probability that the play will terminate at or before the xth game, is represented by as many terms of the following series as there are units in the integer next below $\dfrac{x}{n}$:

$$\frac{x - n + 2}{2^n} - \frac{(x - 2n + 1)}{1 \cdot 2 \cdot 2^{2n}} (x - 2n + 4)$$

$$+ \frac{(x - 3n + 1)(x - 3n + 2)}{1 \cdot 2 \cdot 3 \cdot 2^{3n}} (x - 3n + 6) - \dots$$

The sum of the coefficients of every power of t up to infinity in the expansion of u will represent the probability that the play will terminate if there be no limit assigned to the number of games. But the sum of these coefficients will be equal to the value of u when t is made equal to unity; and this value of u is unity. Hence we infer that the probability of the termination of the play may be made as near to unity as we please by allowing a sufficient number of games.

983. In Laplace's own solution no notice is taken of the fact that equation (1) does not hold for $x = n$. Professor De Morgan remarks in a note to Art. 52 of the *Theory of Probabilities* in the *Encyclopædia Metropolitana*,

Laplace (p. 240) has omitted all allusion to this circumstance ; and the omission is highly characteristic of his method of writing. No one was more sure of giving the result of an analytical process correctly, and no one ever took so little care to point out the various small considerations on which correctness depends.. His *Théorie des Probabilités* is by very much the most difficult mathematical work we have ever met with, and principally from this circumstance: the *Mécanique Céleste* has its full share of the same sort of difficulty; but the analysis is less intricate.

984. We may observe that as Laplace continues his discussion of Waldegrave's problem he arrives at the following equation in Finite Differences,

$$y_{r, z} - y_{r-1, z-1} + \frac{1}{2^n} y_{r, z-1} = 0 \; ;$$

in integrating this, although his final result is correct, his process is unsatisfactory, because it depends upon an error we have already indicated. See Art. 955.

985. Laplace's next problem is that relating to a *run of events* which was discussed by De Moivre and Condorcet; see Arts. 325, 677 : this problem occupies Laplace's pages 247—253.

Let p denote the chance of the happening of the event in a single trial; let $\phi(x)$ denote the probability that in x trials the event will happen i times in succession. Then from equation (1) of Art. 678 by changing the notation we have

$$\phi(x) = p^i + p^{i-1}(1-p)\,\phi(x-i) + p^{i-2}(1-p)\,\phi(x-i+1) + \ldots$$
$$\ldots + p(1-p)\,\phi(x-2) + (1-p)\,\phi(x-1) \ldots\ldots\ldots(1).$$

Laplace takes z_x to denote the probability that the run will finish at the xth trial, and not before; then he obtains

$$z_x = (1 \quad p)\left\{z_{x-1} + pz_{x-2} + p^2 z_{x-3} + \ldots + p^{i-1}z_{x-i}\right\} \ldots\ldots (2).$$

We may deduce (2) thus; it is obvious that

$$z_x = \phi(x) - \phi(x-1);$$

hence in (1) change x into $x-1$ and subtract, and we obtain (2).

Laplace proceeds nearly thus. If the run is first completed at the xth trial the $(x-i)^{\text{th}}$ trial must have been unfavourable, and the following i trials favourable. Laplace then makes i distinct cases.

I. The $(x-i-1)^{\text{th}}$ trial unfavourable.

II. The $(x-i-1)^{\text{th}}$ favourable; and the $(x-i-2)^{\text{th}}$ unfavourable.

III. The $(x-i-1)^{\text{th}}$ and the $(x-i-2)^{\text{th}}$ favourable, and the $(x-i-3)^{\text{th}}$ unfavourable.

IV. The $(x-i-1)^{\text{th}}$, the $(x-i-2)^{\text{th}}$, and the $(x-i-3)^{\text{th}}$ favourable; and the $(x-i-4)^{\text{th}}$ unfavourable.

And so on.

Let us take one of these cases, say IV. Let P_4 denote the probability of this case existing; then will

$$P_4 p^{i-3} = z_{x-4}.$$

For in this case a run of 3 has been obtained, and if this be followed by a run of $i-3$, of which the chance is p^{i-3}, we obtain a run of i ending at the $(x-4)^{\text{th}}$ trial.

Now the part of z_x which arises from this case IV. is $P_4(1-p)\,p^i$; for we require an unfavourable result at the $(x-i)^{\text{th}}$ trial, of which the chance is $1 - p$, and then a run of i. Thus the part of z_x is

$$\frac{z_{x-4}}{p^{i-3}} (1 - p) p^i, \text{ or } p^3 (1 - p) z_{x-4}.$$

We have said that Laplace adopts *nearly* the method we have given; but he is rather obscure. In the method we have given P_4 denotes the probability of the following compound event: no run of i before the $(x - i - 4)^{\text{th}}$ trial, the $(x - i - 4)^{\text{th}}$ trial unfavourable, and then the next three trials favourable. Similarly our P_2 would denote the probability of the following compound event; no run of i before the $(x - i - 2)^{\text{th}}$ trial, the $(x - i - 2)^{\text{th}}$ trial unfavourable, and the next trial favourable. Laplace says, Nommons P' la probabilité qu'il n'arrivera pas au coup $x - i - 2$. Now Laplace does not formally say that there is to be no run of i before the $(x - i - 2)^{\text{th}}$ trial; but this must be understood. Then his P' agrees with our P_2 if we omit the last of the three clauses which form our account of the probability represented by P_2; so that in fact pP' with Laplace denotes the same as P_2 with us.

Laplace gives the integral of the equation (2), and finally obtains the same result as we have exhibited in Art. 325.

986. Laplace then proceeds to find the probability that one of two players should have a run of i successes before the other; this investigation adds nothing to what Condorcet had given, but is more commodious in form. Laplace's result on his page 250 will be found on examination to agree with what we have given in Art. 680, after Condorcet.

Laplace then supplies some new matter, in which he considers the expectation of each player supposing that after failing he deposits a franc, and that the sum of the deposits is taken by him who first has a run of i successes.

987. Laplace's next problem is the following. An urn contains $n + 1$ balls marked respectively $0, 1, \dots n$; a ball is drawn and replaced: required the probability that after i drawings the sum of the numbers drawn will be s. This problem and applications of it occupy pages 253—261. See Arts. 888, 915.

The problem is due to De Moivre; see Arts. 149, 364. Laplace's solution of the problem is very laborious. We will pass to the application which Laplace makes of the result to the subject of the planes of motion of the planets.

By proceeding as in Art. 148, we find that the probability that after i drawings the sum of the numbers drawn will be s is the coefficient of x^s in the expansion of

$$\frac{1}{(n+1)^i}\,(1-x^{n+1})^i\,(1-x)^{-i}.$$

Thus we obtain for the required probability

$$\frac{1}{(n+1)^i}\left\{\frac{\underline{|i+s-1}}{\underline{|i-1}\,\underline{|s}}-\frac{i}{1}\frac{\underline{|i+s-n-2}}{\underline{|i-1}\,\underline{|s-n-1}}\right.$$
$$\left.+\frac{i\,(i-1)}{1.2}\frac{\underline{|i+s-2n-3}}{\underline{|i-1}\,\underline{|s-2n-2}}-\cdots\right\}.$$

If the balls are marked respectively 0, θ, 2θ, 3θ, ... $n\theta$, this expression gives the probability that after i drawings the sum of the numbers drawn will be $s\theta$.

Now suppose θ to become indefinitely small, and n and s to become indefinitely great. The above expression becomes ultimately

$$\frac{1}{\underline{|i-1}}\left\{\left(\frac{s}{n}\right)^{i-1}-\frac{i}{1}\left(\frac{s}{n}-1\right)^{i-1}+\frac{i\,(i-1)}{1.2}\left(\frac{s}{n}-2\right)^{i-1}-\cdots\right\}\frac{1}{n}.$$

Let $\dfrac{s}{n}$ be denoted by x, and $\dfrac{1}{n}$ by dx, so that we obtain

$$\frac{1}{\underline{|i-1}}\left\{x^{i-1}-\frac{i}{1}(x-1)^{i-1}+\frac{i\,(i-1)}{1.2}(x-2)^{i-1}-\cdots\right\}dx;$$

this expression may be regarded as the conclusion of the following problem. The numerical result at a single trial must lie between 0 and 1, and all fractional values are equally probable: determine the probability that after i trials the sum of the results obtained will lie between x and $x+dx$, where dx is indefinitely small.

Hence if we require the probability that after i trials the sum of the results obtained will lie between x_1 and x_2, we must integrate the above expression between the limits x_1 and x_2; thus we obtain

$$\frac{1}{\underline{|i}}\left\{x_2{}^i - \frac{i}{1}(x_2-1)^i + \frac{i\,(i-1)}{1\,.\,2}(x_2-2)^i - \ldots\right\}$$

$$- \frac{1}{\underline{|i}}\left\{x_1{}^i - \frac{i}{1}(x_1-1)^i + \frac{i\,(i-1)}{1\,.\,2}(x_1-2)^i - \ldots\right\}.$$

Each series, like the others in the present Article, is to be continued only so long as the quantities which are raised to the power i are positive.

We might have obtained this result more rapidly by using Art. 364 as our starting point instead of Art. 148.

At the beginning of the year 1801, the sum of the inclinations of the orbits of the ten planets to the ecliptic was 91·4187 French degrees, that is ·914187 of a right angle; suppose that for each planet any inclination between zero and a right angle had been equally likely: required the probability that the sum of the inclinations would have been between 0 and ·914187 of a right angle. By the preceding expression we obtain for the result $\frac{1}{\underline{|10}}(·914187)^{10}$, that is about ·00000011235.

Speaking of this probability, Laplace says:

... Elle est déjà très-petite; mais il faut encore la combiner avec la probabilité d'une circonstance très-remarquable dans le système du monde, et qui consiste en ce que toutes les planètes se meuvent dans le même sens que la terre. Si les mouvemens directs et rétrogrades sont supposés également possibles, cette dernière probabilité est $\left(\frac{1}{2}\right)^{10}$; il faut donc multiplier ·00000011235 par $\left(\frac{1}{2}\right)^{10}$, pour avoir la probabilité que tous les mouvemens des planètes et de la terre seront dirigés dans le même sens, et que la somme de leurs inclinaisons à l'orbite de la terre, sera comprise dans les limites zéro et 91°·4187; on aura ainsi $\dfrac{1·0972}{(10)^{10}}$ pour cette probabilité; ce qui donne $1 - \dfrac{1·0972}{(10)^{10}}$ pour la probabilité que cela n'a pas dû avoir lieu, si toutes les inclinaisons, ainsi que les mouvemens directs et rétrogrades, ont été également faciles. Cette probabilité approche tellement de la certitude, que le résultat observé devient invraisemblable dans cette hypothèse; ce résultat indique donc avec une très-grande probabilité, l'existence d'une cause primitive qui a déterminé les mouvemens des planètes à se rapprocher du plan de l'écliptique, ou plus naturellement, du plan de l'équateur solaire, et à se mouvoir dans le sens de la rotation du soleil...

Laplace then mentions other circumstances which strengthen his conclusion, such as the fact that the motion of the satellites is also in the same direction as that of the planets.

A similar investigation applied to the observed comets does not give any ground for suspecting the existence of a primitive cause which has affected the inclination of their planes of motion to the plane of the ecliptic. See however Cournot's *Exposition de la Théorie des Chances* ... page 270.

Laplace's conclusion with respect to the motions of the planets. has been accepted by very eminent writers on the subject; for example by Poisson: see his *Recherches sur la Prob.* ... page 302. But on the other hand two most distinguished philosophers have recorded their dissatisfaction; see Professor Boole's *Laws of Thought*, page 364, and a note by R. L. Ellis in *The Works of Francis Bacon* ... Vol. I. 1857, page 343.

988. Laplace devotes his pages 262—274 to a very remarkable process and examples of it; see Art. 892. The following is his enunciation of the problem which he solves:

Soient i quantités variables et positives t, $t_1,...t_{i-1}$ dont la somme soit s, et dont la loi de possibilité soit connue; on propose de trouver la somme des produits de chaque valeur que peut recevoir une fonction donnée $\psi(t, t_1, t_2,$ &c.) de ces variables, multipliée par la probabilité correspondante à cette valeur.

The problem is treated in a very general way; the laws of possibility are not assumed to be continuous, nor to be the same for the different variables. The whole investigation is a characteristic specimen of the great powers of Laplace, and of the brevity and consequent difficulty of his expositions of his methods.

Laplace applies his result to determine the probability that the sum of the errors of a given number of observations shall lie between assigned limits, supposing the law of the facility of error in a single observation to be known: Laplace's formula when applied by him to a special case coincides with that which we have given in Art. 567 from Lagrange.

989. An example is given by Laplace, on his page 271, which we may conveniently treat independently of his general investigation, with which he himself connects it. Let there be a number n of points ranged in a straight line, and let ordinates be drawn at these points; the sum of these ordinates is to be equal to s: moreover the first ordinate is not to be greater than the second, the second not greater than the third, and so on. Required the mean value of the rth ordinate.

Let z_1 denote the first ordinate, let $z_1 + z_2$ denote the second, $z_1 + z_2 + z_3$ the third, and so on: thus $z_1, z_2, z_3, \ldots z_n$ are all positive variables, and since the sum of the ordinates is s we have

$$nz_1 + (n-1) z_2 + (n-2) z_3 + \ldots + z_n = s \ldots\ldots\ldots (1).$$

The mean value of the rth ordinate will be

$$\frac{\iiint \ldots\ldots (z_1 + z_2 + \ldots + z_r)\, dz_1 dz_2 \ldots dz_n}{\iiint \ldots\ldots dz_1 dz_2 \ldots dz_n},$$

where the integrations are to be extended over all positive values of the variables consistent with the limitation (1).

Put $nz_1 = x_1$, $(n-1) z_2 = x_2$, and so on. Then our expression becomes

$$\frac{\iiint \ldots \left(\dfrac{x_1}{n} + \dfrac{x_2}{n-1} + \dfrac{x_3}{n-2} + \ldots + \dfrac{x_r}{n-r+1} \right) dx_1 dx_2 \ldots dx_n}{\iiint \ldots dx_1 dx_2 \ldots dx_n},$$

with the limitation

$$x_1 + x_2 + \ldots + x_n = s \ldots\ldots\ldots\ldots\ldots\ldots (2).$$

The result then follows by the aid of the theorem of Lejeune Dirichlet: we shall shew that this result is

$$\frac{s}{n} \left\{ \frac{1}{n} + \frac{1}{n-1} + \frac{1}{n-2} + \ldots + \frac{1}{n-r+1} \right\}.$$

35

For let us suppose that instead of (2) we have the condition that $x_1 + x_2 + \ldots + x_n$ shall lie between s and $s + \Delta s$. Then by the theorem to which we have just referred we have

$$\int\!\!\int\!\!\int \ldots x_m\, dx_1\, dx_2 \ldots dx_n = \frac{(s + \Delta s)^{n+1} - s^{n+1}}{\underline{|n+1}},$$

and

$$\int\!\!\int\!\!\int \ldots\ldots dx_1\, dx_2 \ldots dx_n = \frac{(s + \Delta s)^n - s^n}{\underline{|n}}.$$

Hence by division we obtain

$$\frac{\int\!\!\int\!\!\int \ldots x_m\, dx_1\, dx_2 \ldots dx_n}{\int\!\!\int\!\!\int \ldots dx_1\, dx_2 \ldots dx_n} = \frac{(s + \Delta s)^{n+1} - s^{n+1}}{(s + \Delta s)^n - s^n} \cdot \frac{1}{n+1}.$$

The limit of this expression when Δs is indefinitely diminished is $\frac{s}{n}$. Then by putting for m in succession the values $1, 2, \ldots r$, we obtain the result.

Laplace makes the following application of the result. Suppose that an observed event must have proceeded from one of n causes $A, B, C, \ldots$; and that a tribunal has to judge from which of the causes the event did proceed.

Let each individual arrange the causes in what he considers the order of probability, beginning with the least probable. Then to the rth cause on his list we must consider that he assigns the numerical value

$$\frac{1}{n}\left\{\frac{1}{n} + \frac{1}{n-1} + \frac{1}{n-2} + \ldots + \frac{1}{n-r+1}\right\}.$$

The sum of all the values belonging to the same cause, according to the arrangement of each member of the tribunal, must be taken; and the greatest sum will indicate in the judgment of the tribunal the most probable cause.

990. Another example is also given by Laplace, which we will treat independently. Suppose there are n candidates for an office, and that an elector arranges them in order of merit; let a denote the maximum merit: required the mean value of the merit of a candidate whom the elector places r^{th} on his list.

Let $t_1,\ t_2,\ \dots\ t_n$ denote the merits of the candidates, beginning with the most meritorious. The problem differs from that just discussed, because there is now no condition corresponding to the *sum* of the ordinates being given ; the elector may ascribe any merits to the candidates, consistent with the conditions that the merits are in order, none being greater than that which immediately precedes it, and no merit being greater than a.

The mean value of the merit of the rth candidate will be

$$\frac{\iiint \dots t_r\, dt_1\, dt_2 \dots dt_n}{\iiint \dots dt_1\, dt_2 \dots dt_n}.$$

The integrations are to be taken subject to the following conditions : the variables are to be all positive; a variable t_m is never to be greater than the preceding variable t_{m-1}, and no variable is to be greater than a. Laplace's account of the conditions is not intelligible ; and he *states* the result of the integration without explaining how it is obtained. We may obtain it thus.

Put $t_n = x_n,\ t_{n-1} = t_n + x_{n-1},\ t_{n-2} = t_{n-1} + x_{n-2},\ \dots$; then the above expression for the mean value becomes

$$\frac{\iiint \dots (x_n + x_{n-1} + \dots + x_r)\, dx_1\, dx_2 \dots dx_n}{\iiint \dots dx_1\, dx_2 \dots dx_n},$$

with the condition that all the variables must be positive, and that $x_1 + x_2 + \dots + x_n$ must not be greater than a. Then we may shew in the manner of the preceding Article that the result is

$$\frac{(n - r + 1)\, a}{n + 1}.$$

Laplace suggests, in accordance with this result, that each elector should ascribe the number n to the candidate whom he thinks the best, the number $n - 1$ to the candidate whom he thinks the next, and so on. Then the candidate should be elected who has the greatest sum of numbers. Laplace says,

Ce mode d'élection serait sans doute le meilleur, si des considérations étrangères au mérite n'influaient point souvent sur le choix des électeurs, même les plus honnêtes, et ne les déterminaient point à placer aux derniers rangs, les candidats les plus redoutables à celui qu'ils préfèrent ; ce qui donne un grand avantage aux candidats d'un mérite médiocre. Aussi l'expérience l'a-t-elle fait abandonner aux établissemens qui l'avaient adopté.

It would be interesting to know where this mode of managing elections had been employed. The subject had been considered by Borda and Condorcet ; see Arts. 690, 719, 806.

991. Thus we close our account of the second Chapter of Laplace's work which we began in Art. 970 ; the student will see that comparatively a small portion of this Chapter is originally due to Laplace himself.

992. Laplace's Chapter III. is entitled *Des lois de la probabilité, qui résultent de la multiplication indéfinie des événemens :* it occupies pages 275—303.

993. The first problem is that which constitutes James Bernoulli's theorem. We will reproduce Laplace's investigation.

The probability of the happening of an event at each trial is p; required the probability that in a given number of trials the number of times in which the event happens will lie between certain assigned limits.

Let $q = 1 - p$ and $\mu = m + n$; then the probability that the event will happen m times and fail n times in μ trials is equal to a certain term in the expansion of $(p + q)^\mu$, namely

$$\frac{\lfloor \mu}{\lfloor m \lfloor n} p^m q^n.$$

Now it is known from Algebra that if m and n vary subject to the condition that $m + n$ is constant, the greatest value of the above term is when $\dfrac{m}{n}$ is as nearly as possible equal to $\dfrac{p}{q}$, so that m and n are as nearly as possible equal to μp and μq respectively. We say as *nearly as possible*, because μp is not necessarily an integer, while m is. We may denote the value of m by $\mu p + z$, where z is some proper fraction, positive or negative; and then $n = \mu q - z$.

The rth term, counting onwards, in the expansion of $(p+q)^{\mu}$ after $\dfrac{\lfloor \mu}{\lfloor m \lfloor n} p^m q^n$ is $\dfrac{\lfloor \mu}{\lfloor m-r \lfloor n+r} p^{m-r} q^{n+r}$.

We shall now suppose that m and n are large numbers, and transform the last expression by the aid of Stirling's Theorem; see Arts. 333, 962. We have

$$\lfloor \mu = \mu^{\mu + \frac{1}{2}} e^{-\mu} \sqrt{(2\pi)} \left\{ 1 + \frac{1}{12\mu} + \ldots \right\},$$

$$\frac{1}{\lfloor m-r} = (m-r)^{r-m-\frac{1}{2}} e^{m-r} \frac{1}{\sqrt{(2\pi)}} \left\{ 1 - \frac{1}{12(m-r)} - \ldots \right\},$$

$$\frac{1}{\lfloor n+r} = (n+r)^{-n-r-\frac{1}{2}} e^{n+r} \frac{1}{\sqrt{(2\pi)}} \left\{ 1 - \frac{1}{12(n+r)} - \ldots \right\}.$$

We shall transform the term $(m-r)^{r-m-\frac{1}{2}}$. Its logarithm is

$$\left(r - m - \frac{1}{2} \right) \left\{ \log m + \log \left(1 - \frac{r}{m} \right) \right\},$$

and $\qquad \log \left(1 - \frac{r}{m} \right) = -\frac{r}{m} - \frac{r^2}{2m^2} - \frac{r^3}{3m^3} - \ldots\ldots$

We shall suppose that r^2 does not surpass μ in order of magnitude, and we shall neglect fractions of the order $\dfrac{1}{\mu}$; we shall thus neglect such a term as $\dfrac{r^4}{m^3}$, because m is of the order μ. Thus we have approximately

$$\left(r - m - \frac{1}{2} \right) \left\{ \log m + \log \left(1 - \frac{r}{m} \right) \right\}$$

$$= \left(r - m - \frac{1}{2} \right) \log m + r + \frac{r}{2m} - \frac{r^2}{2m} - \frac{r^3}{6m^2};$$

and then, passing from the logarithms to the numbers,

$$(m-r)^{r-m-\frac{1}{2}} = m^{r-m-\frac{1}{2}} e^{r-\frac{r^2}{2m}} \left(1 + \frac{r}{2m} - \frac{r^3}{6m^2} \right).$$

Similarly

$$(n+r)^{-n-r-\frac{1}{2}} = n^{-n-r-\frac{1}{2}} e^{-r-\frac{r^2}{2n}} \left(1 - \frac{r}{2n} + \frac{r^3}{6n^2}\right).$$

Thus we have approximately

$$\frac{\lfloor \mu}{\lfloor m-r \rfloor \lfloor n+r} = \frac{\mu^{\mu+\frac{1}{2}} e^{-\frac{\mu r^2}{2mn}}}{m^{m-r+\frac{1}{2}} n^{n+r+\frac{1}{2}} \sqrt{(2\pi)}} \left\{1 + \frac{r(n-m)}{2mn} - \frac{r^3}{6m^2} + \frac{r^3}{6n^2}\right\}.$$

Now suppose that the values of m and n are those which we have already assigned as corresponding to the greatest term of the expansion of $(p+q)^\mu$, then

$$p = \frac{m-z}{\mu}, \quad q = \frac{n+z}{\mu};$$

thus we have approximately

$$p^{m-r} q^{n+r} = \frac{m^{m-r} n^{n+r}}{\mu^\mu} \left(1 + \frac{\mu r z}{mn}\right).$$

Therefore finally we have approximately for the rth term *after* the greatest

$$\frac{e^{-\frac{\mu r^2}{2mn}} \sqrt{\mu}}{\sqrt{(2\pi mn)}} \left\{1 + \frac{\mu r z}{mn} + \frac{r(n-m)}{2mn} - \frac{r^3}{6m^2} + \frac{r^3}{6n^2}\right\}.$$

We shall obtain the approximate value of the rth term *before* the greatest by changing the sign of r in the above expression; by adding the values of the two terms we have

$$\frac{2\sqrt{\mu}}{\sqrt{(2\pi mn)}} e^{-\frac{\mu r^2}{2mn}}.$$

If we take the sum of the values of this expression from $r=0$ to $r=r$, we obtain approximately the sum of twice the greatest term of a certain binomial expansion together with the r terms which precede and the r terms which follow the greatest term; subtract the greatest term, and we have the approximate value of the sum of $2r+1$ terms of a binomial expansion which include the greatest term as their middle term.

Now by Euler's theorem, given in Art. 334,

$$\Sigma y = \int y \, dr - \frac{1}{2} y + \frac{1}{12} \frac{dy}{dr} - \dots$$

Here $y = \frac{2\sqrt{\mu}}{\sqrt{(2\pi mn)}} e^{-\frac{\mu r^2}{2mn}}$, and the differential coefficients of y with respect to r will introduce the factor $\frac{\mu r}{2mn}$, and its powers;

and $\frac{\mu r}{2mn}$ is of the order $\frac{1}{\sqrt{\mu}}$ at most, so that when multiplied by the constant factor in y we obtain a term of the oraer $\frac{1}{\mu}$. Thus as far as we need proceed,

$$\Sigma y = \int y\,dr - \frac{1}{2}y + \frac{1}{2}Y,$$

where both the symbols Σ and $\int$ are supposed to indicate operations commencing with $r = 0$, and $\frac{1}{2}Y$ denotes the greatest term of the binomial expansion, that is the value of $\frac{1}{2}y$ when $r = 0$. The expression Σy denotes as usual the sum of the values of y up to that corresponding to $r - 1$; adding the value of y corresponding to r we obtain

$$\int y\,dr + \frac{1}{2}y + \frac{1}{2}Y;$$

subtract the greatest term of the binomial, and thus we have

$$\int y\,dr + \frac{1}{2}y.$$

Put $\qquad \tau = \frac{r\sqrt{\mu}}{\sqrt{(2mn)}}$; thus we obtain finally

$$\frac{2}{\sqrt{\pi}} \int_0^\tau e^{-t^2}\,dt + \frac{\sqrt{\mu}}{\sqrt{(2\pi mn)}} e^{-\tau^2}.$$

This expression therefore is the approximate value of the sum of $2r + 1$ terms of the expansion of $(p+q)^\mu$, these terms including the greatest term as their middle term. In the theory of probability the expression gives the probability that the number of times in which the event will happen in μ trials will lie between $m - r$ and $m + r$, both inclusive, that is between

$$\mu p + z - \frac{\tau\sqrt{(2mn)}}{\sqrt{\mu}} \text{ and } \mu p + z + \frac{\tau\sqrt{(2mn)}}{\sqrt{\mu}} ;$$

or, in other words, the expression gives the probability that the ratio of the number of times in which the event happens to the whole number of trials will lie between

$$p + \frac{z}{\mu} - \frac{\tau \sqrt{(2mn)}}{\mu \sqrt{\mu}} \text{ and } p + \frac{z}{\mu} + \frac{\tau (2mn)}{\mu \sqrt{\mu}}.$$

If μ be very large we may neglect z in comparison with μp or μq; and then $mn = \mu^2 pq$ approximately, so that we obtain the following result: If the number of trials, μ, be very large, the probability that the ratio of the number of times in which the event happens to the whole number of trials will lie between

$$p - \frac{\tau \sqrt{(2pq)}}{\sqrt{\mu}} \text{ and } p + \frac{\tau \sqrt{(2pq)}}{\sqrt{\mu}}$$

is

$$\frac{2}{\sqrt{\pi}} \int_0^\tau e^{-t^2} dt + \frac{1}{\sqrt{(2\pi\mu pq)}} e^{-\tau^2}.$$

994. The result which has just been obtained is one of the most important in the whole range of our subject. There are two points to be noticed with respect to the result.

In the first place, it is obvious that supposing τ to be constant we may by sufficiently increasing μ render the limits

$$p - \frac{\tau \sqrt{(2pq)}}{\sqrt{\mu}} \text{ and } p + \frac{\tau \sqrt{(2pq)}}{\sqrt{\mu}}$$

as close as we please, while the corresponding probability is always greater than $\frac{2}{\sqrt{\pi}} \int_0^\tau e^{-t^2} dt$.

In the second place, it is known that the value of $\frac{2}{\sqrt{\pi}} \int_0^\tau e^{-t^2} dt$ approaches very near to unity for even moderate values of τ. Tables of the value of this expression will be found in the works of Professor De Morgan cited in Arts. 268 and 485, and in that of Galloway cited in Art. 753. The following extract will sufficiently illustrate the rapid approach to unity: the first column gives values of τ, and the second column the corresponding values of the expression $\frac{2}{\sqrt{\pi}} \int_0^\tau e^{-t^2} dt$.

·5	·5204999
1·0	·8427008
1·5	·9661052
2·0	·9953223
2·5	·9995930
3·0	·9999779

995. With respect to the history of the result obtained in Art. 994, we have to remark that James Bernoulli began the investigation; then Stirling and De Moivre carried it on by the aid of the theorem known by Stirling's name; and lastly, the theorem known by Euler's name gave the mode of expressing the finite summation by means of an integral. See Arts. 123, 334, 335, 423. But it will be seen that practically we use only the first term of the series given in Euler's theorem, in fact no more than amounts to evaluating an integral by a rough approximate quadrature. Thus the result given by Laplace was within the power of mathematicians as soon as Stirling's Theorem had been published.

Laplace, in his introduction, page XLII, speaking of James Bernoulli's theorem says,

Ce théorème indiqué par le bon sens, était difficile à démontrer par l'Analyse. Aussi l'illustre géomètre Jacques Bernoulli qui s'en est occupé le premier, attachait-il une grande importance à la démonstration qu'il en a donnée. Le calcul des fonctions génératrices, appliqué à cet objet, non-seulement démontre avec facilité ce théorème ; mais de plus il donne la probabilité que le rapport des événemens observés, ne s'écarte que dans certaines limites, du vrai rapport de leurs possibilités respectives.

Laplace's words ascribe to the theory of generating functions the merit which should be shared between the theorems known by the names of Stirling and Euler.

We may remark that in one of his memoirs Laplace had used a certain process of summation not connected with Euler's theorem : see Art. 897.

996. Laplace gives the following example of the result obtained in Art. 993.

Suppose that the probability of the birth of a boy to that of the birth of a girl be as 18 to 17 : required the probability that in 14000 births the number of boys will fall between 7363 and 7037.

Here

$$p = \frac{18}{35}, \quad q = \frac{17}{35}, \quad m = 7200, \quad n = 6800, \quad r = 163 :$$

the required probability is ·994303.

The details of the calculation will be found in Art. 74 of the *Theory of Probabilities* in the *Encyclopædia Metropolitana.*

997. We have now to notice a certain inverse application which Laplace makes of James Bernoulli's theorem : this is a point of considerable importance to which we have already alluded in Art. 125, and which we must now carefully discuss.

In Art. 993 it is supposed that p is given, and we find the probability that the ratio of the number of times in which the event happens to the whole number of trials will lie between assigned limits. Suppose however that p is not known *a priori*, but that we have *observed* the event to happen m times and to fail n times in μ trials. Then Laplace *assumes* that the expression given in Art. 993 will be the probability that $p - \dfrac{m}{n}$ lies between

$$- \frac{\tau \sqrt{(2mn)}}{\mu \sqrt{\mu}} \quad \text{and} \quad + \frac{\tau \sqrt{(2mn)}}{\mu \sqrt{\mu}} ;$$

that is, Laplace takes for this probability the expression

$$\frac{2}{\sqrt{\pi}} \int_0^\tau e^{-t^2} dt + \frac{\sqrt{\mu}}{\sqrt{(2\pi mn)}} e^{-\tau^2} \ldots\ldots\ldots\ldots\ldots(1).$$

He draws an inference from the formula, and then says, on his page 282,

On parvient directement à ces résultats, en considérant p comme une variable qui peut s'étendre depuis zéro jusqu'à l'unité, et en déterminant, d'après les événemens observés, la probabilité de ses diverses valeurs, comme on le verra lorsque nous traiterons de la probabilité des causes, déduite des événemens observés.

Accordingly we find that Laplace does in effect return to the subject; see his pages 363—366.

In the formula which we have given in Art. 697, suppose $a = 0$, and $b = 1$; then if the event has been observed to happen m times and to fail n times out of $m + n$ trials, the probability that the chance at a single trial lies between α and β is

$$\frac{\int_{\alpha}^{\beta} x^m (1 - x)^n \, dx}{\int_{0}^{1} x^m (1 - x)^n \, dx}.$$

Let

$$\alpha = \frac{m}{\mu} - \frac{\tau \sqrt{(2mn)}}{\mu \sqrt{\mu}}, \qquad \beta = \frac{m}{\mu} + \frac{\tau \sqrt{(2mn)}}{\mu \sqrt{\mu}},$$

where $\mu = m + n$; then we shall shew, by using Laplace's method of approximation, that the probability is nearly

$$\frac{2}{\sqrt{\pi}} \int_{0}^{\tau} e^{-t^2} \, dt \quad \dots\dots\dots\dots\dots\dots (2).$$

For with the notation of Art. 957 we have $y = x^m (1 - x)^n$; the value of x which makes y a maximum is found from the equation

$$\frac{m}{x} - \frac{n}{1 - x} = 0,$$

so that

$$a = \frac{m}{m + n}$$

Then

$$t^2 = \log \frac{Y}{(a + \theta)^m (1 - a - \theta)^n}$$

$$= \log \frac{Y}{a^m (1 - a)^n} - m \log \left(1 + \frac{\theta}{a}\right) - n \log \left(1 - \frac{\theta}{1 - a}\right)$$

$$= \frac{\theta^2}{2} \left\{ \frac{m}{a^2} + \frac{n}{(1 - a)^2} \right\} - \frac{\theta^3}{3} \left\{ \frac{m}{a^3} - \frac{n}{(1 - a)^3} \right\} + \cdots$$

Thus, approximately,

$$t^2 = \frac{\theta^2}{2} \left\{ \frac{m}{a^2} + \frac{n}{(1 - a)^2} \right\} = \frac{\theta^2 (m + n)^3}{2mn}.$$

Therefore

$$\frac{\int_{\alpha}^{\beta} x^m (1-x)^n \, dx}{\int_0^1 x^m (1-x)^n dx} = \frac{Y \int_{-\tau}^{\tau} e^{-t^2} dt}{Y \int_{-\infty}^{\infty} e^{-t^2} \, dt}$$

$$= \frac{1}{\sqrt{\pi}} \int_{-\tau}^{\tau} e^{-t^2} dt = \frac{2}{\sqrt{\pi}} \int_0^{\tau} e^{-t^2} \, dt.$$

We have thus two results, namely (1) and (2): the former is obtained by what we may call an assumed inversion of James Bernoulli's theorem, and the latter we may say depends on Bayes's theorem. It will be seen that the two results are not quite consistent; the difference is not practically very important, but it is of interest theoretically.

The result (2) is in effect given by Laplace on his page 366; he does not however make any remark on the difference between this result and that which we find on his page 282.

On page 209 of his *Recherches...sur la Prob.* Poisson gives the result (1) which he obtains by the same assumption as Laplace. But on his page 213 Poisson gives a different result, for he finds in effect that the probability that the chance at a single trial lies between

$$\frac{m}{\mu} - \frac{v\sqrt{(2mn)}}{\mu\sqrt{\mu}} \quad \text{and} \quad \frac{m}{\mu} - \frac{(v+dv)\sqrt{(2mn)}}{\mu\sqrt{\mu}}$$

is $V \, dv$,

where $V = \dfrac{1}{\sqrt{\pi}} e^{-v^2} - \dfrac{2(m-n)v^3}{\sqrt{(2\pi\mu mn)}} e^{-v^2} \ldots\ldots\ldots (3)$.

This is inconsistent with Poisson's page 209; for if we take the integral $\int V dv$ between the limits $-\tau$ and $+\tau$ for v it reduces to $\dfrac{2}{\sqrt{\pi}} \int_0^{\tau} e^{-t^2} dt$, so that we arrive at the result (2), and not at the result (1). It is curious that Poisson makes no remark on the difference between his pages 209 and 213; perhaps he regarded his page 209 as supplying a first approximation, and his page 213 as a more correct investigation.

Poisson's result (3) is deduced by him in his *Recherches...sur la Prob.* from the same *kind* of assumption as that by which he and

Laplace arrived at the result (1) ; but the assumption is used in such a way as to diminish very decidedly the apprehension of any erroneous consequences : the assumption, so to speak, is made to extend over an indefinitely small interval instead of over a finite interval.

Poisson had however previously considered the question in his *Mémoire sur la proportion des naissances des deux sexes;* this memoir is published in the *Mémoires...de l'Institut,* Vol. IX, 1830 ; there he uses Bayes's theorem, and proceeds as we have done in establishing (2), but he carries the approximation further: *he arrives at the result* (3). See page 271 of the memoir.

Thus the result (3) is demonstrable in two ways, namely, by the assumed inversion of James Bernoulli's theorem, and by Bayes's theorem. As Poisson in his latest discussion of the question adopted the inversion of James Bernoulli's theorem, we may perhaps infer that he considered the amount of assumption thus involved to be no greater than that which is required in the use of Bayes's theorem. See Art. 552.

In a memoir published in the *Cambridge Philosophical Transactions,* Vol. VI. 1837, Professor De Morgan drew attention to the circumstance that Laplace and Poisson had arrived at the result (1) by *assuming* what we have called an inversion of James Bernoulli's theorem ; and he gave the investigation which, as we have said, depends on Bayes's theorem. Professor De Morgan however overlooked the fact that Laplace had also implicitly given the result (2),. and that Poisson had arrived at the result (3) by both methods. It will be found on examining page 428 of the volume which contains Professor De Morgan's memoir, that his final result amounts to changing v^3 into v in the second term of the value of V in Poisson's result (3). Poisson, however, is correct ; the disagreement between the two mathematicians arises from the fact that the approximations to the values of μ and ν which Professor De Morgan gives towards the top of the page under consideration are not carried far enough for the object he has in view.

In the Treatise on Probability by Galloway, which is contained in the *Encyclopædia Britannica,* reference is expressly made to Professor De Morgan's memoir, without any qualifying remark ;

this is curious, for the Treatise may be described as an abridgement of Poisson's *Recherches...sur la Prob.*, and Poisson himself
refers to his memoir of 1830; so that it might have been expected
that some, if not all, of our conclusions would have presented
themselves to Galloway's attention.

998.   Laplace discusses in his pages 284—286 the following
problem.   An urn contains a large number, n, of balls, some white,
and the rest black; at each drawing a ball is extracted and replaced by a black ball: required the probability that after r
drawings there will be x white balls in the urn.

999.   The remainder of the Chapter, forming pages 287—303,
is devoted to investigations arising from the following problem.
There are two urns, A and B, each containing n balls, some white
and the rest black; there are on the whole as many white balls as
black balls.   A ball is drawn out from each urn and put into the
other urn; and this operation is repeated r times.   Required the
probability that there will then be x white balls in the urn A.

This problem is formed on one which was originally given by
Daniel Bernoulli; see Arts. 417, 587, 807, 921.

Let $z_{x,r}$ denote the required probability; then Laplace obtains
the following equation:

$$z_{x,r+1} = \left(\frac{x+1}{n}\right)^2 z_{x+1,r} + \frac{2x}{n}\left(1 - \frac{x}{n}\right)z_{x,r} + \left(1 - \frac{x-1}{n}\right)^2 z_{x-1,r}.$$

This equation however is too difficult for exact solution, and so
Laplace mutilates it most unsparingly.   He supposes n to be very
large, and he says that we have then approximately

$$z_{x+1,r} = z_{x,r} + \frac{dz_{x,r}}{dx} + \frac{1}{2}\frac{d^2z_{x,r}}{dx^2},$$

$$z_{x-1,r} = z_{x,r} - \frac{dz_{x,r}}{dx} + \frac{1}{2}\frac{d^2z_{x,r}}{dx^2},$$

$$z_{x,r+1} = z_{x,r} + \frac{dz_{x,r}}{dr}.$$

Let $x = \dfrac{n + \mu\sqrt{n}}{2}$, $r = nr'$, $z_{x,r} = U$; then he says that neglecting terms of the order $\dfrac{1}{n^2}$ the equation becomes

$$\frac{dU}{dr'} = 2U + 2\mu\frac{dU}{d\mu} + \frac{d^2U}{d\mu^2}.$$

It is difficult to see how Laplace establishes this; for if we adopt his expressions for $z_{x+1,r}$, $z_{x-1,r}$, and $z_{x,r+1}$, the equation becomes

$$\frac{dU}{dr'} = 2\left(1 + \frac{1}{n}\right)U + 2\mu\left(1 + \frac{2}{n}\right)\frac{dU}{d\mu}$$

$$+ \left(1 + \frac{\mu^2}{n} + \frac{4}{n} + \frac{4}{n^2}\right)\frac{d^2U}{d\mu^2};$$

and thus the error seems to be of the order $\dfrac{1}{n}$, or even larger, since μ^2 *may* be as great as n.

1000. Laplace proceeds to integrate his approximate equation by the aid of definite integrals. He is thus led to investigate some auxiliary theorems in definite integrals, and then he passes on to other theorems which bear an analogy to those which occur in connexion with what are called *Laplace's Functions.* We will give two of the auxiliary theorems, demonstrating them in a way which is perhaps simpler than Laplace's.

To shew that, if i is a positive integer,

$$\int_{-\infty}^{\infty}\int_{-\infty}^{\infty} e^{-s^2-\mu^2}(s + \mu\sqrt{-1})^i\, ds\, d\mu = 0.$$

Transform the double integral by putting

$$s = r\cos\theta, \ \mu = r\sin\theta;$$

we thus obtain

$$\int_{0}^{\infty}\int_{0}^{2\pi} e^{-r^2}(\cos i\theta + \sqrt{-1}\sin i\theta)\, r^{i+1}\, dr\, d\theta.$$

It is obvious that the positive and negative elements in this integral balance each other, so that the result is zero.

Again to shew that, if i and q are positive integers and q less than i,

$$\int_{-\infty}^{\infty} \int_{-\infty}^{\infty} e^{-s^2-\mu^2} \mu^q (s + \mu \sqrt{-1})^i \, ds \, d\mu = 0.$$

Transforming as before we obtain

$$\int_0^{\infty} \int_0^{2\pi} e^{-r^2} (\cos i\theta + \sqrt{-1} \sin i\theta) \sin^q \theta \, r^{q+i+1} \, dr \, d\theta.$$

Now $\sin^q \theta$ may be expressed in terms of sines or of cosines of multiples of θ, according as q is odd or even, and the highest multiple of θ will be $q\theta$. And we know that if m and n are unequal integers we have

$$\int_0^{2\pi} \sin m\theta \cos n\theta \, d\theta = 0,$$

$$\int_0^{2\pi} \cos m\theta \cos n\theta \, d\theta = 0,$$

$$\int_0^{2\pi} \sin m\theta \sin n\theta \, d\theta = 0 ;$$

thus the required result is obtained.

Laplace finally takes the same problem as Daniel Bernoulli had formerly given; see Art. 420. Laplace forms the differential equations, supposing any number of vessels; and he gives without demonstration the solutions of these differential equations: the demonstration may be readily obtained by the modern method of separating the symbols of operation and quantity.

1001. Laplace's Chapter IV. is entitled, *De la probabilité des erreurs des résultats moyens d'un grand nombre d'observations, et des résultats moyens les plus avantageux:* this Chapter occupies pages 304—348.

This Chapter is the most important in Laplace's work, and perhaps the most difficult; it contains the remarkable theory which is called the *method of least squares.* Laplace had at an early period turned his attention to the subject of the mean to be taken of the results of observations; but the contents of the present Chapter occur only in his later memoirs. See Arts. 874, 892, 904, 917, 921.

Laplace's processes in this Chapter are very peculiar, and it is scarcely possible to understand them or feel any confidence in their results without translating them into more usual mathematical language. It has been remarked by R. Leslie Ellis that, "It must be admitted that there are few mathematical investigations less inviting than the fourth Chapter of the *Théorie des Probabilités*, which is that in which the method of least squares is proved." *Cambridge Phil. Trans.* Vol. VIII. page 212.

In the *Connaissance des Tems* for 1827 and for 1832 there are two most valuable memoirs by Poisson on the probability of the mean results of observations. These memoirs may be described as a commentary on Laplace's fourth Chapter. It would seem from some words which Poisson uses at the beginning— j'ai pensé que les remarques que j'ai eu l'occasion de faire en l'étudiant,—that his memoirs form a kind of translation, which he made for his own satisfaction, of Laplace's investigations. Poisson embodied a large part of his memoirs in the fourth Chapter of his *Recherches sur la Prob....*

We shall begin our account of Laplace's fourth Chapter by giving Poisson's solution of a very general problem, as we shall then be able to render our analysis of Laplace's processes more intelligible. But at the same time it must be remembered that the merit is due almost entirely to Laplace; although his processes are obscure and repulsive, yet they contain all that is essential in the theory: Poisson follows closely in the steps of his illustrious guide, but renders the path easier and safer for future travellers.

1002. Suppose that a series of s observations is made, each of which is liable to an error of unknown amount; let these errors be denoted by $\epsilon_1, \epsilon_2, \ldots \epsilon_s$. Let E denote the sum of these errors, each multiplied by an assigned constant, say

$$E = \gamma_1\epsilon_1 + \gamma_2\epsilon_2 + \gamma_3\epsilon_3 + \ldots + \gamma_s\epsilon_s:$$

required the probability that E will lie between assigned limits.

Suppose that each error is susceptible of various values, positive or negative, and that these values are all multiples of a given quantity ω. These values will be assumed to lie between $\alpha\omega$ and $\beta\omega$, both inclusive; here α and β will be positive or negative integers, or zero, and we shall suppose that α is algebraically

greater than β, so that $\alpha - \beta$ is positive. The chance of an assigned error will not be assumed the same at each observation.
If n be any integer comprised between α and β we shall denote
the chance of an error $n\omega$ at the first observation by N_1, at the
second observation by N_2, at the third observation by N_3, and
so on. Let ϖ be a factor such that all the products $\varpi\gamma_1$, $\varpi\gamma_2$,
$\varpi\gamma_3$, ... $\varpi\gamma_s$ are integers; such a factor can always be found either
exactly or to any required degree of approximation. Let

$$Q_i = \Sigma N_i t^{\varpi\gamma_i n\omega},$$

where Σ denotes a summation with respect to n for all values
from β to α, both inclusive; and let

$$T = Q_1 Q_2 \ldots Q_s :$$

then the probability that ϖE will be exactly equal to $m\omega$, where
m is a given integer, is the coefficient of $t^{m\omega}$ in the development
of T according to powers of t; or, which is the same thing, the
probability is equal to the term independent of t in the development of $T t^{-m\omega}$.

For t^ω put $e^{\theta\sqrt{-1}}$, and denote by X what T becomes; then the
required probability is equal to

$$\frac{1}{2\pi} \int_{-\pi}^{\pi} X e^{-m\theta\sqrt{-1}} d\theta.$$

Let λ and μ be two given integers, such that $\lambda - \mu$ is positive;
then the probability that ϖE will lie between $\lambda\omega$ and $\mu\omega$, both
inclusive, may be derived from the last expression by putting for
m in succession the values μ, $\mu+1$, $\mu+2$, ... λ, and adding the results. Since the sum of the values of $e^{-m\theta\sqrt{-1}}$ is

$$\frac{\sqrt{-1}}{2\sin\frac{1}{2}\theta}\left\{e^{-(\lambda+\frac{1}{2})\theta\sqrt{-1}} - e^{-(\mu-\frac{1}{2})\theta\sqrt{-1}}\right\},$$

the required probability is equal to

$$\frac{\sqrt{-1}}{4\pi}\int_{-\pi}^{\pi}\left\{e^{-(\lambda+\frac{1}{2})\theta\sqrt{-1}} - e^{-(\mu-\frac{1}{2})\theta\sqrt{-1}}\right\}\frac{X d\theta}{\sin\frac{1}{2}\theta};$$

we shall denote this probability by P.

Let us now suppose that ω is indefinitely small, and that λ and μ are infinite; and let

$$\lambda\omega = (c + \eta)\,\varpi, \quad \mu\omega = (c - \eta)\,\varpi, \quad \varpi\theta = \omega x.$$

The limits of the integration with respect to x will be $\pm\infty$. Also we have

$$d\theta = \frac{\omega}{\varpi}\,dx, \quad \sin\frac{1}{2}\theta = \frac{\omega x}{2\varpi}.$$

Thus, neglecting $\pm\frac{1}{2}$ compared with λ and μ, we obtain

$$P = \frac{1}{\pi}\int_{-\infty}^{\infty} Xe^{-cx\sqrt{-1}}\sin\eta x\,\frac{dx}{x} \quad\ldots\ldots\ldots\ldots (1).$$

This expression gives the probability that ϖE will lie between $(c+\eta)\,\varpi$ and $(c-\eta)\,\varpi$, that is, the probability that E will lie between $c+\eta$ and $c-\eta$.

Since we suppose ω indefinitely small we consider that the error at each observation may have any one of an infinite number of values; the chance of each value will therefore be indefinitely small. Let

$$\alpha\omega = a, \quad \beta\omega = b, \quad n\omega = z\,;$$

then

$$t^{\varpi\gamma_i n\omega} = e^{\varpi\gamma_i n\theta\sqrt{-1}} = e^{\gamma_i n\omega x\sqrt{-1}} = e^{\gamma_i xz\sqrt{-1}}.$$

Let

$$N_i = \omega f_i(z)\,;$$

thus Q_i becomes

$$\int_b^a f_i(z)\,e^{\gamma_i xz\sqrt{-1}}\,dz\,;$$

and for X in (1) we must put the new form which we thus obtain for the product

$$Q_1 Q_2 Q_3 \cdots Q_s.$$

Assume

$$\int_b^a f_i(z)\cos\gamma_i xz\,dz = \rho_i\cos r_i,$$

$$\int_b^a f_i(z)\sin\gamma_i xz\,dz = \rho_i\sin r_i\,;$$

then

$$Q_i = \rho_i e^{r_i\sqrt{-1}}.$$

Let

$$Y = \rho_1\rho_2\rho_3\cdots\rho_s,$$

$$y = r_1 + r_2 + r_3 + \ldots + r_s\,;$$

then
$$X = Ye^{v\sqrt{-1}}.$$

Substitute in (1) and we obtain

$$P = \frac{1}{\pi} \int_{-\infty}^{\infty} Y \cos (y - cx) \sin \eta x \, \frac{dx}{x}$$

$$+ \frac{\sqrt{-1}}{\pi} \int_{-\infty}^{\infty} Y \sin (y - cx) \sin \eta x \, \frac{dx}{x}.$$

The elements in the second integral occur in pairs of equal numerical value and of opposite signs, while the elements in the first integral occur in pairs of equal numerical value and of the same sign. Thus

$$P = \frac{2}{\pi} \int_{0}^{\infty} Y \cos (y - cx) \sin \eta x \frac{dx}{x} \quad \dots\dots\dots\dots \text{(2)}.$$

Since each error is supposed to lie between a and b we have

$$\int_{b}^{a} f_i(z) \, dz = 1.$$

Hence it follows that $\rho_i = 1$ when $x = 0$; and we shall now shew that when x has any other value ρ_i is less than unity.

For
$$\rho_i^2 = \left\{ \int_{b}^{a} f_i(z) \cos \gamma_i xz \, dz \right\}^2 + \left\{ \int_{b}^{a} f_i(z) \sin \gamma_i xz \, dz \right\}^2 ;$$

that is
$$\rho_i^2 = \int_{b}^{a} f_i(z) \cos \gamma_i xz \, dz \int_{b}^{a} f_i(z') \cos \gamma_i xz' \, dz'$$

$$+ \int_{b}^{a} f_i(z) \sin \gamma_i xz \, dz \int_{b}^{a} f_i(z') \sin \gamma_i xz' \, dz'$$

$$= \int_{b}^{a} \int_{b}^{a} f_i(z) f_i(z') \cos \gamma_i x (z - z') \, dz \, dz' ;$$

and this is less than

$$\int_{b}^{a} \int_{b}^{a} f_i(z) f_i(z') \, dz \, dz',$$

that is less than

$$\int_{b}^{a} f_i(z) \, dz \int_{b}^{a} f_i(z') \, dz',$$

that is less than unity.

Up to this point the investigation has been exact: we shall now proceed to approximate. Suppose s to be a very large number; then Y is the product of a very large number of factors, each of which is less than unity except when $x = 0$. We may infer that Y will always be small except when x is very small; and we shall find an approximate value of Y on the supposition that x is small.

Let
$$\int_b^a z f_i(z)\, dz = k_i,$$

$$\int_b^a z^2 f_i(z)\, dz = k_i',$$

$$\int_b^a z^3 f_i(z)\, dz = k_i'',$$

$$\int_b^a z^4 f_i(z)\, dz = k_i''',$$

$$\dots\dots\dots\dots\dots$$

Then we shall have in converging series
$$\rho_i \cos r_i = 1 - \frac{x^2 \gamma_i^2 k_i'}{\underline{2}} + \frac{x^4 \gamma_i^4 k_i'''}{\underline{4}} - \dots\dots,$$

$$\rho_i \sin r_i = x \gamma_i k_i - \frac{x^3 \gamma_i^3 k_i''}{\underline{3}} + \dots\dots\dots\dots.$$

Let $\quad \frac{1}{2}(k_i' - k_i^2) = h_i^2;$ then we obtain

$$\rho_i = 1 - x^2 \gamma_i^2 h_i^2 + \dots\dots\dots,$$

$$r_i = x \gamma_i k_i + \dots\dots\dots\dots.$$

Hence $\qquad \log \rho_i = -x^2 \gamma_i^2 h_i^2 + \dots\dots\dots\dots;$

therefore $\qquad \rho_i = e^{-x^2 \gamma_i^2 h_i^2}$ approximately.

Let κ^2 stand for $\Sigma \gamma_i^2 h_i^2$, and l for $\Sigma \gamma_i k_i$, each summation extending for the values of i from 1 to s inclusive. Then approximately

$$Y = e^{-\kappa^2 x^2}, \quad y = lx.$$

Thus (2) becomes

$$P = \frac{2}{\pi} \int_0^\infty e^{-\kappa^2 x^2} \cos(lx - cx) \sin \eta x \, \frac{dx}{x} \quad\dots\dots (3).$$

The approximate values which have been given for Y and y can only be considered to be near the truth when x is very small;

but no serious error will arise from this circumstance, because the true value of Y and the approximate value are both very small when x is sensibly different from zero. We may put (3) in the form

$$P = \frac{2}{\pi} \int_0^\infty \left\{ \int_{-\eta}^{\eta} \cos\,(lx - cx + xv)\,dv \right\} e^{-\kappa^2 x^2}\,dx;$$

then by changing the order of integration, and using a result given in Art. 958, we obtain

$$P = \frac{1}{2\kappa\sqrt{\pi}} \int_{-\eta}^{\eta} e^{-\frac{(l-c+v)^2}{4\kappa^2}}\,dv \,\ldots\ldots\ldots\ldots (4).$$

This is therefore approximately the probability that E will lie between $c - \eta$ and $c + \eta$.

It is necessary to shew that the quantity which we have denoted by κ^2 is really positive; this is the case since h_i^2 is really positive, as we shall now shew. From the definition of h_i^2 in conjunction with the equation $\int_b^a f_i\,(z)\,dz = 1$, we have

$$2h_i^2 = \int_b^a z^2 f_i\,(z)\,dz \int_b^a f_i\,(z')\,dz' - \int_b^a z f_i\,(z)\,dz \int_b^a z' f_i\,(z')\,dz'$$

$$= \int_b^a \int_b^a (z^2 - zz')f\,(z) f_i\,(z')\,dz\,dz'.$$

And so also

$$2h_i^2 = \int_b^a \int_b^a (z'^2 - zz') f_i\,(z) f_i\,(z')\,dz\,dz'.$$

Hence, by addition,

$$4h_i^2 = \int_b^a \int_b^a (z - z')^2 f_i\,(z) f_i\,(z')\,dz\,dz'.$$

Thus $4h_i^2$ is essentially a positive quantity which cannot be zero, for every element in the double integral is positive.

It is usual to call $f_i\,(z)$ the function which gives the *facility of error* at the ith observation; this means that $f_i(z)\,dz$ expresses the chance that the error will lie between z and $z + dz$.

If the function of the facility of error be the same at every observation we shall denote it by $f(z)$; and then dropping those suffixes which are no longer necessary, we have

$$k = \int_b^a z f(z)\, dz, \quad k' = \int_b^a z^2 f(z)\, dz,$$

$$h^2 = \frac{1}{2}(k' - k^2),$$

$$\kappa^2 = h^2 \Sigma \gamma_i^2, \quad l = k \Sigma \gamma_i.$$

Such is the solution which we have borrowed from Poisson; he presents his investigation in slightly varying forms in the places to which we have referred: we have not adopted any form exclusively but have made a combination which should be most serviceable for the object we have in view, namely, to indicate the contents of Laplace's fourth Chapter. Our notation does not quite agree with that which Poisson has employed in any of the forms of his investigation; we have, for example, found it expedient to interchange Poisson's a and b.

We may make two remarks before leaving Poisson's problem.

I. We have supposed that the error at each observation lies between the same limits, a and b; but the investigation will apply to the case in which the limits of error are different for different observations. Suppose, for example, it is known at the first observation that the error must lie between the limits a_1 and b_1, which are *within* the limits a and b. Then $f_1(z)$ will be a function of z which must be taken to vanish for all values of z between b and b_1 and between a_1 and a.

Thus in fact it is only necessary to suppose that a and b are so chosen, that no error at any observation can be algebraically greater than a or less than b.

II. Poisson shews how to proceed one step further in the approximation. We took $y = lx$; we have more closely $y = lx - l_1 x^3$, where

$$l_1 = \frac{1}{6} \Sigma \gamma_i^3 \{k_i'' - 3k_i k_i' + 2k_i^3\}.$$

Hence, approximately,

$$\cos(y - cx) = \cos(lx - cx) + l_1 x^3 \sin(lx - cx).$$

Therefore (2) becomes

$$P = \frac{2}{\pi} \int_0^\infty e^{-\kappa^2 x^2} \cos (lx - cx) \sin \eta x \frac{dx}{x}$$

$$+ \frac{2l_1}{\pi} \int_0^\infty e^{-\kappa^2 x^2} \sin (lx - cx) \, x^2 \sin \eta x \, dx.$$

We formerly transformed the first term in this expression of P; it is sufficient to observe that the second term may be derived from the first by differentiating three times with respect to l and multiplying by l_1; so that a transformation may be obtained for the second term similar to that for the first term.

1003. Laplace gives separately various cases of the general result contained in the preceding Article. We will now take his first case.

Let $\gamma_1 = \gamma_2 = \ldots = \gamma_s = 1$. Suppose that the function of the facility of error is the same at every observation, and is a constant; and let the limits of error be $\pm a$. Then

$$\int_{-a}^a f(z) \, dz = 1.$$

If C denote the constant value of $f(z)$ we have then

$$2aC = 1.$$

Here
$$k = 0, \quad k' = \frac{2Ca^3}{3} = \frac{a^2}{3}, \quad h^2 = \frac{a^2}{6},$$

$$l = 0, \quad \kappa^2 = h^2 \Sigma \gamma_i^2 = sh^2 = \frac{sa^2}{6}.$$

Let $c = 0$; then by equation (4) of the preceding Article the probability that the sum of the errors at the s observations will lie between $-\eta$ and η

$$= \frac{\sqrt{6}}{2a\sqrt{(s\pi)}} \int_{-\eta}^\eta e^{-\frac{3v^2}{2sa^2}} \, dv = \frac{\sqrt{6}}{a\sqrt{(s\pi)}} \int_0^\eta e^{-\frac{3v^2}{2sa^2}} \, dv.$$

Let $\frac{v^2}{sa^2} = t^2$; then the probability that the sum of the errors will lie between $-\tau a \sqrt{s}$ and $\tau a \sqrt{s}$

$$= \frac{\sqrt{6}}{\sqrt{\pi}} \int_0^\tau e^{-\frac{3t^2}{2}} \, dt.$$

This will be found to agree with Laplace's page 305.

1004. We take Laplace's next case.

Let $\gamma_1 = \gamma_2 = \ldots = \gamma_s = 1$. Let the limits of error be $\pm a$; suppose that the function of the facility of error is the same at every observation, and that positive and negative errors are equally likely: thus $f(-x) = f(x)$.

Here $\quad k = 0, \quad h^2 = \frac{1}{2}k', \quad l = 0, \quad \kappa^2 = \frac{s}{2}k'.$

By equation (4) of Article 1002 the probability that the sum of the errors at the s observations will lie between $-\eta$ and η is

$$\frac{2}{\sqrt{(2sk'\pi)}} \int_0^\eta e^{-\frac{v^2}{2sk'}}\, dv.$$

This will be found to agree with Laplace's page 308.

We have $\quad k' = \int_{-a}^{a} z^2 f(z)\, dz = 2\int_0^a z^2 f(z)\, dz,$

and $\quad 1 = \int_{-a}^{a} f(z)\, dz = 2\int_0^a f(z)\, dz;$

hence if $f(z)$ always decreases as z increases from 0 to a we see, as in Art. 922, that k' is less than $\frac{a^2}{3}$.

1005. Laplace next considers the probability that the sum of the errors in a large number of observations will lie between certain limits, the sign of the error being disregarded, that is all errors being treated as positive; the function of the facility of error is supposed to be the same at every observation.

Since all errors are treated as positive, we in fact take negative errors to be impossible; we must therefore put $b = 0$ in Poisson's problem.

Take $\gamma_1 = \gamma_2 = \ldots = \gamma_s = 1$. Then

$$l = sk, \quad \kappa^2 = \frac{s}{2}(k' - k^2).$$

Take $c = l$; then, by equation (4) of Art. 1002, the probability that the sum of the errors will lie between $l-\eta$ and $l+\eta$ is

$$\frac{2}{\sqrt{\{2s\pi\,(k'-k^2)\}}} \int_0^\eta e^{-\frac{v^2}{2s(k'-k^2)}}\, dv.$$

This will be found to agree with Laplace's page 311.

For an example suppose that the function of the facility of error is a constant, say C; then since

$$\int_0^a f(z)\, dz = 1,$$

we have $$aC = 1.$$

Thus $$k = \frac{a}{2}, \quad k' = \frac{a^2}{3}, \quad k' - k^2 = \frac{a^2}{12}.$$

Therefore the probability that the sum of the errors will lie between $\frac{sa}{2} - \eta$ and $\frac{sa}{2} + \eta$ is

$$\frac{2\sqrt{6}}{a\sqrt{(s\pi)}} \int_0^\eta e^{-\frac{6v^2}{sa^2}}\, dv.$$

1006. Laplace next investigates the probability that the sum of the squares of the errors will lie between assigned limits, supposing the function of the facility of error to be the same at every observation, and positive and negative errors equally likely. In order to give the result we must first generalise Poisson's problem.

Let $\phi_i(z)$ denote any function of z: required the probability that

$$\phi_1(\epsilon_1) + \phi_2(\epsilon_2) + \dots + \phi_s(\epsilon_s)$$

will lie between the limits $c - \eta$ and $c + \eta$. The investigation will differ very slightly from that in Art. 1002. In that Article we have

$$Q_i = \int_b^a f_i(z)\, e^{\gamma_i x z \sqrt{-1}}\, dz;$$

in the present case the exponent of e instead of being $\gamma_i x z \sqrt{-1}$, will be $x\phi_i(z)\sqrt{-1}$. The required probability will be found to be

$$\frac{1}{2\kappa\sqrt{\pi}} \int_{-\eta}^{\eta} e^{-\frac{(l-c+v)^2}{4\kappa^2}}\, dv;$$

where $$l = \Sigma \int_b^a \phi_i(z) f_i(z)\, dz,$$

and $$2\kappa^2 = \Sigma \int_b^a \left\{\phi(z)\right\}^2 f_i(z)\, dz - \Sigma \left\{\int_b^a \phi_i(z) f_i(z)\, dz\right\}^2.$$

The summations extend for all values of i from 1 to s, both inclusive.

It is not necessary that $\phi_i(z)$ should be restricted to denote the same function of z for all the values of i; Poisson however finds it sufficient for his purpose to allow this restriction.

Suppose now, for example, that $\phi_i(z) = z^2$ for all the values of i; and let the function of the facility of error be the same at every observation. Then, taking $b = 0$, as in the preceding Article,

$$l = s \int_0^a z^2 f(z)\, dz,$$

$$2\kappa^2 = s \int_0^a z^4 f(z)\, dz - s \left\{\int_0^a z^2 f(z)\, dz\right\}^2.$$

Take $c = l$; then the probability that the sum of the squares of the errors will lie between $l - \eta$ and $l + \eta$ is

$$\frac{1}{\kappa \sqrt{\pi}} \int_0^\eta e^{-\frac{v^2}{4\kappa^2}}\, dv.$$

This will be found to agree with Laplace's page 312.

1007. Laplace proceeds in his pages 313—321 to demonstrate the advantage of the method of least squares in the simplest case, that is when one unknown element is to be determined from observations; see Art. 921. This leads him to make an investigation similar to that which we have given in Art. 1002 from Poisson: Laplace however assumes that the function of the facility of error is the same at every observation, and that positive and negative errors are equally likely, and thus his investigation is less general than Poisson's.

Laplace and Poisson agree closely in their application of the investigation to the method of least squares: we will follow the latter.

In a system of observations the quantity given by the observation is in general not the element which we wish to determine, but some function of that element. We suppose that we already know the approximate value of the element, and that the required correction is so small that we may neglect its square and higher powers. Let the correction be represented by u; let A_i be the approximate value of the function at the ith observation, and $A_i + uq_i$ its corrected value. Let B_i be the value of the function given by observation, ϵ_i the unknown error of this observation. Then we shall have

$$B_i + \epsilon_i = A_i + uq_i.$$

Put δ_i for $B_i - A_i$, so that δ_i is the excess of the observed value above the approximate value of the function; thus we have

$$\epsilon_i = uq_i - \delta_i.$$

A similar equation will be furnished by each of the s observations. All the quantities of which q_i and δ_i are the types will be known, and all those of which ϵ_i is the type will be unknown. We wish to obtain from the system of equations the best value of u.

Form the sum of all such equations as the preceding, each multiplied by a factor of which γ_i is the type. Thus we obtain

$$\Sigma\gamma_i\epsilon_i = u\Sigma\gamma_iq_i - \Sigma\gamma_i\delta_i \dots\dots\dots\dots (1).$$

Then by equation (4) of Art. 1002 the probability that $\Sigma\gamma_i\epsilon_i$ will lie between $l - \eta$ and $l + \eta$ is

$$\frac{1}{\kappa\sqrt{\pi}} \int_0^\eta e^{-\frac{v^2}{4\kappa^2}} \, dv,$$

where l and κ have the values assigned in that Article.

Put $\dfrac{v^2}{4\kappa^2} = t^2$; thus the probability that $\Sigma\gamma_i\epsilon_i$ will lie between $l - 2\tau\kappa$ and $l + 2\tau\kappa$ is

$$\frac{2}{\sqrt{\pi}} \int_0^\tau e^{-t^2} \, dt \dots\dots\dots\dots\dots\dots(2).$$

If in (1) we put l for $\Sigma\gamma_i\epsilon_i$ we obtain

$$u = \frac{\Sigma\gamma_i\delta_i}{\Sigma\gamma_iq_i} + \frac{l}{\Sigma\gamma_iq_i} \quad\dots\dots\dots\dots\dots\dots (3),$$

and there is therefore the probability assigned in (2) that the error in the value of u will lie between

$$-\frac{2\tau\kappa}{\Sigma\gamma_iq_i} \text{ and } \frac{2\tau\kappa}{\Sigma\gamma_iq_i}.$$

Supposing then that τ remains constant, the error to be apprehended will be least when $\dfrac{\kappa}{\Sigma\gamma_iq_i}$ is least; and therefore the factors of which γ_i is the type must be taken so as to make this expression as small as possible. Put for κ its value; and then the expression becomes

$$\frac{\sqrt{(\Sigma\gamma_i^2h_i^2)}}{\Sigma\gamma_iq_i}.$$

We then make this expression a minimum by the rules of the Differential Calculus, and we find that the factors must be determined by equations of which the type is

$$\gamma_i = \frac{\nu q_i}{h_i^2},$$

where ν is a coefficient which is constant for all the factors.

With these values of the factors, equation (3) becomes

$$u = \frac{\Sigma\dfrac{q_i\delta_i}{h_i^2}}{\Sigma\dfrac{q_i^2}{h_i^2}} + \frac{\Sigma\dfrac{q_ik_i}{h_i^2}}{\Sigma\dfrac{q_i^2}{h_i^2}} \quad\dots\dots\dots\dots\dots (4);$$

and the limits of the error for which there is the probability assigned in (2) become

$$\pm\frac{2\tau}{\sqrt{\left(\Sigma\dfrac{q_i^2}{h_i^2}\right)}}.$$

If the function of the facility of error is the same at every observation the quantities of which h_i is the type are all equal, and so are those of which k_i is the type. Thus (4) becomes

$$u = \frac{\Sigma q_i \delta_i}{\Sigma q_i^2} + \frac{k \Sigma q_i}{\Sigma q_i^2} \dots\dots\dots\dots\dots (5);$$

and the limits of error become

$$\pm \frac{2\tau h}{\sqrt{(\Sigma q_i^2)}}.$$

If we suppose also that positive and negative errors are equally likely, we have $k = 0$, as in Art. 1004. Thus (5) becomes

$$u = \frac{\Sigma q_i \delta_i}{\Sigma q_i^2} \dots\dots\dots\dots\dots\dots (6).$$

This agrees with Laplace's result.

Laplace also presents another view of the subject. Suppose that $\psi(x)\, dx$ represents the chance that an error will lie between x and $x + dx$; then $\int_0^\infty x \psi(x)\, dx$ may be called the mean value of the positive error to be apprehended—*la valeur moyenne de l'erreur à craindre en plus*. Laplace compares an error with a loss at play, and multiplies the amount of the error by the chance of its happening, in the same way as we multiply a gain or loss by the chance of its happening in order to obtain the advantage or disadvantage of a player. Laplace then examines how the mean value of the error to be apprehended may be made as small as possible.

In equation (4) of Art. 1002 put $c = \eta$; and suppose positive and negative errors equally likely, so that $l = 0$: then the probability that $\Sigma \gamma_i \epsilon_i$ will lie between 0 and 2η

$$= \frac{1}{2\kappa \sqrt{\pi}} \int_{-\eta}^{\eta} e^{-\frac{(\eta - v)^2}{4\kappa^2}}\, dv = \frac{1}{2\kappa \sqrt{\pi}} \int_0^{2\eta} e^{-\frac{v^2}{4\kappa^2}}\, dv.$$

Thus the probability that $\Sigma \gamma_i \epsilon_i$ will lie between 0 and τ is

$$\frac{1}{2\kappa \sqrt{\pi}} \int_0^\tau e^{-\frac{v^2}{4\kappa^2}}\, dv,$$

and therefore the probability that $\Sigma\gamma_i\epsilon^i$ will lie between τ and $\tau + d\tau$ is

$$\frac{1}{2\kappa\sqrt{\pi}}\, e^{-\frac{\tau^2}{4\kappa^2}}\, d\tau.$$

This then is the probability that the error in u will lie between $\frac{\tau}{\Sigma\gamma_i q_i}$ and $\frac{\tau+d\tau}{\Sigma\gamma_i q_i}$; and therefore the probability that the error in u will lie between x and $x+dx$ is

$$\frac{\Sigma\gamma_i q_i}{2\kappa\sqrt{\pi}}\, e^{-\frac{x^2(\Sigma\gamma_i q_i)^2}{4\kappa^2}}\, dx.$$

This then is what we denoted above by $\psi(x)\,dx$; and we obtain therefore

$$\int_0^\infty x\psi(x)\,dx = \frac{\kappa}{\Sigma\gamma_i q_i.\sqrt{\pi}},$$

which is least when $\frac{\kappa}{\Sigma\gamma_i q_i}$ is least. This leads to the same result as before. The mean value of the positive error to be apprehended becomes $\frac{h}{\sqrt{(\pi\Sigma q_i^2)}}$.

Since $\epsilon_i = uq_i - \delta_i$ we have

$$\Sigma\epsilon_i^2 = \Sigma\,(uq_i - \delta_i)^2.$$

If we were to find u from the condition that the sum of the squares of the errors shall be as small as possible, we should obtain by the Differential Calculus

$$u = \frac{\Sigma q_i\delta_i}{\Sigma q_i^2},$$

which coincides with (6); so that the result previously obtained for u is the same as that assigned by the condition of making the sum of the squares of the errors as small as possible. It will be remembered that (6) was obtained by assuming that the function of the facility of error is the same at every observation, and that positive and negative errors are equally likely. The result in (4) does not involve these assumptions. It will be found that the value of u in (4) is the same as we should obtain by seeking the minimum value of

$$\Sigma \frac{(uq_i - \delta_i - k_i)^2}{h_i^2},$$

that is the minimum value of

$$\Sigma \left(\frac{\epsilon_i - k_i}{h_i}\right)^2.$$

1008. It is very important to observe how much is demonstrated with respect to the results (4), (5), and (6) of the preceding Article. There is nothing to assure us that we thus obtain the *most probable value* of u, in the strict sense of these words; neither Laplace nor Poisson makes such an assertion: they speak of the method as the *most advantageous method*, as the *method which ought to be preferred*.

Let us compare this method with another which would perhaps appear the most natural, namely that in which each of the factors $\gamma_1, \gamma_2, \ldots$ is taken equal to unity.

In the preceding Article we arrived at the following result,

$$u = \frac{\Sigma q_i \delta_i}{\Sigma q_i^2} + \frac{k \Sigma q_i}{\Sigma q_i^2} \ldots\ldots\ldots\ldots\ldots (5).$$

Now suppose that instead of giving to the factors $\gamma_1, \gamma_2, \ldots$ the values assigned in the preceding Article we take each of them equal to unity; then the quantity l of the preceding Article becomes Σk_i, that is sk if we suppose the function of the facility of error to be the same at each observation. Hence instead of (5) we shall have

$$u = \frac{\Sigma \delta_i}{\Sigma q_i} + \frac{sk}{\Sigma q_i} \ldots\ldots\ldots\ldots\ldots (7).$$

Now (5) is preferable to (7) because it was shewn in the preceding Article that, corresponding to a given probability, the limits of the error in (5) are less than the limits of the error in (7). In fact the limits of the error in (5) are $\pm \frac{2\tau h}{\sqrt{(\Sigma q_i^2)}}$, and in (7) they are $\pm \frac{2\tau h \sqrt{s}}{\Sigma q_i}$; and the result that the former limits are less than the latter is equivalent to the known algebraical theorem that

$$(\Sigma q_i)^2 \text{ is less than } s\Sigma q_i^2.$$

Moreover suppose that we neglect the second term on the right-hand side of (5) and of (7), and thus arrive at

$$u = \frac{\Sigma q_i \delta_i}{\Sigma q_i^2} \ldots\ldots (6), \qquad u = \frac{\Sigma \delta_i}{\Sigma q_i} \ldots\ldots (8);$$

then there is another reason why (6) is preferable to (8); for, by virtue of the algebraical theorem just quoted, the term which is neglected in arriving at (6), is less than the term which is neglected in arriving at (8).

1009. It was shewn in Art. 1007 that there is the probability (2) that the limits of the error in (6) are $\pm \dfrac{2\tau h}{\sqrt{(\Sigma q_i^2)}}$. This involves an unknown quantity h. Laplace proposes to obtain an approximate value of h from the observations themselves. It is shewn in Art. 1006 that there is a certain probability that the sum of the squares of the errors will lie between $l - \eta$ and $l + \eta$. Assume l for the value of the sum of the squares of the errors; thus

$$\Sigma \epsilon_i^2 = l = s \int_0^a z^2 f(z)\, dz = 2sh^2.$$

Therefore approximately

$$h^2 = \frac{\Sigma \epsilon_i^2}{2s} = \frac{\Sigma (uq_i - \delta_i)^2}{2s};$$

and with the value of u from (6) of Art. 1007, we obtain

$$h^2 = \frac{(\Sigma q_i^2)(\Sigma \delta_i^2) - (\Sigma q_i \delta_i)^2}{2s \Sigma q_i^2}.$$

Thus the mean value of the positive error to be apprehended, which was found in Art. 1007 to be $\dfrac{h}{\sqrt{(\pi \Sigma q_i^2)}}$, becomes

$$\frac{\sqrt{\{(\Sigma q_i^2)(\Sigma \delta_i^2) - (\Sigma q_i \delta_i)^2\}}}{\Sigma q_i^2 \sqrt{(2\pi s)}}.$$

This agrees with Laplace's page 322.

37

1010. Laplace now proceeds in his pages 322—329 to the case where *two* unknown elements are to be determined from a large number of observations; see Art. 923. Laplace arrives at the conclusion that the method of least squares is advantageous because the results which it gives coincide with those obtained by making the mean values of the positive errors to be apprehended as small as possible; the investigation is very laborious. The same assumptions are made as we have stated at the end of Art. 1007.

Laplace considers that he has thus established the method of least squares for any number of unknown quantities, for he asserts, on his page 327, ... *il est visible que l'analyse précédente peut s'étendre à un nombre quelconque d'élémens.* This assertion, however, seems very far from being obvious. ᶦ

Poisson has not considered this part of the subject; on account of its importance I shall now supply investigations by which the conclusions obtained in Art. 1007 will be extended to the case of more than one unknown element. I shall give, as in Art. 1007, two modes of arriving at the result; Laplace himself omits the first, and he presents the second in a form extremely different from that which will be here adopted. In drawing up the next Article I have obtained great assistance from the memoir by R. L. Ellis cited in Art. 1001.

1011. Suppose that instead of *one* element to be determined by the aid of observations we have any number of elements; suppose that approximate values of these elements are known, and that we have to find the small correction which each element requires. Denote these corrections by x, y, z, ... Then the general type of the equations furnished by the aid of observations will be

$$\epsilon_i = a_i x + b_i y + c_i z + \ldots - q_i \ldots\ldots\ldots\ldots\ldots\ldots(1).$$

Here ϵ_i is unknown, while a_i, b_i, c_i, ... q_i are known. Multiply (1) by γ_i, and then form the sum of the products for all values of i, which we suppose to be from 1 to s, both inclusive. And let the factors γ_1, γ_2, ... γ_s be taken subject to the conditions

$$\cdot \Sigma \gamma_i b_i = 0, \quad \Sigma \gamma_i c_i = 0, \ldots\ldots\ldots\ldots\ldots\ldots(2);$$

thus we obtain

$$x = \frac{\Sigma \gamma_i q_i}{\Sigma \gamma_i a_i} + \frac{\Sigma \gamma_i \epsilon_i}{\Sigma \gamma_i a_i} \quad \dots \dots \dots (3).$$

Now we know from equation (4) of Art. 1002 that there is the probability

$$\frac{2}{\sqrt{\pi}} \int_0^\tau e^{-t^2} dt \quad \dots \dots \dots (4),$$

that $\Sigma \gamma_i \epsilon_i$ will lie between $l - 2\tau\kappa$ and $l + 2\tau\kappa$, where, as before, $l = \Sigma \gamma_i k_i$. Put l for $\Sigma \gamma_i \epsilon_i$; thus (3) becomes

$$x = \frac{\Sigma \gamma_i q_i}{\Sigma \gamma_i a_i} + \frac{l}{\Sigma \gamma_i a_i} \quad \dots \dots \dots (5);$$

and there is the probability (4) that the error in the value of x, when determined by (5), will lie between

$$\pm \frac{2\tau\kappa}{\Sigma \gamma_i a_i}.$$

We propose then to make $\frac{\kappa}{\Sigma \gamma_i a_i}$ as small as possible, the factors being taken consistent with the limitations (2).

Since it is obvious that we want not the absolute values of the factors $\gamma_1, \gamma_2, \gamma_3, \dots$, but only the *ratio* which they bear to any arbitrary magnitude, we shall not really change the problem if we impose the condition $\Sigma \gamma_i a_i = 1$. Thus, since $\kappa^2 = \Sigma \gamma_i^2 h_i^2$, we require that $\Sigma \gamma_i^2 h_i^2$ shall be a minimum consistent with the conditions

$$\Sigma \gamma_i a_i = 1, \quad \Sigma \gamma_i b_i = 0, \quad \Sigma \gamma_i c_i = 0 \quad \dots \dots \dots (6).$$

Hence, by the Differential Calculus, we have

$$\Sigma \gamma_i h_i^2 d\gamma_i = 0,$$
$$\Sigma a_i d\gamma_i = 0,$$
$$\Sigma b_i d\gamma_i = 0,$$
$$\dots \dots \dots \dots$$

Therefore by the use of arbitrary multipliers $\lambda, \mu, \nu, \dots$ we obtain a set of s equations of which the type is

$$\gamma_i h_i^2 = \lambda a_i + \mu b_i + \nu c_i + \quad \dots \dots \dots (7).$$

37.—2

Let j_i stand for $\frac{1}{h_i^2}$; then from (7) we can deduce the follow-ing system of equations:

$$
\left.
\begin{aligned}
1 &= \lambda \Sigma a_i^2 j_i + \mu \Sigma a_i b_i j_i + \nu \Sigma a_i c_i j_i + \dots \\
0 &= \lambda \Sigma a_i b_i j_i + \mu \Sigma b_i^2 j_i + \nu \Sigma b_i c_i j_i + \dots \\
0 &= \lambda \Sigma a_i c_i j_i + \mu \Sigma b_i c_i j_i + \nu \Sigma c_i^2 j_i + \dots
\end{aligned}
\right\} \dots\dots\dots\dots(8).
$$

To obtain the first of equations (8) we multiply (7) by $a_i j_i$, and then sum for all values of i paying regard to (6); to obtain the second of equations (8) we multiply (7) by $b_i j_i$ and sum; to obtain the third of equations (8) we multiply (7) by $c_i j_i$ and sum; and so on. The number of equations (8) will thus be the same as the number of conditions in (6), and therefore the same as the number of arbitrary multipliers λ, μ, ν, ... Thus equations (8) will determine λ, μ, ν, ...; and then from (5) we have

$$
x = \Sigma \gamma_i q_i + l \dots\dots\dots\dots\dots\dots(9).
$$

We shall now shew how this value of x may practically be best calculated.

Take s equations of which the type is

$$
a_i x' + b_i y' + c_i z' + \dots\dots\dots\dots\dots = q_i + k_i.
$$

First multiply by $a_i j_i$ and sum for all values of i; then mul-tiply by $b_i j_i$ and sum; then multiply by $c_i j_i$ and sum; and so on: thus we obtain the following system

$$
\left.
\begin{aligned}
x' \Sigma a_i^2 j_i + y' \Sigma a_i b_i j_i + z' \Sigma a_i c_i j_i + \dots &= \Sigma (q_i + k_i) a_i j_i \\
x' \Sigma a_i b_i j_i + y' \Sigma b_i^2 j_i + z' \Sigma b_i c_i j_i + \dots &= \Sigma (q_i + k_i) b_i j_i \\
x' \Sigma a_i c_i j_i + y' \Sigma b_i c_i j_i + z' \Sigma c_i^2 j_i + \dots &= \Sigma (q_i + k_i) c_i j_i
\end{aligned}
\right\} \dots\dots(10).
$$

Now we shall shew that if x' be deduced from (10) we shall have $x' = \Sigma \gamma_i q_i + l$, and therefore $x = x'$.

For multiply equations (10) in order by λ, μ, ν, ... and add; then by (8)

$$x' = \lambda\Sigma\,(q_i + k_i)\,a_i j_i + \mu\Sigma\,(q_i + k_i)\,b_i j_i + \nu\Sigma\,(q_i + k_i)c_i j_i + \ldots$$
$$= \Sigma\,(q_i + k_i).j_i\,\{\lambda a_i + \mu b_i + \nu c_i + \ldots\}$$
$$= \Sigma\gamma_i\,(q_i + k_i) \quad \text{by (7)}.$$

The advantage of using equations (10) is twofold; in the first place we determine x', and thence x, by a systematic process, and in the next place we see that the equations (10) are *symmetrical* with respect to x', y', z', ...: thus if we had proposed to find y, or z, or any of the other unknown quantities instead of x, we should, by proceeding in the same manner as we have already, arrive at the same system (10). Hence the same advantage which we have shewn by the Theory of Probability to belong to the value of x by taking it equal to x', will belong to the value of y by taking it equal to y', and to the value of z by taking it equal to z', and so on. In fact it is obvious that if we had begun by investigating the value of y instead of the value of x the conditions (6) would have been changed in such a manner as to leave the proportion of the factors γ_1, γ_2, γ_3, ... unchanged; and thus we might have anticipated that a *symmetrical* system of equations like (10) could be formed.

We have thus shewn how to obtain the most advantageous values for the required quantities x, y, z, ...

Suppose now that we wished to find the values of x', y', z', ... which render the following expression a minimum,

$$\Sigma j_i\,\{a_i x' + b_i y' + c_i z' + \ldots - q_i - k_i\}^2;$$

it will be found that we arrive at the equations (10) for determining x', y', z' ... Hence the values which have been found for x, y, z, ... give a minimum value to the following expression

$$\Sigma j_i\,(\epsilon_i - k_i)^2 \text{ that is } \Sigma\left(\frac{\epsilon_i - k_i}{h_i}\right)^2.$$

If k_i be zero, and h_i constant, for all values of i, the values which have been found for x, y, z, ... render the sum of the squares of the errors a minimum: as in Art. 1007 these conditions will hold if the function of the facility of error is the same at every observation, and positive and negative errors are equally likely.

Thus we have completed one mode of arriving at the result, and we shall now pass on to the other.

If we proceed as in the latter part of Art. 1007 we shall find that the probability that the error in the value of x, when it is determined by (5), lies between t and $t + dt$ is

$$\frac{\Sigma \gamma_i a_i}{2\kappa \sqrt{\pi}} e^{-\frac{t^2 (\Sigma \gamma_i a_i)^2}{4\kappa^2}} dt \dots\dots\dots\dots\dots(11).$$

For put $c = \eta$ in equation (4) of Art. 1002. Then the probability that $\Sigma \gamma_i \epsilon_i$ will lie between 0 and 2η

$$= \frac{1}{2\kappa \sqrt{\pi}} \int_{-\eta}^{\eta} e^{-\frac{(l - \eta + v)^2}{4\kappa^2}} dv = \frac{1}{2\kappa \sqrt{\pi}} \int_{0}^{2\eta} e^{-\frac{(l - v)^2}{4\kappa^2}} dv.$$

Thus the probability that $\Sigma \gamma_i \epsilon_i$ will lie between τ and $\tau + d\tau$ is

$$\frac{1}{2\kappa \sqrt{\pi}} e^{-\frac{(l - \tau)^2}{4\kappa^2}} d\tau,$$

and therefore the probability that $\Sigma \gamma_i \epsilon_i$ will lie between $l + \tau'$ and $l + \tau' + d\tau'$ is

$$\frac{1}{2\kappa \sqrt{\pi}} e^{-\frac{\tau'^2}{4\kappa^2}} d\tau'.$$

This is therefore the probability that the error in the value of x when determined by (5) will lie between

$$\frac{\tau'}{\Sigma \gamma_i a_i} \quad \text{and} \quad \frac{\tau' + d\tau'}{\Sigma \gamma_i a_i}.$$

And therefore the probability that the error in the value of x when determined by (5) will lie between t and $t + dt$ is given by (11).

The mean value of the positive error to be apprehended in the value of x will be obtained by multiplying the expression in (11) by t and integrating between the limits 0 and ∞ for t. Thus, since $\Sigma \gamma_i a_i = 1$, we obtain $\dfrac{\kappa}{\sqrt{\pi}}$ for the result; and therefore if we proceed to make this mean error as small as possible we obtain the same values as before for the factors $\gamma_1, \gamma_2, \gamma_3, \dots$

It will be interesting to develop the value of κ. Multiply equation (7) by γ_i, and sum for all values of i; thus by (6) we obtain

$$\kappa^2 = \lambda.$$

Suppose then we have *two* unknown quantities, x and y; we find from (8)

$$\lambda = \frac{\Sigma b_i^2 j_i}{(\Sigma a_i^2 j_i)\,(\Sigma b_i^2 j_i) - (\Sigma a_i b_i j_i)^2},$$

and the mean error for x will be $\dfrac{\sqrt{\lambda}}{\sqrt{\pi}}$.

The mean error to be apprehended for y may be deduced from that for x by interchanging a_i with b_i.

If there are *three* unknown quantities we may deduce the mean error from that which has just been given in the case of *two* unknown quantities by the following rule:

change $\Sigma a_i^2 j_i$ into $\Sigma a_i^2 j_i - \dfrac{(\Sigma a_i c_i j_i)^2}{\Sigma c_i^2 j_i}$,

change $\Sigma b_i^2 j_i$ into $\Sigma b_i^2 j_i - \dfrac{(\Sigma b_i c_i j_i)^2}{\Sigma c_i^2 j_i}$,

change $\Sigma a_i b_i j_i$ into $\Sigma a_i b_i j_i - \dfrac{(\Sigma a_i c_i j_i)\,(\Sigma b_i c_i j_i)}{\Sigma c_i^2 j_i}$.

To establish this rule we need only observe that if we have *three* equations (8) we may begin the solution of them by expressing ν from the last equation in terms of λ and μ, and substituting in the first and second.

By a similar rule we can deduce the mean error in the case of four unknown quantities from that in the case of three unknown quantities: and so on.

The rule is given by Laplace on his page 328, without any demonstration. He assumes however the function of the facility of error to be the same at every observation so that j_i is constant for all values of i; and he takes, as in Art. 1009,

$$h_i^2 = \frac{\Sigma \epsilon_i^2}{2s}.$$

1012. Laplace gives on his pages 329—332 an investigation which approaches more nearly in generality to that which we have supplied in Art. 1007 than those which we have hitherto noticed in the fourth Chapter of the *Théorie ... des Prob.*; see Art. 917. Laplace takes the same function of the facility of error at every observation, but he does not assume that positive and negative errors are equally likely, or have equal ranges.

1013. Laplace says, on his page 333, that hitherto he has been considering observations not yet .made; but he will now consider observations that have been already made.

Suppose that observations assign values a_1, a_2, a_3, ... to an unknown element; let $\phi(z)$ be the function of the facility of an error z, the function being supposed the same at every observation. Let us now determine the probability that the true value of the element is x, so that the errors are $a_1 - x$, $a_2 - x$, $a_3 - x$, ... at the various observations.

Let $\qquad P = \phi(a_1 - x) \cdot \phi(a_2 - x) \cdot \phi(a_3 - x) \cdot \ldots$

Then, by the ordinary principles of inverse probability, the probability that the true value lies between x and $x + dx$ is

$$\frac{P dx}{\int P dx},$$

the integral in the denominator being supposed to extend over all the values of which x is susceptible.

Let H be such that, with the proper limits of integration,

$$H \int P dx = 1,$$

and let $\qquad y = H\phi(a_1 - x) \cdot \phi(a_2 - x) \cdot \phi(a_3 - x) \cdot \ldots$

Laplace conceives that we draw the curve of which the ordinate is y corresponding to the abscissa x. He says that the value which we ought to take as the mean result of the observations is that which renders the mean error a minimum, every error being considered positive. He shews that this corresponds to the point the ordinate of which bisects the area of the curve just drawn; that is the mean result which he considers the best is such that the true result is equally likely to exceed it or to fall short of it. See Arts. 876, 918.

Laplace says, on his page 335,

Des géomètres célèbres ont pris pour le milieu qu'il faut choisir, celui qui rend le résultat observé, le plus probable, et par conséquent l'abscisse qui répond à la plus grande ordonnée de la courbe ; mais le milieu que nous adoptons, est évidemment indiqué par la théorie des probabilités.

This extract illustrates a remark which we have already made in Art. 1008, namely that strictly speaking Laplace's method does not profess to give the *most probable* result but one which he considers the most advantageous.

1014. Laplace gives an investigation in his pages 335—340 which amounts to solving the following problem : if we take the *average* of the results furnished by observations as the *most probable* result, and assume that positive and negative errors are equally likely and that the function of the facility of error is the same at every observation, what function of the facility of error is implicitly assumed ?

Let the function of the facility of an error z be denoted by $e^{-\psi(z^2)}$, which involves only the assumption that positive and negative errors are equally likely. Hence the value of y in the preceding Article becomes

$$He^{-\sigma},$$

where $\quad \sigma = \psi(x - a_1)^2 + \psi(x - a_2)^2 + \psi(x - a_3)^2 + \dots$

To obtain the most probable result we must determine x so that σ shall be a minimum ; this gives the equation

$$(x - a_1)\,\psi'(x - a_1)^2 + (x - a_2)\,\psi'(x - a_2)^2$$
$$+ (x - a_3)\,\psi'(x - a_3)^2 + \dots = 0.$$

Now let us assume that the *average* result is always the *most probable* result ; suppose that out of s observations i coincide in giving the result a_1, and $s - i$ coincide in giving the result a_2 ; the preceding equation becomes

$$i\,(x - a_1)\,\psi'(x - a_1)^2 + (s - i)(x - a_2)\,\psi'(x - a_2)^2 = 0.$$

The average value in this case is

$$\frac{ia_1 + (s - i)\,a_2}{s}.$$

Substitute this value of x in the equation, and we obtain

$$\psi'\left\{\frac{s - i}{s}(a_1 - a_2)\right\}^2 = \psi'\left\{\frac{i}{s}(a_1 - a_2)\right\}^2.$$

This cannot hold for all values of $\frac{i}{s}$ and $a_1 - a_2$ unless $\psi'(z)$ be independent of z; say $\psi'(z) = c$.

Hence $\psi(z) = cz + c'$, where c and c' are constants.

Thus the function of the facility of error is of the form Ce^{-cz^2}; and since an error must lie between $-\infty$ and ∞, we have

$$C \int_{-\infty}^{\infty} e^{-cz^2} \, dz = 1;$$

therefore
$$C = \frac{\sqrt{c}}{\sqrt{\pi}}.$$

The result given by the method of least squares, in the case of a single unknown quantity, is the same as that obtained by taking the average. For if we make the following expression a minimum

$$(x - a_1)^2 + (x - a_2)^2 + \ldots + (x - a_s)^2$$

we obtain
$$x = \frac{a_1 + a_2 + \ldots + a_s}{s}.$$

Hence the assumption in the preceding investigation, that the average of the results furnished by observations will be the most probable result, is equivalent to the assumption that the method of least squares will give the most probable result.

1015. Laplace devotes his pages 340—342 to shewing, as he says, that in a certain case the method of least squares becomes necessary. The investigation is very simple when divested of the cumbrous unsymmetrical form in which Laplace presents it.

Suppose we require to determine an element from an assemblage of a large number of observations of various kinds. Let there be s_1 observations of the first kind, and from these let the value a_1 be deduced for the unknown quantity; let there be s_2 observations of the second kind, and from these let the value a_2 be deduced for the unknown quantity; and so on.

Take x to represent a hypothetical value of the unknown quantity. Assume positive and negative errors to be equally likely; then by Art. 1007 the probability that the error of the result deduced from the first set of observations will lie between $x - a_1$ and $x + dx - a_1$ is $\frac{\beta_1}{\sqrt{\pi}} e^{-\beta_1^2(x - a_1)^2} dx$.

Here β_1^2 stands for $\frac{(\Sigma\gamma_i q_i)^2}{4\Sigma\gamma_i^2 h_i^2}$, and the value of β_1 will therefor depend on the values of the factors γ_1, γ_2, ... which we employ; for example we may take each of these factors equal to unity, which amounts to adopting the *average* of the results of observation ; or we may take for these factors the system of values which we have called the most advantageous system : if we adopt the latter we find $\beta_1^2 = \frac{1}{4}\Sigma\frac{q_i^2}{h_i^2}$.

Similarly the probability that the error of the result deduced from the second set of observations will lie between $x - a_2$ and $x + dx - a_2$ is $\frac{\beta_2}{\sqrt{\pi}} e^{-\beta_2^2(x-a_2)^2} dx$.

And so on for the other sets of observations.

Hence we shall find, in the manner of Art. 1013, that the probability that x is the true value of the unknown quantity is proportional to

$$e^{-\sigma},$$

where $\sigma = \beta_1^2(x-a_1)^2 + \beta_2^2(x-a_2)^2 + \beta_3^2(x-a_3)^2 + \ldots$

Now determine x so that this probability shall have its greatest value; σ must be a minimum, and we find that

$$x = \frac{\beta_1^2 a_1 + \beta_2^2 a_2 + \beta_3^2 a_3 + \ldots}{\beta_1^2 + \beta_2^2 + \beta_3^2 + \ldots}.$$

We may say then that Laplace obtains this result by deducing a value of the unknown quantity from each set of observations, and then seeking for the *most probable* inference. If a_1, a_2, a_3, ... are determined by the most advantageous method, this result is similar in form to that which is given in Art. 1007, if we suppose that positive and negative errors are equally likely, and that one function of facility of error applies to the first set of observations, another function to the second set, and so on. For the numerator of the value of x just given corresponds with the $\Sigma\frac{q_i\delta_i}{h_i^2}$, and the denominator with the $\Sigma\frac{q_i^2}{h_i^2}$ of Art. 1007.

1016. Laplace gives some remarks on his pages 343—348 relative to another method of treating errors, namely, that which consists in making the sum of the $2n^{th}$ powers of the errors a minimum, n being supposed indefinitely great. He explains this method for the case of one unknown quantity, and he refers to the *Mécanique Céleste, Livre* III. for the case in which there is more than one unknown quantity. The section intended of *Livre* III. must be the 39th, in which Laplace gives some rules as in the present place, but without connecting his rules with the consideration of infinite powers of the errors. Another method is given in the next section of the *Mécanique Céleste* which Dr Bowditch in a note on the passage ascribes to Boscovich: Laplace takes up this method in the second Supplement to the *Théorie...des Prob.*, where he calls it the *method of situation*.

1017. Laplace gives on his pages 346—348 some account of the history of the methods of treating the results of observations. Cotes first proposed a rule for the case in which a single element was to be determined. His rule amounts to taking

$$\gamma_1 = \gamma_2 = \ldots = \gamma_s = 1$$

in Art. 1007, so that

$$u = \frac{\Sigma \delta_i}{\Sigma q_i}.$$

Laplace says that the rule was however not employed by mathematicians until Euler employed it in his first memoir on Jupiter and Saturn, and Mayer in his investigations on the libration of the moon. Legendre suggested the method of least squares as convenient when any number of unknown quantities had to be found; Gauss had however previously used this method himself and communicated it to astronomers. Gauss was also the first who endeavoured to justify the method by the Theory of Probability.

We have seen that Daniel Bernoulli, Euler, and Lagrange had studied the subject: see Arts. 424, 427, 556. Lambert and Boscovich also suggested rules on the subject; see the article *Milieu* of the *Encyclopédie Méthodique* and Dr Bowditch's translation of the *Mécanique Céleste*, Vol. II. pages 434, 435.

The titles of some other memoirs on the subject of least squares will be found at the end of the Treatise on Probability in the *Encyclopædia Britannica;* we would also refer the student to the work by the Astronomer Royal *On the Algebraical and Numerical Theory of Errors of Observations and the combination of Observations.*

1018. Laplace's fifth Chapter is entitled *Application du Calcul des Probabilités, à la recherche des phénomènes et de leurs causes:* it occupies pages 349—362.

The example with which Laplace commences will give a good idea of the object of this Chapter. Suppose that observations were made on 400 days throughout which the height of the barometer did not vary 4 millimetres; and that the sum of the heights at nine in the morning exceeded the sum of the heights at four in the afternoon by 400 millimetres, giving an average excess of one millimetre for each day: required to estimate the probability that this excess is due to a constant cause.

We must examine what is the probability of the result on the supposition that it is *not* due to any constant cause, but arises from accidental perturbations and from errors of observation.

By the method of Art. 1004, supposing that it is equally probable that the daily algebraical excess of the morning result over the afternoon result will be positive or negative, the probability that the sum of s excesses will exceed the positive quantity c

$$= \frac{1}{\sqrt{(2k's\pi)}} \int_0^\infty e^{-\frac{v^2}{2sk'}} dv$$

$$= \frac{1}{\sqrt{\pi}} \int_\tau^\infty e^{-t^2} dt, \quad \text{where } \tau = \frac{c}{\sqrt{(2sk')}}.$$

Hence the probability that the sum will be algebraically less than c is

$$1 - \frac{1}{\sqrt{\pi}} \int_\tau^\infty e^{-t^2} dt.$$

Now, as in Art. 1004, we may take $\dfrac{a^2}{3}$ as the greatest value

of k', so that the least value of τ is $\dfrac{c\sqrt{3}}{a\sqrt{(2s)}}$; also $a = 4$, $c = 400$,

$s = 400$: thus the least value of τ is $\dfrac{5\sqrt{3}}{\sqrt{2}}$, that is $\sqrt{(37\cdot 5)}$.

Hence $1 - \dfrac{1}{\sqrt{\pi}}\displaystyle\int_{\tau}^{\infty} e^{-t^2}dt$ is found to be very nearly equal to
unity. We may therefore regard it as nearly certain that the
sum of the excesses would fall below 400 if there were no constant
cause: that is we have a very high probability for the existence of
a constant cause.

1019. Laplace states that in like manner he had been led
by the theory of probabilities to recognise the existence of constant causes of various results in physical astronomy obtained by
observation; and then he had proceeded to verify the existence
of these constant causes by mathematical investigations. The
remarks on this subject are given more fully in the *Introduction*,
pages LVII—LXX; see Art. 938.

1020. Laplace on his pages 359—362 solves Buffon's problem,
which we have explained in Art. 650.

Suppose that there is one set of parallel lines; let a be the
distance of two consecutive straight lines of the system, and $2r$
the length of the rod: then the chance that the rod will fall

across a line is $\dfrac{4r}{\pi a}$. Hence, by Art. 993, if the rod be thrown

down a very large number of times we may be certain that the
ratio of the number of times in which the rod crosses a line

to the whole number of trials will be very nearly $\dfrac{4r}{\pi a}$: we might

therefore determine by experiment an approximate value of π.

Laplace adds ... et il est facile de voir que le rapport $\dfrac{8r}{a\pi}$ qui,

pour un nombre donné de projections, rend l'erreur à craindre la
plus petite, est l'unité... Laplace seems to have proceeded thus.
Suppose p the chance of the event in one trial; then, by Art. 993,

the probability that in μ trials the number of times in which the event happens will lie between

$$p\mu - \tau \sqrt{2\mu p\,(1-p)} \text{ and } p\mu + \tau \sqrt{2\mu p\,(1-p)}$$

is approximately $\dfrac{2}{\sqrt{\pi}} \displaystyle\int_0^\tau e^{-t^2}\,dt.$

Hence to make the limits as close as possible we must have $p\,(1-p)$ as small as possible, and thus $p = \dfrac{1}{2}$. This, we say, appears to have been Laplace's process. It is however wrong; for $p\,(1-p)$ is a *maximum* and not a *minimum* when $p = \dfrac{1}{2}$. Moreover we have not to make $\tau \sqrt{2\mu p\,(1-p)}$ as small as possible, but the ratio of this expression to $p\mu$. Hence we have to make $\dfrac{\sqrt{p\,(1-p)}}{p}$ as small as possible; that is we must make $\dfrac{1}{p} - 1$ as small as possible: therefore p must be as great as possible. In the present case $p = \dfrac{4r}{\pi a}$; we must therefore make this as great as possible: now in the solution of the problem $2r$ is assumed to be not greater than a, and therefore we take $2r = a$ as the most favourable length of the rod.

Laplace's error is pointed out by Professor De Morgan in Art. 172 of the *Theory of Probabilities* in the *Encyclopædia Metropolitana*. The most curious point however has I believe hitherto been unnoticed, namely, that Laplace had the correct result in his first edition, where he says ...et il est facile de voir que le rapport $\dfrac{2r}{a}$ qui, pour un nombre donné de projections, rend l'erreur à craindre la plus petite, est l'unité... The original leaf was cancelled, and a new leaf inserted in the second and third editions, thus causing a change from truth to error. See Art. 932.

Laplace solves the second part of Buffon's problem correctly, in which Buffon himself had failed; Laplace's solution is much less simple than that which we have given in Art. 650.

1021. Laplace's sixth Chapter is entitled *De la probabilité des causes et des événemens futurs, tirée des événemens observés:* it occupies pages 363—401.

The subject of this Chapter had engaged Laplace's attention from an early period, and to him we must principally ascribe the merit of the important extension thus given to the Theory of Probability, due honour being at the same time reserved for his predecessor Bayes. See Arts. 851, 868, 870, 903, 909.

Let *x* denote the chance, supposed unknown, of a certain simple event; let *y* denote the chance of a certain compound event depending in an assigned manner on this simple event: then *y* will be a known function of *x*. Suppose that this compound event has been observed; then the probability that the chance of the simple event lies between α and β is

$$\frac{\int_{\alpha}^{\beta} y\, dx}{\int_{0}^{1} y\, dx} \cdot$$

This is the main formula of the present Chapter: Laplace applies it to examples, and in so doing he evaluates the integrals by his method of approximation.

In like manner if the compound event depends on *two* independent simple events, the probability that the chance of one lies between α and β and the chance of the other between α′ and β′ is

$$\frac{\int_{\alpha'}^{\beta'}\int_{\alpha}^{\beta} y\, dx'\, dx}{\int_{0}^{1}\int_{0}^{1} y\, dx'\, dx} \cdot$$

1022. The examples in the present Chapter of Laplace's work exhibit in a striking way the advantage of his method of approximation; but as they present no novelty nor difficulty of principle we do not consider it necessary to reproduce any of them in detail.

1023. Laplace makes a remark on his page 366 which may deserve a brief examination. He says that if we have to take the integral $\int e^{-t^2}\, dt$ between the limits $-\tau$ and τ' we may for an approximation take the integral between the limits 0 and $\sqrt{\left(\dfrac{\tau^2 + \tau'^2}{2}\right)}$,

and double the result: he says this amounts to neglecting the square of $\tau'^2 - \tau^2$. We may put the matter in the following form: suppose that a and b are positive, and we require x such that

$$\int_0^a e^{-t^2}\, dt + \int_0^b e^{-t^2}\, dt = 2 \int_0^x e^{-t^2}\, dt.$$

Suppose a less than b; then in fact we require that

$$\int_a^x e^{-t^2}\, dt = \int_x^b e^{-t^2}\, dt.$$

Laplace, in effect, tells us that we should take $x = \sqrt{\left(\dfrac{a^2 + b^2}{2}\right)}$ as an approximation. He gives no reason however, and the more natural approximation would be to take $x = \dfrac{1}{2}(a+b)$, and this is certainly a better approximation than his. For since the function e^{-t^2} decreases as t increases, the true value of x is less than $\dfrac{1}{2}(a+b)$, while Laplace's approximation is greater than $\dfrac{1}{2}(a+b)$.

1024. Laplace discusses on his pages 369—376 a problem relating to play; see Art. 868. A and B play a certain number of matches; to gain a match a player must win two games out of three; having given that A has gained i matches out of a large number n, determine the probability that A's skill lies within assigned limits. If a player wins the first and second games of a match the third is not played, being unnecessary; hence if n matches have been played the number of games must lie between $2n$ and $3n$: Laplace investigates the most probable number of games.

1025. Laplace discusses in his pages 377—380 the problem which we have enunciated in Art. 896. The required probability is

$$\frac{\displaystyle\int_{\frac{1}{2}}^1 x^p (1-x)^q\, dx}{\displaystyle\int_0^1 x^p (1-x)^q\, dx},$$

where p and q have the values derived from observations during

38

40 years; these values are given in Art. 902. Laplace finds that the probability is approximately

$$1 - \frac{1 - \cdot0030761}{\mu},$$

where μ is a very large number, its logarithm being greater than 72. Thus Laplace concludes that the probability is at least equal to that of the best attested facts in history.

With respect to a formula which occurs in Laplace's solution see Art. 767. With respect to an anomaly observed at Vitteaux see Arts. 768, 769.

1026. Laplace discusses in his pages 381—384 the problem which we have noticed in Art. 902.

He offers a suggestion to account for the observed fact that the ratio of the number of births of boys to girls is larger at London than at Paris.

1027. Laplace then considers the probability of the results founded on tables of mortality: he supposes that if we had observations of the extent of life of an infinite number of infants the tables would be perfect, and he estimates the probability that the tables formed from a finite number of infants will deviate to an assigned extent from the theoretically perfect tables. We shall hereafter in Art. 1036 discuss a problem like that which Laplace here considers.

1028. A result which Laplace indicates on his page 390 suggests a general theorem in Definite Integrals, which we will here demonstrate.

Let $u^2 =$

$$a_1^2 z_1^2 + a_2^2 (z_2 - b_1 z_1)^2 + a_3^2 (z_3 - b_2 z_2)^2 + \ldots + a_n^2 (z_n - b_{n-1} z_{n-1})^2;$$

let e^{-u^2} be integrated with respect to each of the $n-1$ variables $z_1, z_2, \ldots z_{n-1}$, between the limits $-\infty$ and ∞: then the result will be

$$\frac{\gamma \pi^{\frac{n-1}{2}}}{a_1 a_2 \ldots a_{n-1} a_n} e^{-\gamma^2 z_n^2},$$

where
$$\frac{1}{\gamma^2} = \frac{1}{a_n^2} + \frac{b_{n-1}^2}{a_{n-1}^2} + \frac{b_{n-2}^2 b_{n-1}^2}{a_{n-2}^2} + \ldots + \frac{b_1^2 b_2^2 \ldots b_{n-1}^2}{a_1^2}.$$

Let us consider first the integration with respect to z_1; we have

$$a_1^2 z_1^2 + a_2^2 (z_2 - b_1 z_1)^2 = (a_1^2 + a_2^2 b_1^2) z_1^2 - 2a_2^2 b_1 z_1 z_2 + a_2^2 z_2^2$$

$$= (a_1^2 + a_2^2 b_1^2) \left(z_1 - \frac{a_2^2 b_1 z_2}{a_1^2 + a_2^2 b_1^2} \right)^2 + a_2^2 z_2^2 - \frac{a_2^4 b_1^2 z_2^2}{a_1^2 + a_2^2 b_1^2}$$

$$= (a_1^2 + a_2^2 b_1^2) t^2 + \frac{a_1^2 a_2^2 z_2^2}{a_1^2 + a_2^2 b_1^2},$$

where $\quad t = z_1 - \dfrac{a_2^2 b_1 z_2}{a_1^2 + a_2^2 b_1^2}.$

The limits of t will be $-\infty$ and ∞; integrate with respect to t: thus we remove z_1 entirely, and obtain the factor

$$\frac{\sqrt{\pi}}{\sqrt{(a_1^2 + a_2^2 b_1^2)}},$$

and instead of the first two terms in u^2 we have the single term

$$\frac{a_1^2 a_2^2 z_2^2}{a_1^2 + a_2^2 b_1^2}.$$

We integrate next with respect to z_2; thus we shall remove z_2 entirely, and introduce the factor

$$\frac{\sqrt{\pi}}{\sqrt{\left(\dfrac{a_1^2 a_2^2}{a_1^2 + a_2^2 b_1^2} + a_3^2 b_2^2 \right)}};$$

and instead of the first three terms in u^2 we shall have the single term

$$\frac{a_1^2 a_2^2 a_3^2 z_3^2}{a_1^2 + a_2^2 b_1^2} \left\{ \frac{a_1^2 a_2^2}{a_1^2 + a_2^2 b_1^2} + a_3^2 b_2^2 \right\}^{-1}.$$

Thus we have now on the whole the factor

$$\frac{(\sqrt{\pi})^2 \lambda}{a_1 a_2 a_3},$$

where $\qquad \dfrac{1}{\lambda^2} = \dfrac{1}{a_3^2} + \dfrac{b_2^2}{a_2^2} + \dfrac{b_1^2 b_2^2}{a_1^2};$

and the first three terms in u^2 are replaced by the single term $\lambda^2 z_3^2$.

We integrate next with respect to z_3; thus we shall remove z_3 entirely, and introduce the factor

$$\frac{\sqrt{\pi}}{\sqrt{(\lambda^2 + a_4^2 b_3^2)}}, \text{ that is, } \frac{\sqrt{\pi}}{\lambda a_4 \sqrt{\left(\frac{1}{a_4^2} + \frac{b_3^2}{\lambda^2}\right)}},$$

$$\text{say } \frac{\mu \sqrt{\pi}}{\lambda a_4},$$

$$\text{where } \frac{1}{\mu^2} = \frac{1}{a_4^2} + \frac{b_3^2}{\lambda^2}:$$

and the first four terms in u^2 are replaced by the single term

$$\frac{\lambda^2 a_4^2 z_4^2}{\lambda^2 + a_4^2 b_3^2}, \text{ that is, by } \mu^2 z_4^2.$$

By proceeding in this way it is obvious that we shall arrive at the assigned result.

1029. Laplace devotes his pages 391—394 to a problem which we have indicated in Art. 911. The problem resembles that which we have noticed in Art. 1027, and the mode of solution will be illustrated hereafter in Art. 1036.

The problems which Laplace considers in his pages 385—394 relate to the probabilities of *future* events ; and thus these pages are strangely out of their proper place : they should have *followed* the discussion which we are about to analyse in our next Article, and which begins thus, *Considérons maintenant la probabilité des événemens futurs, tirée des événemens observés.*

1030. Laplace considers in his pages 394—396 the important subject of the probability of future events deduced from observed events : see Arts. 870, 903, 909.

Retaining the notation of Art. 1021, suppose that z, which is a known function of x, represents the chance of some compound future event depending on the simple event of which x represents the chance : then the whole probability, P, of this future event will be given by

$$P = \frac{\int_0^1 y z \, dx}{\int_0^1 y \, dx}.$$

Laplace then suggests approximations for the integrals in the above expression. We will reproduce the substance of his remarks. In Art. 957 we have

$$t^2 = \log Y - \log \phi (a + \theta)$$

$$= \log Y - \log \left\{ \phi (a) + \theta \phi' (a) + \frac{\theta^2}{2} \phi'' (a) + ... \right\}$$

$$= - \frac{\theta^2}{2} \frac{\phi'' (a)}{\phi (a)} + ... ;$$

for $Y = \phi (a)$, and $\phi' (a) = 0$, by hypothesis.

Thus approximately

$$t = \theta \sqrt{\left\{ - \frac{1}{2} \frac{\phi'' (a)}{\phi (a)} \right\}}.$$

Hence if y vanishes when $x = 0$ and when $x = 1$, we have approximately

$$\int_0^1 y\, dx = \frac{Y^{\frac{3}{2}} \sqrt{(2\pi)}}{\sqrt{\left(- \dfrac{d^2 Y}{da^2} \right)}}.$$

Similarly if we suppose that yz is a maximum when $x = a'$, and that then $yz = Y'Z'$, we have

$$\int_0^1 yz\, dx = \frac{(Y'Z')^{\frac{3}{2}} \sqrt{(2\pi)}}{\sqrt{\left(- \dfrac{d^2 Y'Z'}{da'^2} \right)}}.$$

Suppose that z is a function of y, say $z = \phi (y)$, then yz is a maximum when y is a maximum, so that $a' = a$; and since $\dfrac{dY}{da} = 0$, we find that

$$\frac{d^2 Y'Z'}{da'^2} = \left\{ \phi (Y) + Y\phi' (Y) \right\} \frac{d^2 Y}{da^2}.$$

Hence we have approximately

$$P = \frac{\phi (Y)}{\sqrt{\left\{ 1 + \dfrac{Y\phi' (Y)}{\phi (Y)} \right\}}}.$$

1031. Laplace discusses on his pages 397—401 the following problem. It has been observed during a certain number of years at Paris that more boys than girls are annually baptised: determine the probability that this superiority will hold during a century. See Art. 897.

Let p be the observed number of baptisms of boys during a certain number of years, q the observed number of baptisms of girls, $2n$ the annual number of baptisms. Let x represent the chance that an infant about to be born and baptised will be a boy.

Let $(x + 1 - x)^{2n}$ be expanded in a series

$$x^{2n} + 2nx^{2n-1}(1-x) + \frac{2n(2n-1)}{1.2} x^{2n-2}(1-x)^2 + \dots ;$$

then the sum of the first n terms of this series will represent the probability that in a year the number of baptisms of boys will predominate.

Denote this sum by ζ; then ζ^i will be the probability that the superiority will be maintained during i years.

Hence we put $x^p(1-x)^q$ for y and ζ^i for z in the formula of the preceding Article, and obtain

$$P = \frac{\int_0^1 x^p (1-x)^q \zeta^i \, dx}{\int_0^1 x^p (1-x)^q \, dx}.$$

Laplace applies his method of approximation with great success to evaluate the integrals. He uses the larger values of p and q given in Art. 902; and he finds that $P = \cdot 782$ approximately.

1032. Laplace's seventh Chapter is entitled *De l'influence des inégalités inconnues qui peuvent exister entre des chances que l'on suppose parfaitement égales*: it occupies pages 402—407.

The subject of this Chapter engaged the attention of Laplace at an early period; see Arts. 877, 881, 891. Suppose the chance of throwing a head with a coin is either $\frac{1+\alpha}{2}$ or $\frac{1-\alpha}{2}$, but it is as likely to be one as the other. Then the chance of throwing n heads in succession will be

$$\frac{1}{2}\left\{\left(\frac{1+\alpha}{2}\right)^n + \left(\frac{1-\alpha}{2}\right)^n\right\},$$

that is, $\dfrac{1}{2^n}\left\{1 + \dfrac{n(n-1)}{1.2}\alpha^2 + \dfrac{n(n-1)(n-2)(n-3)}{\underline{4}}\alpha^4 + \dots\right\}.$

Thus there is an advantage in undertaking to throw n heads in succession beyond what there would be if the coin were perfectly symmetrical.

Laplace shews how we may diminish the influence of the want of symmetry in a coin.

Let there be two coins A and B; let the chances of head and tail in A be p and q respectively, and in B let them be p' and q' respectively: and let us determine the probability that in n throws the two coins shall always exhibit *the same faces*. The chance required is $(pp' + qq')^n$.

Suppose that

$$p = \frac{1+a}{2}, \quad q = \frac{1-a}{2},$$

$$p' = \frac{1+a'}{2}, \quad q' = \frac{1-a'}{2};$$

then $\quad (pp' + qq')^n = \frac{1}{2^n}(1 + aa')^n.$

But as we do not know to which faces the want of symmetry is favourable, the preceding expression might also be $\frac{1}{2^n}(1 - aa')^n$ by interchanging the forms of p and q or of p' and q'. Thus the true value will be

$$\frac{1}{2}\left\{\frac{1}{2^n}(1 + aa')^n + \frac{1}{2^n}(1 - aa')^n\right\},$$

that is

$$\frac{1}{2^n}\left\{1 + \frac{n(n-1)}{1.2}a^2a'^2 + \frac{n(n-1)(n-2)(n-3)}{\lfloor 4}a^4a'^4 + \ldots\right\}.$$

It is obvious that this expression is nearer to $\frac{1}{2^n}$ than that which was found for the probability of securing n heads in n throws with a single coin.

1033. Laplace gives again the result which we have noticed in Art. 891. Suppose p to denote A's skill, and q to denote B's skill; let A have originally a counters and B have originally b counters. Then A's chance of ruining B is

$$\frac{p^b(p^a - q^a)}{p^{a+b} - q^{a+b}}.$$

Laplace puts for p in succession $\frac{1}{2}(1+a)$ and $\frac{1}{2}(1-a)$, and takes half the sum. Thus he obtains for A's chance

$$\frac{1}{2} \frac{\{(1+a)^a - (1-a)^a\}\{(1+a)^b + (1-a)^b\}}{(1+a)^{a+b} - (1-a)^{a+b}}.$$

Laplace says that it is easy to see that, supposing a less than b, this expression is always greater than $\dfrac{a}{a+b}$, which is its limit when $\alpha = 0$. This is the same statement as is made in Art. 891, but the proof will be more easy, because the transformation there adopted is not reproduced.

Put $$\frac{1+\alpha}{1-\alpha} = x,$$

and $$u = \frac{(x^a - 1)(x^b + 1)}{x^{a+b} - 1}.$$

We have to shew that u continually increases as x increases from 1 to ∞, supposing that a is less than b. It will be found that

$$\frac{1}{u}\frac{du}{dx} = \frac{ax^a(x^{2b} - 1) - bx^b(x^{2a} - 1)}{x(x^a - 1)(x^b + 1)(x^{a+b} - 1)}.$$

We shall shew that this expression cannot be negative.

We have to shew that

$$\frac{x^b - x^{-b}}{b} - \frac{x^a - x^{-a}}{a}$$

cannot be negative.

This expression vanishes when $x = 1$, and its differential coefficient is $(x^{b-1} - x^{a-1})(1 - x^{-a-b})$, which is positive if x lie between 1 and ∞; therefore the expression is positive if x lie between 1 and ∞.

Laplace says that if the players agree to double, triple, ... their respective original numbers of counters the advantage of A will continually increase. This may be easily shewn. For change a into ka and b into kb: we have then to shew that

$$\frac{(x^{ka} - 1)(x^{kb} + 1)}{x^{ka+kb} - 1}$$

continually increases with k. Let $x^k = y$; then we have to shew that

$$\frac{(y^a - 1)\,(y^b + 1)}{y^{a+b} - 1}$$

continually increases as y increases from unity: and this is what we have already shewn.

1034. Laplace's eighth Chapter is entitled *Des durées moyennes de la vie, des mariages et des associations quelconques:* it occupies pages 408—418.

Suppose we have found from the tables of mortality the mean duration of the life of n infants, where n is a very large number. Laplace proposes to investigate the probability that the deviation of this result from what may be considered to be the true result will lie within assigned limits : by the *true* result is meant the result which would be obtained if n were infinite. Laplace's analysis is of the same kind as that in his fourth Chapter.

1035. Laplace then examines the effect which would be produced on the laws of mortality if a particular disease were extinguished, as for example the small-pox. Laplace's investigation resembles that of Daniel Bernoulli, as modified by D'Alembert : see Arts. 402, 405, 483.

We will give Laplace's result. In Art. 402, we have arrived at the equation

$$\frac{dq}{dx} = \frac{q}{n} - \frac{1}{mn},$$

where $q = \frac{\xi}{s}$. Put i for $\frac{1}{n}$, and r for $\frac{1}{m}$; and let i and r not be assumed constant. Thus we have

$$\frac{dq}{dx} = iq - ir.$$

Let v denote $e^{-\int i\,dx}$; thus

$$\frac{d}{dx}\,qv = -irv;$$

therefore

$$qv = \text{constant} - \int irv\,dx.$$

The constant is unity, if we suppose the lower limit of the integral to be 0, for q and v are each unity when $x = 0$; thus

$$qv = 1 - \int irv\, dx.$$

The differential equation obtained in Art. 405 becomes when expressed in our present notation

$$\frac{1}{z}\frac{dz}{dx} - \frac{1}{\xi}\frac{d\xi}{dx} = \frac{ir}{q} = \frac{irv}{1 - \int irv\, dx};$$

therefore, by integration,

$$\frac{z}{\xi} = \frac{\text{constant}}{1 - \int irv\, dx}.$$

As before the constant is unity; thus

$$z = \frac{\xi}{1 - \int irv\, dx}.$$

This result agrees with that on Laplace's page 414.

Laplace intimates that this would be an advantageous formula if i and r were constants; but as these quantities may vary, he prefers another formula which he had previously investigated, and which we have given from D'Alembert in Art. 483. He says that by using the data furnished by observation, it appears that the extinction of the small-pox would increase by three years the mean duration of life, provided this duration be not affected by a diminution of food owing to the increase of population.

1036. Laplace discusses in his pages 415—418 the problem of the mean duration of marriages which had been originally started by Daniel Bernoulli; see Arts. 412, 790.

Laplace's investigation is very obscure: we will examine various ways in which the problem may be treated.

Suppose μ men aged A years to marry μ women of the same age, μ being a large number: determine the probability that at the end of T years there will remain an assigned number of unbroken couples. The law of mortality is assumed to be the same for men as for women; and we suppose that the tables shew that out of $m_1 + n_1$ persons aged A years, m_1 were alive at the end of T years, m_1 and n_1 being large.

One mode of solving the proposed problem would be as follows.

Take $\dfrac{m_1}{m_1 + n_1}$ as the chance that a specified individual will be alive at the end of T years; then $\left(\dfrac{m_1}{m_1 + n_1}\right)^2$ will be the chance that a specified pair will be alive, and we shall denote this by p. Therefore the chance that at the end of T years there will be ν unbroken couples, out of the original μ couples, is

$$\frac{\lfloor \mu}{\lfloor \mu - \nu \lfloor \nu} p^\nu (1 - p)^{\mu - \nu}.$$

This is rigorous on the assumption that $\dfrac{m_1}{m_1 + n_1}$ is exactly the chance that a specified individual will be alive at the end of T years: the assumption is analogous to what we have called an inverse use of James Bernoulli's theorem; see Art. 997.

Or we may solve the problem according to the usual principles of inverse probability as given by Bayes and Laplace. Let x denote the chance, supposed unknown, that an individual aged A years will be alive at the end of T years. We have the observed event recorded in the tables of mortality, that out of $m_1 + n_1$ persons aged A years, m_1 were alive at the end of T years. Hence the quantity denoted by y in Art. 1030 is

$$\frac{\lfloor m_1 + n_1}{\lfloor m_1 \lfloor n_1} x^{m_1} (1 - x)^{n_1},$$

and the quantity denoted by z is

$$\frac{\lfloor \mu}{\lfloor \mu - \nu \lfloor \nu} (x^2)^\nu (1 - x^2)^{\mu - \nu};$$

therefore $\quad P = \dfrac{\lfloor \mu}{\lfloor \mu - \nu \lfloor \nu} \dfrac{\displaystyle\int_0^1 x^{m_1} (1 - x)^{n_1} (x^2)^\nu (1 - x^2)^{\mu - \nu}\, dx}{\displaystyle\int_0^1 x^{m_1} (1 - x)^{n_1}\, dx}.$

Laplace however adopts neither of the above methods; but forms a mixture of them. His process may be described thus: Take the first form of solution, but use Bayes's theorem to determine the value of p, instead of putting p equal to $\left(\dfrac{m_1}{m_1+n_1}\right)^2$.

We will complete the second solution. The next step ought to consist in evaluating strictly the integrals which occur in the expression for P; we shall however be content with some rough approximations which are about equivalent to those which Laplace himself adopts.

Assume, in accordance with Art. 993, that

$$\frac{\lfloor \mu}{\lfloor \nu \rfloor \lfloor \mu - \nu} (x^2)^\nu (1 - x^2)^{\mu-\nu} = \frac{e^{-\frac{r^2}{2\mu x^2 (1-x^2)}}}{\sqrt{2\pi \mu x^2 (1-x^2)}},$$

where r is supposed to be not large, and to be such that nearly

$$\nu = x^2\mu - r, \quad \mu - \nu = (1 - x^2)\mu + r.$$

Thus
$$P = \frac{\displaystyle\int_0^1 \frac{x^{m_1}(1-x)^{n_1}}{\sqrt{2\pi\mu x^2(1-x^2)}} e^{-\frac{r^2}{2\mu x^2(1-x^2)}}\, dx}{\displaystyle\int_0^1 x^{m_1}(1-x)^{n_1}\, dx}.$$

Then, as in Arts. 957, 997, we put

$$x^{m_1}(1-x)^{n_1} = Y e^{-t^2},$$

$$x = a + \frac{t\sqrt{(2m_1 n_1)}}{(m_1+n_1)^{\frac{3}{2}}}, \quad \text{nearly,}$$

where
$$a = \frac{m_1}{m_1 + n_1}.$$

And finally we have approximately

$$P = \frac{e^{-\frac{r^2}{2\mu a^2(1-a^2)}}}{\sqrt{2\pi\mu a^2(1-a^2)}}.$$

Then we have to effect a summation for different values of r, like that given in Art. 993. The result is that there is approximately the probability

$$\frac{2}{\sqrt{\pi}} \int_0^\tau e^{-t^2}\,dt + \frac{1}{\sqrt{2\pi\mu a^2\,(1-a^2)}}\,e^{-\tau^2},$$

that the number of unbroken couples will lie between

$$\mu a^2 - \tau\,\sqrt{2\mu a^2\,(1-a^2)} \text{ and } \mu a^2 + \tau\,\sqrt{2\mu a^2\,(1-a^2)}.$$

This substantially agrees with Laplace, observing that in the third line of his page 418 the equation ought to be simplified by the consideration that p' has been assumed very great; so that the equation becomes

$$k^2 = \frac{1}{2n\phi^2\,(1-\phi^2)}.$$

See Art. 148 of the *Theory of Probabilities* in the *Encyclopædia Metropolitana.*

There is still another way in which the problem may be solved. We may take it as a result of observation that out of μ_1 marriages of persons aged A years there remained ν_1 unbroken couples at the end of T years, and we require the consequent probability that out of μ marriages now contracted between persons aged A years ν unbroken couples will remain at the end of T years. Then as in Art. 1030 we obtain

$$P = \frac{\lfloor\mu}{\lfloor\nu\,\lfloor\mu - \nu} \; \frac{\int_0^1 x^{\nu_1+\nu}\,(1-x)^{\mu_1-\nu_1+\mu-\nu}\,dx}{\int_0^1 x^{\nu_1}\,(1-x)^{\mu_1-\nu_1}\,dx}.$$

The result will be like that which we have found by the second method, having $\dfrac{\nu_1}{\mu_1}$ instead of a^2. Practically $\dfrac{\nu_1}{\mu_1}$ may be nearly equal to a^2, but they must not be confounded in theory, being obtained from different data. The last mode is simpler in theory than the second, but it assumes that we have from observation data which bear more immediately on the problem.

1037. Laplace's ninth Chapter is entitled *Des bénéfices dépendans de la probabilité des événemens futurs:* it occupies pages 419—431.

Suppose that a large number of trials, s, is to be made, and that at each trial one of two cases will happen; suppose that in one case a certain sum of money is to be received, and in the other case a certain other sum : determine the expectation.

Laplace applies an analysis of the same kind as in his fourth Chapter ; we shall deduce the required result from the investigation in Art. 1002. We supposed in Art. 1002 that all values of a certain variable z were possible, and that $f_i(z)$ denoted the chance at the ith trial that the value would lie between z and $z + \delta z$. Suppose however that only *two* values are possible which we may denote by ζ_i and ξ_i ; then we must suppose that $f_i(z)$ vanishes for all values of z except when z is very nearly equal to ζ_i or to ξ_i, and we may put

$$\int_b^a f_i(z)\, dz = p_i + q_i,$$

where p_i stands for the part of the integral arising from values of z nearly equal to ζ_i, and q_i stands for the part of the integral arising from values of z nearly equal to ξ_i ; and thus

$$p_i + q_i = 1.$$

Again, $\int_b^a z f_i(z)\, dz$ will reduce to two terms arising from values of z nearly equal to ζ_i and ξ_i respectively, so that we shall have

$$\int_b^a z f_i(z)\, dz = \zeta_i p_i + \xi_i q_i.$$

Similarly,

$$\int_b^a z^2 f_i(z)\, dz = \zeta_i^2 p_i + \xi_i^2 q_i.$$

Suppose now in Art. 1002 that $\gamma_1 = \gamma_2 = \ldots = \gamma_s = 1$; then

$$l = \Sigma k_i = \Sigma\,(\zeta_i p_i + \xi_i q_i) ;$$
$$2\kappa^2 = \Sigma\,(k_i' - k_i^2)$$
$$= \Sigma\,\{\zeta_i^2 p_i + \xi_i^2 q_i - (\zeta_i p_i + \xi_i q_i)^2\}$$
$$= \Sigma\,\{(\zeta_i^2 p_i + \xi_i^2 q_i)(p_i + q_i) - (\zeta_i p_i + \xi_i q_i)^2\}$$
$$= \Sigma\, p_i q_i\,(\zeta_i - \xi_i)^2.$$

And there is, by Art. 1002, the probability $\dfrac{2}{\sqrt{\pi}} \displaystyle\int_0^\tau e^{-t^2} dt$ that $\Sigma\epsilon$ will lie between

$$\Sigma\left(\zeta_i p_i + \xi_i q_i\right) - 2\tau\kappa \quad \text{and} \quad \Sigma\left(\zeta_i p_i + \xi_i q_i\right) + 2\tau\kappa.$$

There has been no limitation as to the sign of ζ_i or ξ_i.

This result will be found to agree with that given by Laplace on his page 423; he had previously, on his page 420, treated the particular case in which the function $f_i(z)$ is supposed the same at every trial, so that the suffix i becomes unnecessary, and the result simplifies in the manner which we have explained towards the end of Art. 1002.

1038. An important consequence follows so naturally from the investigation in the preceding Article, that in order to explain it we will interrupt our analysis of Laplace. Suppose that $\zeta_i = 1$ and $\xi_i = 0$, for all values of i: thus

$$l = \Sigma p_i, \quad 2\kappa^2 = \Sigma p_i q_i;$$

and $\Sigma\epsilon_i$ becomes equal to the number of times in which an event happens out of s trials, the chance of the happening of the event being p_i at i^{th} trial. Thus we have the probability $\dfrac{2}{\sqrt{\pi}} \displaystyle\int_0^\tau e^{-t^2} dt$ that the number of times will lie between

$$\Sigma p_i - \tau\sqrt{2\Sigma p_i q_i} \quad \text{and} \quad \Sigma p_i + \tau\sqrt{2\Sigma p_i q_i}.$$

This is an extension of James Bernoulli's theorem to the case in which the chance of the event is not constant at every trial; if we suppose that p_i *is* independent of i we have a result practically coincident with that in Art. 993. This extension is given by Poisson, who attaches great importance to it; see his *Recherches sur la Prob.* ..., page 246.

1039. If instead of *two* values at the ith trial as in Art. 1037, we suppose a larger number, the investigation will be similar to that already given. Denote these values by $\zeta_i, \xi_i, \chi_i \dots$; we shall have

$$l = \Sigma \ (\zeta_i p_i + \xi_i q_i + \chi_i w_i + \dots),$$

where $p_i + q_i + w_i + \dots = 1$;

$$2\kappa^2 = \Sigma \left\{ \zeta_i^2 p_i + \xi_i^2 q_i + \chi_i^2 w_i + \dots - (\zeta_i p_i + \xi_i q_i + \chi_i w_i + \dots)^2 \right\}.$$

Laplace himself takes the particular case in which the function $f_i(z)$ is supposed the same at every trial; see his pages 423—425.

1040. Laplace proceeds to a modification of the problem just considered, which may be of more practical importance. Nothing is supposed known *a priori* respecting the chances, but data are taken from observations. Suppose we have observed that in μ_1 trials a certain result has been obtained ν_1 times: if μ more trials are made determine the expectation of a person who is to receive ζ each time the result is obtained, and to forfeit ξ each time the result fails.

The analysis now is like that which we have given at the end of Art. 1036. There is the probability $\dfrac{2}{\sqrt{\pi}} \displaystyle\int_0^\tau e^{-t^2}\, dt$ that the number of times the result is obtained will lie between

$$\frac{\mu\nu_1}{\mu_1} - \frac{\tau \sqrt{2\mu\nu_1 \ (\mu_1 - \nu_1)}}{\mu_1} \text{ and } \frac{\mu\nu_1}{\mu_1} + \frac{\tau \sqrt{2\mu\nu_1 \ (\mu_1 - \nu_1)}}{\mu_1}.$$

But if the result is obtained σ times in μ trials the advantage is

$$\sigma\zeta - (\mu - \sigma)\, \xi, \text{ that is, } \sigma\, (\zeta + \xi) - \mu\xi.$$

Hence there is the probability above assigned that the advantage will lie between

$$\mu \left\{ \frac{\nu_1}{\mu_1}\, \zeta - \frac{\mu_1 - \nu_1}{\mu_1}\, \xi \right\} \pm \frac{\tau\, (\zeta + \xi)}{\mu_1} \ \sqrt{2\mu\nu_1 \ (\mu_1 - \nu_1)}.$$

This will be found to agree substantially with Laplace's page 425.

1041. Laplace passes on to questions connected with life insurances: he shews that the stability of insurance companies depends on their obtaining a very large amount of business. It has been pointed out by Bienaymé, that if the consideration of compound interest is neglected we shall form too high an estimate of the stability of insurance companies; see Cournot's *Exposition de la Théorie des Chances*...page 333 : see also page 143 of the same work for a formula by Bienaymé connected with the result given in Art. 1038.

1042. Laplace's tenth Chapter is entitled *De l'espérance morale:* it occupies pages 432—445. This Chapter may be described as mainly a reproduction of the memoir by Daniel Bernoulli, which we have analysed in Arts. 377—393 ; Laplace himself names his predecessor. Laplace adds the demonstration to which we have referred in Art. 388 ; see his pages 436, 437. Laplace also applies the theory of moral expectation to an example connected with life annuities ; see his pages 442—444.

The following example in inequalities is involved in Laplace's page 444. If $a_1, a_2, a_3, \ldots$ and $b_1, b_2, b_3, \ldots$ are series both in increasing or both in decreasing order of magnitude

$$\frac{a_1^2 b_1 + a_2^2 b_2 + a_3^2 b_3 + \ldots + a_n^2 b_n}{a_1 b_1 + a_2 b_2 + a_3 b_3 + \ldots + a_n b_n}$$

is greater than

$$\frac{a_1^2 + a_2^2 + a_3^2 + \ldots + a_n^2}{a_1 + a_2 + a_3 + \ldots + a_n} ;$$

for if we multiply up and bring all the terms together, we find that the result follows from the fact that $\sigma_r a_s (a_r - a_s)(b_r - b_s)$ is positive.

Hence too if one of the two series is in increasing and one in decreasing order of magnitude the inequality becomes inverted.

1043. Laplace's eleventh Chapter is entitled *De la Probabilité des témoignages:* it occupies pages 446—461.

We have given sufficient indication of the main principle of the Chapter in Art. 735 ; see also Art. 941.

Laplace's process on his page 457, although it leads to no error in the case he considers, involves an unjustifiable assumption ; see Poisson, *Recherches sur la Prob.*...page 112. See also pages 3 and 364 of Poisson's work for criticisms bearing on Laplace's eleventh Chapter.

1044. Laplace's pages 464—484 are headed *Additions;* see Arts. 916, 921. There are three subjects discussed.

I. Laplace demonstrates Wallis's theorem, and he gives an account of the curious way in which the theorem was discovered, although it cannot be said to have been demonstrated by its discoverer.

II. Laplace demonstrates a formula for $\Delta^n s^i$ which he had formerly obtained by a bold assumption ; see Arts. 916, 966.

III. Laplace demonstrates the formula marked (p) on page 168 of the *Théorie...des Prob.;* see Art. 917.

1045. The first Supplement to the *Théorie...des Prob.* is entitled *Sur l'application du Calcul des Probabilités à la Philosophie Naturelle;* it occupies 34 pages: see Art. 926. The title of the Supplement does not seem adapted to give any notion of the contents.

1046. We have seen in Art. 1009 that in Laplace's theory of the errors of observations a certain quantity occurs the value of which is not known *a priori,* but which may be approximately determined from the observations themselves. Laplace proposes to illustrate this point, and to shew that this approximation is one which we need not hesitate to adopt : see pages 7—11 of the first Supplement. It does not appear to me however that much conviction could be gained from Laplace's investigation.

A very remarkable theorem is enunciated by Laplace on page 8 of the first Supplement. He gives no demonstration, but says in his characteristic way, L'analyse du n° 21 du seconde Livre conduit à ce théorème général.... The theorem is as follows: Suppose, as in Art. 1011, that certain quantities are to be determined by the aid of observations; for simplicity we will assume that there are three quantities x, y, z. Let values be found for these quantities by the most advantageous method, and denote these values by x_1, y_1, z_1, respectively. Put

$$x = x_1 + \xi, \quad y = y_1 + \eta, \quad z = z_1 + \zeta.$$

Then Laplace's theorem asserts that the probability of the simultaneous existence of ξ, η, ζ, as values of the errors of the quantities to be determined, is proportional to $e^{-\sigma}$, where

$$\sigma = \frac{1}{4\kappa^2} \Sigma \, (a_i\xi + b_i\eta + c_i\zeta)^2.$$

I am compelled to omit the demonstration of this theorem for want of space; but I shall endeavour to publish it on some other occasion.

1047. Laplace next supposes that six elements are to be determined from a large number of observations by the most advantageous method. He arranges the algebraical work in what he considers a convenient form, supposing that we wish to determine for each variable the mean value of the error to be apprehended, or to determine the probability that the error will lie within assigned limits ; see pages 11—19 of the first Supplement. He then, on his pages 21—26, makes a numerical application, and arrives at the result to which we have already referred in Art. 939.

1048. Laplace observes that all his analysis rests on the assumption that positive and negative errors are equally likely, and he now proposes to shew that this limitation does not practically affect the value of his results: see his pages 19—21. Here again however it does not appear to me that much conviction would be gained from Laplace's investigation.

1049. The first Supplement closes with a section on the Probability of judgments; it is connected with the eleventh Chapter : see Art. 1043.

1050. The second Supplement is entitled *Application du Calcul des Probabilités aux opérations géodésiques :* it occupies 50 pages: see Art. 927. This Supplement is dated February 1818.

This Supplement is very interesting, and considering the subject and the author it cannot be called difficult. Laplace shews how the knowledge obtained from measuring a *base of verification* may be used to correct the values of the elements of the triangles of a survey. He speaks favourably of the use of *repeating circles;* see his pages 5, 8, 20. He devotes more space than the subject seems to deserve to discuss an arbitrary method proposed by

Svanberg for deducing a result from observations made with a repeating circle: see Laplace's pages 32—35.

Laplace explains a method of treating observations which he calls the *method of situation*, and which he considers may in some cases claim to be preferable to the *most advantageous method* explained in his fourth Chapter. This method of situation had been given in the *Mécanique Céleste, Livre* III., but without receiving a special name: see Art. 1016. Laplace gives an investigation to determine when the *method of situation* should be preferred to the *most advantageous method*, and an investigation of the value of a combination of the two methods.

1051. The third Supplement is entitled *Application des formules géodésiques de probabilité, à la méridienne de France;* it occupies 36 pages: see Art. 928.

Laplace begins by giving a numerical example of some of the formulæ in the second Supplement. In his pages 7—15 he gives what he calls a simple example of the application of the geodesic formulæ. He takes a system of isosceles triangles, having their bases all parallel to a given line, and he finds the errors in lengths arising from errors in the angles. The investigation is like that in the second Supplement.

Laplace devotes his pages 16—28 to discussions respecting the error in level in large trigonometrical surveys.

Pages 29—36 contain what Laplace calls *Méthode générale du calcul des probabilités, lorsqu'il y a plusieurs sources d'erreurs.*

1052. Here we close our account of the *Theorie Analytique des Probabilités*. After every allowance has been made for the aid which Laplace obtained from his predecessors there will remain enough of his own to justify us in borrowing the words applied to his Theory of the Tides by a most distinguished writer, and pronouncing this also "to be one of the most splendid works of the greatest mathematician of the past age."

For remarks which will interest a student of Laplace's work I may refer to the first page in the Appendix to De Morgan's *Essay on Probabilities*...in the Cabinet Cyclopædia; to the *History of the Science* which forms the introduction to Galloway's Treatise published in the *Encyclopædia Britannica;* to the work of Gouraud, pages 107—128; and to various passages in Dugald Stewart's *Works edited by Hamilton,* which will be found by consulting the General Index in the Supplementary volume.

· Some observations by Poisson will find an appropriate place here: they occur in the *Comptes Rendus*...Vol. II. page 396.

Sans doute Laplace s'est montré un homme de génie dans la mécanique céleste; c'est lui qui a fait preuve de la sagacité la plus pénétrante pour découvrir les causes des phénomènes; et c'est ainsi qu'il a trouvé la cause de l'accélération du mouvement de la Lune et celle des grandes inégalités do Saturne et de Jupiter, qu'Euler et Lagrange avaient cherchées infructueusement. Mais on peut dire que c'est encore plutôt dans le calcul des probabilités qu'il a été un grand géomètre; car ce sont les nombreuses applications qu'il a faites de ce calcul qui ont donné naissance au calcul aux différences finies partielles, à sa méthode pour la réduction de certaines intégrales en séries, et à ce qu'il a nommé la *théorie des fonctions génératrices.* Un des plus beaux ouvrages de Lagrange, son Mémoire de 1775, a aussi pour occasion, et en partie pour objet, le calcul des probabilités. Croyons donc qu'un sujet qui a fixé l'attention de pareils hommes est digne de la nôtre; et tâchons, si cela nous est possible, d'ajouter quelque chose à ce qu'ils ont trouvé dans une matière aussi difficile et aussi intéressante.

APPENDIX.

1053. THIS Appendix gives a notice of some writings which came under my attention during the printing of the book, too late to be referred to their proper places.

1054. John de Witt's tract which was mentioned in the fifth Chapter has been recovered in modern times, and printed in an English translation. See *Contributions to the History of Insurance...* by Frederick Hendriks, Esq. in the *Assurance Magazine*, Vol. II. 1852, page 231. For some remarks on John de Witt's hypothesis as to the rate of mortality, see page 393 of the same volume.

Many interesting and valuable memoirs connected with the history of Insurance and kindred subjects will be found in the volumes of the *Assurance Magazine*.

1055. A memoir on our subject occurs in the *Actorum Eruditorum...Supplementa.* Tomus IX. Lipsiæ, 1729. The memoir is entitled, *Johannis Rizzetti Ludorum Scientia, sive Artis conjectandi elementa ad alias applicata:* it occupies pages 215—229 and 295—307 of the volume.

It appears from page 297 of the memoir that Daniel Bernoulli had a controversy with Rizzetti and Riccati relating to some problems in chances; I have found no other reference to this controversy. Rizzetti cites the *Exercitationes Mathematicæ* of Daniel Bernoulli; I have not seen this book myself, which appears to have been published in 1724.

The chief point in dispute may be said to be the proper definition of *expectation*. Suppose that A and B play together; let A stake the sum a, and B stake the sum b; suppose that there are $m + n + p$ equally likely cases, in m of them A is to take both the stakes, in n of them B is to take both the stakes, and in p of them each takes his own stake. Then according to the ordinary principles we estimate the expectation of A at

$$\frac{m(a+b)+pa}{m+n+p},$$

so far as it depends upon the game which is to be played. Or if we wish to take account of the fact that A has already paid down the sum a, we may take for the expectation

$$\frac{m(a+b)+pa}{m+n+p} - a, \text{ that is, } \frac{mb-na}{m+n+p}.$$

Rizzetti however prefers another definition; he says that A has m chances out of $m+n+p$ of gaining the sum b; so that his expectation is $\frac{mb}{m+n+p}$. Rizzetti tries to shew that the ordinary definition employed by Montmort and Daniel Bernoulli leads to confusion and error; but these consequences do not really follow from the ordinary definition but from the mistakes and unskilfulness of Rizzetti himself.

The memoir does not give evidence of any power in the subject. Rizzetti considers that he demonstrates James Bernoulli's famous theorem by some general reasoning which mainly rests on the axiom, Effectus constans et immutabilis pendet a causa constante, et immutabili. On his page 224 he gives what he considers a short investigation ·of a problem discussed by Huygens and James Bernoulli; see Arts. 33, 103: but the investigation is unsatisfactory, and shews that Rizzetti did not clearly understand the problem.

1056. I am indebted for a reference to the memoir noticed in the preceding Article to Professor De Morgan,,who derived it from Kahle, *Bibliothecæ Philosophiæ Struvianæ*...Göttingen, 1740. 2 Vols. 8vo. Vol. I. p. 295. Professor De Morgan supplied me from the same place with references to the following works which I have not been so fortunate as to obtain.

Andrew Rudiger, *De sensu falsi et veri, lib. I. cap. xii. et lib. III.*: no further description given.

Kahle himself. *Elementa logicæ probabilium, methodo mathematica*...Halæ Magdeburgicæ, 1735, 8vo.

1057. The work which we have quoted at the beginning of Art. 347 contains some remarks on our subject; they form part of the *Introduction à la Philosophie*, and occur on pages 82—93 of the second volume. It appears from page XLVII of the first volume that this work was first published by 's Gravesande in 1736. The remarks amount to an outline of the mathematical Theory of Probability. It is interesting to observe that 's Gravesande gives in effect an example of the inverse use of James Bernoulli's theorem; see his page 85: the example is of the kind which we have used for illustration in Art. 125.

1058. The result attributed to Euler in Art. 131 is I find really due to John Bernoulli. See *Johannis Bernoulli...Opera Omnia, Tomus Quartus*, 1742, p. 22. He says,

Atque ita satisfactum est ardenti desiderio Fratris mei, qui agnoscens summæ hujus pervestigationem difficiliorem esse quam quis putaverit, ingenue fassus est, omnem suam industriam fuisse elusam : *Si quis inveniat,* inquit, *nobisque communicet, quod industriam nostram elusit hactenus, magnas de nobis gratias feret.* Vid. Tractat. *de Seriebus infinitis,* p. 254. Utinam Frater superstes esset.

1059. An essay on Probability was written by the celebrated Moses Mendelsohn; it seems to have been published in his *Philosophische Schriften* in 1761. I have read it in the edition of the *Philosophische Schriften* which appeared at Berlin in 1771, in two small volumes. The essay occupies pages 243—283 of the second volume.

Mendelsohn names as writers on the subject, Pascal, Fermat, Huygens, Halley, Craig, Petty, Montmort, and De Moivre. Mendelsohn cites a passage from the work of 's Gravesande, which amounts to an example of James Bernoulli's theorem ; and Mendelsohn gives what he considers to be a demonstration of the theorem, but it is merely brief general reasoning.

The only point of interest in the memoir is the following. Suppose an event A has happened simultaneously, or nearly so, with an event B; we are then led to enquire whether the concurrence is accidental or due to some causal connexion. Mendelsohn says that if the concurrence has happened n times the probability that there is a causal connexion is $\dfrac{n}{n+1}$; but he gives no intimation of the way in which he obtains this result. He takes the following illustration : suppose a person to drink coffee, and to be attacked with giddiness ; the concurrence may be accidental or there may be some causal connexion : if the concurrence has been observed n times the probability is $\dfrac{n}{n+1}$ that the giddiness will follow the drinking of coffee.

If we apply the theorem of Bayes and Laplace, and suppose that an event has happened n times, the probability that it will happen at the next trial is $\dfrac{n+1}{n+2}$; see Art. 848. It is certainly curious that Mendelsohn's rule should agree so nearly with this result when n is large, but it is apparently only an accidental coincidence, for there is nothing in Mendelsohn's essay which suggests that he had much knowledge of the subject or any great mathematical power : we cannot therefore consider that he in any way anticipated Bayes.

Mendelsohn makes his rule serve as the foundation of some remarks on the confidence which we repose on the testimony of our senses, referring especially to the scepticism of Hume. Mendelsohn also touches on the subjects of Free Will and the Divine Foreknowledge ; but as it appears to me without throwing any light on these difficult problems.

I was aware that Mendelsohn had written on Probability from the occurrence of his name in Art. 840, but I assumed that his essay would not contain any matter bearing on the mathematical theory, and so I omitted to examine it. I supply the omission at the request of the late Professor Boole ; he had seen a reference to Mendelsohn in some manuscripts left by Dr Bernard, formerly teacher of Hebrew in the University of Cambridge, and, in consequence of this reference, expressed a wish that I would report on the character of the essay.

1060. I take from Booksellers' Catalogues the titles of four works which I have never seen.

Thubeuf. Élémens et principes de la royale Arithmétique aux jettons, etc. 12mo. Paris, 1661.

Marpurg, F. W., Die Kunst, sein Glück spielend zu machen. Hamburg, 1765. 4to.

Fenn, (I.) Calculations and formulæ for determining the Advantages or Disadvantages of Gamesters,...1772. 4to.

Frömmichen Ueber Lehre d. Wahrscheinl. Braunschw.1773. 4to.

1061. I had overlooked a passage in Montucla which bears on the point noticed in Art. 990; see Montucla, page 421. It seems that a mode of election suggested by Condorcet was for some time adopted at Geneva. The defects of the mode were indicated in a work by Lhuilier entitled, *Examen du mode d'élection proposé en février* 1793, *à la Convention nationale de France, et adopté à Genève* (1794, en 8°).

1062. A very curious application of the Theory of Probability was stated by Waring; see his *Meditationes Algebraicæ*, 3rd Edition, 1782, pages xi, 69, 73. For example, he gives a rule for ascertaining the number of imaginary roots in an equation, and says: Hæc methodus in quadraticis æquationibus verum præbet numerum impossibilium radicum : in cubicis autem ejus probabilitas inveniendi impossibiles radices non videtur majorem habere rationem ad probabilitatem fallendi quam 2 : 1.

I owe this reference to the kindness of Professor Sylvester in sending me a copy of his remarkable memoir in the *Philosophical Transactions* for 1864, on the *Real and Imaginary roots of Algebraical Equations*. Professor Sylvester had independently made the same kind of application ; see page 580 of the volume, where he says: "Like myself, too, in the body of the memoir Waring has given theorems of probability in connexion with rules of this kind, but without any clue to his method of arriving at them. Their correctness may legitimately be doubted."

CHRONOLOGICAL LIST OF AUTHORS.

The figures refer to the pages of the Volume.

INDEX.

The figures refer to the pages of the Volume.

THE END.